Lecture Notes in Computer Science   **9411**

Commenced Publication in 1973
Founding and Former Series Editors:
Gerhard Goos, Juris Hartmanis, and Jan van Leeuwen

More information about this series at http://www.springer.com/series/7407

Emilio Di Giacomo · Anna Lubiw (Eds.)

Graph Drawing
and Network Visualization

23rd International Symposium, GD 2015
Los Angeles, CA, USA, September 24–26, 2015
Revised Selected Papers

 Springer

Editors
Emilio Di Giacomo
Università degli Studi di Perugia
Perugia
Italy

Anna Lubiw
School of Computer Science
University of Waterloo
Waterloo, ON
Canada

ISSN 0302-9743   ISSN 1611-3349   (electronic)
Lecture Notes in Computer Science
ISBN 978-3-319-27260-3   ISBN 978-3-319-27261-0   (eBook)
DOI 10.1007/978-3-319-27261-0

Library of Congress Control Number: 2015955866

LNCS Sublibrary: SL1 – Theoretical Computer Science and General Issues

Springer Cham Heidelberg New York Dordrecht London

Printed on acid-free paper

Springer International Publishing AG Switzerland is part of Springer Science+Business Media
(www.springer.com)

Preface

This volume contains the papers presented at the 23rd International Symposium on Graph Drawing and Network Visualization (GD 2015), which took place September 24–26, 2015, in Los Angeles, California, USA. The conference was hosted by California State University at Northridge, with Csaba Tóth as chair of the Organizing Committee. A total of 86 participants from 12 countries attended the conference.

This year the symposium added "Network Visualization" to its name to better emphasize the focus of the conference both on the combinatorial and algorithmic aspects of graph drawing, and on the design of visualization systems and interfaces.

Paper submissions were divided into three tracks plus a poster track: Track 1 for combinatorial and algorithmic aspects; Track 2 for experimental, applied, and network visualization aspects; and Track 3 for shorter notes and demos. All tracks were handled by a single Program Committee. The total number of submissions was 77 papers and nine posters. At least three Program Committee members reviewed each submission and the Program Committee then accepted 42 papers and eight posters, for acceptance rates of 24/42 in Track 1, 11/22 in Track 2, 7/13 in Track 3, and 8/9 posters. In addition to the papers, these proceedings include a two-page description of each poster.

GD 2015 was preceded by a two-day graduate workshop on "Recent Trends in Graph Drawing: Curves, Graphs, and Intersections." A report about the workshop is included in the proceedings.

There were two invited talks at GD 2015. Herbert Edelsbrunner of the Institute of Science and Technology, Austria, talked about "Shape, Homology, Persistence, and Stability." Kwan-Liu Ma of the University of California at Davis, USA, talked about "Emerging Topics in Network Visualization." Abstracts of both talks are included in these proceedings.

Springer sponsored awards for best paper in each of Track 1 and Track 2, plus a best presentation award and a best poster award. The Program Committee voted to give the best paper award in Track 1 to "Drawing Graphs Using a Small Number of Obstacles," by M. Balko, J. Cibulka, and P. Valtr, and in Track 2 to "An Incremental Layout Method for Visualizing Online Dynamic Graphs," by T. Crnovrsanin, J. Chu, and K.-L. Ma. The participants of the conference voted to give the best presentation award to M. Löffler for his presentation of the paper "Realization of Simply Connected Polygonal Linkages and Recognition of Unit Disk Contact Trees" and the best poster award to P. Angelini, G. Da Lozzo, G. Di Battista, F. Frati, M. Patrignani, and I. Rutter for their poster entitled "On the Relationship Between Map Graphs and Clique Planar Graphs."

Following tradition, the 22nd Annual Graph Drawing Contest was held during the conference. The contest had two parts, each with two categories: Creative Topics (Graph Classes and Tic Tac Toe) and Live Challenge (Automatic Category and Manual Category). Awards were made in each of the four categories. A report about the contest is included in the proceedings.

Many people and organizations contributed to the success of GD 2015. We thank the Program Committee members and the additional reviewers for carefully reviewing the submitted papers and posters and for putting together a strong and interesting program. Thanks to all the authors for choosing GD 2015 as the publication venue for their research.

We warmly thank the Organizing Committee, Bernardo Ábrego, Silvia Fernández-Merchant, Csaba Tóth, and all the volunteers from the California State University at Northridge, who put a lot of time and effort into the organization of GD 2015. This year's Contest Committee was chaired by Maarten Löffler, Utrecht University. We thank the committee for preparing interesting and challenging problems.

GD 2015 thanks its sponsors, "diamond" sponsor California State University at Northridge, "gold" sponsors Tom Sawyer Software and yWorks, "silver" sponsor Microsoft, and "bronze" sponsor Springer. Their generous support helps ensure the continued success of this conference.

The 24th International Symposium on Graph Drawing and Network Visualization (GD 2016) will take place September 19–21, 2016, in Athens, Greece. Yifan Hu and Martin Nöllenberg will co-chair the Program Committee, and Antonios Symvonis will chair the Organizing Committee.

October 2015

Emilio Di Giacomo
Anna Lubiw

Organization

Program Committee

Carla Binucci	University of Perugia, Italy
Prosenjit Bose	Carleton University, Canada
Giuseppe Di Battista	Roma Tre University, Italy
Emilio Di Giacomo (Co-chair)	University of Perugia, Italy
Vida Dujmović	University of Ottawa, Canada
Tim Dwyer	Monash University, Australia
Fabrizio Frati	Roma Tre University, Italy
Michael Goodrich	University of California, Irvine, USA
Nathalie Henry Riche	Microsoft Research, USA
Yifan Hu	Yahoo Labs, USA
Michael Kaufmann	University of Tübingen, Germany
Andreas Kerren	Linnaeus University, Sweden
Anna Lubiw (Co-chair)	University of Waterloo, Canada
Tamara Munzner	University of British Columbia, Canada
Stephen North	Infovisible LLC, USA
Martin Nöllenburg	Karlsruhe Institute of Technology, Germany
Yoshio Okamoto	University of Electro-Communications, Japan
Ignaz Rutter	Karlsruhe Institute of Technology, Germany
Maria Saumell	University of West Bohemia, Czech Republic
Marcus Schaefer	DePaul University, USA
Heidrun Schumann	University of Rostock, Germany
Geza Toth	Rényi Institute, Hungary
Jarke van Wijk	Eindhoven University of Technology, The Netherlands
Alexander Wolff	University of Würzburg, Germany

Organizing Committee

Bernardo Ábrego	California State University at Northridge, USA
Silvia Fernández-Merchant	California State University at Northridge, USA
Csaba D. Tóth (Chair)	California State University at Northridge, USA

Graph Drawing Contest Committee

Philipp Kindermann	University of Würzburg, Germany
Maarten Löffler (Chair)	Utrecht University, The Netherlands
Lev Nachmanson	Microsoft Research, USA
Ignaz Rutter	Karlsruhe Institute of Technology, Germany

Additional Reviewers

Aichholzer, Oswin
Angelini, Patrizio
Bekos, Michael
Bläsius, Thomas
Bruckdorfer, Till
Da Lozzo, Giordano
Di Bartolomeo, Marco
Di Donato, Valentino
Didimo, Walter
van Dijk, Thomas C.
Feng, Wendy
Fink, Martin
Fulek, Radoslav
Gansner, Emden
Grilli, Luca
Hasunuma, Toru
Hernandez, Gregorio
Kainen, Paul
Khoury, Marc
Kieffer, Steven
Kindermann, Philipp
Klein, Karsten
Kleist, Linda
Kobourov, Stephen
Kucher, Kostiantyn
Kusters, Vincent
Lee, Bongshin
Lipp, Fabian
Liu, Qingsong
Löffler, Maarten
Mchedlidze, Tamara
Mondal, Debajyoti

Montecchiani, Fabrizio
Morin, Pat
Nayyeri, Amir
Niedermann, Benjamin
Ozeki, Kenta
Park, Ji-Won
Patrignani, Maurizio
Pizzonia, Maurizio
Prutkin, Roman
Radermacher, Marcel
Raftopoulou, Chrysanthi
Richter, Bruce
Roselli, Vincenzo
Schreiber, Falk
Sheffer, Adam
Shermer, Thomas
Shi, Conglei
Smorodinsky, Shakhar
Song, Qi
Spisla, Christiane
Strash, Darren
Ueckerdt, Torsten
van den Elzen, Stef
van Renssen, Andr
Verbeek, Kevin
Vesonder, Gregg
Yamanaka, Katsuhisa
Yang, Yalong
Yoghourdjian, Vahan
Zielke, Christian
Zimmer, Björn

Sponsors

Diamond Sponsor

Gold Sponsors

Silver Sponsor

Bronze Sponsor

Invited Talks

Shape, Homology, Persistence, and Stability

Herbert Edelsbrunner

Institute of Science and Technology, Austria

Abstract. My personal journey to the fascinating world of geometric forms started more than 30 years ago with the invention of alpha shapes in the plane. It took about 10 years before we generalized the concept to higher dimensions, we produced working software with a graphics interface for the three-dimensional case. At the same time, we added homology to the computations. Needless to say that this foreshadowed the inception of persistent homology, because it suggested the study of filtrations to capture the scale of a shape or data set. Importantly, this method has fast algorithms. The arguably most useful result on persistent homology is the stability of its diagrams under perturbations.

Emerging Topics in Network Visualization

Kwan-Liu Ma

University of California at Davis, USA

Abstract. Visualizing networks commonly found in a wide variety of applications, such as bioinformatics, computer security, social networks, telecommunication, transportation systems, etc., can lead to important insights. While visualizing small, static networks is relatively easy to do, larger and more complex networks present many challenges. In particular, real-world network data are almost all time-varying, and effective techniques for visualizing and analyzing networks evolving over time are lacking. I will discuss emerging topics in network visualization using research results that my group has produced as examples.

Contents

Drawings with Crossings

Polygons and Convexity

Drawing Graphs on Point Sets

Posters

Large and Dynamic Graphs

GraphMaps: Browsing Large Graphs
as Interactive Maps

Lev Nachmanson[1]([✉]), Roman Prutkin[2], Bongshin Lee[1],
Nathalie Henry Riche[1], Alexander E. Holroyd[1], and Xiaoji Chen[3]

[1] Microsoft Research, Redmond, WA, USA
{levnach,bongshin,nath,holroyd}@microsoft.com
[2] Karlsruhe Institute of Technology, Karlsruhe, Germany
roman.prutkin@kit.edu
[3] Microsoft, Redmond, WA, USA
missx@xbox.com

Abstract. Algorithms for laying out large graphs have seen significant progress in the past decade. However, browsing large graphs remains a challenge. Rendering thousands of graphical elements at once often results in a cluttered image, and navigating these elements naively can cause disorientation. To address this challenge we propose a method called GraphMaps, mimicking the browsing experience of online geographic maps.

GraphMaps creates a sequence of layers, where each layer refines the previous one. During graph browsing, GraphMaps chooses the layer corresponding to the zoom level, and renders only those entities of the layer that intersect the current viewport. The result is that, regardless of the graph size, the number of entities rendered at each view does not exceed a predefined threshold, yet all graph elements can be explored by the standard zoom and pan operations.

GraphMaps preprocesses a graph in such a way that during browsing, the geometry of the entities is stable, and the viewer is responsive. Our case studies indicate that GraphMaps is useful in gaining an overview of a large graph, and also in exploring a graph on a finer level of detail.

1 Introduction

Graphs are ubiquitous in many different domains such as information technology, social analysis or biology. Graphs are routinely visualized, but their large size is often a barrier. The difficulty comes not from the layout which can be calculated very fast. (For example, by using Brandes and Pich's algorithm [9] a graph with several thousand nodes and links can be laid out in a few seconds on a regular personal computer.) Rather, viewing and browsing these large graphs is problematic. Firstly, rendering thousands of graphical elements on a computer might take a considerable time and may result in a cluttered image if the graph is dense. Secondly, navigating thousands of elements rendered naively disorients the user.

© Springer International Publishing Switzerland 2015
E. Di Giacomo and A. Lubiw (Eds.): GD 2015, LNCS 9411, pp. 3–15, 2015.
DOI: 10.1007/978-3-319-27261-0_1

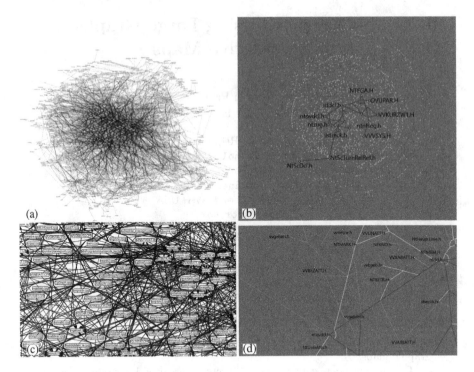

Fig. 1. A graph (https://github.com/ekoontz/graphviz/blob/master/rtest/graphs/
b100.dot.) with 1436 nodes and 5806 edges. (a) The full view with a standard method
which draws all nodes and edges regardless of the zoom. (b) The full view rendered by
GraphMaps. (c) A view with the zoom close to 9.13 with the standard viewing. (d) A
view with zoom 9.26 with GraphMaps.

Our intention is to provide a graph browsing experience similar to that of
online geographic maps, for example, Bing or Google Maps. We propose a set
of requirements for such a visualization and introduce a method, GraphMaps,
fulfilling these requirements. GraphMaps renders a graph as an interactive map
by displaying only the most essential elements for the current view. We allow fast
interactions using standard pan and zoom operations. The drawing is visually
stable, in the sense that during these operations, nodes do not change their
relative positions, and edges do not change their geometry. To the best of our
knowledge, GraphMaps is the first method having these properties. Figure 1
illustrates the method.

Related Work. The problem of visualizing large graphs has been extensively
addressed in the literature, but here we discuss only the approaches most relevant
to ours. Most research efforts have concentrated on reducing the number of
visual elements to make node-link diagrams readable. We mention three different
approaches.

Aggregation techniques group vertices and edges of the graph together to obtain
a smaller graph [10]. Most techniques compute a hierarchical partitioning and

offer interaction to explore different branches of the tree. Early work by Eades and Feng [11] proposes 3-dimension visualization to navigate in this tree. Abello et al. [2] use treemaps and fisheye view to show a combination of the hierarchy levels. Later research [1] demonstrated that such hierarchy-based techniques can scale up to very large graphs (16 million edges and 200,000 nodes). Similar approaches attempt to give more clues about the content of the aggregates. Balzer and Deussen [6] represent aggregates by 3-dimensional shapes, whose sizes convey the number of vertices, with bundled edges whose thickness indicates the density of the connection. Zinmaier et al. [26] utilize the GPU to create an aggregated image of a large graph, using heatmaps to convey the number of vertices and edges in the aggregates. Van den Elzen and van Wijk [12] recently introduced a system for interactive exploration of large graphs via manual or automatic selections and aggregations.

While these techniques can scale up to very large graphs, they have several disadvantages. Aggregating nodes involves a loss of information concerning intra- and inter-connectivity. Spatial stability is another issue. The drawing may change dramatically when several entities collapse into one, potentially disorienting the user.

Multiscale techniques allow users to explore the partition hierarchy at different depths. These techniques aim at disambiguating the topology induced by aggregating vertices and edges together. Auber et al. [5] propose a clustered multiscale technique, for which the interiors of the aggregates are shown at a finer scale. However, aggregated edges are shown between clusters, risking misinterpretation. Henry et al. [18] propose a hybrid technique that can only represent one level of clustering. In a similar spirit van Ham and van Wijk [15] propose an aggregate method in which users can expand one aggregate at a time. Henry et al. [17] attempt to indicate inter-aggregate connectivity by duplicating elements, but their solution only works for a single level of clustering.

A different technique by Koren et al. [13] aims at smoothly integrating the level of detail, as opposed to discrete partitioning of the graph. The authors build a hierarchy of graphs and, for each viewpoint, construct a smaller graph by "borrowing" parts of the corresponding hierarchy levels and adjusting the layout of this smaller graph. The strength of this technique is that it avoids potentially misleading partitioning of the graph. However, there is a lack of stability: a small change in viewport may lead to a large change in the viewed graph. The fisheye technique of the paper may also add a spatial distortion, further disrupting the user's mental map.

Filtering techniques approach the visualization of large graphs by filtering the elements rendered in the view. For example, SocialAction [22] provides a set of measures to rank vertices and edges, rendering node-link diagrams with manageable sizes. A related technique by Perer and van Ham [16] proposes to build a filtered node-link diagram based on the queries made by the user, via the concept of degree-of-interest. The principal disadvantage of these techniques is the lack of overview of the entire graph. The progressive rendering approach proposed by Auber et al. [3,4] renders the node-link diagram entities in order of their importance. The rendering stops when the view changes. Given enough non-interaction time, all entities intersecting the viewport are rendered. In contrast to the previous filtering techniques, the benefit of this approach is to reveal the key features of the graph first. However, the user does not directly control the level of detail, which potentially disrupts the experience.

GMaps by Gansner et al. [14] also uses the map metaphor to draw graphs. Its main focus is on representing clusters of vertices as countries with map-like borders and coloring. The entire graph is drawn on top of the map as a node-link diagram with straight lines. When zooming in, labels of less important nodes appear gradually.

Design Rationale Motivated by Online Maps. Exploring online geographic maps is probably the most common scenario for browsing large graphs. Millions of people every day browse maps on their cellular phones or computers for finding a location or driving directions. We decided to search for key ideas used in interactive geographic maps that could be applied to browsing general graphs. One insight is that showing everything at all times is counterproductive. In a digital map on the top level we only see major cities and major roads connecting them. Objects on finer levels of detail, like smaller roads, are not shown explicitly. They may be hinted by using, for example, pre-rendered bitmap tiles. When we zoom in, other, less significant features appear and become labeled. Online maps can answer search queries such as finding a route from source to destination or showing a point of interest close to the mouse position.

Design goals identified are as follows:

1. The method should be able to reveal most details of the graph by using only the zoom in, zoom out, and pan operations. As we zoom in, more vertices and edges should appear according to their importance. Interactions such as node or edge highlighting or search by label should help discover further details.
2. During these operations, the user's mental map must be preserved. In particular, vertex positions and edge trajectories should not change between zoom levels.
3. In order to limit visual clutter, the number of rendered visual elements at each view should not exceed some predefined bound.

2   Method Description

The input to the algorithm is a graph with given node positions; the edge routes are not part of the input. The output is a set of *layers* containing nodes and edge routes. Let $G = (V, E)$ be the input graph, where V is the set of nodes and E the set of edges. The input also includes an ordering of V. This ordering should reflect the relative importance of the vertices. If such an ordering is not provided then we can sort the nodes, for example, by using PageRank [21], by node degree, or by shortest-path betweenness [8]. Finding a good order reflecting the node importance is a separate problem which is outside the scope of this research. Here we look at the node order as input and consider $V = [v_1, \ldots, v_N]$ to be an array. Before giving a detailed description of the algorithm we describe its high level steps.

We build the **layer** 0, denoted by L_0, as follows. For some number $k_0 > 0$ we assign nodes $v_1, \ldots, v_{k_0}$ to L_0 and route all edges $(v_m, v_n) \in E$ with $m, n \leq k_0$. Suppose we have already built L_{i-1} containing vertices v_j, for $j \leq k_{i-1}$. Then, if $k_{i-1} < N$, that is we have vertices that are not assigned to a layer yet, for a number $k_i \geq k_{i-1}$ we assign nodes $v_1, \ldots, v_{k_i}$ to L_i and route all edges $(v_m, v_n) \in E$ with $m, n \leq k_i$. Otherwise we are done. Note that a node can be assigned to several consecutive layers. To achieve the assignment we define a function z from V to the set $\{2^0, 2^1, 2^2, \ldots\}$. The value $z(v)$ we call the **zoom level** of the node. For $n \in \mathbb{N}_0$, the layer L_n contains node v if and only if $z(v) \leq 2^n$. For each layer an edge is represented by a set of straight line segments called **rails**. We define function z on rails too, but the layer assignment rule is different for rails; a rail r belongs to L_n iff $z(r)$ is equal to 2^n.

Algorithm 1. Setting node zoom levels with rail quota

SetNodeRailZoomLevels()

---

1 **assignedNodes** = $\emptyset$, **maximalRails** = $\emptyset$
2 tileSize = BoundingBox(G), nodeSize = InitialNodeSize, $n = 0$
3 **while** |**assignedNodes**| < |V| **do**
4   **ProcessLayer()**
5   tileSize = tileSize * 0.5, nodeSize = nodeSize * 0.5, $n = n + 1$

ProcessLayer()

---

6 initialize **tileMap** with **assignedNodes** and **maximalRails**
7 **candidateNodes** = TryAddingNodesUntilNodeQuotaFull(n)
8 **prevLayerRails** = rails of L_{n-1}, or $\emptyset$ if $n = 0$
9 M = GenerateMesh(**assignedNodes** $\cup$ **candidateNodes**, **prevLayerRails**)
10 **prevLayerRailsUpdated** = SegmentsOfMeshOn(M, **prevLayerRails**)
11 z(**prevLayerRailsUpdated**) $= 2^n$
12 **foreach** $v \in$ **candidateNodes do**
13   **rails**(v) = RailsOnEdgeRoutes(v, **assignedNodes**, M)
14   **maximalRailsOfV** = FindMaximalRails(**rails**(v))
15   **if** *adding all* **maximalRailsOfV** *to L_n exceeds rail quota* **then return**
16   set $z(v) = z$(**rails**(v)) $= 2^n$
17   update **tileMap** with v and **maximalRailsOfV**
18   **maximalRails** = **maximalRails** $\cup$ **maximalRailsOfV**
19   **assignedNodes** = **assignedNodes** $\cup \{v\}$

Calculation of Layers. Algorithm 1 computes the function z on the nodes and extends it to the rails. The flow of the algorithm is illustrated in Fig. 2.

Let B be the bounding box of G with width w and height h. For $i, j, n \in \mathbb{N}_0$ we define T_{ij}^n as the rectangle with width $w_n = w/2^n$, height $h_n = h/2^n$ and the

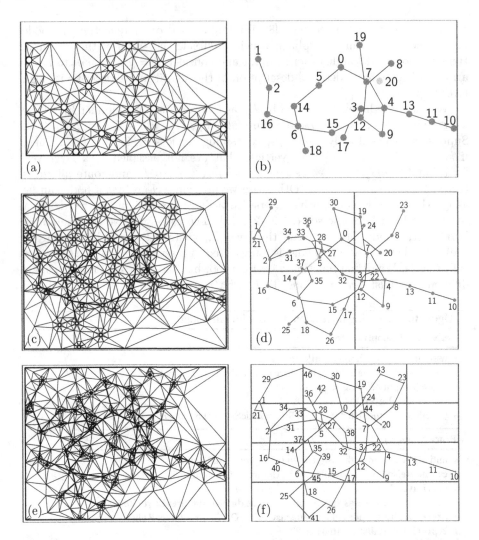

Fig. 2. Graph *abstract.dot*, $Q_N = 80$, $Q_R = 180$. (a) The mesh containing node boundaries of layer 0 (thick). (b) Nodes of layer 0 (green) and rails of edges routed using the mesh in (a) (black). Adding node 20 would exceed the node quota. (c) The mesh containing rails and node boundaries of layer 0 and boundaries of candidate nodes for layer 1 (thick). (d) Node 37 has inserted nodes 6 and 28 as neighbors. The edges incident to Node 37 are routed through red rails, which are new maximal rails. After adding the red rails the upper left tile would intersect more than $Q_R/4 = 45$ maximal rails. (e) Mesh containing rails and node boundaries of layer 1 and candidates for layer 2. (f) All nodes and edges are added to layer 2 without exceeding the quotas (Color figure online).

bottom left corner with coordinates $x = u + i \cdot w_n$ and $y = v + j \cdot h_n$, where (u, v) is the left bottom corner of B. We call T_{ij}^n a **tile**. The algorithm is driven by positive integers Q_N and Q_R, which we call **node** and **rail quota**, respectively.

We say that tile T_{ij}^n *exceeds node quota* Q_N if it intersects more than $Q_N/4$ nodes of layer n.

To work with the rail quota Q_R we need the following definition. For a set of rails R and a rail $r \in R$ we call r **maximal in** R if r is not a sub-segment of any other rail in R. During the algorithm we maintain the set of maximal rails among the set of rails already assigned to layers and count intersections between the tiles and the maximal rails only. The union of all maximal rails will always form the same set of points as the union of all rails created so far. Tile T_{ij}^n *exceeds rail quota* if it intersects more than $Q_R/4$ rails which are maximal among all rails of layer n and below. Assume both Q_N and Q_R are divisible by 4.

The outer loop of Algorithm 1 in line 3 works as follows. Starting with $n = 0$, each call to **ProcessLayer** in line 4 tries to greedily assign the nodes to the current layer. Each such attempt starts with the first unassigned node in V. Procedure **ProcessLayer** terminates if adding the next node in V and its edges incident to already assigned nodes would exceed the node or rail quota of some tile.

After calling **ProcessLayer** tile dimensions and the node size become twice smaller, and a new attempt starts for $n + 1$ in line 4. The algorithm stops when all nodes are assigned to a layer. Figure 2 illustrates Algorithm 1 for graph *abstract.dot*[1] with 47 nodes, labeled from 0 to 46 according to their order in V.

In line 6 **tileMap** is a map from $\mathbb{N}_0^2$ to $\mathbb{N}_0^2$. If for some n we have **tileMap** $(i, j) = (r, k)$, then r nodes in L_n and k maximal rails intersect T_{ij}^n.

Consider **ProcessLayer** for $n = 0$. For this case the domain of **tileMap** is $\{(0, 0)\}$. The sets **assignedNodes** and **maximalRails** are empty, and there is only one tile $T_{0,0}^0$, which has the size of B (blue in Figs. 2a and 2b). After executing line 7 the set **candidateNodes** contains the first $Q_N/4$ nodes of V (green in Fig. 2b). The boundaries of these nodes are represented by regular polygons (thick in Fig. 2a) and used to generate a triangular mesh M. The mesh is a constrained triangulation in a sense that any straight line segment of the input can still be traced in M although it can be split into several segments. The edges with both endpoints in **candidateNodes** are routed on M.

In the $i + 1$-th iteration of the loop in line 12 the algorithm tries to add node i, while nodes $0, \ldots, i - 1$ have already been added to L_0, and **tileMap** $(0,0)=(i, k)$, where $k \leq Q_R/4$ is the number of rails used by edges routed so far. All these rails are maximal rails by construction.

In line 13 the routes of edges from node i to nodes $0, \ldots, i-1$ are computed as shortest paths on M, and the set **rails**(v) is the set of all rails of these routes. In line 14 we find **maximalRailsOfV**, the rails from **rails**(v) which are maximal with respect to the set **maximalRails** $\cup$ **rails**(v). In the case of $n = 0$ they are all the rails of **rails**$(v)\setminus$ **maximalRails**. For $n \geq 1$, these are the rails from **rails**(v) covered by no rail from **maximalRails**. In Fig. 2d, such maximal rails for node 37 are drawn red.

If $T_{0,0}^0$ still contains no more than $Q_R/4$ rails after adding **maximalRailsOfV**, then node i is added to L_0. Otherwise, **ProcessLayer** terminates.

---

[1] https://github.com/ekoontz/graphviz/blob/master/rtest/graphs/abstract.dot.

In Fig. 2b, all $Q_N/4 = 20$ candidate nodes and the rails on the corresponding edges could be added to L_0.

The procedure works similarly for $n \geq 1$. One notable difference is that rails from L_{n-1} are passed as input to the mesh generator in addition to the boundaries of the appropriate nodes in line 9. For more details we refer to the in the full version [19]. Figures 2c, ..., 2f show **ProcessLayer** for $n = 1, 2$.

Using the Layers During the Visualization. Let H be a rectangle. We denote by $w(H)$ the width of H and by $h(p)$ the height of H. Recall that B is the bounding box of G. Then the *zoom level of H to B* is the value $l(H) = \min\{\frac{w(B)}{w(H)}, \frac{h(B)}{h(H)}\}$.

Let K be the transformation matrix from the graph to the user window W. Then the rectangle $P = K^{-1}(W)$, where K^{-1} is the inverse of K, is the current viewport.

To decide which elements of G are displayed to the user, we find the *zoom level $Z = l(P)$* and set the layer index $n = \max(0, \lfloor \log_2 Z \rfloor)$. Finally, the elements displayed to the user are all the nodes and rails of layer L_n intersecting P. We show in the full version [19] that, by following this strategy, we render at most Q_N nodes, and the rendered rails can be exactly covered by at most Q_R maximal rails.

Edge Routing and Overlap Removal. Consider the nodes of L_0. To construct a graph on which the edges are routed, we first create a regular polygon for each vertex. Then, we generate a triangular mesh using the *Triangle* mesh generator by Shewchuk [23]. By inserting additional vertices Triangle creates meshes with a lower-bounded minimum angle, which implies the upper-bounded vertex degree. Each edge between a pair of L_0 nodes is then assigned the corresponding Euclidean shortest path in the mesh, which is computed using the A^* algorithm. Mesh segments lying on such paths become rails of L_0 and the remaining mesh segments are discarded.

We now proceed with edge routing for L_n for $n \geq 1$. Consider Procedure **ProcessLayer**. Since the initial node placement did not take edge trajectories into account, at the beginning of the procedure some unassigned nodes might overlap the entities of L_{n-1}. We move these nodes away from their initial positions to resolve these overlaps, but this might create overlaps with the nodes that are not assigned to a layer yet.

The overlap removal process happens before line 7. We follow the metro map labeling method of Wu et al. [25]. All line segments and bounding boxes of fixed nodes are drawn on a monochromatic bitmap and the image is *dilated* by the diameter of a node on L_n. To define a position for a candidate node v at which it does not overlap already placed nodes or rails, we find a free pixel p in the image, ideally close to the initial location of v. We draw a dilated v at p and proceed with the next candidate, etc.

To generate a graph for edge routing on L_n, we use the bounding polygons of nodes from **candidateNodes** and the nodes of L_{n-1}, and the rails of L_{n-1},

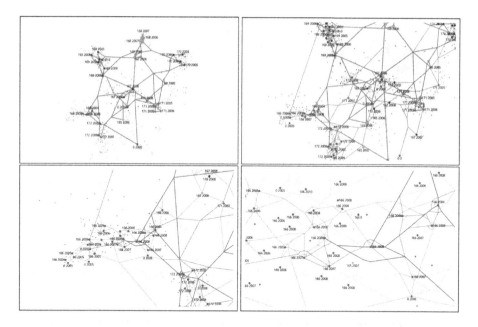

Fig. 3. Caltech graph on four different zoom levels visualized using our approach.

as the input segments for Triangle. Already routed edges maintain their trajectories, while edges incident to a node not belonging to L_{n-1} are routed over the triangulation created by Triangle in line 13. To create the bundling effect by reusing existing rails, we slightly reduce their weights during routing.

Pre-rendered Tiles. To help users gain spatial orientation, we hint the nodes which are not yet visible at the current zoom level, but will appear if we zoom in further. We create and store on the disk the images of some graph nodes and use them as the background. The images are generated very fast and are loaded and unloaded dynamically by a background thread to keep the visualization responsive. See [19] for more details.

Interaction. We define several interactions in addition to the zoom and pan. Clicking on a node (even if it is hinted, but not visible yet) highlights all edges incident to it and unhides all adjacent nodes. The highlighted elements are always shown regardless of the zoom level. Clicking on a rail highlights the most important edge passing through it, and unhides the edge endpoints. Additionally, nodes can be searched by substrings of their labels.

3 Experiments

Visualizing and Evaluating Clusterings. The test graph of our first experiment is the *Caltech* graph used by Nocaj et al. [20]. It is the graph of Facebook

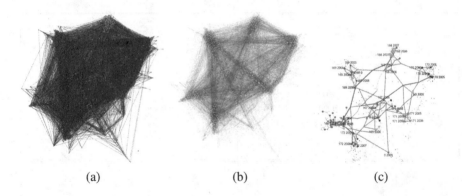

(a)   (b)   (c)

Fig. 4. (a) Caltech graph visualized using Gephi. (b), (c): showing neighbors of node "165 2007" (leftmost orange node surrounded by green nodes) using Gephi and our tool (Color figure online).

friendships at California Institute of Technology from September 2005 and contains 769 nodes and 16k edges [24]. The nodes are labeled by class year and residence, or house, of the corresponding student, and colored by the house. Label 0 denotes missing data. A computer science researcher, with focus on clustering algorithms, used GraphMaps to browse this graph. He was interested in discovering the connectivity structure of the graph, e.g., which houses or years have strong ties. The researcher's tool of choice for visualizing clusterings was *Gephi* [7]. The node layout was computed by a force-directed layout algorithm applied to the Simmelian backbone of the graph, as proposed by Nocaj et al. [20]. The result is shown in Fig. 4a. The same initial layout was used as input for GraphMaps. The resulting drawing on four different zoom levels is shown in Fig. 3.

The user noted that the view in Fig. 4a was too dense and gave no insight into the graph connectivity. On the other hand, he found our result in Fig. 3 less cluttered. The user mentioned that when looking at the drawing created by GraphMaps, for some pairs of nodes one may think that an edge between them exists, when, in fact, it does not. However, by using additional interactions besides zoom and pan, e.g., edge highlighting, the connectivity can be understood.

One interaction mode that the user tested for both tools was selecting all neighbors of a node. In Gephi, when hovering the mouse over a node, all non-incident edges and non-adjacent nodes are grayed out; see Fig. 4b. In our method, when clicking on a node, routes of all its incident edges are highlighted and, additionally, all adjacent nodes are shown, regardless of the zoom level; see Fig. 4c. According to the user, both methods provided satisfactory results. He noted that GraphMaps, by using edge bundling, provides a tidier picture than Gephi.

For dense graphs, like Caltech, the user would prefer to view the neighbors of a node in GraphMaps. The user commented that, contrary to Gephi, GraphMaps exposes the most important nodes and their labels in a readable fashion.

Experiments with Other Graphs. In the video at http://1drv.ms/1IsBEVh we demonstrate browsing the graph "composers"[2] with GraphMaps. The nodes of the graph represent the articles on Wikipedia on composers, and the edges represent Internet links between the articles. We show the user interactions that help us to explore the graph.

When browsing the graph of InfoVis coauthors, created from ACM data, another user was able to notice two groups of coauthors, one connected to Peter Eades, and another one to Ulrik Brandes and Michael Goodrich. By selecting all direct neighbors of Peter Eades, the user was able to see that only one member of the second group, Roberto Tamassia, has a paper with Peter Eades; see the corresponding figure in the full version [19]. Further analysis showed that, according to the data set, Roberto Tamassia, is the only author with coauthors from both groups. GraphMaps enabled the user to gain insights on the graph structure.

Running Time. GraphMaps processes a graph with 1463 nodes and 5806 edges for 1 min, a graph with 3405 nodes and 13832 edges for 130 seconds, and a graph with 38395 nodes and 85763 for less than 6 hours. The experiments were done on an HP-Z820 with Intel Xeon CPU E5-2690 under Windows 8.1. The required memory was 16 GB. The current bottleneck in performance is the edge routing. We hope to speed up the edge routing by using parallel processing.

The Sources of GraphMaps. GraphMaps is implemented in MSAGL, which is available as Open Source at github.com/Microsoft/automatic-graph-layout.

4   Discussion

The users of GraphMaps appreciate its aesthetics and the similarity to browsing online maps. GraphMaps helps in gaining the first impression of the graph structure and, in spite of the fact that precise knowledge of the connectivity cannot be obtained with GraphMaps by zooming and panning alone, additional interactions allow answering the queries as, for example, finding if two nodes are direct neighbors. A current shortcoming of GraphMaps is that the direction of the edges is lost. It happens for other methods as well, when edges are bundled. Solving this issue is a possible future work item. The labeling algorithm needs improvement, since it does not always respect the node ranking and does not always utilize free space well enough.

Future Work. Currently we cannot guarantee that our layer generation algorithm always reaches the end, although, in all our experiments it did. Creating a version of the algorithm which provably stops, or, even better, guarantees that

---

[2] http://www.graphdrawing.de/contest2011/topic2-2011.html.

the number of generated layers is within predefined bounds, is a very interesting problem.

Another problem is finding a node placement that works nicely with the node ranking to improve the distribution of nodes among levels. Ideally, such a layout algorithm is aware of the edge routing too, and avoids the overlap removal step.

Acknowledgements. We are grateful to Roberto Sonnino for the useful discussions on the rendering of the tile images in a background thread, and to Itzhak Benenson for sharing with us his ideas on the visualization style.

References

1. Abello, J., van Ham, F., Krishnan, N.: Ask-graphview: a large scale graph visualization system. IEEE Trans. Vis. Comput. Graph. **12**(5), 669–676 (2006)
2. Abello, J., Kobourov, S.G., Yusufov, R.: Visualizing large graphs with compoundfisheye views and treemaps. In: Pach, J. (ed.) GD 2004. LNCS, vol. 3383, pp. 431–441. Springer, Heidelberg (2005)
3. Auber, D.: Using Strahler numbers for real time visual exploration of huge graphs. In: Computer Vision and Graphics (ICCVG'02), pp. 56–69 (2002)
4. Auber, D.: Tulip - a huge graph visualization framework. In: Graph Drawing Software, pp. 105–126 (2004)
5. Auber, D., Chiricota, Y., Jourdan, F., Melançon, G.: Multiscale visualization of small world networks. In: IEEE Symposium on Information Visualization (INFOVIS'03), pp. 75–81 (2003)
6. Balzer, M., Deussen, O.: Level-of-detail visualization of clustered graph layouts. In: Asia-Pacific Symposium on Information Visualisation (APVIS'07), pp. 133–140. IEEE (2007)
7. Bastian, M., Heymann, S., Jacomy, M.: Gephi: an open source software for exploring and manipulating networks. In: International AAAI Conference on Weblogs and Social Media (ICWSM'09) (2009)
8. Brandes, U.: A faster algorithm for betweenness centrality. J. Math. Sociol. **25**(2), 163–177 (2001)
9. Brandes, U., Pich, C.: Eigensolver methods for progressive multidimensional scaling of large data. In: Kaufmann, M., Wagner, D. (eds.) GD 2006. LNCS, vol. 4372, pp. 42–53. Springer, Heidelberg (2007)
10. Brunel, E., Gemsa, A., Krug, M., Rutter, I., Wagner, D.: Generalizing geometric graphs. In: Speckmann, B. (ed.) GD 2011. LNCS, vol. 7034, pp. 179–190. Springer, Heidelberg (2011)
11. Eades, P., Feng, Q.-W.: Multilevel visualization of clustered graphs. In: North, S.C. (ed.) GD 1996. LNCS, vol. 1190, pp. 101–112. Springer, Heidelberg (1997)
12. van den Elzen, S., van Wijk, J.: Multivariate network exploration and presentation: from detail to overview via selections and aggregations. IEEE Trans. Vis. Comput. Graph. **20**(12), 2310–2319 (2014)
13. Gansner, E.R., Koren, Y., North, S.C.: Topological fisheye views for visualizing large graphs. IEEE Trans. Vis. Comput. Graph. **11**(4), 457–468 (2005)
14. Gansner, E., Hu, Y., Kobourov, S.: Gmap: visualizing graphs and clusters as maps. In: IEEE Pacific Visualization Symposium (PacificVis'10), pp. 201–208. IEEE (2010)

15. van Ham, F., van Wijk, J.: Interactive visualization of small world graphs. In: IEEE Symposium on Information Visualization (INFOVIS'04), pp. 199–206 (2004)
16. van Ham, F., Perer, A.: Search, show context, expand on demand: supporting large graph exploration with degree-of-interest. IEEE Trans. Vis. Comput. Graph. **15**(6), 953–960 (2009)
17. Henry, N., Bezerianos, A., Fekete, J.D.: Improving the readability of clustered social networks using node duplication. IEEE Trans. Vis. Comput. Graph. **14**(6), 1317–1324 (2008)
18. Henry, N., Fekete, J.D., McGuffin, M.J.: NodeTrix: a hybrid visualization of social networks. IEEE Trans. Vis. Comput. Graph. **13**(6), 1302–1309 (2007)
19. Nachmanson, L., Prutkin, R., Lee, B., Riche, N.H., Holroyd, A.E., Chen, X.: Graphmaps: Browsing large graphs as interactive maps. CoRR arXiv:1506.06745 (2015)
20. Nocaj, A., Ortmann, M., Brandes, U.: Untangling hairballs. In: Duncan, C., Symvonis, A. (eds.) GD 2014. LNCS, vol. 8871, pp. 101–112. Springer, Heidelberg (2014)
21. Page, L., Brin, S., Motwani, R., Winograd, T.: The PageRank citation ranking: Bringing order to the web. Technical report 1999–66, Stanford InfoLab (1999)
22. Perer, A., Shneiderman, B.: Balancing systematic and flexible exploration of social networks. IEEE Trans. Vis. Comput. Graph. **12**(5), 693–700 (2006)
23. Shewchuk, J.R.: Delaunay refinement algorithms for triangular mesh generation. Comput. Geom. Theory Appl. **22**(1–3), 21–74 (2002)
24. Traud, A.L., Kelsic, E.D., Mucha, P.J., Porter, M.A.: Comparing community structure to characteristics in online collegiate social networks. SIAM Rev. **53**(3), 526–543 (2011)
25. Wu, H.-Y., Takahashi, S., Lin, C.-C., Yen, H.-C.: A zone-based approach for placing annotation labels on metro maps. In: Dickmann, L., Volkmann, G., Malaka, R., Boll, S., Krüger, A., Olivier, P. (eds.) SG 2011. LNCS, vol. 6815, pp. 91–102. Springer, Heidelberg (2011)
26. Zinsmaier, M., Brandes, U., Deussen, O., Strobelt, H.: Interactive level-of-detail rendering of large graphs. IEEE Trans. Vis. Comput. Graph. **18**(12), 2486–2495 (2012)

An Incremental Layout Method for Visualizing Online Dynamic Graphs

Tarik Crnovrsanin[✉], Jacqueline Chu, and Kwan-Liu Ma

University of California, Davis, USA
{tecrnovr,sjchu}@ucdavis.edu, ma@cs.ucdavis.edu

Abstract. Graphs provide a visual means for examining relation data and force-directed methods are often used to lay out graphs for viewing. Making sense of a dynamic graph as it evolves over time is challenging, and previous force-directed methods were designed for static graphs. In this paper, we present an incremental version of a multilevel multi-pole layout method with a refinement scheme incorporated, which enables us to visualize online dynamic networks while maintaining a mental map of the graph structure. We demonstrate the effectiveness of our method and compare it to previous methods using several network data sets.

Keywords: Dynamic graphs · Streaming data · Graph layout

1 Introduction

In many fields of study, from biology to chemistry to sociology, software engineering and cyber security, an essential task is to identify and understand relationships of interest among different entities. Graphs in the form of nodes and links are commonly used to represent such relations. Graph drawing is an indispensable tool for visually studying the relationships. Many techniques have been introduced for aesthetically and efficiently laying out static graphs [11,13,14,16,17], but a large class of real-world applications involve graphs that change over time [8].

Visualizing dynamic graphs is often done by animating over the sequence of graphs [3,9,10,22] or by displaying selected ones side-by-side as small multiples [25]. Finding the best way to visualize dynamic graphs remains a challenging research topic. When laying out dynamic graphs for visual analysis, the primary goal is to ensure the stability of the layout [5,10,15,18] and preserve the mental map [1,21–23].

Most previous dynamic graph algorithms address the problem of laying out offline graphs consisting of the entire sequence of graphs. With prior knowledge of the complete time sequence, we can best optimize the layout for animation and specific analysis goals [4,8,9,19]. For online, real-time monitoring or analysis applications, however, the graph is constantly updated and how the graph might change over time cannot be predicted. Making optimal layouts for such evolving graphs is an even more challenging problem, which has received limited attention

© Springer International Publishing Switzerland 2015
E. Di Giacomo and A. Lubiw (Eds.): GD 2015, LNCS 9411, pp. 16–29, 2015.
DOI: 10.1007/978-3-319-27261-0_2

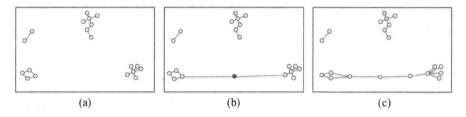

Fig. 1. An undesirable limitation. A graph has many disconnected components (a), and a node is introduced linking two components together (b). One layout method [10] allows the new node (in orange) and its neighbors to move after the new node is added. If these nodes cannot reach their ideal position in a single time step, they are affixed to the same positions (c) until new nodes or edges are later introduced into the same neighborhood (Color figure online).

in existing research [6,10,12,19]. One reason is that online dynamic graph data were not readily available, but the situation has begun to change with the rapid growth of mobile, online, and real-time monitoring applications. Consequently, demands for the ability to understand online dynamic graph data have arisen in various fields.

We have examined previous online dynamic graph layout methods and found they have some undesirable limitations in layout quality or the connectivity of the graph. Some are too expensive to use for real-time applications. In order to speed up the process of laying out an online dynamic graph, a commonly employed approach is to anchor large portions of the graph and allow only a small subset of the graph to move; nevertheless, this speedup comes with several tradeoffs. One tradeoff is that when a new node or an edge is introduced, only that node and its neighbors are allowed to move at that instance. In most cases, nodes are placed near their ideal spots. If two disconnected components merge, the nodes usually cannot reach their ideal spots at once, as depicted in Fig. 1. In addition, linking disconnected graphs may lead to edge crossings. Parts of the graph would stay in suboptimal positions, unless new nodes or edges are added to the same neighborhood to allow the layout algorithm to fix this problem.

In this paper, we present an incremental version of the multilevel multi-pole layout method that is suitable for visualizing online dynamic graphs. Our work makes the following contributions to online dynamic graph drawing:

- The incremental layout method reduces the computation cost while best characterizing inherent network structure and maintaining graph readability.
- Our refinement technique reduces edge crossings and long edges by using the nodes' energy to determine correct placement.
- The refinement technique can be applied independently or in tandem with an existing force-directed layout method.
- The layout is fast because our implementation for both the layout and refinement calculations are GPU-accelerated.

We evaluated our methods using several dynamic graph data sets, including ones from real-world online applications, and compared the layouts with those produced by previous methods. The test results demonstrate the effectiveness and usability of our method.

2   Related Work

Online dynamic graphs are series of graphs in which time steps are not known ahead of time. Lee et al. [19] was one of the earliest to work with online graphs. The algorithm preserves the mental map while generating aesthetically pleasing graphs. The drawback is that the algorithm is slow, recalculating the full layout at each time step. Brandes and Wagner [6] instead used Bayesian decision theory to generate the graph. Their work characterizes the tradeoff between dynamic stability and local quality using conditional probabilities. Frishman and Tal [10] created a novel force-directed algorithm that can handle large graphs. Their GPU implementation provides a 17 times speedup over the CPU version. Gorochowski et al. [12] used the age of the node to stabilize the graph. The age was calculated based on when the node appeared and how much movement it saw through its life. Che et al. [7] proposed a novel layout algorithm that enforces graph component shapes by using Laplacian constrained distance embedding. However, the Gorochowski et al. and Frishman and Tal algorithms do not address the disconnected component problem mentioned in the introduction.

Our layout method addresses this problem by using a novel refinement technique that gradually alleviates areas of high energy. Energy is defined as the amount of force applied to a node. In Sect. 4, our evaluation shows that our method produces more aesthetic graphs at the cost of more movement in the graph. This movement is necessary to reduce long edges and edge crossings that occur.

3   An Incremental Algorithm

Our incremental algorithm is a modified version of FM^3, which is a fast, multlievel, multi-pole, force-directed layout method. What makes FM^3 fast is that it does not calculate all the repulsion forces, which is the most expensive operation of any force-directed calculation. For a single time step, given the finest level of the graph $G = (V, E) = G_0$, FM^3 reduces computation by partitioning and collapsing G_0 until reaching a prescribed number of nodes. This subset of nodes represents the coarsest level, K. A force calculation is applied to this graph G_K, where the resulting node positions are used as the initial layout for the next finer graph, G_{K-1}. These steps are repeated until the original graph G_0 is drawn. More details of FM^3 can be found in Godiyal [11], which our GPU-accelerated implementation is based on.

FM^3 is not designed for online dynamic graphs drawing. To make it incremental, we need to:

1. Include an initial layout construction step
2. Add a merging step, which includes placing new nodes and selecting nodes to move

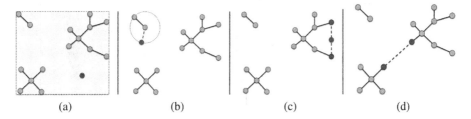

(a)   (b)   (c)   (d)

Fig. 2. The figure shows how our algorithm assigns positions to new nodes. A dashed node and edge indicate a new node and edge, respectively. Nodes colored in orange represent nodes that are flagged to move by our algorithm. (a) A node with no edges is placed randomly inside the bounding box of the graph. (b) A node connected to a positioned node is placed at a desired length, dl, from the positioned node. (c) A node connected to at least two positioned nodes is placed at the centroid of the position nodes. (d) When an edge is added or removed between two positioned nodes, our algorithm flags both nodes to move (Color figure online).

3. Modify the multilevel calculation step
4. Pick a specific force model for the force calculation step
5. Add an animation step for smooth transition of the layout rendering.

We do not modify the multi-pole calculation step. We describe each of these five steps in more detail below.

Initial Layout Construction: For the initial layout, L_0, we use standard FM^3 layout with a degree metric for the selection of the super nodes, which is described in the Multilevel Calculation section.

Merging: This stage attempts to place new nodes at their ideal positions by using affixed nodes from the previous layout. Initial node placement is imperative because error is introduced when previously positioned nodes are at their suboptimal positions. This error propagates across layouts, making it difficult to correct in subsequent time steps.

Our approach minimizes this error by assigning coordinates to new nodes in the following manner. Positioned nodes from L_{i-1} are copied over to L_i. If a new node v is not connected to any other positioned node, v is placed in a random position within the bounds of the graph, as shown in Fig. 2a. If v is connected to one positioned node u, v is placed randomly around u at a distance dl. dl is the desired length between two connected nodes in our spring-based energy model, as seen in Fig. 2b. If v is connected to at least two positioned nodes, v is placed at the geometric center of all the connected nodes, shown in Fig. 2c. All affected nodes are flagged to move.

In our merging stage, the insertion or deletion of edges affects node placement. If an edge is inserted between two new nodes, u and v, node u is randomly placed inside the bounding box, similar to Fig. 2a, and node v is placed randomly around u at a radius of dl, equivalent to Fig. 2b. Both nodes are selected to move. Also, our method moves affixed nodes when a new edge is introduced to another node–whether new or affixed–namely, when node u is connected and node v is

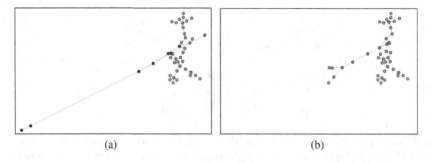

(a)   (b)

Fig. 3. Energy levels are mapped from yellow to red, where red represents high energy. Refinement allows only nodes with high energy (a) to move until they reach a low energy state, which is represented in yellow (b) (Color figure online).

not. Since it is not restricted by other nodes, node v is randomly placed around node u at a distance dl as if it were a new node and is marked to move. Another instance of node placement is the change of connectivity between positioned nodes u and v. When an edge is removed, the two affected nodes are flagged to move because their current positions are invalid and should move closer to their respective components. After adding an edge, we flag both nodes to move, shown in Fig. 2d, to minimize overlapping edges in case these components are distant from one another.

Multilevel Calculation: In FM^3, the process of picking a super node–a single node that represents a large set of nodes from the finer levels–is done randomly or by indexing [11]. When dealing with multiple levels from the coarsening of G_0, our method is more deterministic when selecting a super node than FM^3's multilevel approach. A super node is selected by having the highest degree. A new node will have a low chance of becoming a super node, but the likelihood increases when its degree increases.

Having a multilevel representation of the graph alleviates the computation time. In incremental layout methods, including ours, only nodes within a certain vicinity have their forces calculated. Also, coarser graph levels have cheaper computation compared to the original graph because force calculations are done on the super nodes. Starting from the coarsest level, the super node's resulting movement is used to interpolate the movement of its adjacent nodes at the next finer graph level, until the finest level G_0 is reached. In our method, when we calculate L_i, we compute the layout 250 times at the coarsest level. The number of iterations to compute the layout linearly decays per level until we reach the finest level. At the finest level, we compute the layout 30 times.

Our method uses a contribution factor to restrict the super nodes' range of movement. This solves the problem of nodes at coarser levels of the graph having greater range of movement than those at finer levels. Without the contribution factor, large disparities of movement occur due to simulated annealing. This causes suboptimal node positioning, which ultimately degrades the final graph level at G_0. The contribution factor is determined by how many nodes are allowed to move under the super node. For instance, if there is only one node that is allowed to move under a super node, then the super node will only move slightly.

Force Calculation: Our repulsive forces are modeled as

$$F^{rep} = \frac{C * (u - v)}{\|u - v\|^3} \tag{1}$$

to achieve a greater spreading of disconnected graph components. Our spring forces can be modeled as [11]

$$F^{spring} = \|u - v\| * log\left(\frac{\|u - v\|}{dl}\right) * (u - v) \tag{2}$$

where C is the repulsive constant and dl is the desired length between two nodes. In practice, we found that C as 4.0 and dl as 0.055 works best with our implementation.

Animation: Animation is employed to display the graph changes between L_{i-1} to L_i. Existing nodes smoothly transition into their new positions from L_{i-1} to L_i. New nodes do not exist in L_{i-1} and must be introduced into L_i.

By default, we use Graph Diaries [2], an animation mode that uses different stages to emphasize graph changes, such as deletion, movement, and addition.

3.1   Refinement Method

Our method allows nodes to reposition themselves if high energy, which is characterized by long edges and edge crossings, exists between their components. This occurrence is not adequately addressed by previous methods. We minimize this effect by refining a subset of the graph which not only reduces the cost of refinement, but shortens long edges–a result of minimizing the total energy in these components.

We expect that refinement is best used when it runs independently from the layout method. However, a layout method may not have a sufficient time window to apply refinement in between time steps. A possible option is to incorporate refinement directly into the layout method. However, such integrated refinement has limited opportunities to fix the graph as it is only called once before the main layout algorithm is executed.

We describe our implementation, which makes refinement a viable option for existing layout methods. In our refinement technique, we compute the layout for the finest level of the graph. Although the original graph, G_0, does not leverage the multilevel algorithm, we run the layout step for a subset of the graph that has been marked to have high energy. In addition, refinement runs the layout step for a set number of iterations. This is an adjustable parameter, in which reducing the number of steps trades quality for speed. In our implementation, we have set this number to 20. We modify the temperature factor, defined in FM^3, to "anneal" nodes to their final positions. This factor affects the mental map's quality [24] and complements our force model. In our system, we set temperature to 1.0.

Ideally, we want an approach that will gradually modify the graph, but only move high energy nodes. This reduces the overall energy in the system. We calculate the energy per node by deriving the relation $F = \nabla En$ [10]. Given two nodes, positioned at u and v, the repulsive energy is calculated by

$$En^{rep} = \frac{-C}{\|u - v\|} \tag{3}$$

The spring energy is calculated by

$$En^{spring} = \frac{1}{9} * (\|u - v\|^3 * (log\left(\frac{\|u - v\|}{dl}\right) - 1) + dl^3) \tag{4}$$

The total energy for node v is computed by summing over all edges connected to v and all v and u node pairs: $En(v) = En^{rep} + En^{spring}$.

$$En(v) = \sum_{u,v \in V, u \neq v} \frac{-C}{\|u - v\|} + \sum_{u:(u,v) \in E} \|u - v\| * log\left(\frac{\|u - v\|}{dl}\right) * (u - v) \tag{5}$$

Calculating the energy for nodes takes $O(N^2 + E)$ time, where N is the number of nodes and E is the number of edges. The cost comes from the all-pair computation. Computing a single iteration of the layout is $O(N * log(N) + E)$, making the computation of energy more expensive. In most cases, the cost of one energy computation is cheaper than the cost of computing the entire layout. It is possible to achieve the same cost for the energy computation by leveraging FM^3's multi-pole method to estimate the energy instead. Since FM^3 uses a kd-tree for traversal [11], this adds another $O(N * log(N))$ cost for the multi-pole estimation. The estimation will be faster with large graphs.

Once we quantify the energy for individual nodes, we need to determine when a node's energy is high in relation to the entire system. Every introduced node or edge increases the total energy of the system, making it difficult to define high energy. A simple approach is to subtract graph G_k from G_{k-1} to see which nodes have high energy. However, this is only conclusive for the current time step and nodes that gradually increase in energy over time will not be detected.

Instead, we take the mean of the nodes' energy, μ, and compare it against each node, yielding a definition of high energy. The mean scales with the total energy, U^{Total}, and allows us to compare the individual nodes. Thus, we define a node to have high energy when $\frac{abs(U^{Total} - \mu)}{\mu} > K$, where K is a user-defined constant. In our implementation, we define K to be 1.

4   Evaluation

In this section, we evaluate our layout method visually and use a series of metrics to examine the stability, quality and time of our layout method for comparison

Table 1. Comparison of layout methods using energy, Δ position, and time. Lowest quantities are in bold. Results characterize the graphs' state throughout the observed session. Energy is the total energy in the system, Δ position is the change in nodes' position, and time is measured in seconds.

Layout method	McFarland			Stack overflow-live			Stack overflow			Facebook		
	Energy	Δpos	Time	Energy	Δpos	Time	Energy	Δpos	Time	Energy	Δpos	Time
Pinning	1584	2.48	0.020	137651	**119**	0.067	1457k	**151**	**0.084**	14803k	**308**	0.208
Aging	**25.12**	0.747	0.021	546130	272	0.061	113188k	658	0.085	186310k	1019	**0.131**
Our Layout	1159	**0.745**	**0.008**	**24764**	350	**0.059**	**862k**	658	**0.084**	**9724k**	3042	0.133

against existing methods. We apply our refinement technique to these methods to show the benefits of relieving high energy areas when nodes are placed in suboptimal positions. We discuss the details of the metrics used to characterize the graphs' state. We use a combination of real and synthesized data sets that vary in both size and the number of time steps.

4.1 Layout Methods

The evaluation of our layout method is done by comparing it against two advanced online dynamic layout methods called "pinning", by Frishman and Tal [10], and "aging", by Gorochowski et al. [12]. Pinning reduces calculation by allowing recently updated nodes and their neighbors to move. Nodes closer to the recently updated node have wider range of movement. Aging uses an "aging factor" that is quantified by a relationship between the node's age and how much its immediate neighborhood has changed over time. Nodes that are younger, or experience a large amount of change around them, have more freedom to move. We could not find existing implementation of these algorithms, so we implemented them according to their respective papers. Any assumptions made when implementing these methods can be found in the Appendix (http://vis.cs.ucdavis.edu/papers/tarik_incremental_appendix.pdf).

4.2 Data Sets

We use four data sets with varying size and velocity. The first data set is taken from McFarland's study [20] which documents student interaction in a classroom. The visualization of this graph shows clusters that expand, shrink, and split over time. This data set is our smallest graph, with 20 nodes and 82 time steps. We use the McFarland data set for direct comparison against pinning and aging algorithms since their results are shown in Gorochowski et al's work.

The second and third data sets are from Stack Overflow, a forum where individuals post questions about programming. Users not only answer questions, but also provide feedback to the questions and supplied answers. Users are rewarded points when they post popular questions, answers, or comments. The first Stack Overflow data set is a one-month trial run of the collection in November 2014. The data set starts with 304 nodes and 606 edges and expands to 4000 nodes and 5000 edges. The second data set, Stack Overflow Live, is a live feed of the site.

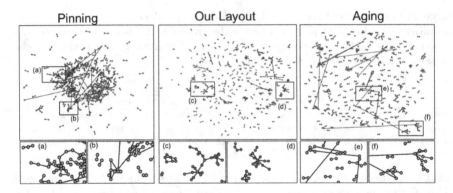

Fig. 4. Visualization of the Stack Overflow-Live data using pinning, our layout method with independent refinement, and aging at the same instance. Pinning tends to have nodes near the center due to its central attractive force, whereas aging and our layout have nodes spread out across the viewing space. Pinning and aging generate long edges and edge crossings (a,b,e,f)–characteristics which degrade the graph over time. With refinement, our method relieves this problem by shifting parts of the graph to lower the system's energy (c,d).

At the time of the measuring for generating Table 1, the data set started with 80 nodes and 80 edges and ended with 638 nodes and 964 edges. Both data sets are characterized by many small, independent components that merge together over time.

The last data set is from Facebook and is acquired from a website hosting collections of streaming graph data sets [26]. This data set starts with 822 nodes and 1160 edges and expands to 1268 nodes and 2004 edges. The Facebook data set focuses on connections between individuals. The data set, an example of a small world graph, is characterized by one large cluster and many small clusters.

4.3   Metrics

Stability is synonymous to the preservation of the mental map. Stability measures the amount of change in a graph by quantifying the change in position for all nodes or the distance a node moved. New nodes' change in position is 0 at the first time step they are introduced.

Timing is measured before and after every layout computation call. We use the average time across layout computations to assess the speed of layout methods. The speed of our refinement technique is difficult to measure because it runs when the layout is waiting for new data. Therefore, it is not part of the layout step and can be considered free as it is not taking away from the computation of the layout.

Selecting a quality metric to evaluate dynamic layouts is difficult. There have been few studies looking at the importance of preserving the mental map in dynamic layouts [23,24]. We define quality as the measurement of energy, where low energy produces aesthetically-pleasing graphs–nodes are placed at optimal

Table 2. Comparison of pinning with and without refinement, using energy, Δ position, and time to measure the performance. Lowest quantities are in bold. Results characterize the graphs' state throughout the observed session. Energy is the total energy in the system, Δ position is the change in nodes' position, and time is measured in seconds.

Layout method	McFarland			Stack oveflow-live			Stack overflow			Facebook		
	Energy	Δpos	Time	Energy	Δpos	Time	Energy	Δpos	Time	Energy	Δpos	Time
Pinning	1584	**2.48**	**0.020**	138k	**119**	**0.067**	1457k	**151**	0.084	14803k	**308**	**0.208**
Pinning+ref	**671**	6.92	**0.020**	**42.9k**	294	0.082	**43.8k**	317	**0.0124**	**76.6k**	409	0.231

edge lengths from each other, making the graph's structure easy to comprehend. We use the total energy of the system to match the metrics used by Frishman and Tal [10] and Gorochowski et al. [12]. Since our refinement technique uses our energy model to determine which nodes have high energy, we simply sum the energy for all nodes as such

$$En^{total} = \sum_{i=1...n} En(i) \tag{6}$$

where n is the total number of nodes.

To ensure fair comparisons of layout quality, all layout implementations use the same force model. Aging naturally uses our force model, since it is built upon our layout method. Our pinning implementation uses our spring-system force model.

4.4   Analysis of Our Layout Method

The results of our study are given in Tables 1 and 2, Figs. 4 and 5, and http://vis.cs.ucdavis.edu/Videos/Incr.mp4.

The evaluation is conducted on a Macbook Pro laptop. It has an Nvidia GeForce GT 750M graphics card, a 2.3 GHz Intel Core i7 processor, and 16 gigabytes of RAM.

Quality, Stability, and Timing Comparisons: Table 1 is the quantitative comparison amongst our layout method, pinning, and aging. Figure 4 shows a visual comparison of the three layouts for Stack Overflow-Live data set. In the pinning results, a distinct ring of nodes forms. The ring is a consequence of the pinning implementation, which places new nodes with no edges around this ring. Nodes are spread out in aging and our layout method because nodes are placed randomly inside the bounding box.

From Table 1, we can see in most cases our layout has the lowest energy. We observe around 1.5 to 5.5 times improvement over pinning and 19 to 133 times for aging. The low energy is attributed to the layout gradually repairing itself. This translates visually, where our layout method better handles merging of distant components than the other two methods. Our layout reduces long edges or edge crossings, whereas these problems are evident in the other two layouts due to their high energy.

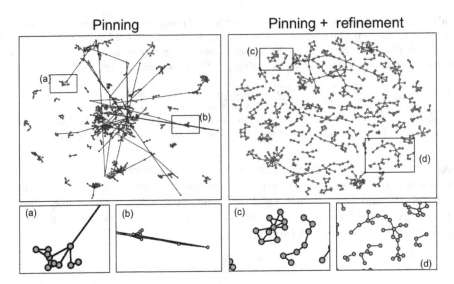

Fig. 5. Visualization of the Stack Overflow data, comparing solely pinning and pinning with independent refinement at the same instance. Many of the same trends found in Fig. 4 are observed in this visualization. Pinning suffers from long edges and edge crossings (a,b), which are fixed when refinement is added (c,d).

The layout's stability is analyzed using an average Δ position. In general, the pinning layout has the smallest average Δ position because it uses pinning weights to minimize node movement in order to produce a stable layout. Our layout has a higher Δ position because nodes are shifting into a better position to reduce energy. Aging also suffers from large Δ position. This is explained by the layout attempting to shorten long edges. In all accounts, our refinement technique increases node movement in favor of gradually fixing the graph, evident in Fig. 4.

Across the layout implementations, there are small differences in speed when computing layouts. Based on our results, additional force calculations do not necessarily increase computation time. This is likely attributed to how nodes are partitioned and bottlenecks found in the GPU. The GPU may not be fully utilized when running the force calculations. For each node, a GPU thread is created for each kd-tree leaf for force calculations. Depending on the implementation, a kd-tree can have leaves that vary in size from 4 to 20 nodes. A bottleneck occurs when the GPU is waiting for kd-tree leaf nodes that take longer to process.

Refinement with Pinning: Table 2 shows the results of applying our refinement technique to pinning. With refinement, pinning has 3 to 200 times lower energy. As expected, the refinement version takes longer to calculate than pinning by itself. However, this extra time is negligible, as refinement is meant to run while the layout is idle. Similar to Table 1, pinning with refinement has higher Δ position than pinning. From Fig. 5, we can see that extra movement is used to fix long edges and spread out nodes.

Figure 5 shows the benefits of our refinement technique. We can see that long edges or edge crossings are less evident on the right figure. The added benefit is that refinement helps spread out the nodes in each component, making it easier to see the structure.

The previous layout methods used in our evaluation have unique benefits. Pinning maintains graph stability using pinning weights to restrict node movement. Aging provides an anchor point for graph exploration by moving nodes based on their evolutionary changes. However, our layout algorithm places nodes at their optimal positions by considering each node's energy. Our refinement technique identifies high energy components in the graph and reduces the system's energy by gradually moving nodes to a lower energy state. Our results show that our refinement technique can be used to improve existing layout methods with respect to both layout quality and aesthetics, creating graph drawings that best visualize the relations between involved entities.

5 Conclusion

We have presented an incremental layout method and a refinement technique for visualizing online dynamic graphs that is used to create stable and aesthetic layouts. First, we have shown how to convert FM^3 into an incremental multilevel multi-pole algorithm. Second, our refinement technique is used to identify high energy nodes and move them to a low energy state. The refinement technique can be used in tandem or separately from the layout method. Lastly, we are able to employ a GPU to accelerate the layout and refinement technique. An empirical evaluation with metrics shows that our method helps improve the stability and aesthetic appeal of layouts.

Acnowledgments. This research is sponsored in part by the U.S. National Science Foundation via grants NSF DRL-1323214 and NSF IIS-1320229, the U.S. Department of Energy through grant DE-FC02-12ER26072, and also the UC Davis RISE program.

References

1. Archambault, D., Purchase, H.C., Pinaud, B.: Animation, small multiples, and the effect of mental map preservation in dynamic graphs. IEEE Trans. Vis. Comput. Graph. **17**(4), 539–552 (2011)
2. Bach, B., Pietriga, E., Fekete, J.D.: GraphDiaries: animated transitions and temporal navigation for dynamic networks. IEEE Trans. Vis. Comput. Graph. **20**(5), 740–754 (2014)
3. Boitmanis, K., Brandes, U., Pich, C.: Visualizing internet evolution on the autonomous systems level. In: Hong, S.-H., Nishizeki, T., Quan, W. (eds.) GD 2007. LNCS, vol. 4875, pp. 365–376. Springer, Heidelberg (2008)
4. Brandes, U., Fleischer, D., Puppe, T.: Dynamic spectral layout of small worlds. In: Healy, P., Nikolov, N.S. (eds.) GD 2005. LNCS, vol. 3843, pp. 25–36. Springer, Heidelberg (2006)

5. Brandes, U., Mader, M.: A quantitative comparison of stress-minimization approaches for offline dynamic graph drawing. In: Speckmann, B. (ed.) GD 2011. LNCS, vol. 7034, pp. 99–110. Springer, Heidelberg (2011)

6. Brandes, U., Wagner, D.: A bayesian paradigm for dynamic graph layout. In: Di Battista, G. (ed.) GD 1997. LNCS, vol. 1353, pp. 236–247. Springer, Heidelberg (1997)

7. Che, L., Liang, J., Yuan, X., Shen, J., Xu, J., Li, Y.: Laplacian-based dynamic graph visualization. In: Visualization Symposium (PacificVis), 2015 IEEE Pacific, pp. 69–73 (2015)

8. Diehl, S., Görg, C.: Graphs, they are changing. In: Goodrich, M.T., Kobourov, S.G. (eds.) GD 2002. LNCS, vol. 2528, pp. 23–31. Springer, Heidelberg (2002)

9. Erten, C., Harding, P.J., Kobourov, S.G., Wampler, K., Yee, G.: Graphael: graph animations with evolving layouts. In: Liotta, G. (ed.) GD 2003. LNCS, vol. 2912, pp. 98–110. Springer, Heidelberg (2004)

10. Frishman, Y., Tal, A.: Online dynamic graph drawing. IEEE Trans. Vis. Comput. Graph. **14**(4), 727–740 (2008)

11. Godiyal, A., Hoberock, J., Garland, M., Hart, J.C.: Rapid multipole graph drawing on the GPU. In: Tollis, I.G., Patrignani, M. (eds.) GD 2008. LNCS, vol. 5417, pp. 90–101. Springer, Heidelberg (2009)

12. Gorochowski, T., di Bernardo, M., Grierson, C.: Using aging to visually uncover evolutionary processes on networks. IEEE Trans. Vis. Comput. Graph. **18**(8), 1343–1352 (2012)

13. Hachul, S., Jünger, M.: An experimental comparison of fast algorithms for drawing general large graphs. In: Healy, P., Nikolov, N.S. (eds.) GD 2005. LNCS, vol. 3843, pp. 235–250. Springer, Heidelberg (2006)

14. Harel, D., Koren, Y.: A fast multi-scale method for drawing large graphs. In: Marks, J. (ed.) GD 2000. LNCS, vol. 1984, pp. 183–196. Springer, Heidelberg (2001)

15. Hu, Y., Kobourov, S.G., Veeramoni, S.: Embedding, clustering and coloring for dynamic maps. In: Visualization Symposium (PacificVis), 2012 IEEE Pacific, pp. 33–40 (2012)

16. Khoury, M., Hu, Y., Krishnan, S., Scheidegger, C.: Drawing large graphs by low-rank stress majorization. Comput. Graph. Forum **31**(3pt1), 975–984 (2012)

17. Koren, Y., Carmel, L., Harel, D.: Drawing huge graphs by algebraic multigrid optimization. Multiscale Model. Simul. **1**, 645–673 (2003)

18. Kumar, G., Garland, M.: Visual exploration of complex time-varying graphs. IEEE Trans. Vis. Comput. Graph. **12**(5), 805–812 (2006)

19. Lee, Y.Y., Lin, C.C., Yen, H.C.: Mental map preserving graph drawing using simulated annealing. In: Proceedings of the 2006 Asia-Pacific Symposium on Information Visualisation, APVis 2006, Vol. 60, pp. 179–188 (2006)

20. Mcfarland, D.: Student resistance: how the formal and informal organization of classrooms facilitate everyday forms of student defiance. Am. J. Sociol. **107**(3), 612–678 (2001)

21. Misue, K., Eades, P., Lai, W., Sugiyama, K.: Layout adjustment and the mental map. J. Vi. Lang. Comput. **6**(2), 183–210 (1995)

22. North, S.C.: Incremental layout in dynadag. In: Brandenburg, F.J. (ed.) GD 1995. LNCS, vol. 1027, pp. 409–418. Springer, Heidelberg (1996)

23. Purchase, H.C., Hoggan, E., Görg, C.: How important is the "Mental Map"? – an empirical investigation of a dynamic graph layout algorithm. In: Kaufmann, M., Wagner, D. (eds.) GD 2006. LNCS, vol. 4372, pp. 184–195. Springer, Heidelberg (2007)

24. Purchase, H.C., Samra, A.: Extremes are better: investigating mental map preservation in dynamic graphs. In: Stapleton, G., Howse, J., Lee, J. (eds.) Diagrams 2008. LNCS (LNAI), vol. 5223, pp. 60–73. Springer, Heidelberg (2008)
25. Tufte, E.R.: Envisioning Information. Graphic Press, Cheshire (1990)
26. Yao, Y.: Collection and streaming of graph datasets. http://www.eecs.wsu.edu/yyao/StreamingGraphs.html

Drawing Large Graphs by Multilevel Maxent-Stress Optimization

Henning Meyerhenke[1], Martin Nöllenburg[2], and Christian Schulz[1]([⊠])

[1] Institute of Theoretical Informatics, Karlsruhe Institute of Technology (KIT),
Karlsruhe, Germany
{meyerhenke,christian.schulz}@kit.edu
[2] Algorithms and Complexity Group, TU Wien, Vienna, Austria
noellenburg@ac.tuwien.ac.at

Abstract. Drawing large graphs appropriately is an important step for the visual analysis of data from real-world networks. Here we present a novel multilevel algorithm to compute a graph layout with respect to a recently proposed metric that combines layout stress and entropy. As opposed to previous work, we do not solve the linear systems of the maxent-stress metric with a typical numerical solver. Instead we use a simple local iterative scheme within a multilevel approach. To accelerate local optimization, we approximate long-range forces and use shared-memory parallelism. Our experiments validate the high potential of our approach, which is particularly appealing for dynamic graphs. In comparison to the previously best maxent-stress optimizer, which is sequential, our parallel implementation is on average 30 times faster already for static graphs (and still faster if executed on one thread) while producing a comparable solution quality.

1 Introduction

Drawing large networks (or graphs, we use both terms interchangeably) with hundreds of thousands of nodes and edges has a variety of relevant applications. One of them can be interactive visualization, which helps humans working on graph data to gain insights about the properties of the data. If a very large high-end display is not available for such purpose, a hierarchical approach allows the user to select an appropriate zoom level [1]. Moreover, drawings of large graphs can also be used as a preprocessing step in high-performance applications [22].

One very promising class of layout algorithms in this context is based on the *stress* of a graph. Such algorithms can for instance be used for drawing graphs with fixed distances between vertex pairs, provided *a priori* in a distance matrix [13]. More recently, Gansner et al. [12] proposed a similar model that includes besides the stress an additional entropy term (hence its name *maxent-stress*). While still using shortest path distances, this model often results in more satisfactory layouts for large networks. The optimization problem can be cast

---

This is a short version of the technical report (TR) [27].

© Springer International Publishing Switzerland 2015
E. Di Giacomo and A. Lubiw (Eds.): GD 2015, LNCS 9411, pp. 30–43, 2015.
DOI: 10.1007/978-3-319-27261-0_3

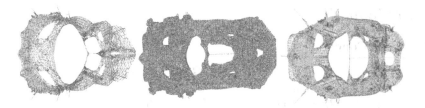

Fig. 1. Drawings of bcsstk31. Left to right: PivotMDS [5], Maxent [12], MulMent (new).

as solving Laplacian linear systems successively. Since each right-hand side in this succession depends on the previous solution, many linear systems need to be solved until convergence – more details can be found in Sect. 2.3.

Motivation. We want to employ this maxent-stress model for drawing large networks quickly. Yet, solving many large Laplacian linear systems can be quite costly. A conjugate gradient solver (used in [12]) is easy to implement but has superlinear running time. Solvers with provably nearly-linear running time exist but are not yet competitive with established methods in practice (see [18] for an experimental comparison). Multigrid methods [24, 26] for Laplacian systems may seem appealing in this context, but their setup phase building the multigrid hierarchy can be expensive for large graphs.

Gansner et al. [12] also suggested (but did not use) a simpler iterative refinement procedure for solving their optimization problem. This procedure would be slow to converge if used unmodified. However, if designed and implemented appropriately, it has the potential for fast convergence even on large graphs. Moreover, as already observed in [12], it has high potential for parallelism and should work well on dynamic graphs by profiting from previous solutions.

Outline and Contribution. The main contribution of this paper is to make the alternative iterative local optimizer suggested by Gansner et al. [12] (for details on this and other related work see Sect. 2) usable and fast in practice. To this end, we design and implement a multilevel algorithm tailored to large networks with unit target edge lengths (see Sect. 3). The employed coarsening algorithm for building the multilevel hierarchy can control the trade-off between the number of hierarchy levels and convergence speed of the local optimizer. One property of the local optimizer we exploit is its high degree of parallelism. Further acceleration is obtained by approximating long-range forces. To this end, we use coarser representatives stored in the multilevel hierarchy.

Our experimental results in Sect. 4 show that force approximation rarely affects the layout *quality* significantly – in terms of maxent-stress values as well as visual quality, also see Fig. 1 and TR [27]. The parallel implementation of our multilevel algorithm MulMent with force approximation is, however, on average 30 times faster than the reference implementation [12] – and even our sequential approximate algorithm is faster than the reference. A contribution besides higher speed is that, in contrast to [12], our approach does not require input coordinates to optimize the maxent-stress measure.

2   Preliminaries

2.1   Basic Concepts

Consider an undirected, connected graph $G = (V, E, c, \omega, d)$ with node weights $c : V \to \mathbb{R}_{\geq 0}$, edge weights $\omega : E \to \mathbb{R}_{\geq 0}$, target edge lengths $d : E \to \mathbb{R}_{> 0}$, $n = |V|$, and $m = |E|$. Often the function d models the required distance between two adjacent vertices. By default, our initial inputs will have unit edge length $d \equiv 1$ as well as unit node weight and edge weight $c \equiv 1$, $\omega \equiv 1$. However, we will encounter weighted problems in the course of our multilevel algorithm. Let $N(v) := \{u : \{v, u\} \in E\}$ denote the set of neighbors of v. A clustering of a graph is a set of *blocks* (= clusters) of nodes $\{V_1, \ldots, V_k\}$ that partition V, i.e., $V_1 \cup \cdots \cup V_k = V$ and $V_i \cap V_j = \emptyset$ for $i \neq j$. A *layout* of a graph is represented as a coordinate vector x, where x_v is the two-dimensional coordinate of vertex v. Since edges are drawn as straight-line segments between their incident nodes, x is sufficient to define the complete graph layout.

2.2   Related Work

Most general-purpose layout algorithms for arbitrary undirected graphs are based on physical analogies and can be grouped, according to Hu and Shi [19], into two main classes: algorithms in the *spring-electrical model* and algorithms in the *stress model*. Both classes of algorithms often yield aesthetically pleasing graph layouts that emphasize symmetries and avoid edge crossings at least in sparse graphs. Recent surveys of algorithms in these models are given by Hu and Shi [19] and by Kobourov [23].

In the spring-electrical model, first presented by Eades in 1984 [8], the analogy is to represent nodes as electrically charged particles that repel each other while edges are represented as springs exerting attraction forces to adjacent nodes. A graph layout is then seen as a physical system of forces and the goal is to find an optimal layout corresponding to a minimum energy state. Spring-electrical algorithms are also known as *spring embedders*, with the algorithm by Fruchterman and Reingold [10] being one of the most widely used spring embedder algorithms. It simulates the physical system of attractive and repulsive forces and iteratively moves each node into the direction of the resulting force. Each iteration requires, however, a quadratic number of force computations due to the repulsive forces between all pairs of nodes, which limits the scalability of the original approach. A faster approximative force calculation method based on quadtrees, aggregating especially the long-range forces, has been proposed by Barnes and Hut [3] and yields running times of $O(n \log n)$ under certain assumptions.

The (full) stress model is closely related to multidimensional scaling [25], and was introduced in graph drawing by Kamada and Kawai [21]. It is based on defining ideal distances d_{uv} not only between adjacent vertices but between all vertex pairs $(u, v) \in V \times V$ and then minimizing the layout stress $\sum_{u \neq v} w_{uv} (\|x_u - x_v\| - d_{uv})^2$, where w_{uv} is a weight factor typically chosen as $w_{uv} = 1/d_{uv}^2$. Often, the distance d_{uv} between adjacent nodes is set to 1, while the distance of non-adjacent nodes is the shortest-path distance in the graph.

Solving this model is typically done by iteratively solving a series of linear systems [13]. The need to compute all-pairs shortest paths and to store a quadratic number of distances again defeats the scalability of this original approach for large graphs. One of the fastest algorithms for approximatively solving the stress model instead is PivotMDS [5], which requires distance calculations from each vertex only to a small set of $k \ll n$ suitably chosen pivot vertices.

The stress model prescribes target distances not only for edges but for all vertex pairs. While this is a reasonable approach, it still brings artificial information into the layout process. An interesting alternative has been proposed by Gansner et al. [12]. Their algorithm (called *Maxent*) uses the sparse stress model, which only contains the stress terms for the edges of the graph. In order to deal with the remaining degrees of freedom in the layout, they suggest using the maximum entropy principle instead. Since our algorithm is closely related to Maxent, we discuss the latter in more detail in Sect. 2.3.

A general approach for speeding up layout computations for large graphs is the *multilevel technique*, which has been used in the spring-electrical [16,29,32] and in the stress model [11]. A multilevel algorithm computes a sequence of increasingly coarse but structurally related graphs as abstractions of the original graph. Starting from a layout of the coarsest graph, incremental refinement steps using the previous layout as a scaffold eventually produce a layout of the entire input graph, where the refinement steps are fast due to the good initial layouts. Hachul and Jünger [15] performed an extensive experimental evaluation of state-of-the-art layout algorithms for large graphs, including multilevel algorithms, and Bartel et al. [4] experimentally compared different combinations of coarsening, placement, and layout methods for the generic multilevel approach.

In addition to sequential algorithms for drawing large graphs, there is previous research in parallel layout algorithms, particularly using a graphics processing unit (GPU). Frishman and Tal [9] presented a multilevel force-based layout algorithm and implemented it using GPU-based parallelization. Ingram et al. [20] also exploit parallel GPU computations and presented a multilevel stress-based layout algorithm. Godiyal et al. [14] implemented a fast multipole algorithm on the GPU.

2.3 Maxent-Stress Optimization

Gansner et al. [12] proposed the maxent-stress model that combines a sparse stress model with an entropy term to resolve the degrees of freedom for non-adjacent vertex pairs. The entropy term itself is optimized when all nodes are spread out uniformly, similar to the repulsive forces in the spring-electrical model. Gansner et al. [12] showed that the maxent-stress model performs well on several measures of layout quality in distance-based embeddings and avoids typical shortcomings of other stress models, particularly for non-rigid graphs. Formally, the maxent-stress $M(x)$ of a layout x is defined[1] as

---

[1] In fact, Gansner et al. define a slightly more general model that considers the stress term for arbitrary supersets $S \supseteq E$ and allows variations of the entropy term. Our algorithm also works for the general model; to simplify the description, we restrict ourselves to the default model.

$$M(x) = \sum_{\{u,v\}\in E} w_{uv}(\|x_u - x_v\| - d_{uv})^2 - \alpha \sum_{\{u,v\}\notin E} \ln \|x_u - x_v\|, \qquad (1)$$

where d_{uv} is the target distance between nodes u and v and w_{uv} is a weight factor typically chosen as $w_{uv} = 1/d_{uv}^2$. Throughout the paper, we use this as a weight factor. The scaling factor α is used to modulate the strength of the entropy term and is gradually reduced in the implementation.

Gansner et al. minimize the maxent-stress using a technique that repeatedly solves Laplacian linear systems that additionally include a repulsive force vector which is approximated following the quadtree method of Barnes and Hut [3].

Alternatively, they proposed (but did not implement) the following local iterative force-based scheme to solve the maxent-stress model:

$$x_u \leftarrow \frac{1}{\rho_u} \sum_{\{u,v\}\in E} w_{uv}\left(x_v + d_{uv}\frac{x_u - x_v}{\|x_u - x_v\|}\right) + \frac{\alpha}{\rho_u} \sum_{\{u,v\}\notin E} \frac{x_u - x_v}{\|x_u - x_v\|^2}, \qquad (2)$$

where $\rho_u = \sum_{\{u,v\}\in E} w_{uv}$. Note that sometimes we use the abbreviation $r(u,v) := \frac{x_u - x_v}{\|x_u - x_v\|^2}$ and shortly call these values *r-values*.

3   Multilevel Maxent-Stress Optimization

As mentioned, a successful (meta)heuristic for graph drawing (and other optimization problems on large graphs) is the multilevel approach. We also employ this approach for maxent-stress optimization for several other reasons: (i) Some graphs (such as road networks) feature a hierarchical structure, which can be exploited to some extent by a multilevel approach and (ii) the computed hierarchy may be useful later on for multiscale visualization.

Before going into the details, we briefly sketch our algorithmic approach: The method for creating the graph hierarchy is based on fast graph clustering with controllable cluster sizes. Each cluster computed on one hierarchy level is contracted into a new supervertex for the next level. After computing an initial layout on the coarsest hierarchy level, we improve the drawing on each finer level by iterating Eq. (2). Additionally, this process exploits the hierarchy and draws vertices that are densely connected with each other (i.e. which are in the same cluster) close to each other.

3.1   Coarsening and Initial Layout

To compute the clustering we adapt size-constrained label propagation (SCLaP) [28], an algorithm originally developed for coarsening and local improvement during multilevel graph partitioning. SCLaP itself is based on the graph clustering algorithm *label propagation* [30]. The latter starts with a single-ton clustering (i.e. each node is a cluster). The algorithm then works in rounds. Roughly speaking, in each round the algorithm visits all nodes in random order and assigns each node to the predominant cluster in its neighborhood. This way, cluster IDs (= labels) propagate through the graph and nodes in a dense cluster usually agree on a common label.

However, clusters with unconstrained sizes are not desirable here since they would hamper convergence of the local improvement phase. The trade-off between this convergence speed and the number of hierarchy levels needs to be chosen properly for a fast overall running time. That is why SCLaP constrains cluster sizes, i.e. it introduces an upper bound $U := \max(\max_v c(v), W)$ on the cluster sizes (W is specified below), where constraining on the maximum node weight favors uniform coarsening. Consequently, in each SCLaP round, nodes are assigned to the predominant cluster that is not overloaded after the label change.

In our implementation, based on preliminary experiments, we set the parameter W to $\min(b^h, \frac{|V|}{f})$, where b and f are tuning parameters and h is the level in the hierarchy that we are currently working on. The intuition behind this choice is that we want the contraction process not to be too strong on the fine levels in order to allow fast convergence of local improvement algorithms, whereas we allow stronger contractions on coarser levels. If the contracted graph is not more than 10 % smaller than the graph on the current level, we decrease the value of f and set it to $0.7f$.

While the original label propagation algorithm repeats the process until convergence, SCLaP performs at most ℓ rounds, where ℓ is a tuning parameter. One round of the algorithm can be implemented to run in $\mathcal{O}(n + m)$ time.

Contracting a clustering works as follows: each block of the clustering is contracted into a single node. The weight of the node is set to the sum of the weight of all nodes in the original block. There is an edge between two nodes u' and v' in the contracted graph if the two corresponding blocks in the clustering are adjacent to each other in G, i.e. block u' and block v' are connected by at least one edge. The weight of an edge (u', v') is set to the sum of the weight of edges that run between block u' and block v' of the clustering.

Initial Layout. The process of computing a size-constrained clustering and contracting it is repeated recursively. Then an initial layout is drawn, meaning that each of the two nodes of the coarsest graph is assigned to a position. We place the vertices such that the distance is optimal. The optimal distance of the two vertices is defined and motivated in the next section.

3.2   Uncoarsening and Local Improvement

When the initial layout has been computed, the solution is successively prolongated to the next finer level, where a local maxent-stress minimizer is used to improve the layout. For undoing the contraction, nodes that have been in a cluster are drawn at a random position around the location of its coarse representative. More precisely, let v be a (fine) vertex that is represented by the coarse supervertex v' at $P = (x, y)$. We place v at a random position in a circle around P with radius $r := \sqrt{c(v')}$. We do this by picking an angle uniformly at random in $[0, 2\pi]$ and a distance to P uniformly at random in $[0, r]$. These two values are then used as a polar coordinate for v with respect to the origin P.

Local Improvement. Our local improvement tries to minimize the maxent-stress on each level of the hierarchy based on Eq. (2). Note, however, that simply iterating Eq. (2) on each level is not sensible since coarse vertices represent a multitude of vertices. These vertices need space to be drawn on the next finer level. Now let u and v be two vertices on the same fixed level. We adjust distances d_{uv} on the current level in the hierarchy under consideration to $\sqrt{c(u)} + \sqrt{c(v)}$ with the intuition that vertices represented by u should be drawn in a circle around u with radius $\sqrt{c(u)}$ (similarly for v).

As Gansner et al. [12], we adjust the value of α in Eq. (2) during the process. Since we want to approximate the maxent-stress, the value should be small. However, it cannot be too small initially since one would only solve a sparse stress model in this case. Hence, following Gansner et al. [12], we set α to one initially and gradually reduce it by $\alpha := 0.3 \cdot \alpha$ until $\alpha_{\min} = 0.008$ is reached.

We call a single update step of the coordinates of all vertices using Eq. (2) an *iteration*. Multiple iterations with the same value of α are called *round*. The current iteration uses the coordinates that have been computed in the previous iteration. We perform at most a iterations with the same value of α in one round. Then we reduce α as described above. If the relative change $||x^{\ell+1} - x^{\ell}||/||x^{\ell}||$ in the layout is smaller than some threshold ϵ, we directly reduce the value of α and continue with the next round.

Faster Local Improvement. The local optimization algorithm presented above has a theoretical running time of $\mathcal{O}(n^2)$ per iteration. To speed this up, one can use approximations for the distances in the entropy term in Eq. (2). We do this by taking the cluster structure computed during coarsening into account: Let $V_1 \cup \ldots \cup V_k$ be the corresponding clustering and $M : V \to V' = \{1, \ldots, k\}$ be the mapping that maps a node $v \in V$ to its coarse representative. The first term in Eq. (2) is computed as before and the second term is approximated by using the coordinates of the corresponding coarse vertex. As formula the second term written without the multiplicative factor $\frac{\alpha}{\rho_u}$ becomes

$$\sum_{\substack{u \neq v \\ M(u) = M(v)}} r(u, v) + \sum_{\substack{v' \in V' \\ v' \neq M(u)}} \nu(v') \frac{x_u - x'_{v'}}{||x_u - x'_{v'}||^2} - \sum_{\{u,v\} \in E} r(u, v), \qquad (3)$$

where x' maps a coarse vertex to its coordinates and $\nu(v')$ is the *number of nodes* that the coarse vertex represents on the *current* finer level. Note that this is different from the vertex weight $c(v')$ which represents the number of nodes that the coarse vertex represents on the *finest* level. Roughly speaking, we reduced the necessary amount of computation to add up the values of r by summing up the correct values of r for all vertices that are in a sense *close* and using approximations for vertices that are far away. In our context, a vertex is close if it is in the same cluster as the currently processed vertex. If a vertex is not close, we use the coordinate of its coarse representative instead. We avoid unnecessary computation by scaling the approximated value of r with the number $\nu(v')$ of vertices it represents and adding approximated value of r only once. The last term in Eq. (3) subtracts values of r for $\{u, v\} \in E$ that have been added in good faith in the first two summations.

Note that if M is the identity, then the term in Eq. (3) is the same as in the original Eq. 2. In this case the first two summations add up the r-values for all pairs of vertices and the last sum subtracts the r-values for pairs that are in E.

After the update of the vertices on the current level, we update the coordinates of the vertices on the coarser level used for approximation. We set the coordinate of a vertex v' on the coarser level to the barycenter of the vertices represented by v'.

Note that one obtains even faster algorithms by using a coarser version of the graph that is *multiple levels beneath* the current level in the graph hierarchy. That means instead of using the next coarser graph, we use the contracted graph which is $h > 1$ levels beneath the current graph in the hierarchy – if there is such. Otherwise, we use the coarsest graph in the hierarchy. Obviously this yields a trade-off between solution quality and running time. Also note that this introduces an additional error. To see this, let the coarser vertices that have the same coarse representative on the level used for approximating values of r be called $\mathcal{M}$-vertices (merged vertices). Now, for a vertex on the current level, the r-values of $\mathcal{M}$-vertices are not accounted for in Eq. (3). Hence, we look at the parameter h carefully in Sect. 4 and evaluate its impact on running time and solution quality. We call our algorithms MulMent and denote by MulMent$_h$ the algorithm that uses an h-level approximation of the r-values. With $h = 0$ we denote the quadratic-time algorithm. A rough analysis in TR [27] yields:

Proposition 1. *Under the assumption of equal cluster sizes, the running time of one iteration of algorithm MulMent$_h$, $h \geq 0$, is $\mathcal{O}(m + n^{\frac{h+2}{h+1}})$, respectively.*

Properly implemented, multilevel algorithms lead to fast convergence of their local optimizers. Moreover, the overall work performed by the multilevel approach is only a constant factor times the one on the finest level. This leads us to the initial appraisal that the same asymptotic running times may hold for the respective complete algorithms.

Shared-Memory Parallelization. Our shared-memory parallelization of an iteration of the local optimizer uses OpenMP and works as follows: Since new coordinates of the vertices in the same iteration can be computed independently, we use multiple threads to do so. The relative change in the layout $||x^{\ell+1} - x^\ell|| / ||x^\ell||$ can be computed in parallel using a reduce operation. Parallelism is also used analogously when working on different levels for the distance approximations in the entropy term. Other parts of the overall algorithm could potentially be parallelized, too – such as coarsening. However, already on medium sized graphs coarsening consumes less than 5 % of the algorithm's overall running time. Moreover, the relative running time of coarsening decreases even more with increasing graph size so that the effort does not seem worth it.

4   Experimental Evaluation

Methodology. We implemented[2] the algorithm described above using C++. Parallelization of our algorithm has been done using OpenMP. We compiled our programs using g++ 4.9 -O3 and OpenMP 3.1. Executables for Pivot-MDS (PMDS) [5] and MaxEnt (GHN, for clarity we use the author names as acronym) [12] have been kindly provided by Yifan Hu. When comparing layouts computed by different algorithms, we evaluate two metrics. The first metric is the full stress measure, $F(x) = \sum_{u,v \in V} w_{uv}(||x_u - x_v|| - d_{uv})^2$, and the second one is the maxent-stress function $M(x)$ as defined in Eq. (1) at the final penalty level of $\alpha = 0.008$. The latter is of primary importance since that is what GHN and Mul-Ment optimize for. The implementations PMDS and GHN sometimes compute vertices that are on the same position. Hence, we add small random noise to the coordinates of these layouts in order to be able to compute the maxent-stress. More precisely, for each of the components of the 2D-coordinate of a node, we randomly add or subtract a random value from the interval $[10^{-7}, 10^{-4}]$. This changes the full stress measure by less than 10^{-4} percent on average. We follow the methodology of Gansner et al. [12] and scale the layout of all algorithms to minimize the stress to be fair to all methods: We find a scalar s such that $\sum_{u,v \in V} w_{uv}(s||x_u - x_v|| - d_{uv})^2$ is minimized for a given layout x.

Machine. Our machine has four Octa-Core Intel Xeon E5-4640 (Sandy Bridge) processors (32 cores, 64 with hyperthreading active) which run at a clock speed of 2.4 GHz. It has 512 GB local memory, 20 MB L3-Cache and 8x256 KB L2-Cache. Unless otherwise mentioned, our algorithms use all 64 cores (hyperthreading) of that machine. Since PMDS and GHN are sequential algorithms, they use one core of that machine.

Algorithm Configuration. After an extensive evaluation of the parameters, we fixed the cluster coarsening parameters f to 20 and b to 2. The initial value of the penalty parameter α is set to 1. We perform at most $a = 2$ iterations with the same value of α, while it has not reached its minimum value of 0.008. When it has reached its minimum value, we iterate until the relative error $||x^{\ell+1} - x^\ell||/||x^\ell||$ is smaller than 0.0001. Yet, our experiments indicate that our algorithm is not very sensitive about the choice of these parameters. We evaluate the influence of the approximation level h in Sect. 4.1.

Instances. We use the instances 1138_bus, USpowerGrid, bcsstk31, commanche and luxembourg employed in [12] and extend the set to include larger instances. We excluded the graphs gd, qh882 and lp_ship04l from [12] from our experiments since the graphs are either not undirected or the corresponding matrix is rectangular. Most of the instances taken from [12] are available at the Florida Sparse

---

[2] We released the implementation of our algorithms as open source in the KaDraw (Karlsruhe Graph Drawing) framework available at http://algo2.iti.kit.edu/kadraw/.

Matrix Collection [6]. The graphs 3elt, bcsstk31, fe_pwt and auto are available at the Walshaw benchmark archive [31]. The graphs del$X$ are Delaunay triangulations of 2^X random points in the unit square [17]. Moreover, the graphs nyc and luxembourg are road networks. These graphs have been taken from the benchmark set of the 9th and 10th DIMACS Implementation Challenge [2,7]. A summary of the basic properties of these instances can be found in the technical report version of this paper [27]. In any case, we draw the largest connected component if the graph has more than one. We assume unit length distance for all graphs.

4.1   Influence of Coarse Graph Approximation and Scalability

In this section, we investigate the influence of the parameter h on layout quality and running time (algorithmic speedup) as well as the scalability of our algorithms with varying number of threads (parallel speedup). We perform detailed experiments on our medium sized networks (using 64 threads) and present parallel speedups on the largest graphs auto and del20. We report absolute running times and parallel speedups for the graph del20 in Fig. 2 and present detailed data for the medium size networks as well as more plots in [27]. We do not report layout quality metrics for auto and del20 since the size of the network makes it infeasible to compute them and the result of the algorithm is independent of the number of threads used.

We now investigate the influence of the parameter h. In general, the larger the graphs get, the larger the algorithmic speedups obtained with increasing h. On the smallest graph in this collection, we obtain an algorithmic speedup of about 3 with $h = 6$ (fe_pwt) over MulMent$_0$. On the largest two instances in this section, we obtain an algorithmic speedup of 30 with $h = 9$ (auto) and of 122 with $h = 10$ (del20). In addition, the precise choice of the parameter does not seem to have a very large impact on solution quality on these graphs. This is also due to the size of the networks. The graphs on which full stress measure slightly increases are luxembourg and bcsstk31 (7 % and 15 % respectively – see [27]). The metric actually under consideration, maxent-stress, always remains comparable. On all instances under consideration, we observe a locally optimal value for h in terms of running time. It is around seven and seems to get larger with increasing graph size. This is due to the fact that too large values of h provide less precision and slower convergence.

On del20, the scalability with the number of threads is almost perfect for small values of h. With enabled hyperthreading, we achieve slightly *superlinear* speedups for MulMent$_0$. As less work has to be done for increasing h, speedups get smaller. The smallest speedup on this graph has been observed for MulMent$_{10}$. In this case, we achieve a speedup of 11.5 using 64 threads over MulMent$_{10}$ using one thread. With even larger h speedups increase again. The parallel scalability on auto is similar.

Another interesting way to look at the data is the overall speedup – algorithmic and parallel speedup combined – achieved over MulMent$_0$ using only one thread. The largest overall speedup is obtained by MulMent$_{10}$ using 64 threads.

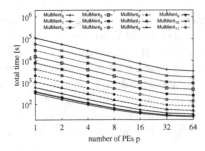 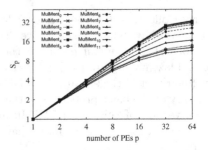

Fig. 2. Running times and parallel speedups of our algorithms on del20.

In this case, the overall speedup is larger than 4000 – reducing the running time of the algorithm from 30 hours to 27 seconds. Speedups over PMDS and GHN are found in the next section.

4.2 Comparison to Other Drawing Algorithms

We now compare MulMent to the two implementations PMDS [5] and GHN [12]. We do this on all networks but only report quality metrics for small and medium sized graphs since it is infeasible to compute quality metrics for the large graphs. We report detailed data in [27].

Most importantly, although MulMent sometimes performs a few percent worse than GHN, the maxent-stress of all layouts is more or less *similar*. PMDS performs slightly worse in this metric. Intriguingly, the alternative full stress metric is consistently better on small networks for MulMent than the results obtained by PMDS (except for $h = 10$). On the other hand, full stress obtained by our algorithms is comparable to the layout computed by GHN on four out of nine instances. On the three largest medium sized networks, we obtain worse full stress than PMDS and GHN. However, this is not astonishing since our algorithm does not optimize for full stress – in *contrast* to PMDS. And GHN at least starts with a PMDS solution and improves maxent-stress afterwards.

Our implementations of MulMent$_{7,10}$ are always faster than GHN, both of them a factor 30 on average. Also, MulMent$_{7,10}$ outperform even PMDS in terms of running time as soon as the graphs get large enough (medium and large sized graphs). On the large graphs, MulMent$_{10}$ is a factor of 2 to 3 faster than PMDS and a factor of 32 to 63 faster than GHN. In addition, MulMent$_{7,10}$ are also several times faster than GHN when using one thread only (see TR [27]).

4.3 Dynamic Networks

One of the main advantages of the iterative scheme is its ability to use an existing layout for computing a new one, e. g. for a graph that has changed over time. We perform experiments with dynamic graphs obtained by modifying our medium sized networks. Often one is interested in drawing graphs that have more or less good locality. Hence, we define a random model that modifies the edges of a graph by removing random edges and inserting edges between vertices that are not too far apart.

To be more precise, we start with an input graph G and perform a breadth first search from a random start node to compute a random spanning tree. We then remove $x\%$ undirected non-tree edges at random in the beginning. Note that this ensures that the graph stays connected. Afterwards, we insert $x\%$ new edges as follows. We pick a random node and insert an undirected edge to a random node that has distance $1 < d \leq \mathcal{D}$ in the original graph G, where $\mathcal{D}$ is a tuning parameter. We denote the graph that results out of this process as Q.

We compute two layouts of Q. The first one updates coordinates given by an initial layout of G (update algorithm). The second layout is computed by our algorithm from scratch (scratch algorithm), i.e. discarding the initial layout. In the first case, we start directly at the penalty level $\alpha = 0.008$ and only update coordinates on the finest level of the hierarchy. We compute the graph hierarchy as before but stop the coarsening process after the computation of h levels. Coordinates of the vertices on the approximation level are set to the middle point of the vertices in the corresponding cluster initially.

We vary $x \in \{1, 5\}$, $\mathcal{D} \in \{2, 16\}$ and $h \in \{0, 7\}$, and present detailed data in [27]. As expected, the *running time* of the update algorithm (t_{dyn}) is always smaller than the running time of the scratch algorithm (t_{scratch}). As MulMent$_7$ performs less work than MulMent$_0$, algorithmic speedups are always larger for the latter. For $h = 0$, the update algorithm is a factor of 4 faster than the scratch algorithm on average. On the other hand, for $h = 7$ the update algorithm saves about 50% time on average over the scratch algorithm. *Solution quality* is not influenced much. On average, the full stress measure of the update algorithm is 9% larger and maxent-stress improves by 1% compared to the scratch algorithm. The increase in full stress is mostly due to the Delaunay instance and $\mathcal{D} = 16$, in which the full stress of the layout of the update algorithm is a factor of two larger. The algorithmic speedup does not seem to be largely influenced by $\mathcal{D}$. However, we expect that much larger values of $\mathcal{D}$ will decrease the speedup of the update algorithm over the scratch algorithm.

5   Conclusions

We have presented a new multilevel algorithm for iteratively and approximatively optimizing the maxent-stress model, a model proposed by Gansner et al. [12] to avoid typical pitfalls of other stress models. From the experimental evaluation we conclude that our parallel algorithm produces layouts with similar visual quality and maxent-stress values as the reference implementation [12]. At the same time it is on average 30 times faster, even more for dynamic graphs. Moreover, our algorithm is even up to twice as fast as the fastest stress-based algorithm PivotMDS [5]. It thus combines the high speed of PivotMDS with the high visual quality of Maxent in a single algorithm, at least if a multicore system is available.

Currently our method is only capable of handling constant edge lengths. This requirement is due to the way coarse vertices are placed and later interpolated to a finer level. In future work we would like to eliminate this limitation.

Acknowledgements. Financial support by DFG is acknowledged (DFG grants ME 3619/3-1 and SA 933/10-1). We thank Yifan Hu for providing us the codes from [12].

References

1. Abello, J., van Ham, F., Krishnan, N.: ASK-GraphView: a large scale graph visualization system. IEEE Trans. Vis. Comput. Graph. **12**(5), 669–676 (2006)
2. Bader, D.A., Meyerhenke, H., Sanders, P., Schulz, C., Kappes, A., Wagner, D.: A benchmark set for graph clustering and graph partitioning. In: Encyclopedia of Social Network Analysis and Mining (2014)
3. Barnes, J., Hut, P.: A hierarchical $O(n \log n)$ force-calculation algorithm. Nature **324**, 446–449 (1986)
4. Bartel, G., Gutwenger, C., Klein, K., Mutzel, P.: An experimental evaluation of multilevel layout methods. In: Brandes, U., Cornelsen, S. (eds.) GD 2010. LNCS, vol. 6502, pp. 80–91. Springer, Heidelberg (2011)
5. Brandes, U., Pich, C.: Eigensolver methods for progressive multidimensional scaling of large data. In: Kaufmann, M., Wagner, D. (eds.) GD 2006. LNCS, vol. 4372, pp. 42–53. Springer, Heidelberg (2007)
6. Davis, T.: The university of florida sparse matrix collection (2008). http://www.cise.ufl.edu/research/sparse/matrices
7. Demetrescu, C., Goldberg, A.V., Johnson, D.S.: The Shortest Path Problem: 9th DIMACS Implementation Challenge, vol. 74. AMS (2009)
8. Eades, P.: A heuristic for graph drawing. Congressus Numerantium **42**, 146–160 (1984)
9. Frishman, Y., Tal, A.: Multi-level graph layout on the GPU. IEEE Trans. Vis. Comput. Graph. **13**(6), 1310–1319 (2007)
10. Fruchterman, T.M.J., Reingold, E.M.: Graph drawing by force-directed placement. Softw. Pract. Exp. **21**(11), 1129–1164 (1991)
11. Gajer, P., Kobourov, S.G.: GRIP: Graph drawing with intelligent placement. J. Graph Algorithms Appl. **6**(3), 202–224 (2002)
12. Gansner, E.R., Hu, Y., North, S.C.: A maxent-stress model for graph layout. IEEE Trans. Vis. Comput. Graph. **19**(6), 927–940 (2013)
13. Gansner, E.R., Koren, Y., North, S.C.: Graph drawing by stress majorization. In: Pach, J. (ed.) GD 2004. LNCS, vol. 3383, pp. 239–250. Springer, Heidelberg (2005)
14. Godiyal, A., Hoberock, J., Garland, M., Hart, J.C.: Rapid multipole graph drawing on the GPU. In: Tollis, I.G., Patrignani, M. (eds.) GD 2008. LNCS, vol. 5417, pp. 90–101. Springer, Heidelberg (2009)
15. Hachul, S., Jünger, M.: Large-graph layout algorithms at work: an experimental study. J. Graph Algorithms Appl. **11**(2), 345–369 (2007)
16. Hachul, S., Jünger, M.: Drawing large graphs with a potential-field-based multilevel algorithm. In: Pach, J. (ed.) GD 2004. LNCS, vol. 3383, pp. 285–295. Springer, Heidelberg (2005)
17. Holtgrewe, M., Sanders, P., Schulz, C.: Engineering a scalable high quality graph partitioner. In: IEEE Parallel and Distributed Computing (IPDPS 2010), pp. 1–12. IEEE (2010)
18. Hoske, D., Lukarski, D., Meyerhenke, H., Wegner, M.: Is nearly-linear the same in theory and practice? a case study with a combinatorial laplacian solver. In: Bampis, E. (ed.) SEA 2015. LNCS, vol. 9125, pp. 205–218. Springer, Heidelberg (2015)

19. Hu, Y., Shi, L.: Visualizing large graphs. Wiley Interdisc. Rev. Comput. Stat. **7**(2), 115–136 (2015)
20. Ingram, S., Munzner, T., Olano, M.: Glimmer: multilevel MDS on the GPU. IEEE Trans. Vis. Comput. Graph. **15**(2), 249–261 (2009)
21. Kamada, T., Kawai, S.: An algorithm for drawing general undirected graphs. Inf. Process. Lett. **31**, 7–15 (1989)
22. Kirmani, S., Raghavan, P.: Scalable parallel graph partitioning. In: High Performance Computing, Networking, Storage and Analysis (S 2013), pp. 51:1–51:10. ACM (2013)
23. Kobourov, S.G.: Force-directed drawing algorithms. In: Tamassia, R. (ed.) Handbook of Graph Drawing and Visualization, Chap. 12, pp. 383–408. CRC Press, Boca Raton (2013)
24. Koutis, I., Miller, G.L., Tolliver, D.: Combinatorial preconditioners and multilevel solvers for problems in computer vision and image processing. Comput. Vis. Image Underst. **115**(12), 1638–1646 (2011)
25. Kruskal, J.B.: Multidimensional scaling by optimizing goodness of fit to a nonmetric hypothesis. Psychometrika **29**(1), 1–27 (1964)
26. Livne, O.E., Brandt, A.: Lean algebraic multigrid (LAMG): fast graph Laplacian linear solver. SIAM J. Sci. Comput. **34**(4), B499–B522 (2012)
27. Meyerhenke, H., Nöllenburg, M., Schulz, C.: Drawing large graphs by multilevel maxent-stress optimization. CoRR, arXiv:1506.04383 (2015)
28. Meyerhenke, H., Sanders, P., Schulz, C.: Partitioning complex networks via size-constrained clustering. In: Gudmundsson, J., Katajainen, J. (eds.) SEA 2014. LNCS, vol. 8504, pp. 351–363. Springer, Heidelberg (2014)
29. Quigley, A., Eades, P.: FADE: graph drawing, clustering, and visual abstraction. In: Marks, J. (ed.) GD 2000. LNCS, vol. 1984, pp. 197–210. Springer, Heidelberg (2001)
30. Raghavan, U.N., Albert, R., Kumara, S.: Near linear time algorithm to detect community structures in large-scale networks. Phy. Rev. E **76**(3), 036106 (2007)
31. Soper, A.J., Walshaw, C., Cross, M.: A combined evolutionary search and multilevel optimisation approach to graph-partitioning. J. Global Optim. **29**(2), 225–241 (2004)
32. Walshaw, C.: A multilevel algorithm for force-directed graph-drawing. J. Graph Algorithms Appl. **7**(3), 253–285 (2003)

A Million Edge Drawing for a Fistful of Dollars

Alessio Arleo, Walter Didimo$^{(\boxtimes)}$, Giuseppe Liotta,
and Fabrizio Montecchiani

Università degli Studi di Perugia, Perugia, Italy
alessio.arleo@studenti.unipg.it,
{walter.didimo,giuseppe.liotta,fabrizio.montecchiani}@unipg.it

Abstract. In this paper we study the problem of designing a graph drawing algorithm for large graphs. The algorithm must be simple to implement and the computing infrastructure must not require major hardware or software investments. We report about the experimental analysis of a simple implementation of a spring embedder in Giraph, a vertex-centric open source framework for distributed computing. The algorithm is tested on real graphs of up to 1 million edges by using a cheap PaaS (Platform as a Service) infrastructure of Amazon. We can afford drawing graphs with about one million edges in about 8 min, by spending less than 1 USD per drawing for the cloud computing infrastructure.

1 Introduction

Classical force-directed algorithms, like *spring embedders*, are by far the most popular graph drawing techniques (see, e.g., [4,10]). One of the key components of this success is the simplicity of their implementation and the effectiveness of the resulting drawings. Spring embedders make the final user only a few lines of code away from an effective layout of a network. They model the graph as a physical system, where vertices are equally-charged electrical particles that repeal each other and edges are modeled as springs that give rise to attractive forces. Computing a drawing corresponds to finding an equilibrium state of the force system by a simple iterative approach (see, e.g., [5,6]).

The main drawback of spring embedders is that they are relatively expensive in terms of computational resources, which gives rise to scalability problems even for graphs with a few thousands vertices. To overcome this limit, sophisticated variants of force-directed algorithms have been proposed; they include *hierarchical space partitioning*, *multidimensional scaling* techniques, *multi-scale* techniques, and *stress-majorization* approaches (see, e.g., [1,8,10] for a survey). Also, both centralized and parallel multi-scale force-directed algorithms that use the power of graphical processing units (GPU) are described [7,9,14,18]. They scale to graphs with some million edges, but their implementation is not easy

Research supported in part by the MIUR project AMANDA "Algorithmics for MAssive and Networked DAta", prot. 2012C4E3KT_001. We are grateful to Clint Eastwood and Sergio Leone for inspiring the title of this work with their movies.

E. Di Giacomo and A. Lubiw (Eds.): GD 2015, LNCS 9411, pp. 44–51, 2015.
DOI: 10.1007/978-3-319-27261-0_4

and the required infrastructure could be expensive in terms of hardware and maintenance.

A few works concentrate on designing relatively simple parallel implementations of classical spring embedders. Mueller et al. [13] and Chae et al. [2] propose a graph visualization algorithm that uses multiple large displays. Vertices are evenly distributed on the different displays, each associated with a different processor, which is responsible for computing the positions of its vertices; scalability experiments are limited to graphs with some thousand vertices. Tikhonova and Ma [15] present a parallel force-directed algorithm that scales well to graphs with some hundred thousand edges. It is important to remark that all the above algorithms are mainly *parallel*, rather than *distributed*, force-directed techniques. Their basic idea is to partition the set of vertices among the processors and to keep data locality as much as possible throughout the computation.

Motivated by the increasing availability of scalable cloud computing services, we study the problem of adapting a simple spring embedder to a distributed architecture. We want to use such an algorithm on a cheap PaaS (Platform as a Service) infrastructure to compute drawings of graphs with million edges. We design, engineer, and experiment a distributed Fruchterman-Reingold (FR) spring embedder [6] in the open source *Giraph* framework [3], on a small Amazon cluster of 10 computers, each equipped with 4 vCPUs (http://aws.amazon.com/en/elasticmapreduce/). Giraph is a popular computing framework for distributed graph algorithms, used for instance by Facebook to analyze social networks (http://giraph.apache.org/).

Our distributed algorithm is based on the *"Think-Like-A-Vertex (TLAV)"* design paradigm [12] and its performance is experimentally tested on a set of real-world graphs whose size varies from tens of thousand to one million edges. The experiments measure not only the execution time and the visual complexity, but also the cost in dollars on the Amazon PaaS infrastructure. For example, computing a drawing on a graph of our test suite with 1,049,866 edges required about 8 min, which corresponds to less than 1$ payed to Amazon. The parallel algorithm described by Tikhonova and Ma [15] needs about 40 min for a graph of 260,385 edges, on 32 processors of the PSC's BigBen Cray XT3 cluster.

The remainder of the paper is organized as follows. Section 2 describes the algorithmic pipeline of our distributed spring embedder. The experimental analysis is the subject of Sect. 3. Conclusions and future work can be found in Sect. 4.

2 A Vertex Centric Spring Embedder

The vertex-centric programming model is a paradigm for distributed processing frameworks to address computing challenges on large graphs. The main idea behind this model is to "Think-Like-A-Vertex" (TLAV), which translates to implementing distributed algorithms from the perspective of a vertex rather than the whole graph. Such an approach improves locality, demonstrates linear scalability, and provides a natural way to express and compute many iterative graph algorithms [12]. TLAV approaches overcome the limits of other popular

Fig. 1. Algorithm pipeline of CLINT: G is the input graph and Γ the computed drawing.

distributed paradigms like MapReduce, which are often poor-performing for iterative graph algorithms.

The first published implementation of a TLAV framework was Google's Pregel [11], based on the Bulk Synchronous Programming model (BSP) [16]. *Giraph* is a popular open-source TLAV framework built on the Apache Hadoop infrastructure [3]. In Giraph, the computation is split into *supersteps* executed iteratively and synchronously. A superstep consists of two processing steps: (*i*) a vertex executes a user-defined vertex function based on both local vertex data, and on data coming from adjacent vertices; (*ii*) the results of such local computation are sent to the vertex neighbors along its incident edges. Supersteps always end with a synchronization barrier which guarantees that messages sent in a given superstep are received at the beginning of the next superstep. The whole computation ends after a fixed number of supersteps has occurred or when all vertices are inactive (i.e., no message has been sent).

Our distributed spring embedder algorithm, called CLINT, is simple and is based on the fact that a classical force-directed approach can be naturally reinterpreted according to a vertex-centric paradigm. CLINT is designed to run on a cluster of computers within a TLAV framework and consists of the following algorithmic pipeline (see also Fig. 1).

Pruning: For the sake of efficiency, we first remove all vertices of degree one from the graph, which will be reinserted at the end of the computation by means of an ad-hoc technique. This operation can be directly performed while loading the graph. The number of degree-one vertices adjacent to a vertex v is stored as a local information of v, to be used throughout the computation.

Partitioning: We then partition the vertex set into subsets, each assigned to a computing unit, also called *worker* in Giraph; in the distributed architecture, each computer may have more than one worker. The default partitioning algorithm provided by Giraph aims at making the partition sets of uniform size, but it does not take into account the graph topology (it is based on applying a hash function). As a result a default Giraph partition may have a very high number of edges that connect vertices of different partition sets; this would negatively affect the communication load between different computing units. To cope with this problem, we used a partitioning algorithm by Vaquero *et al.*, called Spinner [17], which creates balanced partition sets by exploiting the graph topology. It is based on iterative vertex migrations, relying only on local information.

Layout: Recall that classic spring embedders split the computation into a set of iterations. In each iteration every vertex updates its position based on the positions of all other vertices. The computation ends after a fixed number of iterations has occurred or when the positions of the vertices become sufficiently stable. The design of a distributed spring embedder within the TLAV paradigm must consider the following architectural constraints: (i) each vertex can exchange messages only with its neighbors, (ii) each vertex can locally store a small amount of data, and (iii) the communication load, i.e., the total number of messages and length sent at a particular superstep, should not be too high, for example linear in the number of edges of the graph. These three constraints together do not allow for simple strategies to make a vertex aware of the positions of all other vertices in the graph, and hence a distributed spring embedding approach must use some locality principle. We exploit the experimental evidence that in a drawing computed by a spring embedder: (a) the graph theoretic distance between two vertices is a good approximation for their geometric distance; (b) the fact that the repulsive forces between a vertex u and a vertex v tend to be less influential as the distance between u and v increases. See, e.g., [10]. Hence, we find it reasonable to adopt a locality principle where the force acting on each vertex v only depends on its k-*neighborhood* $N_v(k)$, i.e., the set of vertices whose graph theoretic distance from v is at most k, where k is a suitably defined constant. The attractive and repulsive forces acting on a vertex are defined according to the FR spring embedder model [6]. In our distributed implementation, each drawing iteration consists of a sequence of Giraph supersteps.

An iteration works as follows. By means of a controlled flooding technique, every vertex v knows the position of each vertex in $N_v(k)$. In the first superstep, vertex v sends a message to its neighbors. The message contains the coordinates of v, its unique identifier, and an integer number, called *TTL (Time-To-Live)*, equal to k. In the second superstep, v processes the received messages and uses them to compute the attractive and repulsive forces with respect to its adjacent vertices. Then, v uses a data structure H_v (a hash set) to store the unique identifiers of its neighbors. The TTL of each received message is decreased by one unit, and the message is broadcasted to v's neighbors. In superstep i ($i > 2$), vertex v processes the received messages and, for each message, v first checks whether the sender u is already present in H_v. If this is not the case, v uses the message to compute the repulsive force with respect to u, and then u is added to H_v. Otherwise, the forces between u and v had already been computed in some previous superstep. In both cases, the TTL of the message is decreased by one unit, and if the TTL is still greater than zero, the message is broadcasted. When no message is sent, the coordinates of each vertex are updated and the iteration is ended.

Reinsertion: After a drawing of the pruned graph has been computed, we reinsert the degree-one vertices by means of an ad-hoc technique. The general idea is to reinsert in a region close to v its adjacent vertices of degree one. Namely, each angle around v formed by two consecutive edges will host a number of vertices that is proportional to its extent. To reduce the crossings, the edges incident to the reinserted vertices are assigned a length of one tenth of the ideal spring length. We found experimentally that this solution gives good results on graphs with many one-degree vertices.

Table 1. Benchmark of real-world complex networks and results of our experiments.

	$|V|$	$|E|$	δ	CLINT- $k = 2$			CLINT- $k = 3$			FR	
				Time [sec.]	$	CR (10^6)	Time [sec.]	$	CR (10^6)	Time [sec.]	CR(10^6)
add32	4,960	9,462	28	30.3	0.04	0.26	40.2	0.06	0.24	6.9	0.1
ca-GrQc	5,242	14,496	17	112.6	0.16	2.08	128.9	0.18	1.2	4.9	1.85
grund	15,575	17,427	15	36.1	0.05	0.34	46.3	0.07	0.19	71	0.35
pGp-giantcompo	10,680	48,632	17	35.1	0.05	3.11	72.4	0.1	2.02	32.2	1.8
p2p-Gnutella04	10,876	39,994	9	39.2	0.06	73.8	122	0.17	59.5	40	12.3
ca-CondMat	23,133	93,497	14	179.1	0.25	146.8	525.4	0.74	100.2	59	77.9
p2p-Gnutella31	62,586	147,892	11	58.5	0.08	694.4	323.5	0.46	545.4	-	-
amazon0302	262,111	899,792	32	203.2	0.29	5,267.4	1,228.7	1.74	4,213.9	-	-
com-amazon	334,863	925,872	44	278.9	0.39	3,314.6	946.8	1.34	3,130.3	-	-
com-DBLP	317,080	1,049,866	21	508.5	0.72	11,978.7	-	-	-	-	-

The pipeline described above is applied independently to each connected component of the input graph. The layouts of the different components are then conveniently arranged in a matrix, so to avoid overlap. The pre-processing phase that computes the connected components of the graph is a distributed adaptation of a classical BFS algorithm, still based on a simple flooding technique. We experimentally observed that, when the graph consists of many connected components, the time required for such a pre-processing step could be higher than 70 % of the total time.

We conclude this section with the analysis of the time complexity of CLINT. Let G be a graph with n vertices and maximum vertex degree Δ. Recall that k is the integer value used to initialize the TTL of each message. Then the local function computed by each vertex costs $O(\Delta^k)$, since each vertex needs to process (in constant time) one message for each of its neighbors at distance at most k, which are $O(\Delta^k)$. Moreover, let c be the number of computing units. Assuming that each of them handles (approximately) n/c vertices, we have that each superstep costs $O(\Delta^k)\frac{n}{c}$. Let s be the maximum number of supersteps that CLINT performs (if no equilibrium is reached before), then the time complexity is $O(\Delta^k)s\frac{n}{c}$. If we assume that c and s are two constants in the size of the graph, then we have $O(\Delta^k)n$, which, in the worst case, corresponds to $O(n^{k+1})$.

3   Experiments

We experimentally studied the performances of CLINT. We took into account two main experimental hypotheses: **H1.** For small values of k ($k \leq 2$), CLINT can draw graphs up to one million edges in a reasonable time, on a cloud computing platform whose cost per hour is cheap. This hypothesis is motivated by the fact that, for a small k, the amount of data stored at each vertex should be relatively small and the message traffic load should be limited. **H2.** When the diameter of the graph is not too high, small increases of k may give rise to relatively high improvements of the drawing quality. Nevertheless, small diameters may cause a dramatic increase of the running time even for small changes of k, because the data stored at each vertex might significantly grow.

To test our hypotheses, we performed the experiments on a benchmark of 10 real graphs with up to 1 million edges, taken from the Sparse Matrix Collection of the University of Florida (http://www.cise.ufl.edu/research/sparse/matrices/) and from the Stanford Large Networks Dataset Collection (http://snap.stanford.edu/data/index.html). Previous experiments on the subject use a comparable number of real graphs (see, e.g., [15]). On each graph, we ran CLINT with $k \in \{2,3\}$. Every computation ended after at most 100 iterations (corresponding to a few hundreds Giraph supersteps). The experiments were executed on the Amazon EC2 infrastructure, using a cluster of 10 memory-optimized instances (R3.xlarge) with 4 vCPUs and 30.5 GiB RAM each. The cost per hour to use this infrastructure is about 5 USD. Table 1 reports some experimental data. Each row refers to a different network, with the networks ordered according to increasing number of edges. The columns report the number $|V|$ of vertices and $|E|$ of edges, the network diameter δ, the running time of CLINT, the Amazon cost for the drawing, and the number of crossings in the drawing.

Concerning **H1**, the data suggest that this hypothesis is not disproved. The computation of the biggest network of our test suite, consisting of more than one million edges, took about 8 min with $k = 2$. Most of this time was required for sending messages among the different Giraph workers. The cloud computing infrastructure cost of the computation is less than 1 USD. On graph com-DBLP, the computation for $k = 3$ failed due to a lack of storage resources, which means that more than 10 workers are necessary in this case. On the other hand, for the 4 smallest networks of the test suite we were able to compute the layout up to $k = 5$. The running time on ca-GrQc and ca-CondMat is higher than that spent on other graphs of similar size; more than 70 % of this time was needed to compute the (many) connected components of these graphs.

Concerning **H2**, we report the quality of the drawings in terms of number of edge crossings. The improvement passing from $k = 2$ to $k = 3$ varies from 6 % (on com-amazon) to 44 % (on grund). As expected, the improvement is usually higher on networks with relatively small diameter. Also the increase of the running time, going from $k = 2$ to $k = 3$, is usually more severe for small diameters. For example, graph p2p-Gnutella04 has half the size of graph ca-CondMat, and its diameter is also much smaller; nevertheless, the increase of running time passing from $k = 2$ to $k = 3$ on p2p-Gnutella04 is higher (211 %) than on ca-CondMat (193 %). Again, the increase of time on graph amazon0302 (whose diameter is 32) is almost twice that on graph com-amazon (whose diameter is 44), although the latter is bigger than the former. Hence, also hypothesis **H2** is not disproved.

In addition to the above experiments, we ran a centralized version of the FR algorithm against our benchmark on an Intel i7 3630QM laptop, with 2.4 GHz and 8 GB of RAM. Namely, we ran the optimized FR implementation available in the OGDF library (http://www.ogdf.net/). This algorithm was able to complete the computation for the 6 smaller graphs of the test suite. The last two columns report the time and the number of crossings of the centralized FR computations. In the average, the drawings computed by CLINT for $k = 3$ have about 1.8 times the number of crossings of those computed by FR. In some cases however,

CLINT performed better than FR (see ca-GrQC and grund) or similarly (see pGp-giantcompo). About the running time, CLINT is often slower than the centralized FR, due to the time required by the flooding techniques for exchanging messages and by the fix infrastructure cost of the distributed environment, which is better amortized over the computation of bigger instances.

We also tried to estimate the *strong scalability* of CLINT, that is, how the running time varies on a given instance when the number of workers increases. For each graph we ran CLINT also with 6 and 8 workers. For the largest graphs and for $k = 2$, passing from 6 to 8 workers improves the running time of about 20 %, while passing from 8 to 10 workers causes a further decrease of about 10 %. These percentages increase for $k = 3$. On the smaller graphs, the benefit of using more workers is evident from $k \geq 4$.

4   Conclusions and Future Research

We described and experimented the first TLAV distributed spring embedder. Our results are promising, but more experiments would help to find better trade-offs between values of k, running time, drawing quality, and number of workers in the PaaS. Future work includes: (a) Developing TLAV versions of multi-scale force-directed algorithms, able to compute several million edge graphs on a common cloud computing service; this would improve running times and drawing quality. (b) Designing a vertex-centric distributed service to interact with the visualizations of very large graphs; a TLAV drawing algorithm should be one of the core components of such a service.

References

1. Brandes, U., Pich, C.: Eigensolver methods for progressive multidimensional scaling of large data. In: Kaufmann, M., Wagner, D. (eds.) GD 2006. LNCS, vol. 4372, pp. 42–53. Springer, Heidelberg (2007)
2. Chae, S., Majumder, A., Gopi, M.: Hd-graphviz: highly distributed graph visualization on tiled displays. In: ICVGIP 2012, pp. 43:1–43:8. ACM (2012)
3. Ching, A.: Giraph: large-scale graph processing infrastructure on hadoop. In: Hadoop Summit (2011)
4. Di Battista, G., Eades, P., Tamassia, R., Tollis, I.G.: Graph Drawing. Prentice Hall, Upper Saddle River, NJ (1999)
5. Eades, P.: A heuristic for graph drawing. Congr. Numerant. **42**, 149–160 (1984)
6. Fruchterman, T.M.J., Reingold, E.M.: Graph drawing by force-directed placement. Software, Practice and Experience **21**(11), 1129–1164 (1991)
7. Godiyal, A., Hoberock, J., Garland, M., Hart, J.C.: Rapid multipole graph drawing on the GPU. In: Tollis, I.G., Patrignani, M. (eds.) GD 2008. LNCS, vol. 5417, pp. 90–101. Springer, Heidelberg (2009)
8. Hachul, S., Jünger, M.: Drawing large graphs with a potential-field-based multilevel algorithm. In: Pach, J. (ed.) GD 2004. LNCS, vol. 3383, pp. 285–295. Springer, Heidelberg (2005)
9. Ingram, S., Munzner, T., Olano, M.: Glimmer: multilevel MDS on the GPU. IEEE Trans. Vis. Comput. Graph. **15**(2), 249–261 (2009)

10. Kobourov, S.G.: Force-directed drawing algorithms. In: Tamassia, R. (ed.) Handbook of Graph Drawing and Visualization, pp. 383–408. CRC Press, Boca Raton (2013)
11. Malewicz, G., Austern, M.H., Bik, A.J., Dehnert, J.C., Horn, I., Leiser, N., Czajkowski, G.: Pregel: a system for large-scale graph processing. In: SIGMOD 2010, pp. 135–146. ACM (2010)
12. McCune, R.R., Weninger, T., Madey, G.: Thinking like a vertex: a survey of vertex-centric frameworks for large-scale distributed graph processing. ACM Comput. Surv. 1(1), 1–35 (2015)
13. Mueller, C., Gregor, D., Lumsdaine, A.: Distributed force-directed graph layout and visualization. In: EGPGV 2006, pp. 83–90. Eurographics (2006)
14. Sharma, P., Khurana, U., Shneiderman, B., Scharrenbroich, M., Locke, J.: Speeding up network layout and centrality measures for social computing goals. In: Salerno, J., Yang, S.J., Nau, D., Chai, S.-K. (eds.) SBP 2011. LNCS, vol. 6589, pp. 244–251. Springer, Heidelberg (2011)
15. Tikhonova, A., Ma, K.: A scalable parallel force-directed graph layout algorithm. In: EGPGV 2008, pp. 25–32. Eurographics (2008)
16. Valiant, L.G.: A bridging model for parallel computation. Commun. ACM 33(8), 103–111 (1990)
17. Vaquero, L.M., Cuadrado, F., Logothetis, D., Martella, C.: Adaptive partitioning for large-scale dynamic graphs. In: ICDCS 2014, pp. 144–153. IEEE (2014)
18. Yunis, E., Yokota, R., Ahmadia, A.: Scalable force directed graph layout algorithms using fast multipole methods. In: ISPDC 2012, pp. 180–187. IEEE (2012)

Faster Force-Directed Graph Drawing
with the Well-Separated Pair Decomposition

Fabian Lipp$^{(\boxtimes)}$, Alexander Wolff, and Johannes Zink

Lehrstuhl für Informatik I, Universität Würzburg, Würzburg, Germany
fabian.lipp@uni-wuerzburg.de
http://www1.informatik.uni-wuerzburg.de/en/staff

Abstract. The force-directed paradigm is one of the few generic approaches to drawing graphs. Since force-directed algorithms can be extended easily, they are used frequently. Most of these algorithms are, however, quite slow on large graphs as they compute a quadratic number of forces in each iteration. We speed up this computation by using an approximation based on the well-separated pair decomposition.

We perform experiments on a large number of graphs and show that we can strongly reduce the runtime—even on graphs with less then a hundred vertices—without a significant influence on the quality of the drawings (in terms of number of crossings and deviation in edge lengths).

1 Introduction

Force-directed algorithms are commonly used to draw graphs. They can be used on a wide range of graphs without further knowledge of the graphs' structure. The idea is to define physical forces between the vertices of the graph. These forces are applied to the vertices iteratively until stable positions are reached. The well-known spring-embedder algorithm of Eades [5] models the edges as springs. His approach was refined by Fruchterman and Reingold [9]. Between pairs of adjacent vertices they apply attracting forces caused by springs. To prevent vertices getting too close, they apply repulsive forces between all pairs of vertices.

Generally, force-directed methods are easy to implement and can be extended well. For example, Fink et al. [7] defined additional forces to draw Metro lines in Metro maps as Bézier curves instead of as polygonal chains. Different aesthetic criteria can be balanced by weighing them accordingly. Force-directed algorithms can in principle be used for relatively large graphs with hundreds of vertices and often yield acceptable results. Unfortunately, force-directed methods are rather slow on such graphs. This is caused by the computation of the repulsive force for every vertex pair, which yields a quadratic runtime for each iteration. In this paper, we present a new approach to speed this up.

Previous Work. There are a lot of techniques to speed up force-directed algorithms. For example, Barnes and Hut [1] use a quadtree, a multi-purpose spatial

© Springer International Publishing Switzerland 2015
E. Di Giacomo and A. Lubiw (Eds.): GD 2015, LNCS 9411, pp. 52–59, 2015.
DOI: 10.1007/978-3-319-27261-0_5

data structure, to approximate the forces between the vertex pairs. We will compare our algorithm to theirs subsequently. Another approach is the multilevel paradigm introduced by Walshaw [15]. After contracting dense subgraphs, the resulting coarse graph is laid out. Then the vertices are uncontracted and a layout of the whole graph based on the coarse layout is computed. This can be done over several levels. The multilevel paradigm does not rule out our WSPD-based approach; our approach can be applied to each level.

Various force-directed graph drawing algorithms have been compared before [2,3,8]. We use some of the quality criteria described in the literature to evaluate our algorithm.

Callahan and Kosaraju [4] defined a decomposition for point sets in the plane, the *well-separated pair decomposition* (WSPD). Given a point set P and a number $s > 0$, this decomposition consists of pairs of subsets $(A_i, B_i)_{i=1,\ldots,k}$ of P with two properties. First, for each pair $(p, q) \in P^2$ with $p \neq q$, there is a unique index $i \in \{1, \ldots, k\}$ such that $p \in A_i$ and $q \in B_i$ or vice versa. Second, each pair (A_i, B_i) must be s-well-separated, that is, the distance between the two sets is at least s times the larger of the diameters of the sets. Callahan and Kosaraju showed how to construct a WSPD for a set of n points in $O(n \log n)$ time where the number k of pairs of sets is linear in n.

The WSPD has been used for graph drawing before; Gronemann [10] employed it to speed up the fast multipole multilevel method [11]. While our WSPD is based on the *split tree* [4], Gronemann's is based on a quadtree.

Our Contribution. We use the WSPD in order to speed up the force-directed algorithm of Fruchterman and Reingold (FR). Instead of computing the repulsive forces for every pair of points, we represent every set $A_1, \ldots, A_k, B_1, \ldots, B_k$ in the decomposition by its barycenter and use the barycenter of a set, say A_i, as an approximation when computing the forces between this set and a point in B_i. Thus, an iteration takes us $O(n \log n)$ time, instead of $\Omega(n^2)$ for the classical algorithm.

Additionally, our method is very simple and allows the user to define forces arbitrarily—as long as the total force on a point p is the sum of the forces of point pairs in which p is involved. Hence, our approach can be applied to other force-directed algorithms as well. We don't consider other techniques such as Multidimensional Scaling (MDS) or multi-level algorithms in this paper, as we only want to show that we can speed up a force-directed graph layout algorithm using the WSPD. We guess that this technique can be applied to other algorithms as well. In the above-mentioned fast multipole method, in contrast, the approximation of the repulsive forces is quite complicated (as Hachul and Jünger [11] point out); it requires the expansion of a Laurent series.

2   Algorithm

In this section, we describe our WSPD-based implementation, analyze its asymptotic running time, and give a heuristic speed-up method.

Constructing a WSPD. There are various ways to construct an *efficient* WSPD, that is, a WSPD with a linear number of pairs of sets. We use the *split tree* as described by Callahan and Kosaraju [4] when introducing the WSPD. Our implementation follows the algorithm FastSplitTree in the textbook of Narasimhan and Smid [13, Sect. 9.3.2]. Given n points, this algorithm constructs a linear-size split tree in $O(n \log n)$ time. Given the tree, a WSPD with separation constant s can be built in $O(s^2 n)$ time.

The Force-Directed Algorithm. The general principle of a force-directed algorithm is as follows. In every iteration, the algorithm computes forces on the vertices. These forces depend on the current position of the vertices in the drawing. The forces are applied as an offset to the position of each vertex. The algorithm terminates after a given number of iterations or when the forces get below a certain threshold.

A classical force-directed algorithm such as FR computes, in every iteration, an attractive force for any pair of adjacent vertices and a repulsive force for any pair of vertices. Fruchterman and Reingold [9] use $F_{\text{attractive}}(u, v) = d^2/c$ and $F_{\text{repulsive}}(u, v) = -c^2/d$, where c is a constant describing the ideal edge length and $d = d(u, v)$ is the distance between vertices u and v in the current drawing.

Our modified algorithm is shown in Algorithm 1. We first compute a fair split tree T for the current positions of the vertices of G (which are stored in the leaves of T). Each node μ of T corresponds to the set of vertices in the leaves of the subtree rooted in μ. Bottom-up, we compute the barycenters of the sets corresponding to the nodes of T. From T, we compute a WSPD $(A_i, B_i)_i$ for the current vertex positions. Each set A_i (and B_i) of the WSPD corresponds to a node α_i (and β_i) of T. For each pair (A_i, B_i) of the WSPD, we compute $F_{\text{repulsive}}$ from the barycenter of A_i to the barycenter of B_i (and vice versa), and store the results (in an accumulative fashion) in α_i and β_i. Finally, we traverse T top-down. During the traversal, we add to the force of each node the force of the parent node. When we reach the leaves of T, which correspond to the graph vertices, we have computed the resulting force for each vertex.

Running Time. We denote the number of vertices of the given graph by n and the number of edges by m. In each iteration, the classical algorithm computes the attractive forces in $O(m)$ time and the repulsive forces in $O(n^2)$ time.

We don't modify the computation of the attractive forces. For computing the repulsive forces, the most expensive step is the computation of the split tree T and the WSPD, which takes $O(n \log n)$ time. The barycenters of the sets corresponding to the nodes of T can be computed bottom-up in linear time. The forces between the pairs of the WSPD can also be computed in linear total time. The same holds for the forces acting on the vertices. In total, hence, an iteration takes $O(m + n \log n)$ time.

Improvements. To speed up our algorithm, we compute a new split tree and the resulting WSPD only every few iterations. To be precise, we only recompute it when $\lfloor 5 \log i \rfloor$ changes, where i is the current iteration. Thus, the WSPD may

---

Algorithm 1. WSPD-based force computation for a graph $G = (V, E)$

`// attractive forces for adjacent vertices:`
foreach $e = (u, v) \in E$ **do**
  | u.addForce($F_{\text{attractive}}(u, v)$); v.addForce($F_{\text{attractive}}(v, u)$)

`// approximation of repulsive forces:`
Compute a fair split tree T for the current positions of the vertices (stored in the leaves of T).
For each node μ of T, compute the barycenter $c(\mu)$ of the leaves of the subtree rooted in μ.
Compute a WSPD $(A_i, B_i)_{i=1,\ldots,k}$ from T; each A_i (B_i) corresponds to a node α_i (β_i) of T.
for $i = 1$ **to** k **do**
  | α_i.addForce($|B_i| \cdot F_{\text{repulsive}}(c(\alpha_i), c(\beta_i))$)
  | β_i.addForce($|A_i| \cdot F_{\text{repulsive}}(c(\beta_i), c(\alpha_i))$)

Traverse the split tree top-down to compute the total force for every vertex of G.

---

not be valid for the current vertex positions. This makes the approximation of the forces more inaccurate, but our experiments show that this method does not change the quality of the drawings significantly, while the running time decreases notably (see Figs. 1, 2 and 3).

Implementation. Our Java implementation is based on FRLayout, the FR algorithm implemented in the JUNG library [12]. We slightly optimized the code, which reduced the runtime by a constant factor. Additionally, we removed the frame that bounded the drawing area, as it caused ugly drawings for larger graphs. For our experimental comparison in Sect. 3, we used FRLayout with these modifications. It is this implementation that we then sped up using the WSPD. We recompute the WSPD only every few iterations as described in the previous paragraph. We call the result FR+WSPD. For comparison, we also implemented the quadtree-based speed-up method of Barnes and Hut [1], which we call FR+Quad, and a grid-based approach suggested already by Fruchterman and Reingold [9], which we call FR+Grid. To widen the scope of our study, we included some algorithms implemented in C++ in the *Open Graph Drawing Framework* (OGDF, www.ogdf.net): GEM, FM^3 (with and without multilevel technique, then we call it FM^3 single) of Hachul and Jünger [11], and FRExact (the exact FR implementation in OGDF).

3   Experimental Results

We formulate the following hypotheses which we then test experimentally.

(H1) The quality of the drawings produced by FR+WSPD is comparable to that of FRLayout.
(H2) On sufficiently large graphs, FR+WSPD is faster than FRLayout.

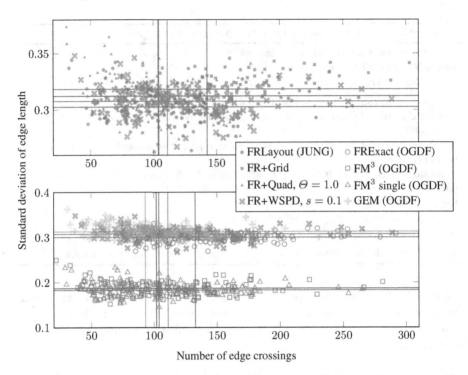

Fig. 1. Standard deviation of the edge length (y-axis) over number of edge crossings (x-axis) for various variants of the algorithm, applied to all 140 Rome graphs with exactly 100 vertices. Top: quality of the unmodified FRLayout algorithm and its variants FR+WSPD, FR+Quad, and FR+Grid. Bottom: FR+WSPD and some algorithms implemented in OGDF. For each algorithm, a vertical and a horizontal line mark its median performance.

We assume these hypotheses due to the favorable properties of the WSPD: the separation property hints at (H1) and the improved time complexity per iteration implies (H2).

We tested our algorithms on two data sets; (i) the Rome graph collection [14] that contains 11528 undirected connected graphs with 10–100 vertices each, and (ii) 40 random graphs that we generated using the EppsteinPowerLawGenerator [6] in JUNG, which yields graphs whose structure is similar to Web graphs. Our graphs had 2,500, 5,000, 7,500, . . . , 100,000 vertices and 2.5 times as many edges. We considered only the largest connected component of each generated graph.

The experiments were performed on an Intel Xeon CPU with 2.67 GHz and 20 GB RAM running Linux. The computer has 16 cores, but we did not parallelize our code. During our experiments, only one core was operating at close-to-full capacity.

We measured the quality of the drawings by (a) the number of edge crossings and (b) the standard deviation of the edge length (normalized by the mean edge length). These criteria have been used before to compare force-directed layout algorithms [3,8].

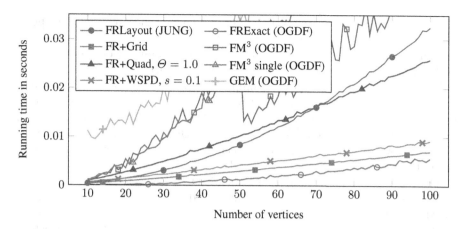

Fig. 2. The runtimes for the different variants of the algorithm as a function of the number of vertices. Each point in the plots represents the mean value of the runtimes on all Rome graphs with the given number of vertices. The markers are used only as a tool to identify the plots.

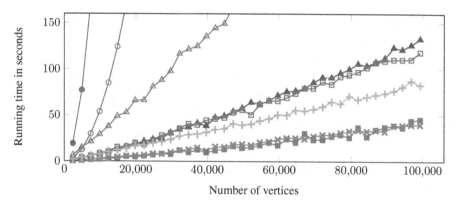

Fig. 3. The runtimes of the algorithms on 40 random graphs. The graphs were generated using the EppsteinPowerLawGenerator [6] in JUNG with $|V| = 2{,}500, 5{,}000, 7{,}500, \ldots, 100{,}000$ and $|E| \approx 2.5 \cdot |V|$. Only the largest connected component of each generated graph was considered. Other than in Fig. 2, each marker represents one of the tested graphs. The legend is the same.

To test hypothesis (H1), we compared the quality of the drawings of FRLayout, FR+WSPD, FR+Quad, and FR+Grid. In order to vary as few parameters as possible, we kept the size of the graphs constant in this part of the study. We used all Rome graphs with exactly 100 vertices. The 140 graphs have, on average, 135 edges.

We first compared the outputs of FR+WSPD for different values $(0.01, 0.1, 1)$ of the separation constant s. The distribution of the results in the plot was roughly the same, that is, the quality of the drawings did not strongly depend on s. Using $s = 1$ was about 30 % slower than $s = 0.1$ or $s = 0.01$.

Similarly, FR+Quad has a parameter Θ that controls how fine the given point set is subdivided. Increasing Θ decreases the running time. Our experiments confirmed what Barnes and Hut [1] observed: only values of Θ close to 1 give results with a similar quality as the unmodified algorithm.

The upper scatterplot in Fig. 1 compares variants of FR based on different speed-up techniques. Compared to FRLayout, FR+Quad is slightly worse in terms of uniformity of edge lengths and FR+WSPD is slightly worse in terms of edge crossings and between FRLayout and FR+Quad in terms of edge lengths. FR+Grid is worse in both measures, especially in the number of edge crossings. Hence, there is support for hypothesis (H1).

The lower scatterplot in Fig. 1 compares FR+WSPD to the above-mentioned OGDF algorithms. In terms of uniformity of edge lengths, there are two clear clusters: the two FM^3 variants are better than the rest. In terms of crossings, GEM is best, followed by the FM^3 variants, and then by FR+WSPD, which surprisingly is *better* than FRExact.

To test hypothesis (H2), we measured the runtimes of the all algorithms on the Rome graphs (Fig. 2) and the random graphs (Fig. 3). In Java, we only measure the time used for the thread running the force-directed algorithm in our Java Virtual Machine; this eliminates the influence of the garbage collector and the JIT compiler on our measurements. In C++, we used an OGDF method for measuring the CPU time. For each graph size, we display the mean runtime over all graphs of that size.

The results are as follows. As expected, FR+WSPD (with $s = 0.1$) is faster than FRLayout on larger graphs. We were surprised, however, to see that FR+WSPD overtakes FRLayout already around $n \approx 30$. FR+WSPD also turned out to be faster than FR+Quad (with $\Theta = 1$) and than FM^3 by a factor of 1.5 to 3. FR+WSPD and FR+Grid are comparable in speed, and twice as fast as GEM. Recall, however, that FR+Grid tends to produce more edge crossings (Fig. 1). Concerning the comparison between Java and C++, FRExact (in C++) is roughly four times faster than FRLayout (in Java).

Conclusion. Our experiments show that the WSPD-based approach speeds up force-directed graph drawing algorithms such as FR considerably without sacrificing the quality of the drawings. The main feature of the new approach is its simplicity. We plan to combine our approach with multi-level techniques in order to draw much larger graphs.

References

1. Barnes, J., Hut, P.: A hierarchical $O(N \log N)$ force-calculation algorithm. Nature **324**(6096), 446–449 (1986)
2. Bartel, G., Gutwenger, C., Klein, K., Mutzel, P.: An experimental evaluation of multilevel layout methods. In: Brandes, U., Cornelsen, S. (eds.) GD 2010. LNCS, vol. 6502, pp. 80–91. Springer, Heidelberg (2011)
3. Brandenburg, F.J., Himsolt, M., Rohrer, C.: An experimental comparison of force-directed and randomized graph drawing algorithms. In: Brandenburg, F.J. (ed.) GD 1995. LNCS, vol. 1027, pp. 76–87. Springer, Heidelberg (1996)

4. Callahan, P.B., Kosaraju, S.R.: A decomposition of multidimensional point sets with applications to k-nearest-neighbors and n-body potential fields. J. ACM **42**(1), 67–90 (1995)
5. Eades, P.: A heuristics for graph drawing. Congr. Numerantium **42**, 146–160 (1984)
6. Eppstein, D., Wang, J.Y.: A steady state model for graph power laws. In: 2nd International Workshop Web Dynamics (2002). http://arxiv.org/abs/cs/0204001
7. Fink, M., Haverkort, H., Nöllenburg, M., Roberts, M., Schuhmann, J., Wolff, A.: Drawing metro maps using Bézier curves. In: Didimo, W., Patrignani, M. (eds.) GD 2012. LNCS, vol. 7704, pp. 463–474. Springer, Heidelberg (2013)
8. Frick, A., Ludwig, A., Mehldau, H.: A fast adaptive layout algorithm for undirected graphs. In: Tamassia, R., Tollis, I.G. (eds.) GD 1994. LNCS, vol. 894, pp. 388–403. Springer, Heidelberg (1995)
9. Fruchterman, T.M.J., Reingold, E.M.: Graph drawing by force-directed placement. Softw. Pract. Exp. **21**(11), 1129–1164 (1991)
10. Gronemann, M.: Engineering the fast-multipole-multilevel method for multicore and SIMD architectures. Master's thesis, Department of Computer Science, TU Dortmund (2009)
11. Hachul, S., Jünger, M.: Drawing large graphs with a potential-field-based multilevel algorithm. In: Pach, J. (ed.) GD 2004. LNCS, vol. 3383, pp. 285–295. Springer, Heidelberg (2005)
12. JUNG: Java Universal Network/Graph Framework. http://jung.sourceforge.net. Accessed 2 September 2015
13. Narasimhan, G., Smid, M.: Geometric Spanner Networks. Cambridge University Press, New York, NY, USA (2007)
14. Rome Graphs. http://graphdrawing.org/data.html, http://www.graphdrawing.org/download/rome-graphml.tgz. Accessed 2 September 2015
15. Walshaw, C.: A multilevel algorithm for force-directed graph-drawing. J. Graph Algorithms Appl. **7**(3), 253–285 (2003)

Crossing Numbers

The Degenerate Crossing Number
and Higher-Genus Embeddings

Marcus Schaefer[1](✉) and Daniel Štefankovič[2]

[1] DePaul University, Chicago, IL 60604, USA
mschaefer@cs.depaul.edu
[2] University of Rochester, Rochester, NY 14627, USA
stefanko@cs.rochester.edu

Abstract. If a graph embeds in a surface with k crosscaps, does it always have an embedding in the same surface in which every edge passes through each crosscap at most once? This well-known open problem can be restated using crossing numbers: the degenerate crossing number, $\mathrm{dcr}(G)$, of G equals the smallest number k so that G has an embedding in a surface with k crosscaps in which every edge passes through each crosscap at most once. The genus crossing number, $\mathrm{gcr}(G)$, of G equals the smallest number k so that G has an embedding in a surface with k crosscaps. The question then becomes whether $\mathrm{dcr}(G) = \mathrm{gcr}(G)$, and it is in this form that it was first asked by Mohar.

We show that $\mathrm{dcr}(G) \leq 6 \, \mathrm{gcr}(G)$, and $\mathrm{dcr}(G) = \mathrm{gcr}(G)$ as long as $\mathrm{dcr}(G) \leq 3$. We can separate dcr and gcr for (single-vertex) graphs with embedding schemes, but it is not clear whether the separating example can be extended into separations on simple graphs. Finally, we show that if a graph can be embedded in a surface with crosscaps, then it has an embedding in that surface in which every edge passes through each crosscap at most twice. This implies that dcr is **NP**-complete.

Keywords: Degenerate crossing number · Non-orientable genus · Genus crossing number

1 Introduction

When defining the crossing number of a graph, one typically requires that at most two edges cross in any point. If $k > 2$ edges cross in a single point, these edges can be perturbed slightly to create $\binom{k}{2}$ crossings of pairs of edges, so multiple crossings in a single point can always be avoided. Günter Rote and M. Sharir, according to Pach and Tóth [10] asked "what happens if multiple crossings are counted only *once*". This led Pach and Tóth to introduce the degenerate crossing number: we allow drawings which are *degenerate* in the sense that more than two edges are allowed to cross in a single point (but which are otherwise standard, in particular, edges have to actually cross, not touch, and self-crossings are not allowed). The *degenerate crossing number* of the drawing is the number of crossing points in the drawing. The *degenerate crossing number*, $\mathrm{dcr}(G)$, of a

© Springer International Publishing Switzerland 2015
E. Di Giacomo and A. Lubiw (Eds.): GD 2015, LNCS 9411, pp. 63–74, 2015.
DOI: 10.1007/978-3-319-27261-0_6

graph G is the smallest degenerate crossing number of any (degenerate) drawing of G. Some papers (e.g. [1]) restrict drawings to be simple, that is, every two edges intersect (or cross, that's not always clearly defined[1]) at most once; to distinguish this variant from dcr we call it the *simple degenerate crossing number*, $\mathrm{dcr}^*(G)$.[2]

If we modify the definition of the degenerate crossing number to allow self-crossings of edges, we obtain the *genus crossing number*, $\mathrm{gcr}(G)$, which was introduced by Mohar [8]. By definition, $\mathrm{gcr}(G) \leq \mathrm{dcr}(G)$. Mohar conjectured that $\mathrm{gcr}(G) = \mathrm{dcr}(G)$. Equality of these two numbers would be particularly interesting, since, as Mohar observes, $\mathrm{gcr}(G) = \widetilde{\gamma}(G)$, where $\widetilde{\gamma}(G)$ is the *non-orientable genus* (or the *minimum crosscap number*) of G, the smallest number k so that G can be embedded on a surface with k crosscaps (we allow the special case of $k = 0$ for planar graphs). Each crossing of multiple edges can be replaced by a crosscap and vice versa, since edges have to cross (and may not touch) in a crossing point. Similarly, $\mathrm{dcr}(G)$ can be viewed (as we did in the abstract) as the smallest number k so that G has an embedding on a surface with k crosscaps so that every edge passes through each crosscap at most once. An edge not being allowed to pass through a crosscap more than once corresponds to prohibiting self-crossings in degenerate drawings in the plane. We view crosscaps as geometric, rather than purely topological objects, a view which we believe makes sense in graph drawing, where we need to visualize objects.[3]

We do not yet know, whether $\mathrm{gcr}(G) = \mathrm{dcr}(G)$ in general, but we can separate them, if we are allowed to equip graphs with an embedding scheme (a fixed rotation at each vertex, and a signature for each edge). In that case, there are graphs for which gcr is 3, but dcr is 4 as we will see in Theorem 4.

Remark 1 (Visualizing Graphs in Higher-Order Surfaces). Whether gcr = dcr or not has consequences for visualizing graphs embeddable in higher-order surfaces in the plane. Typically, such graphs are visualized using a (canonical) polygonal schema. There are polynomial-time algorithms for this task, e.g., see [4] for orientable surfaces, also see [3,5,7]. Assuming that vertices may not lie on the boundary (of the polygonal schema), the question gcr = dcr then becomes: do edges have to pass through the same side of a schema more than once? Many of the visualization algorithms (including [3]) start by contracting the graph to a single-vertex graph with an embedding scheme; for these algorithms, the example in Theorem 4 shows that edges can be forced to cross through the same side more than once.

On the other hand, we can show that $\mathrm{dcr}(G) \leq 6\,\mathrm{gcr}(G)$, so any graph embeddable in a surface with k crosscaps can be embedded in a surface with at most $6k$ crosscaps so that every edge passes through each crosscap at most once. We will

---

[1] The difference is that a shared endpoint counts as an intersection, but not a crossing.

[2] An example in the entry on degenerate crossing number in [12] shows that it matters whether dcr* is defined so as to allow crossings between adjacent edges or not.

[3] Mohar [8] uses a "planarizing system of disjoint 1-sided curves" to define "passing through a crosscap" formally.

establish this in Theorem 2. If we allow an edge to pass through each crosscap just twice, it turns out that every graph can then be embedded in a surface with $\widetilde{\gamma}(G)$ crosscaps (Theorem 5).

1.1 Known Results

Pach and Tóth [10] showed that dcr(G) < |E(G)|. For the simple degenerate crossing number, a crossing lemma is known: $\mathrm{dcr}^*(G) \geq c \cdot |E(G)|^3 / |V(G)|^2$ for $|E(G)| \geq 4|V(G)|$ (and some constant $c > 0$). This was shown by Ackerman and Pinchasei [1], improving an earlier result by Pach and Tóth. We should also mention work by Harborth [6], who may have been the first to study multiple crossings in drawings. His goal is to maximize the number of multiway crossings. For example, he shows that K_{2m} can be drawn with two m-fold crossings; he conjectured that K_{2m} cannot be drawn with three or more m-fold crossings.

2 Tools

We start with some basic facts about (simple) closed curves on a non-orientable surface S. A closed curve C is called *non-separating* if $S - C$ consists of a single piece. Otherwise, C is *separating*. If it is separating, it can be *contractible* (one of the two pieces is homeomorphic to a disk) or *surface-separating*. The *sidedness* of a closed curve is the number of sides it has: it is either *one-sided* (its neighborhood is a Moebius strip) or *two-sided*. A closed curve C in a non-orientable surface *maximal* if $S - C$ is orientable (equivalently, if C passes through every crosscap an odd number of times).[4]

A surface can contain only a small number of different types of closed curves. The following lemma makes this precise.

Lemma 1 (Malnič, Mohar [9, Proposition 4.2.7]). *If G is a graph embedded in a surface S, and $\mathcal{P}$ is a collection of internally disjoint paths between vertices a and b (where $a = b$ is allowed), so that no two of the paths bound a disk in S, then*

$$|\mathcal{P}| \leq \begin{cases} 3\widetilde{\gamma}(S) - 2 & \text{if } \widetilde{\gamma}(S) \geq 2 \\ \widetilde{\gamma}(S) + 1 & \text{otherwise.} \end{cases}$$

Remark 2. We are interested in the case where $a = b$ and there are no surface separating paths; a better upper bound for that case would improve the upper bound in Theorem 4.

We also need some tools from topological graph theory to describe and handle embeddings of graphs on non-orientable surfaces. On orientable surfaces, an embedding can be described by a *rotation system* which prescribes a *rotation* (a clockwise, cyclic ordering) of the ends of all edges incident to a vertex.

---

[4] There seems to be no standard name for curves of this type in the literature. Bojan Mohar suggests "orienting".

On non-orientable surfaces, we also need to prescribe, for every edge, its *signature*, which is a number in $\{-1, 1\}$. A cycle in G is *two-sided* if the signature of its edges multiply to 1, otherwise, it is *one-sided*. A rotation system ρ and signature λ together form an *embedding scheme* (ρ, λ) of a graph on a surface. A drawing of a graph in a surface *realizes* an embedding scheme (ρ, λ), if the rotation at each vertex is as prescribed by ρ, and the sidedness of each cycle is as prescribed by the signatures of the edges. The sidedness of a cycle is determined by the parity of how often the cycle passes through crosscaps. Typical operations on graphs (removing/adding a vertex/edge, contracting an edge) are easily performed on the embedding scheme as well. For details, see [9, Sect. 3.3].

Arguments and algorithms for graph embeddings can often be simplified by replacing an embedded graph with a single-vertex graph with embedding scheme. This is often done for visualizing embeddings of graphs in higher-genus surfaces in the plane (see Remark 1). Note that in a single-vertex graph every edge is a loop, hence a closed curve, and we can talk about the sidedness, which then directly correspond to its signature: a one-sided loop has signature -1, and a two-sided loop signature 1. For a graph G with embedding scheme (ρ, λ) we define $\mathrm{gcr}(G, \rho, \lambda)$ as the smallest number k so that G has an embedding realizing (ρ, λ) on a surface with k crosscaps. Similarly, $\mathrm{dcr}(G, \rho, \lambda)$ is the smallest degenerate crossing number of any drawing of G which realizes the embedding scheme (ρ, λ).

The next lemma shows that as far as gcr is concerned, we can replace a graph with a graph on a single vertex equipped with an embedding scheme. For dcr we can do so for upper bounds only.

Lemma 2. *For every graph G there is a single-vertex graph G' with embedding scheme (ρ, λ) so that $\mathrm{gcr}(G) = \mathrm{gcr}(G', \rho, \lambda)$, and $\mathrm{dcr}(G) \leq \mathrm{dcr}(G', \rho, \lambda)$. Moreover, any embedding of (G', ρ, λ) can be turned into an embedding of G without changing dcr or gcr of the embedding by uncontracting edges.*

Proof. Fix an embedding of G on a surface S with $k = \mathrm{gcr}(G)$ crosscaps. We can assume that G is connected (if it is not, we can extend G to a triangulation of S). Let T be a spanning tree of G. Contract edges of T, merging rotations in the embedding scheme at vertices that are identified and updating signatures of edges. Let G' be the resulting single-vertex graph with embedding scheme (ρ, λ). Then $\mathrm{gcr}(G', \rho, \lambda) \leq \mathrm{gcr}(G)$. If $\mathrm{gcr}(G', \rho, \lambda) < \mathrm{gcr}(G)$ were true, we could undo the operations which turned G into G' (since we maintained the embedding scheme) to find an embedding of G on a surface with less than $\mathrm{gcr}(G)$ crosscaps, which is a contradiction, so $\mathrm{gcr}(G) = \mathrm{gcr}(G', \rho, \lambda)$. The same argument shows that $\mathrm{dcr}(G', \rho, \lambda) < \mathrm{dcr}(G)$ is not possible, so $\mathrm{dcr}(G) \leq \mathrm{dcr}(G', \rho, \lambda)$. ∎

Note that we do not claim that $\mathrm{dcr}(G) \geq \mathrm{dcr}(G', \rho, \lambda)$, the construction we used may force an edge through a crosscap multiple times, so dcr can increase. Lemma 2 allows us to replace a graph with a single-vertex graph when showing that dcr can be bounded in gcr.

Finally, we need some basic techniques to deal with curves in a surface.

Theorem 1 (Weak Hanani-Tutte Theorem for Surfaces [2, 11]). *If G is drawn in a surface so that every pair of edges crosses an even number of times, then G has an embedding on the same surface with the same embedding scheme.*

It is well-known that a handle and two crosscaps are equivalent in the presence of another crosscap. So a graph embeddable on a surface with h handles can be embedded on a surface with $2h + 1$ crosscaps so that every edge is two-sided. The following lemma shows that the odd number of crosscaps is not accidental when restricting to orientable embeddings, where we call an embedding (G, ρ, λ) of a graph G *orientable* if all cycles in G are two-sided (equivalently, multiplying the signatures of edges along each cycle, one always gets 1). Note that if G is a single-vertex graph, then its embedding is orientable, if all loops have signature 1.

Lemma 3. *Suppose k is minimal so that a connected graph (G, ρ) with rotation ρ has an orientable embedding on a surface with k crosscaps. Then either $k = 0$, or $k \geq 3$ and k is odd.*

Proof. Fix an orientable embedding of (G, ρ) in a surface with k crosscaps, where k is minimal. We can assume that G is a single-vertex graph (contract edges of a spanning tree, this leaves the embedding orientable, so $\lambda(e) = 1$ for all loops e now). Suppose k is even. Let c be one of the crosscaps. For any edge that passes oddly through c, push that edge over all crosscaps. Note that pushing an edge over all crosscaps does not change the parity of crossing between any pair of edges since the number of crosscaps is even and every edge initially crosses through an even number of crosscaps oddly, and this remains true. At the end of this operation we have a drawing of G in which every pair of edges crosses an even number of times, and all edges pass through c an even number of times. We can then push all edges off c, again maintaining that every pair of edges crosses evenly. Now, by Theorem 1, (G, ρ, λ) has an orientable embedding in the surface with $k - 1$ crosscaps, so k cannot have been even if it was minimal. If $k = 1$, then an orientable embedding on the projective plane implies that the graph is planar (since every edge passes through the single crosscap an even number of times). ☐

Corollary 1. *If a single-vertex graph (G, ρ) has an orientable embedding on a non-orientable surface with $k \geq 2$ crosscaps, we can add a one-sided loop into its embedding scheme, without changing the surface.*

Proof. Let $k' \leq k$ be minimal so that (G, ρ) has an orientable embedding on the surface with k' crosscaps. If $k' = 0$, then we can add two crosscaps, and a loop that passes through one of them; since $k \geq 2$ this is sufficient. Otherwise, by Lemma 3 we can assume that k' is odd and at least 3. To G add a loop with its ends consecutive in the rotation. Now push this loop once over each crosscap. Since all other loops are two-sided, every pair of edges crosses evenly, so by Theorem 1 the graph embeds in the surface with the same embedding scheme. The loop we added is one-sided and maximal. ☐

3 Removing Self-Crossings

Theorem 2. $\mathrm{dcr}(H) \leq 6\,\mathrm{gcr}(H)$.

So a degenerate drawing with self-crossings can be cleaned of self-crossings at the expense of increasing the degenerate crossing number by a factor of six. We will make use of the following lemma.[5]

Lemma 4. $\mathrm{dcr}(H) \leq 2|E(H)|$.

Proof. Use Lemma 2 to create a single-vertex graph G on vertex v with embedding scheme (ρ, λ) so that $\mathrm{dcr}(H) \leq \mathrm{dcr}(G, \rho, \lambda)$. We proceed by induction on $|E(G)| = |E(H)|$. If $|E(G)| = 0$, there is nothing to show, so G has at least one loop. Pick a loop e whose ends at v are *closest* in the sense, that no other edge begins and ends in the wedge formed by the two ends of e (we direct e to differentiate between the two parts of the rotation system at v enclosed by the ends of v). If we can, we pick e one-sided. Suppose e is one-sided. Let (G', ρ', λ') be the graph obtained from G by reversing the order of the edges enclosed in the wedge formed by e (we "flip" the wedge), changing all their signatures (since every edge has at most one end in the wedge that flips the signature of every edge which has an end in the wedge), and removing e. By induction $\mathrm{dcr}(G', \rho', \lambda') \leq 2|E(G')|$. We can now add a crosscap close to v and pass all edges in the former wedge through that crosscap, reattaching them to v in their original order. This also reestablishes the original signatures of edges in G. Finally, we add back e in its proper place in the rotation, passing it through the crosscap once. By construction, $\mathrm{dcr}(G, \rho, \lambda) \leq 1 + \mathrm{dcr}(G', \rho', \lambda') \leq 2|E(G)|$.

If there is no closest, one-sided loop, e must be two-sided. Let $\tilde{G}$ be the same as G with one modification: let $\tilde{\lambda}(e) = -1$ and proceed as in the first case. We obtain a graph $\tilde{G}'$ so that $\mathrm{dcr}(\tilde{G}, \rho, \tilde{\lambda}) \leq 1 + \mathrm{dcr}(\tilde{G}', \rho', \tilde{\lambda}')$. Now add one additional crosscap passing only edge e through it, making it two-sided again. This shows that $\mathrm{dcr}(G, \rho, \lambda) \leq 1 + \mathrm{dcr}(\tilde{G}, \rho, \tilde{\lambda}) \leq 2 + \mathrm{dcr}(\tilde{G}', \rho', \tilde{\lambda}') \leq 2|E(G)|$. ∎

Proof (of Theorem 2). Let H be a graph with $\mathrm{gcr}(H) = k$. Fix an embedding of H on a surface S with k crosscaps. By Lemma 2 we can transform H into a graph G on a single vertex v with an embedding scheme (ρ, λ) so that $\mathrm{gcr}(H) = \mathrm{gcr}(G, \rho, \lambda)$ and $\mathrm{dcr}(H) \leq \mathrm{dcr}(G, \rho, \lambda)$. We show the result by induction on $|E(G)| = |E(H)|$.

If $|E(G)| \leq 3k$, then the result follows from Lemma 4. So we can assume that $|E(G)| > 3k$. Lemma 1 implies that in this case there are two loops e and f so that $e \cup f$ bounds a disk (e and f are homotopic). Remove the disk (with any loops it may contain) from the surface, and identify e and f. Since this removes at least one edge from G we can apply induction to the resulting graph G'. From G' we can reconstruct an embedding of G by splitting e, f into two loops and reinserting the disk. Any loops in the disk which are not homotopic to e and f can be drawn close to v (so they do not use any crosscaps that e and f may be using). Any loops parallel to e and f use the same crosscaps as e and f, so in the resulting drawing no edge uses any crosscap more than once (note that any such loops have the same signature as e and f, since e and f bound a disk). ∎

---

[5] This approach was suggested by one of the reviewers, and simplifies the original proof.

Since the proof works with single-vertex graphs with embedding schemes, the separation of gcr and dcr for those types of graphs (Theorem 4) implies that the proof approach in Theorem 2 will not yield gcr = dcr, but we can prove equality for small values.

Theorem 3. *If* $\mathrm{dcr}(G) \leq 3$, *then* $\mathrm{gcr}(G) = \mathrm{dcr}(G)$.

For graphs with embedding scheme, this result is sharp, as Theorem 4 shows.

Proof. Since $\mathrm{gcr}(G) \leq \mathrm{dcr}(G)$ it is sufficient to show that if $\mathrm{gcr}(G) \leq 2$, then $\mathrm{dcr}(G) \leq \mathrm{gcr}(G)$. By Lemma 2 it is sufficient to prove the result for single-vertex graphs with embedding scheme: for G there is a single-vertex graph G' and an embedding scheme (ρ, λ) so that $\mathrm{dcr}(G) \leq \mathrm{dcr}(G', \rho, \lambda)$ and $\mathrm{gcr}(G', \rho, \lambda) = \mathrm{gcr}(G)$, so establishing $\mathrm{dcr}(G', \rho, \lambda) \leq \mathrm{gcr}(G', \rho, \lambda)$ will prove the result.

If $\mathrm{gcr}(G', \rho, \lambda) = 0$, there is nothing to prove. If $\mathrm{gcr}(G', \rho, \lambda) = 1$ all loops are either two-sided and contractible or one-sided. Pick a closest loop e (in the sense defined in Lemma 4: every edge has at most one end in the wedge formed by e). If e is one-sided, we can proceed as in Lemma 4, cutting along e, flipping the wedge enclosed by e and changing the signature of all edges in the wedge. The resulting graph is embedded in a plane, and we can add back e so that it, and the edges it encloses cross through the crosscap exactly once. If e is two-sided, the ends of e must be consecutive. We can then remove e from the drawing, inductively draw the remaining graph, and add e back locally without using any crosscaps. If $\mathrm{gcr}(G', \rho, \lambda) = 2$, there may be two-sided loops which are not contractible. However, if there is a closest one-sided loop, or a closest two-sided loop which is contractible, we can proceed as in the case of a single crosscap. Hence, all closest loops are two-sided, and either separating, or maximal. Suppose there is a one-sided loop f. Then the wedge enclosed by f must contain both ends of another loop e. Pick e so it is closest (within the wedge formed by f). Now e cannot be maximal, since the ends of a maximal loop alternate with the ends of a one-sided loop in the rotation. Hence e is separating. But then anything starting inside the wedge formed by e must end within the wedge as well, so since e was chosen to be closest, its ends have to be consecutive in the rotation. We can then remove e, inductively draw the remaining graph, and add e back into the rotation without using any additional crosscaps. We conclude that there is no one-sided loop f, so all loops are two-sided. By Lemma 3, the graph is planar in this case. ∎

A closer look at the proof of Theorems 2 and 3 show that they are purely combinatorial, and the bounds can be implemented algorithmically.

4   Separating dcr and gcr with Embedding Schemes

Theorem 4. *There is a single-vertex graph G with embedding scheme (ρ, λ) for which* $3 = \mathrm{gcr}(G, \rho, \lambda) < \mathrm{dcr}(G, \rho, \lambda) = 4$.

Proof. See the graph pictured in Fig. 1(a). The single vertex is drawn as the outer cycle, to make the picture easier to read. So there are 5 loop edges $e_1, \ldots, e_5$ in this graph, the rotation at v is $e_1, e_2, e_3, e_4, e_5, e_3, e_2, e_1, e_4, e_5$, and the signatures are as in the embedding: $\lambda(e_1) = \lambda(e_3) = \lambda(e_4) = \lambda(e_5) = 1$ and $\lambda(e_2) = -1$. The drawing of G in Fig. 1(a) shows that $\text{gcr}(G, \rho, \lambda) \leq 3$. If $\text{gcr}(G, \rho, \lambda) \leq 2$ were true, then e_2 would have to pass through exactly one of the two crosscaps oddly, say $\otimes_1$. Since the ends of e_4 and e_5 alternate with the ends of e_2, both e_4 and e_5 must also pass through $\otimes_1$ oddly. Since e_4 and e_5 are two-sided, they must then also pass through $\otimes_2$ oddly. But then e_4 and e_5 would be parallel (in the sense that their ends do not alternate), contradicting the fact that their ends alternate in the rotation. Hence, $\text{gcr}(G, \rho, \lambda) = 3$. The embedding in Fig. 1(b) shows that $\text{dcr}(G, \rho, \lambda) \leq 4$, so we are left with the proof that $\text{dcr}(G, \rho, \lambda) \geq 4$. Suppose, for a contradiction, that G can be realized on a surface with three crosscaps so that every edge passes through each crosscap at most once, and the embedding scheme is (ρ, λ), as specified in Fig. 1(a). Then each edge in $\{e_1, e_3, e_4, e_5\}$ passes through an even number of crosscaps. Since none of these edges can be separating (since they would all separate ends of other edges in the rotation), they each pass through two crosscaps. Edge e_2 passes through an odd number of crosscaps. It cannot pass through all three crosscaps, since then all other edges would be parallel to it (as each would share two crosscaps with e_2), but the ends of e_2 alternate with the ends of e_4 and e_5. Hence, e_2 passes through exactly one crosscap, say $\otimes_1$. Since e_3 is parallel to e_2, it must then pass through $\otimes_2$ and $\otimes_3$. Now e_4 and e_5 alternate ends with both e_2 and e_3, so one of them, say e_4, by symmetry, passes through $\otimes_1$ and $\otimes_2$ and e_5 passes through $\otimes_1$ and $\otimes_3$.

Edge	$\otimes_1$	$\otimes_2$	$\otimes_3$
e_2	1	0	0
e_3	0	1	1
e_4	1	1	0
e_5	1	0	1
e_1	0	1	1

Now e_1 is parallel to e_2 and e_3 and passes through two crosscaps, which must therefore be $\otimes_2$ and $\otimes_3$. Now suppose there were such a drawing. Since edges pass through crosscaps at most once, we can think of crosscaps as vertices. But then, there is a path from an end of e_1 to an end of e_3 which passes through $\otimes_2$ and $\otimes_3$ but not through $\otimes_1$. That path now separates the two ends of e_2, since e_2 may only pass through $\otimes_1$.   ∎

Question 1. Can the construction in Theorem 4 be used to construct for every n a single-vertex graph G with embedding scheme (ρ, λ) so that $n \leq \text{dcr}(G, \rho, \lambda) \leq 3/4 \text{gcr}(G, \rho, \lambda)$?

5   Nice Embeddings of Higher Genus Graphs

In this section we consider relaxing the restriction on how often each edge may pass through each crosscap. It turns out that increasing the limit to two is sufficient.

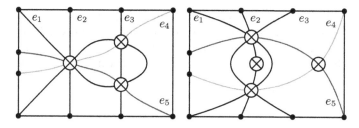

Fig. 1. Graph G with rotation displayed as outer cycle. (*a*) G embedded in a surface with three crosscaps, requiring e_1 to pass through one crossscap twice. (*b*) G embedded in a surface with four crosscaps, each edge passing through each crosscap at most once.

Theorem 5. *If a graph is embeddable in a non-orientable surface S, then it can be embedded in S so that every edge passes through each crosscap at most twice.*

This means, G always has a nearly degenerate drawing in the plane with at most gcr(G) crossings, and in which each edge has at most gcr(G) self-crossings. In the language of topology [8], it means that there is a planarizing system of disjoint one-sided curves each of which intersects every edge of the graph at most twice.

By Theorem 4 the theorem is tight if the graph is given with an embedding scheme (which may not be changed), even if the graph consists of a single vertex.

We will concentrate the proof in a more technical lemma, which may be of interest in its own right. For the proof, we need the *Euler genus*, eg(G, ρ, λ) of an embedded graph, which is defined as $1 + |E| - |F|$, where $|E|$ is the number of edges of G and $|F|$ the number of faces in the embedding scheme (ρ, λ) of G (note that this is a purely combinatorial notion). It's tempting to assume that gcr(G, ρ, λ) = eg(G, ρ, λ), but that is not actually true; take, for example, a single vertex with two two-sided edges alternating at the vertex. The Euler genus of this graph is 2, while it requires 3 crosscaps to realize. The following lemma clarifies the relationship.

Recall that an embedded single-vertex graph (G, ρ, λ) is orientable, if $\lambda(e) = 1$ for all $e \in E(G)$.

Lemma 5. *If (G, ρ, λ) is a single-vertex graph with embedding scheme, then it has an embedding in a surface with eg(G, ρ, λ) crosscaps in which every edge uses every crosscap at most twice, unless (G, ρ, λ) is orientable, in which case such an embedding exists in a surface with eg(G, ρ, λ) + 1 crosscaps.*

We leave the proof of Lemma 5 to the journal version of the paper. The proof can be viewed as a (more sophisticated) extension of the proof of Theorem 3. Since we allow edges to cross through a crosscap twice, the construction becomes simpler, in that we can process one-sided loops, even if they are not closest. The new ingredient needed is a technique for dealing with separating loops. For example, consider the embedding scheme described by $\rho(v) = (abbacdcd)$, and $\lambda(b) = -1$, and $\lambda(a) = \lambda(c) = \lambda(d) = 1$, as illustrated in Fig. 2(*a*). The Euler genus of this graph is 3, and a is a separating loop, splitting the graph into two pieces, one of Euler genus 1, and the other of Euler genus 2. The problem now is that the piece of Euler genus 2 is orientable, and hence needs 3 crosscaps to realize by itself. Hence, some care is needed when merging drawings in this case; the solution in this case is shown in Fig. 2(b). Details will be found in the journal version.

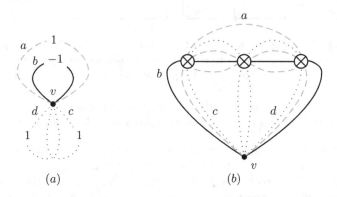

$$(a) \qquad\qquad (b)$$

Fig. 2. (a) Embedding scheme with Euler genus 3; edges are a (red/dashed), b (black), and c, d (blue/dotted). (b) Actual embedding of same scheme on surface with three crosscaps, in which every edge passes through every crosscap at most twice (Color figure online).

Proof (of Theorem 5). Fix an embedding of a graph G on a surface with $k = \widetilde{\gamma}(G)$ crosscaps (without loss of generality, we can assume that it is a minimum genus embedding). By Lemma 2 there is a single-vertex graph G' with embedding scheme (ρ, λ) so that $\mathrm{gcr}(G) = \mathrm{gcr}(G', \rho, \lambda)$. It is sufficient to prove the result for G', since an embedding of G' with embedding scheme (ρ, λ) can be turned back into an embedding of G by uncontracting and deleting edges (in case G was not connected). Since these operations can be done close to the single vertex of G', this does not affect how often edges pass through any crosscap. Hence, we can assume that G is given as a graph on a single vertex v with embedding scheme (ρ, λ).

Since (G, ρ, λ) is an embedding on the surface with k crosscaps, $\mathrm{eg}(G, \rho, \lambda) \leq$ k. If (G, ρ, λ) is not orientable, then the result follows immediately from Lemma 5. If (G, ρ, λ) is orientable, we apply Corollary 1 to extend (G, ρ, λ) to an embedding scheme (G', ρ', λ') which still embeds in the same surface, and is no longer orientable. Since $\mathrm{eg}(G, \rho, \lambda) \leq \mathrm{eg}(G', \rho', \lambda') \leq$ k, and (G', ρ', λ') is not orientable, Lemma 5 gives us an embedding of (G', ρ', λ'), and thereby (G, ρ, λ) in a surface with k crosscaps, in which every edge passes through each crosscap at most twice, completing the proof. ∎

The proof of Theorem 5 is entirely combinatorial, so it can be made algorithmic.

Corollary 2. *Determining the degenerate crossing number is **NP**-complete, even if the graph is cubic.*

Proof. The problem lies in **NP** (since every edge passes through each crosscap at most once we can guess the embedding). On the other hand, Thomassen [9,13] showed that the non-orientable genus problem is **NP**-complete, even for cubic graphs. For a given cubic graph G, let G' be the result of replacing each edge of G with a path of length $2|E(G)|$, and attaching a (local, planar) gadget to each vertex of degree 2, to ensure that G' is cubic. If G has orientable genus at most k, then, by Theorem 5, there is an embedding in which every edge passes through each of the crosscaps at most twice. Since we can assume that $k \leq |E(G)|$ (e.g. [10]), this implies that G' can be embedded so that every edge passes through each crosscap at most once. In other words, the degenerate crossing number of G' equals the non-orientable genus of G, showing that dcr is **NP**-complete.   ∎

6   Open Questions

The main open question which remains is whether dcr(G) = gcr(G); one could weaken this question in various ways, and, for example ask whether dcr(G) ≤ gcr(G) + c for some constant c? Another approach would be to ask whether dcr(G) = gcr(G) if we allow a limited number of self-crossings along each edge. Theorem 5 implies that gcr(G) self-crossings along each edge are sufficient, but can a constant bound be achieved?

Acknowledgments. We would like to thank Bojan Mohar for suggesting the question, and giving us detailed feedback on earlier drafts of this paper. We are also grateful for helpful comments by the anonymous reviewers.

References

1. Ackerman, E., Pinchasi, R.: On the degenerate crossing number. Discrete Comput. Geom. **49**(3), 695–702 (2013)
2. Cairns, G., Nikolayevsky, Y.: Bounds for generalized thrackles. Discrete Comput. Geom. **23**(2), 191–206 (2000)
3. Dey, T.K., Schipper, H.: A new technique to compute polygonal schema for 2-manifolds with application to null-homotopy detection. Discrete Comput. Geom. **14**(1), 93–110 (1995)
4. Duncan, C.A., Goodrich, M.T., Kobourov, S.G.: Planar drawings of higher-genus graphs. J. Graph Algorithms Appl. **15**(1), 7–32 (2011)
5. Erickson, J., Har-Peled, S.: Optimally cutting a surface into a disk. Discrete Comput. Geom. **31**(1), 37–59 (2004). ACM Symposium on Computational Geometry, Barcelona 2002
6. Harborth, H.: Drawings of graphs and multiple crossings. In: Alavi, Y., Chartrand, G., Lick, D.R., Wall, C.E., Lesniak, L. (eds.) Graph Theory with Applications to Algorithms and Computer Science (Kalamazoo, Mich., 1984), pp. 413–421. Wiley-Interscience Publication, New York (1985)

7. Lazarus, F., Pocchiola, M., Vegter, G., Verroust, A.: Computing a canonical polygonal schema of an orientable triangulated surface. In: Proceedings of the Seventeenth Annual Symposium on Computational Geometry (SCG-01), pp. 80–89. ACM Press, New York, 3–5 2001
8. Mohar, B.: The genus crossing number. Ars Math. Contemp. **2**(2), 157–162 (2009)
9. Mohar, B., Thomassen, C.: Graphs on Surfaces. Johns Hopkins Studies in the Mathematical Sciences. Johns Hopkins University Press, Baltimore, MD (2001)
10. Pach, J., Tóth, G.: Degenerate crossing numbers. Discrete Comput. Geom. **41**(3), 376–384 (2009)
11. Pelsmajer, M.J., Schaefer, M., Štefankovič, D.: Removing even crossings on surfaces. European J. Combin. **30**(7), 1704–1717 (2009)
12. Schaefer, M.: The graph crossing number and its variants: a survey. Electron. J. Comb. **20**, 1–90 (2013). Dynamic Survey, #DS21
13. Thomassen, C.: The genus problem for cubic graphs. J. Comb. Theory Ser. B **69**(1), 52–58 (1997)

On Degree Properties of Crossing-Critical Families of Graphs

Drago Bokal[1], Mojca Bračič[1], Marek Derňár[2], and Petr Hliněný[2]([✉])

[1] Faculty of Natural Sciences and Mathematics, University of Maribor,
Maribor, Slovenia
drago.bokal@um.si, mojca.bracic@student.um.si
[2] Faculty of Informatics, Masaryk University, Brno, Czech Republic
m.dernar@gmail.com, hlineny@fi.muni.cz

Abstract. Answering an open question from 2007, we construct infinite k-crossing-critical families of graphs which contain vertices of any prescribed odd degree, for sufficiently large k. From this we derive that, for any set of integers D such that $\min(D) \geq 3$ and $3, 4 \in D$, and for all sufficiently large k there exists a k-crossing-critical family such that the numbers in D are precisely the vertex degrees which occur arbitrarily often in any large enough graph in this family. We also investigate what are the possible average degrees of such crossing-critical families.

Keywords: Crossing number · Tile drawing · Degree-universality · Average degree · Crossing-critical graph

1 Introduction

Reducing the number of crossings in a drawing of a graph is considered one of the most important drawing aesthetics. Consequently, great deal of research work has been invested into understanding what forces the number of edge crossings in a drawing of the graph to be high. There exist strong quantitative lower bounds, such as the famous Crossing Lemma [1,14]. However, the quantitative bounds typically show their strength only in dense graphs, while in the area of graph drawing we often deal with graphs having few edges.

The reasons for sparse graphs to have many crossings in any drawing are structural (there is a lot of "nonplanarity" in them). These reasons can be understood via so called k-*crossing-critical* graphs, which are the subgraph-minimal graphs that require at least k edge crossings (the "minimal obstructions"). While there are only two 1-crossing-critical graphs, up to subdivisions— the Kuratowski graphs K_5 and $K_{3,3}$—it has been known already since Širáň's [19]

D. Bokal—This research was supported by the internationalisation of Slovene higher education within the framework of the Operational Programme for Human Resources Development 2007–2013 and by the Slovenian Research Agency project L7–5459.
M. Derňár and P. Hliněný—This research was supported by the Czech Science Foundation under the project 14-03501S.

E. Di Giacomo and A. Lubiw (Eds.): GD 2015, LNCS 9411, pp. 75–86, 2015.
DOI: 10.1007/978-3-319-27261-0_7

and Kochol's [13] constructions that the structure of crossing-critical graphs is quite rich and nontrivial for any $k \geq 2$.

Although 2-crossing-critical graphs can be reasonably (although not easily) described [5], a full description for any $k \geq 3$ is clearly out of our current reach. Consequently, research has focused on interesting properties shared by all k-crossing-critical graphs (for certain k), successfull attempts include, e.g., [7,8,10, 12,17]. While we would like to establish as many specific properties of crossing-critical graphs as possible, the reality unfortunately seems to be against it. Many desired and conjectured properties of crossing-critical graphs have already been disproved by often complex and sophisticated constructions showing the odd behaviour of crossing-critical families, e.g. [6,9,11,18].

We study properties of *infinite families* of k-crossing-critical graphs, for fixed values of k, since sporadic "small" examples of critical graphs tend to behave very wildly for every $k > 1$. Among the most studied such properties are those related to vertex degrees in the critical families, see [3,6,8,11,18]. Often the research focused on the average degree a k-crossing-critical family may have— this rational number clearly falls into the interval [3, 6] if we forbid degree-2 vertices. It is now known [8] that the true values fall into the open interval (3, 6), and all the rational values in this interval can be achieved [3].

In connection with the proof of bounded pathwidth for k-crossing-critical families [9,10], it turned out to be a fundamental question whether k-crossing-critical graphs have maximum degree bounded in k. The somehow unexpected negative answer was given by Dvořák and Mohar [6]. In 2007, Bokal noted that all the known (by that time) constructions of infinite k-crossing-critical families seem to use only vertices of degrees $3, 4, 6$, and he asked what other degrees can occur frequently (see the definition in Sect. 2) in k-crossing-critical families. Shortly after that Hliněný extended his previous construction [9] to include an arbitrary combination of any even degrees [11], for sufficiently large k.

Though, [11] answered only the easier half of Bokal's question, and it remained a wide open problem of whether there exist infinite k-crossing-critical families whose members contain many vertices of odd degrees greater than 5. Our joint investigation has recently led to an ultimate positive answer.

The contribution and new results of our paper can be summarized as follows:

- In Sect. 2, we review the tools which are commonly used in constructions of crossing-critical families.
- Sect. 3 presents the key new contribution—a construction of crossing-critical graphs with repeated occurrence of any prescribed odd vertex degree (Proposition 3.1 and Theorem 3.2).
- In Sect. 4, we combine the new construction of Sect. 3 with previously known constructions to prove the following: for any set of integers D such that $\min(D) = 3$ and $3, 4 \in D$, and for all sufficiently large k there exists an infinite k-crossing-critical family such that the numbers in D are precisely the vertex degrees which occur frequently in this family (Theorem 4.2).
- We then extend the previous results in Sect. 5 to include also an exhaustive discussion of possible average vertex degrees attained by our degree-restricted crossing-critical families (Theorem 5.1).

– Finally, in the concluding Sect. 6 we pay special attention to 2-crossing-critical graphs, and list some remaining open questions.

2   Preliminaries

We consider *finite multigraphs* without loops by default (i.e., we allow multiple edges unless we explicitly call a graph *simple*), and use the standard graph terminology otherwise. The *degree* of a vertex v in a graph G is the number of edges of G incident to v (cf. multigraphs), and the *average degree* of G is the average of all the vertex degrees of G.

Crossing Number. In a *drawing* of a graph G, the vertices of G are points and the edges are simple curves joining their endvertices. It is required that no edge passes through a vertex, and no three edges cross in a common point. The *crossing number* $\mathrm{cr}(G)$ of a graph G is the minimum number of crossing points of edges in a drawing of G in the plane. For $k \in \mathbb{N}$, we say that a graph G is *k-crossing-critical*, if $\mathrm{cr}(G) \geq k$ but $\mathrm{cr}(G - e) < k$ for each edge $e \in E(G)$.

Note that a vertex of degree 2 in G is not relevant for a drawing of G and for the crossing number, and we will often replace such vertices by edges between their two neighbours. Since also vertices of degree 1 are irrelevant for the crossing number, it is quite common to assume minimum degree 3.

Degree-Universality. The following terms formalize a vague notion that a certain vertex degree occurs frequently or arbitrarily often in an infinite family. For a finite set $D \subseteq \mathbb{N}$, we say that a family of graphs $\mathcal{F}$ is *D-universal*, if and only if, for every integer m there exists a graph $G \in \mathcal{F}$, such that G has at least m vertices of degree d for each $d \in D$. It follows easily that $\mathcal{F}$ has infinitely many such graphs.

Clearly, if $\mathcal{F}$ is D universal and $D' \subseteq D$, then $\mathcal{F}$ is also D'-universal. The family of all sets D, for which a given $\mathcal{F}$ is D-universal, therefore forms a poset under relation $\subseteq$. Maximal elements of this poset are of particular interest, and for "well-behaved" $\mathcal{F}$, these maximal elements are finite and unique. We distinguish this case with the following definition: $\mathcal{F}$ is *D-max-universal*, if it is D-universal, there are only finitely many degrees appearing in graphs of $\mathcal{F}$ that are not in D, and there exists an integer M, such that any degree not in D appears at most M times in any graph of $\mathcal{F}$.

Note that if $\mathcal{F}$ is D-max-universal and D'-max-universal, at the same time, then $D = D'$. It can also be easily seen that if $\mathcal{F}$ is D-max-universal then there exists infinite $\mathcal{F}' \subseteq \mathcal{F}$ such that, for any m, *every* sufficiently large member of $\mathcal{F}'$ has at least m vertices of degree d for each $d \in D$. Though, we do not specifically mention this property in the formal definition.

Tools for Crossing-Critical Graphs. A principal tool used in many constructions of crossing-critical graphs are tiles. They were implicitly used already in the early papers by Kochol [13] and Richter–Thomassen [17], although they were formalized only later in the work of Pinnontoan and Richter [15,16]. In our contribution, we use an extension of their formalization from [3], which we also briefly sketch here.

A *tile* is a triple $T = (G, \lambda, \rho)$ where $\lambda, \rho \subseteq V(G)$ are two disjoint sequences of distinct vertices of G, called the *left and right wall* of T, respectively. A *tile drawing* of T is a drawing of the underlying graph G in the unit square such that the vertices of λ occur in this order on the left side of the square and those of ρ in this order on the right side of it. The *tile crossing number* $\mathrm{tcr}(T)$ of a tile T is the smallest crossing number over all tile drawings of T.

For simplicity, in this brief exposition, we shall assume that all tiles considered in construction of a single graph satisfy $|\lambda| = |\rho| = w$ for a suitable constant $w \geq 2$ depending on the graph (though, a more general treatment is obviously possible). The *join of two tiles* $T = (G, \lambda, \rho)$ and $T' = (G', \lambda', \rho')$ is defined as the tile $T \otimes T' := (G'', \lambda, \rho')$, where G'' is the graph obtained from the disjoint union of G and G', by identifying ρ_i with λ'_i for $i = 1, \ldots, w$. Specially, if $\rho_i = \lambda'_i$ is a vertex of degree 2 (after the identification), we replace it with a single edge in G''. Since the operation $\otimes$ is associative, we can safely define the *join of a sequence of tiles* $\mathcal{T} = (T_0, T_1, \ldots, T_m)$ as $\otimes \mathcal{T} = T_0 \otimes T_1 \otimes \ldots \otimes T_m$. The *cyclization* of a tile $T = (G, \lambda, \rho)$, denoted by $\circ T$, is the ordinary graph obtained from G by identifying λ_i with ρ_i for $i = 1, \ldots, w$. The *cyclization of a sequence of tiles* $\mathcal{T} = (T_0, T_1, \ldots, T_m)$ is $\circ \mathcal{T} := \circ(\otimes \mathcal{T})$. Again, possible degree-2 vertices are replaced with single edges.

Let $T = (G, \lambda, \rho)$ be a tile. The *right-inverted* tile $T^\updownarrow$ is the tile $(G, \lambda, \bar{\rho})$ and the *left-inverted* tile $^\updownarrow T$ is $(G, \bar{\lambda}, \rho)$, where $\bar{\lambda}$ and $\bar{\rho}$ denote the inverted sequences of λ, ρ. For a sequence of tiles $\mathcal{T} = (T_0, \ldots, T_m)$, let $\mathcal{T}^\updownarrow := (T_0, \ldots, T_{m-1}, T_m^\updownarrow)$.

One can easily get the following (cf. [15]): for any tile T, $\mathrm{cr}(\circ T) \leq \mathrm{tcr}(T)$, and for every sequence of tiles $\mathcal{T} = (T_0, T_1, \ldots, T_m)$, $\mathrm{tcr}(\otimes \mathcal{T}) \leq \sum_{i=0}^{m} \mathrm{tcr}(T_i)$. On the other hand, corresponding lower bounds on the crossing number of cyclizations of tile sequences are also possible [3], under additional technical assumptions. A tile $T = (G, \lambda, \rho)$ is *planar* if $\mathrm{tcr}(T) = 0$. T is *perfect* if the following hold:

- $G - \lambda$ and $G - \rho$ are connected;
- for every $v \in \lambda$ there is a path from v to the right wall ρ in G internally disjoint from λ, and for every $u \in \rho$ there is a path from u to the left wall λ in G internally disjoint from ρ;
- for every $0 \leq i < j \leq w$, there is a pair of disjoint paths, one joining λ_i and ρ_i, and the other joining λ_j and ρ_j.

We are particularly interested in the following specialized result:

Theorem 2.1 ([3]). *Let* $T_0, \ldots, T_m$ *be copies of a perfect planar tile* T, *and* $\mathcal{T} = (T_0, \ldots, T_m)$. *Assume that, for some integer* $k \geq 1$, *we have* $m \geq 4k - 2$ *and* $\mathrm{tcr}(\otimes(\mathcal{T}^\updownarrow)) \geq k$. *Then,* $\mathrm{cr}(\circ(\mathcal{T}^\updownarrow)) \geq k$.

To lower-bound the tile crossing number (e.g., for use in Theorem 2.1), we use the following simple tool. A *traversing path* in a tile $T = (G, \lambda, \rho)$ is a path $P \subseteq G$ such that one end of P is in λ and the other in ρ, and P is internally disjoint from $\lambda \cup \rho$. A pair of traversing paths $\{P, Q\}$ is *twisted* if P, Q are disjoint and the mutual order of their ends in λ is the opposite of their order in ρ. Obviously, a twisted pair must induce a crossing in any tile drawing of T. A family of twisted pairs of traversing paths is called a *twisted family*.

Lemma 2.2 ([3]). *Let $\mathcal{F}$ be a twisted family in a tile T, such that no edge occurs in two distinct paths of $\cup\mathcal{F}$. Then, $\mathrm{tcr}(T) \geq |\mathcal{F}|$.*

The second tool for constructing crossing-critical families is the so called *zip product* [2,3], which we introduce in a simplified setting [11]. For $i \in \{1,2\}$, let G_i be a simple graph and let $v_i \in V(G_i)$ be a vertex of degree 3, such that $G_i - v_i$ is connected. We denote the neighbours of v_i by u_j^i for $j \in \{1,2,3\}$. The *zip product of G_1 and G_2 according to v_1, v_2* and their neighbours, is obtained from the disjoint union of $G_1 - v_1$ and $G_2 - v_2$ by adding the three edges $u_1^1 u_1^2$, $u_2^1 u_2^2$, $u_3^1 u_3^2$. The following is true in this special case:

Theorem 2.3 ([4]). *Let G be a zip product of G_1 and G_2 according to degree-3 vertices. Then, $\mathrm{cr}(G) = \mathrm{cr}(G_1) + \mathrm{cr}(G_2)$. Consequently, if G_i is k_i-crossing-critical for $i = 1,2$, then G is $(k_1 + k_2)$-crossing-critical.*

3 Crossing-Critical Families with High Odd Degrees

We first present a new construction of a crossing-critical family containing many vertices of an arbitrarily prescribed odd degree (recall that the question of an existence of such families has been the main motivation for this research).

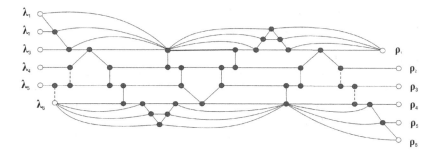

Fig. 1. A tile drawing of the tile $G_{3,4}$. The wall vertices are drawn in white.

The construction defines a graph $G(\ell, n, m)$ with three integer parameters $\ell \geq 1$, $n \geq 3$ and odd $m \geq 3$, as follows. There is a tile $G_{\ell,n}$, with the walls of size $n+\ell-1$, which is illustrated in Fig. 1 and formally defined below. Let $\mathcal{G}(\ell, n, m) = (G_{\ell,n}, {}^\uparrow G_{\ell,n}{}^\uparrow, G_{\ell,n} \ldots, {}^\uparrow G_{\ell,n}{}^\uparrow, G_{\ell,n})$ be a sequence of such tiles of length m, and let $G(\ell, n, m)$ be constructed as the join $\circ(\mathcal{G}(\ell, n, m)^\uparrow)$. In the degenerate case of $\ell = 0$, the graph $G(0, n, m)$ is defined as the "staircase strip" graph from Bokal's [3], and $G(0, n, m)$ will be contained in $G(\ell, n, m)$ as a subdivision for every ℓ.

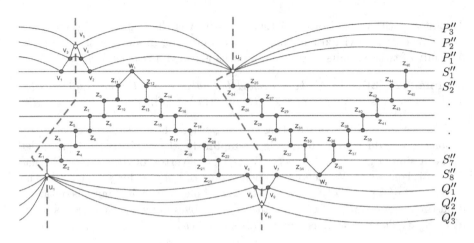

Fig. 2. A fragment of the tile $G_{3,8} = H_{3,8} \otimes {}^{\updownarrow} H_{3,8}{}^{\updownarrow} \otimes H_{3,8}$; defining the one tile $H_{3,8}$ (left, between the dashed margins) and showing the composition $H_{3,8} \otimes {}^{\updownarrow} H_{3,8}{}^{\updownarrow}$ in $G_{3,8}$.

The tile $G_{\ell,n}$ is composed of three copies of a smaller tile $H_{\ell,n}$ such that $G_{\ell,n} = H_{\ell,n} \otimes {}^{\updownarrow} H_{\ell,n}{}^{\updownarrow} \otimes H_{\ell,n}$. A fragment illustrating the join $H_{3,8} \otimes {}^{\updownarrow} H_{3,8}{}^{\updownarrow}$ is presented in Fig. 2. Formally, $H_{\ell,n}$ consists of $2\ell + n$ pairwise edge disjoint paths, grouped into three families $P_1', \ldots, P_\ell'$, $Q_1', \ldots, Q_\ell'$, and $S_1', \ldots, S_n'$, and an additional set F' of $2(n-2)$ edges not on these paths.

- The paths $S_1', \ldots, S_n'$ are pairwise vertex-disjoint except that S_1' shares one vertex with S_2' (w_1 in Fig. 2). The additional $2(n-2)$ edges of F' are in pairs between vertices of the paths S_{i-1}' and S_i' for $i = 3, \ldots, n$, as depicted in Fig. 2 (edges $u_1 z_1, z_2 z_3, \ldots, z_{22} z_{23}$).
- The union $S_1' \cup \ldots \cup S_n' \cup F'$ is (consequently) a subdivision of the aforementioned staircase tile from [3].
- The paths $Q_1', \ldots, Q_\ell'$ all share the bottom-most vertex u_1 of S_n' on the left wall of $H_{\ell,n}$, and are combined in such a way that Q_i', $i = 1, \ldots, \ell$, shares exactly one vertex with Q_{i-1}' (with S_n' for $i = 1$) other than u_1 and this shared vertex is of degree 4, as depicted near the right wall in Fig. 2 (vertices v_6, v_8, v_{10}). The paths $P_1', \ldots, P_\ell'$ analogously share the top-most vertex u_2 of S_1' on the right wall of $H_{\ell,n}$ and are symmetric to Q's.

Let P_i'', Q_i'', S_i'' denote the paths obtained as the union of the three copies of each of P_i', Q_i', S_i' in $G_{\ell,n}$. Then $P_1'', \ldots, P_\ell''$, $Q_1'', \ldots, Q_\ell''$, and $S_1'', \ldots, S_n''$ are all traversing paths of the tile $G_{\ell,n}$. Let P_i, Q_i, S_i denote the corresponding unions of the paths in whole $G(\ell, n, m)$.

The proof of the following basic properties is straightforward, as attentive reader could easily verify from the illustrating pictures of $H_{\ell,n}$ (recall that degree-2 vertices are removed in a tile join).

Proposition 3.1. *For every $\ell \geq 1$ and $n \geq 3$, the tiles $H_{\ell,n}$, and hence also $G_{\ell,n}$, are perfect planar tiles. The graph $G(\ell, n, m)$ has $3m(2\ell + 4n - 8)$ vertices,* *out of which* $3m \cdot 2\ell$ *have degree 4, $3m(4n - 9)$ have degree 3, and remaining $3m$ vertices have degree $2\ell + 3$. The average degree of $G(\ell, n, m)$ is*

$$\frac{5l + 6n - 12}{l + 2n - 4}.$$ $\square$

We conclude with the main desired property of the graph $G(\ell, n, m)$.

Theorem 3.2. *Let $\ell \geq 1$, $n \geq 3$ be integers. Let $k = (\ell^2 + \binom{n}{2} - 1 + 2\ell(n - 1))$ and $m \geq 4k - 1$ be odd. Then the graph $G(\ell, n, m)$ is k-crossing-critical.*

Proof. By using Theorem 2.1 and symmetry, it suffices to prove the following:

(I) $\mathrm{tcr}\big(\otimes \mathcal{G}(\ell, n, m)^{\updownarrow}\big) \geq k$, and
(II) every edge of $G_{\ell,n}$ corresponding to one copy of $H_{\ell,n}$ in it is *critical*, meaning that $\mathrm{tcr}(G_{\ell,n}^{\updownarrow} - e) < k$ for every edge $e \in E(H_{\ell,n}) \subseteq E(G_{\ell,n})$.

Recall the pairwise edge-disjoint traversing paths $P_1, \ldots, P_\ell, Q_1, \ldots, Q_\ell$, and $S_1, \ldots, S_n$ of the composed tile $\otimes \mathcal{G}(\ell, n, m)$. We define the following disjoint sets of pairs of these paths, such that each pair is formed by *vertex-disjoint* paths:

- $\mathcal{A} = \{\{P_i, Q_j\} : 1 \leq i, j \leq \ell\}$ where $|\mathcal{A}| = \ell^2$,
- $\mathcal{B} = \{\{P_i, S_j\} : 1 \leq i \leq \ell, 1 < j \leq n\}$ where $|\mathcal{B}| = \ell(n - 1)$,
- $\mathcal{C} = \{\{Q_i, S_j\} : 1 \leq i \leq \ell, 1 \leq j < n\}$ where $|\mathcal{C}| = \ell(n - 1)$.

Each pair in $\mathcal{A} \cup \mathcal{B} \cup \mathcal{C}$ is twisted in $\otimes \mathcal{G}(\ell, n, m)^{\updownarrow}$, and so these pairs account for at least $|\mathcal{A}| + |\mathcal{B}| + |\mathcal{C}| = 2\ell(n-1) + \ell^2$ crossings in a tile drawing of $\otimes \mathcal{G}(\ell, n, m)^{\updownarrow}$, by Lemma 2.2. Importantly, each of these crossings involves at least one edge of $R = P_1 \cup \ldots \cup P_\ell \cup Q_1 \cup \ldots \cup Q_\ell$. The subgraph $\otimes \mathcal{G}(\ell, n, m) - E(R)$ contains a subdivision of the staircase strip $\otimes \mathcal{G}(0, n, m)$. Hence any tile drawing of $\otimes \mathcal{G}(\ell, n, m)^{\updownarrow}$ contains at least another $\mathrm{tcr}\big(\otimes \mathcal{G}(0, n, m)^{\updownarrow}\big)$ crossings not involving any edges of R. Since $\mathrm{tcr}\big(\otimes \mathcal{G}(0, n, m)^{\updownarrow}\big) \geq \binom{n}{2} - 1$ by [3], we get $\mathrm{tcr}\big(\otimes \mathcal{G}(\ell, n, m)^{\updownarrow}\big) \geq \binom{n}{2} - 1 + 2\ell(n-1) + \ell^2 = k$, thus proving (I).

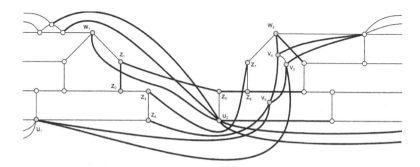

Fig. 3. A fragment of an optimal tile drawing of $G_{2,4}^{\updownarrow}$.

To finish with (II), we investigate the tile drawing in Fig. 3. It is routine to count that a natural generalization of this drawing has precisely $\binom{n-1}{2} + (n-2)\ell +$

$(\ell+1)^2+(\ell+1)(n-3)+\ell = k$ crossings, and so it is optimal. Consequently, every edge which is crossed in Fig. 3 is critical, are so are edges which become crossed after suitable local sliding of some vertex or edge (while preserving optimality) in the picture. This way one can easily verify that all the edges of a copy of $H_{2,4}$ in $G_{2,4}$, up to symmetry, are critical; except possibly three z_3z_4, z_5u_2, z_6z_7. The following local changes in the picture verify criticality also for the latter three edges:

- for z_3z_4, slide the edge z_3z_7 up (above u_2) and the edge w_1u_2 slightly down,
- for z_5u_2, z_6z_7, slide the edge z_3z_7 up (above z_6), the edge w_1u_2 down (below z_4), and the edge z_4v_5 together with the vertex v_5 suitably up.

An extension of this argument to the general case of $G_{\ell,n}$ is again routine.   □

4   Families with Prescribed Frequent Degrees

In order to fully answer the primary question of this paper—about which vertex degrees other than $3, 4, 6$ can occur arbitrarily often in infinite k-crossing-critical families—we start by repeating the three ingredients we have got so far. First, there is a bunch of established critical constructions essentially covering all the even degree cases and degree 3. Second, we have newly covered the cases of any fixed odd degree in Sect. 3. And third, we have got the zip product operation.

Proposition 4.1. *There exist (infinite) families $\mathcal{F}$ of simple, 3-connected, k-crossing-critical graphs such that, in addition, the following holds:*

(a) *([11, Sect. 4].) For every $k \geq 10$ or odd $k \geq 5$, and every rational $r \in (4, 6 - \frac{8}{k+1})$, a family $\mathcal{F}$ which is $\{4,6\}$-max-universal and each member of $\mathcal{F}$ is of average degree exactly r, and another $\mathcal{F}$ which is $\{4\}$-max-universal and of average degree exactly 4. Every graph of the two families has the set of its vertex degrees equal to $\{3, 4, 6\}$ (e.g., degree 3 repeats six times in each).*

(b) *([11, Sects. 3 and 4].) For every $\varepsilon > 0$, any integer $k \geq 5$ and every set D_e of even integers such that $\min(D_e) = 4$ and $6 \leq \max(D_e) \leq 2k-2$, a family $\mathcal{F}$ which is D_e-max-universal, and each graph of $\mathcal{F}$ has the set of its vertex degrees $D_e \cup \{3\}$ and is of average degree from the interval $(4, 4 + \varepsilon)$.*

(c) *([13] for $k = 2$ and [3] for general k, see $G(0, n, m)$.) For every $k = \binom{n}{2} - 1$ where $n \geq 3$ is an integer, a family $\mathcal{F}$ which is $\{3, 4\}$-max-universal and each member of $\mathcal{F}$ is of average degree equal to $3 + \frac{1}{4n-7}$.*

(d) *($G(\ell, 3, m)$ in Theorem 3.2.) For every $k = \ell^2 + 4\ell + 2$ where $\ell \geq 1$ is an integer, a family $\mathcal{F}$ which is $\{3, 4, 2\ell + 3\}$-max-universal and each member of $\mathcal{F}$ is of average degree $5 - \frac{4}{\ell+2}$.*

Using the zip product and Theorem 2.3, we can hence easily combine all the cases of Proposition 4.1 to obtain the following "ultimate" answer:

Theorem 4.2. *Let D be any finite set of integers such that $\min(D) \geq 3$. Then there is an integer $K = K(D)$, such that for every $k \geq K$, there exists a D-universal family of simple, 3-connected, k-crossing-critical graphs. Moreover, if either $3, 4 \in D$ or both $4 \in D$ and D contains only even numbers, then there exists a D-max-universal such family. All the vertex degrees are from $D \cup \{3, 4, 6\}$.*

5 Families with Prescribed Average Degree

In addition to Theorem 4.2, we are going to show that the claimed D-max-universality property can be combined with nearly any feasible rational average degree of the family. The full statement reads:

Theorem 5.1. *Let D be any finite set of integers such that $\min(D) \geq 3$ and $A \subset \mathbb{R}$ an interval. Assume that at least one of the following assumptions holds:*

(a) $D \supseteq \{3,4,6\}$ and $A = (3,6)$,
(b) $D \supsetneq \{3,4\}$ and $A = (3,4]$, or $D = \{3,4\}$ and $A = (3,4)$,
(c) $D \supsetneq \{3,4\}$ and $A = (3, 5 - \frac{8}{b+1})$ where $b \geq 9$ is the largest odd number in D,
(d) $D \supseteq \{4,6\}$ has only even numbers and $A = (4,6)$, or $D = \{4\}$ and $A = \{4\}$.

Then, for every rational $r \in A \cap \mathbb{Q}$, there is an integer $K = K(D,r)$ such that for every $k \geq K$, there exists a D-max-universal family of simple, 3-connected, k-crossing-critical graphs of average degree precisely r.

Due to limited space, we only sketch a proof of the theorem. The basic idea of balancing the average degree in a crossing-critical family is quite simple; assume we have two families $\mathcal{F}_a, \mathcal{F}_b$ of fixed average degrees $a < b$, respectively, and containing some degree-3 vertices. Then, we can use zip product of graphs from the two families to obtain a new family of average degree equal to a convex combination of a and b. This simple scheme, however, has two difficulties:

(I) If one combines graphs $G_1 \in \mathcal{F}_a$ and $G_2 \in \mathcal{F}_b$, the average degree of the disjoint union $G_1 \cup G_2$ is the average of a, b weighted by the sizes of G_1, G_2. Hence we need flexibility in choosing members of $\mathcal{F}_a, \mathcal{F}_b$ of various size.
(II) Moreover, after a zip product of G_1, G_2, the resulting average degree is no longer this weighted average of a, b but a slightly different rational number. We take care of this problem by introducing a special *compensation gadget* whose role is to revert the change in average degree caused by zip product.

Addressing (I); a family of graphs $\mathcal{F}$ is *scalable* if all the graphs in $\mathcal{F}$ have equal average degree and for every $G \in \mathcal{F}$ and every integer a, there exists $H \in \mathcal{F}$ such that $|V(H)| = a|V(G)|$. Furthermore, $\mathcal{F}$ is *D-max-universal scalable* if, additionally, H contains at least a vertices of each degree from D and the number of vertices of degrees not in D is bounded independently of a.

Trivially, the families of Proposition 4.1 (c),(d) are D-max-universal scalable for $D = \{3,4\}$ and $D = \{3,4,2\ell+3\}$, respectively. For families as in Proposition 4.1 (a),(b), the analogous property can be established by a slight modification of the very flexible construction from [11].

Addressing (II); we again exploit the construction from [11], defining a flexible gadget M_m^c as a special case of Proposition 4.1 (a). The graph M_m^c, for any $m \geq 12$ and $0 \leq c \leq m$, is simple, 3-connected, and 5-crossing-critical. The way "compensating by" M_m^c works, is formulated next:

Lemma 5.2. *Let $G_1, \ldots, G_t$ be graphs, each having at least two degree-3 vertices, and $q \in \mathbb{N}$. If H is a graph obtained by arbitrarily using the zip product of all $G_1, \ldots, G_t$ and of M_m^{q+t}, $m \geq \max(q+t, 12)$, then the average degree of H is equal to the average degree of the disjoint union of $G_1, \ldots, G_t$ and M_m^q.*

The next step is to naturally combine available scalable critical families to obtain, with the help of Theorem 2.3 and Lemma 5.2, new families of arbitrary "intermediate" rational average degrees:

Lemma 5.3. *Assume we have simple, D_i-max-universal scalable, 3-connected, k_i-crossing-critical families $\mathcal{F}_i$ of average degree r_i, $i = 1, \ldots, t$, such that $r_1 < r_2$. Then for every $k \geq k_1 + \cdots + k_t + 5$ and any $r \in (r_1, r_2) \cap \mathbb{Q}$, there exists a $(D_1 \cup \cdots \cup D_t)$-max-universal family of simple, 3-connected, k-crossing-critical graphs of average degree exactly r.*

While leaving technical details of these tools to a full paper, we finish with an overview of their case-specific application to Theorem 5.1:

Proof (of Theorem 5.1). The case (d) has already been proved in [11], see Proposition 4.1 (a). In all other cases, let $\mathcal{F}_1$ be the family from Proposition 4.1 (c) such that the parameter n satisfies $r_1 = 3 + \frac{1}{4n-7} < r$ (where $r \in A \cap \mathbb{Q}$, $r > 3$, is the desired fixed average degree).

In the case (a), let $\mathcal{F}_2$ be a family from Proposition 4.1 (a) with average degree equal to arbitrary (but fixed) $r_2 \in (r, 6) \neq \emptyset$, and chosen as scalable. In the case (c), let $\mathcal{F}_2$ be the family from Proposition 4.1 (d) for the parameter ℓ such that $b = 2\ell + 3$; in this case $r_2 = 5 - \frac{8}{b+1} > r$. Finally, consider the remaining sub-cases of (b). If $D = \{3, 4\}$, then let $\mathcal{F}_2$ be the second family from Proposition 4.1 (a) with average degree $r_2 = 4$. If $D \supsetneq \{3, 4\}$, then let $\mathcal{F}_2$ be the family from Proposition 4.1 (b), made scalable and of fixed average degree $r_2 > 4$.

In each of the choices of $\mathcal{F}_1, \mathcal{F}_2$ above, it holds $r_1 < r < r_2$. Furthermore, if needed to fulfill D-max-universality, add more scalable families $\mathcal{F}_3, \ldots$ as in the proof of Theorem 4.2. Theorem 5.1 then follows directly from Lemma 5.3. □

6 Final Remarks

In the previous constructions, we have always assumed that the fixed crossing number k of the families is sufficiently large. One can, on the other hand, ask what happens if we fix a small value of k beforehand (i.e., independently of the asked degree properties).

In this direction, there is the remarkable result of Dvořák and Mohar [6] proving the existence of k-crossing-critical families with unbounded maximum degree for any $k \geq 171$. Unfortunately, since [6] is not really constructive, we do not know anything exact about the degrees occurring in these families. An explicit construction of a k-crossing-critical family with unbounded maximum degree is known only in the projective plane [12] for $k \geq 2$, but that falls outside of the area of interest of this paper.

Fig. 4. Fractions (each of three tiles) of constructions of simple, 3-connected, 2-crossing-critical and D-max-universal families for $D = \{3, 5\}$ (left) and $D = \{3, 6\}$ (right).

It thus appears natural to thoroughly investigate the least non-trivial case of $k = 2$, with help of the remarkably involved characterization result [5][1]. Due to limited space, we can only very briefly survey the obtained results.

Theorem 6.1 *A simple, 3-connected 2-crossing-critical D-max-universal family exists if and only if $\{3\} \subsetneq D \subseteq \{3, 4, 5, 6\}$. Without the simplicity requirement, such a family exists if and only if $D \subseteq \{3, 4, 5, 6\}$, $|D| \geq 2$, and $D \cap \{3, 4\} \neq \emptyset$.*

We remark that it is important that Theorem 6.1 deals with infinite such families (via the universality property) since not all of the (finitely many) sporadic small 2-crossing-critical graphs are explicitly known [5]. Examples of two sub-cases of Theorem 6.1 can be found in Fig. 4.

Theorem 6.2 *A simple, 3-connected, 2-crossing-critical infinite family of graphs with average degree $r \in \mathbb{Q}$ exists if and only if $r \in [3\frac{1}{5}, 4]$. Without the simplicity requirement, such a family exists if and only if $r \in [3\frac{1}{5}, 4\frac{2}{3}]$.*

At last, we return to the statement of Theorem 4.2, which always requires $4 \in D$. On the other hand, from Theorem 6.1 we know that there exist D-max-universal families of simple, 3-connected, 2-crossing-critical graphs for $D = \{3, 5\}$ and $D = \{3, 6\}$ (Fig. 4), e.g., when $4 \notin D$, and these can be generalized to any $k > 2$ by a zip product with copies of $K_{3,3}$.

Hence it is an interesting open question of whether there exists a D-max-universal k-crossing-critical family such that $D \cap \{3, 4\} = \emptyset$. It is unlikely that the answer would be easy since the question is related to another long standing open problem—whether there exists a 5-regular k-crossing-critical infinite family. Related to this is the same question of existence of a 4-regular k-crossing-critical family, which does exist for $k = 3$ [17] and the construction can be generalized to any $k \geq 6$, but the cases $k = 4, 5$ remain open.

Many more questions can be asked in a direct relation to the statement of Theorem 5.1, but we are able to mention only a few of the interesting ones. E.g., if $6 \notin D$, can the average degree of such a family be from the interval $[5, 6)$? Or, assuming $3 \in D$ but $4 \notin D$, for which sets D one can achieve D-max-universality and what are the related average degrees?

We finish with another interesting structural conjecture:

Conjecture 6.3. There is a function $g : \mathbb{N} \to \mathbb{R}^+$ such that, any sufficiently large simple 3-connected k-crossing-critical graph has average degree greater than $3 + g(k)$.

---

[1] Even though this very long manuscript [5] is not published yet, its main result has been known already for many years and it is widely believed to be right.

References

1. Ajtai, M., Chvátal, V., Newborn, M., Szemerédi, E.: Crossing-free subgraphs.In: Theory and Practice of Combinatorics. North-Holland Mathematics Studies, vol.60, pp. 9–12. North-Holland (1982)
2. Bokal, D.: On the crossing numbers of cartesian products with paths. J. Comb. Theory Ser. B **97**(3), 381–384 (2007)
3. Bokal, D.: Infinite families of crossing-critical graphs with prescribed average degree and crossing number. J. Graph Theory **65**(2), 139–162 (2010)
4. Bokal, D., Chimani, M., Leaños, J.: Crossing number additivity over edge cuts. Eur. J. Comb. **34**(6), 1010–1018 (2013)
5. Bokal, D., Oporowski, B., Richter, R.B., Salazar, G.: Characterizing 2-crossing-critical graphs. Manuscript, 171 p. http://arxiv.org/abs/1312.3712 (2013)
6. Dvořák, Z., Mohar, B.: Crossing-critical graphs with large maximum degree. J. Comb. Theory, Ser. B **100**(4), 413–417 (2010)
7. Geelen, J.F., Richter, R.B., Salazar, G.: Embedding grids in surfaces. Eur. J. Comb. **25**(6), 785–792 (2004)
8. Hernández-Vélez, C., Salazar, G., Thomas, R.: Nested cycles in large triangulations and crossing-critical graphs. J. Comb. Theory, Ser. B **102**(1), 86–92 (2012)
9. Hlinecaronný, P.: Crossing-critical graphs and path-width. In: Mutzel, P., Jünger, M., Leipert, S. (eds.) GD 2001. LNCS, vol. 2265, pp. 102–114. Springer, Heidelberg (2002)
10. Hliněný, P.: Crossing-number critical graphs have bounded path-width. J. Comb. Theory Ser. B **88**(2), 347–367 (2003)
11. Hliněný, P.: New infinite families of almost-planar crossing-critical graphs. Electr. J. Comb. **15** R102 (2008)
12. Hliněný, P., Salazar, G.: Stars and bonds in crossing-critical graphs. J. Graph Theory **65**(3), 198–215 (2010)
13. Kochol, M.: Construction of crossing-critical graphs. Discrete Math. **66**(3), 311–313 (1987)
14. Leighton, T.: Complexity Issues in VLSI. Foundations of Computing Series. MIT Press, Cambridge (1983)
15. Pinontoan, B., Richter, R.B.: Crossing numbers of sequences of graphs II: Planar tiles. J. Graph Theory **42**(4), 332–341 (2003)
16. Pinontoan, B., Richter, R.B.: Crossing numbers of sequences of graphs I: General tiles. Australas. J. Comb. **30**, 197–206 (2004)
17. Richter, R.B., Thomassen, C.: Minimal graphs with crossing number at least k. J. Comb. Theory Ser. B **58**(2), 217–224 (1993)
18. Salazar, G.: Infinite families of crossing-critical graphs with given average degree. Discrete Math. **271**(1–3), 343–350 (2003)
19. Širáň, J.: Infinite families of crossing-critical graphs with a given crossing number. Discrete Math. **48**(1), 129–132 (1984)

Genus, Treewidth, and Local Crossing Number

Vida Dujmović[1], David Eppstein[2]([✉]), and David R. Wood[3]

[1] School of Computer Science and Electrical Engineering,
University of Ottawa, Ottawa, Canada
vida.dujmovic@uottawa.ca
[2] Department of Computer Science, University of California,
Irvine, CA, USA
eppstein@uci.edu
[3] School of Mathematical Sciences, Monash University, Melbourne, Australia
david.wood@monash.edu

Abstract. We consider relations between the size, treewidth, and local crossing number (maximum number of crossings per edge) of graphs embedded on topological surfaces. We show that an n-vertex graph embedded on a surface of genus g with at most k crossings per edge has treewidth $O(\sqrt{(g+1)(k+1)n})$ and layered treewidth $O((g+1)k)$, and that these bounds are tight up to a constant factor. As a special case, the k-planar graphs with n vertices have treewidth $O(\sqrt{(k+1)n})$ and layered treewidth $O(k+1)$, which are tight bounds that improve a previously known $O((k+1)^{3/4}n^{1/2})$ treewidth bound. Additionally, we show that for $g < m$, every m-edge graph can be embedded on a surface of genus g with $O((m/(g+1))\log^2 g)$ crossings per edge, which is tight to a polylogarithmic factor.

1 Introduction

Treewidth is a graph parameter that measures how similar a graph is to a tree. It is a key measure of the complexity of a graph and is of fundamental importance in algorithmic graph theory and structural graph theory, especially in Robertson and Seymour's graph minors project. Treewidth is closely related to the size of a smallest *separator*, a set of vertices whose removal splits the graph into connected components of at most $\frac{2n}{3}$ vertices, where n (as always) is the number of vertices in the graph. Graphs of low treewidth necessarily have small separators, and graphs in which every subgraph has a small separator have low treewidth [1,2]. See Sect. 2 for a detailed definition of treewidth.

A graph is *k-planar* if it can be drawn in the plane with at most k crossings on each edge. The *local crossing number* of the graph is the minimum k for which it is k-planar [3, pages 51–53]. An important example is the $p \times q \times r$ grid graph, with vertex set $[p] \times [q] \times [r]$ and all edges of the form $(x, y, z)(x + 1, y, z)$ or $(x, y, z)(x, y + 1, z)$ or $(x, y, z)(x, y, z + 1)$. A suitable linear projection from the natural three-dimensional embedding of this graph to the plane gives a $(r − 1)$-planar drawing, as illustrated in Fig. 1.

© Springer International Publishing Switzerland 2015
E. Di Giacomo and A. Lubiw (Eds.): GD 2015, LNCS 9411, pp. 87–98, 2015.
DOI: 10.1007/978-3-319-27261-0_8

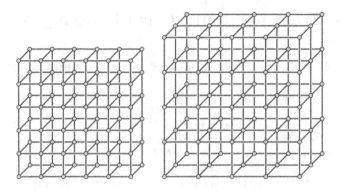

Fig. 1. The $p \times q \times r$ grid graph is $(r-1)$-planar.

The starting point for our work is the following question: what is the maximum treewidth of k-planar graphs? Grigoriev and Bodlaender [4] studied this question and proved an upper bound of $O(k^{3/4}n^{1/2})$. We improve this and give the following tight bound:

Theorem 1. *The maximum treewidth of k-planar n-vertex graphs is*

$$\Theta\left(\min\left\{n, \sqrt{(k+1)n}\right\}\right).$$

More generally, define a graph to be (g,k)-*planar* if it can be drawn in a surface of Euler genus at most g with at most k crossings on each edge.[1] For instance, Guy et al. [5] investigated the local crossing number of toroidal embeddings—in this notation, the $(2,k)$-planar graphs. We again determine an optimal bound on the treewidth of such graphs.

Theorem 2. *The maximum treewidth of (g,k)-planar n-vertex graphs is*

$$\Theta\left(\min\left\{n, \sqrt{(g+1)(k+1)n}\right\}\right).$$

In both these theorems, the $k=0$ case (with no crossings) is well known [6]. We prove our upper bounds by using the concept of *layered treewidth* [7], and we prove matching lower bounds by finding (g,k)-planar graphs without large separators and using the known relations between separator size and treewidth.

Finally, we study the (g,k)-planarity of graphs as a function of their number of edges. For (global) crossing number, it is known that a graph with n vertices and m edges drawn on a surface of genus g (sufficiently small with respect to m) may require $\Omega(\min\{m^2/g, m^2/n\})$ crossings, and it can be drawn with $O((m^2 \log^2 g)/g)$ crossings [8]. In particular, the lower bound implies that some graphs require $\Omega(m/g)$ crossings per edge on average, and therefore also in the worst case. We prove a nearly-matching upper bound:

---

[1] The *Euler genus* is $2h$ for an orientable surface with h handles, and c for a non-orientable surface with c cross-caps.

Theorem 3. *For every graph G with m edges, for every integer $g \geqslant 1$, there is a drawing of G in the orientable surface with at most g handles and with*

$$O\left(\frac{m \log^2 g}{g}\right)$$

crossings per edge.

2   Background

For $\epsilon \in (0, 1)$, a set S of vertices in a graph G is an ϵ-*separator* of G if each component of $G - S$ has at most $\epsilon |V(G)|$ vertices. It is conventional to set $\epsilon = 2/3$ but the precise choice makes no difference to the asymptotic size of a separator.

A *tree-decomposition* of a graph G is given by a tree T whose nodes index a collection $(B_x \subseteq V(G) : x \in V(T))$ of sets of vertices in G called *bags*, such that:

– For every edge vw of G, some bag B_x contains both v and w, and
– For every vertex v of G, the set $\{x \in V(T) : v \in B_x\}$ induces a non-empty (connected) subtree of T.

The *width* of a tree-decomposition is $\max_x |B_x| - 1$, and the *treewidth* $\mathrm{tw}(G)$ of a graph G is the minimum width of any tree decomposition of G. Treewidth was introduced (with a different but equivalent definition) by Halin [9] and tree decompositions were introduced by Robertson and Seymour [10] who proved:

Lemma 1 [10]. *Every graph with treewidth k has a $\frac{1}{2}$-separator of size at most $k + 1$.*

The notion of *layered tree decompositions* is a key tool in proving our main theorems. A *layering* of a graph G is a partition $(V_0, V_1, \ldots, V_t)$ of $V(G)$ such that for every edge $vw \in E(G)$, if $v \in V_i$ and $w \in V_j$, then $|i - j| \leqslant 1$. Each set V_i is called a *layer*. For example, for a vertex r of a connected graph G, if V_i is the set of vertices at distance i from r, then $(V_0, V_1, \ldots)$ is a layering of G, called the *bfs layering* of G starting from r. A *bfs tree* of G rooted at r is a spanning tree of G such that for every vertex v of G, the distance between v and r in G equals the distance between v and r in T. Thus, if $v \in V_i$ then the vr-path in T contains exactly one vertex from layer V_j for $0 \leqslant j \leqslant i$.

The *layered width* of a tree-decomposition $(B_x : x \in V(T))$ of a graph G is the minimum integer ℓ such that, for some layering $(V_0, V_1, \ldots, V_t)$ of G, each bag B_x contains at most ℓ vertices in each layer V_i. The *layered treewidth* of a graph G is the minimum layered width of a tree-decomposition of G. Note that if we only consider the trivial layering in which all vertices belong to one layer, then layered treewidth equals treewidth plus one.

Dujmović, Morin, and Wood [7] introduced layered treewidth and proved the following results, where a graph G is *apex* if $G - v$ is planar for some vertex v:

Theorem 4 (Dujmović, Morin, and Wood [7]).

(a) Every planar graph has layered treewidth at most 3.
(b) Every graph with Euler genus g has layered treewidth at most $2g + 3$.
(c) For every apex graph H, there is a number c such that every H-minor-free graph has layered treewidth at most c.
(d) If a minor-closed class has bounded layered treewidth, then it excludes a fixed apex graph as a minor.

The same characterization by forbidden apex minors was previously known for minor-closed classes of bounded local treewidth [11], establishing the equivalence of bounded local treewidth and bounded layered treewidth in minor-closed classes; however, for families of graphs that are not minor-closed, layered treewidth and local treewidth are distinct. Sergey Norin established the following connection between layered treewidth and treewidth:

Theorem 5 (Norin; see [7]). *Every n-vertex graph with layered treewidth k has treewidth at most $2\sqrt{kn}$.*

Several results that follow depend on expanders; see [12] for a survey.

Lemma 2. *For every $\epsilon \in (0,1)$ there exists $\beta > 0$, such that for all $k \geqslant 3$ and $n \geqslant k+1$ (such that n is even if k is odd), there exists a k-regular n-vertex graph H (called an expander) in which every ϵ-separator in H has size at least βn.*

3 k-Planar Graphs

The following is our first contribution.

Theorem 6. *Every k-planar graph has layered treewidth at most $6(k + 1)$.*

Proof. Let G be k-planar; draw G with at most k crossings per edge, and arbitrarily orient each edge of G. Let G' be the graph obtained from G by replacing each crossing by a new degree-4 vertex. Then G' is planar. By Theorem 4(a), G' has layered treewidth at most 3. That is, there is a tree decomposition T' of G', and a layering $V_0', V_1', \ldots$ of G', such that each bag of T' contains at most three vertices in each layer V_i'. For each vertex v of G', let T_v' be the subtree of T' formed by the bags that contain v.

Let T be the decomposition of G obtained by replacing each occurrence of a dummy vertex x in a bag of T' by the tails of the two edges that cross at x. We now show that T is a tree-decomposition of G. For each vertex v of G, let T_v be the subgraph of T formed by the bags that contain v. Let G_v' be the subgraph of G' induced by v and the division vertices on the edges for which v is the tail. Then G_v' is connected. Thus T_v', which is precisely the set of bags of T' that intersect G_v', form a (connected) subtree of T'. Moreover, for each oriented edge vw of G, if x is the division vertex of vw adjacent to w, then T_x' and T_w' intersect. Since T_v contains T_x', and T_w contains T_w', we have that T_v and T_w intersect. Thus T is a tree-decomposition of G. By construction, each bag of T contains at most six vertices in each layer V_i'.

Note that $\text{dist}_{G'}(v, w) \leqslant k + 1$ for each edge vw of G. Thus, if $v \in V_i'$ and $w \in V_j'$ then $|i - j| \leqslant k+1$. Let V_0 be the union of the first $k+1$ layers restricted to $V(G)$, let V_1 be the union of the second $k+1$ layers restricted to $V(G)$, and so on. That is, for $i \geqslant 0$, let $V_i := V(G) \cap (V_{(k+1)i}' \cup V_{(k+1)i+1}' \cup \cdots \cup V_{(k+1)(i+1)-1}')$. Then $V_0, V_1, \ldots$ is a partition of $V(G)$. Moreover, if $v \in V_i$ and $w \in V_j$ for some edge vw of G, then $|i - j| \leqslant 1$. Thus $V_1, V_2, \ldots$ is a layering of G. Since each layer in G consists of at most $k + 1$ layers in G', the layered treewidth of this decomposition is at most $6(k + 1)$. □

Theorem 4 does not imply Theorem 6, because 1-planar graphs may contain arbitrarily large complete graph minors. For example, the $n \times n \times 2$ grid graph is 1-planar, and contracting the i-th row in the front grid with the i-th column in the back grid gives a K_n minor.

Theorems 5 and 6 imply the upper bound in Theorem 1:

Theorem 7. *Every k-planar n-vertex graph has treewidth at most $2\sqrt{6(k + 1)n}$.*

We now prove the corresponding lower bound.

Theorem 8. *For $1 \leqslant k \leqslant \frac{3}{2}n$ there is a k-planar graph on n vertices with treewidth at least $c\sqrt{kn}$ for some constant $c > 0$.*

Proof. Let G be a cubic expander with n vertices. Then G has treewidth at least ϵn for some constant $\epsilon > 0$ (see for example Grohe and Marx [13]). Consider a straight-line drawing of G. Clearly, each edge is crossed less than $|E(G)| = \frac{3}{2}n$ times. Subdivide each edge of G at most $\frac{3n}{2k}$ times to produce a k-planar graph G' with n' vertices, where $n' \leqslant n + \frac{3n}{2}\frac{3n}{2k} < \frac{4n^2}{k}$. Subdivision does not change the treewidth of a graph. Thus G' has treewidth at least $\epsilon n \geqslant \frac{\epsilon}{2}\sqrt{kn'}$. □

Combining the bound of Theorem 7 with the trivial upper bound $\text{tw}(G) \leqslant n$ for $k \geq n$ shows that the maximum treewidth of k-planar n-vertex graphs is $\Theta(\min\{n, \sqrt{kn}\})$ for arbitrary k and n. This completes the proof of Theorem 1.

4 (g, k)-Planar Graphs

Recall that a graph is (g, k)-planar if it can be drawn in a surface of Euler genus at most g with at most k crossings on each edge. The proof method used in Theorem 6 in conjunction with Theorem 4(b) leads to the following theorem.

Theorem 9. *Every (g, k)-planar graph G has layered treewidth at most $(4g + 6)(k + 1)$.*

Proof. Consider a drawing of G with at most k crossings per edge on a surface Σ of Euler genus g. Arbitrarily orient each edge of G. Let G' be the graph obtained from G by replacing each crossing by a new degree-4 vertex. Then G' is embedded in Σ with no crossings, and thus has Euler genus at most g. By Theorem 4(b), G' has layered treewidth at most $2g + 3$. That is, there is a tree decomposition T' of G', and a layering $V'_0, V'_1, \ldots$ of G', such that each bag of T' contains at most $2g + 3$ vertices in each layer V'_i. For each vertex v of G', let T'_v be the subtree of T' formed by the bags that contain v.

Let T be the decomposition of G obtained by replacing each occurrence of a dummy vertex x in a bag of T' by the tails of the two edges that cross at x. We now show that T is a tree-decomposition of G. For each vertex v of G, let T_v be the subgraph of T formed by the bags that contain v. Let G'_v be the subgraph of G' induced by v and the division vertices on the edges for which v is the tail. Then G'_v is connected. Thus T'_v, which is precisely the set of bags of T' that intersect G'_v, form a (connected) subtree of T'. Moreover, for each oriented edge vw of G, if x is the division vertex of vw adjacent to w, then T'_x and T'_w intersect. Since T_v contains T'_x, and T_w contains T'_w, we have that T_v and T_w intersect. Thus T is a tree-decomposition of G. By construction, each bag of T contains at most $4g + 6$ vertices in each layer V'_i.

Note that $\text{dist}_{G'}(v, w) \leqslant k + 1$ for each edge vw of G. Thus, if $v \in V'_i$ and $w \in V'_j$ then $|i - j| \leqslant k + 1$. Let V_0 be the union of the first $k + 1$ layers restricted to $V(G)$, let V_1 be the union of the second $k + 1$ layers restricted to $V(G)$, and so on. That is, for $i \geqslant 0$, let $V_i := V(G) \cap (V'_{(k+1)i} \cup V'_{(k+1)i+1} \cup \cdots \cup V'_{(k+1)(i+1)-1})$. Then $V_0, V_1, \ldots$ is a partition of $V(G)$. Moreover, if $v \in V_i$ and $w \in V_j$ for some edge vw of G, then $|i - j| \leqslant 1$. Thus $V_1, V_2, \ldots$ is a layering of G. Since each layer in G consists of at most $k + 1$ layers in G', the layered treewidth of this decomposition is at most $(4g + 6)(k + 1)$. □

Theorems 9 and 5 imply:

Theorem 10. *Every n-vertex (g, k)-planar graph has treewidth at most*

$$2\sqrt{(4g + 6)(k + 1)n}.$$

We now show that this bound is tight up to a constant factor.

Theorem 11. *For all $g, k \geqslant 0$ and infinitely many n there is an n-vertex (g, k)-planar graph with treewidth $\Omega(\sqrt{(g + 1)(k + 1)n})$.*

The proof of this result depends on the separation properties of the $p \times q \times r$ grid graph (which is $(r - 1)$-planar). The next two results are not optimal, but have simple proofs and are all that is needed for the main proof that follows.

Lemma 3. *For $q \geqslant (\frac{1}{1-\epsilon})r$, every ϵ-separator of the $q \times r$ grid graph has size at least r.*

Proof. Let S be a set of at most $r - 1$ vertices in the $q \times r$ grid graph. Some row R avoids S, and at least $q - r + 1$ columns avoid S. The union of these columns with R induces a connected subgraph with at least $(q - r + 1)r > \epsilon qr$ vertices. Thus S is not an ϵ-separator. □

Lemma 4. *For $p \geqslant q \geqslant (\frac{1}{1-\epsilon})r$, every ϵ-separator of the $p \times q \times r$ grid graph has size at least $(\frac{1-\epsilon}{1+\epsilon})qr$.*

Proof. Let G be the $p \times q \times r$ grid graph. Let $n := |V(G)| = pqr$. Let S be an ϵ-separator of G. Let $A_1, \ldots, A_c$ be the components of $G - S$. Thus $|A_i| \leqslant \epsilon n$. For $x \in [p]$, let $G_x := \{(x, y, z) : y \in [q], z \in [r]\}$ called a *slice*. Say G_x *belongs* to A_i and A_i *owns* G_x if $|A_i \cap G_x| \geqslant \frac{1+\epsilon}{2}qr$. Clearly, no two components own the same slice. First suppose that at least two components each own a slice. That is, G_v belongs to A_i and G_w belongs to A_j for some $v < w$ and $i \neq j$. Let $X := \{(y, z) : (v, y, z) \in G_v, (w, y, z) \in G_w\}$. Then $|X| \geqslant 2(\frac{1+\epsilon}{2})qr - qr = \epsilon qr$. For each $(y, z) \in X$, the 'straight' path $(v, y, z), (v + 1, y, z), \ldots, (w, y, z)$ contains some vertex in S. Since these paths are pairwise disjoint, $|S| \geqslant |X| \geqslant \epsilon qr \geqslant \frac{1-\epsilon}{1+\epsilon}qr$ (since $\epsilon > \frac{1}{2}$). Now assume that at most one component, say A_1, owns a slice. Say A_1 owns t slices. Thus $t(\frac{1+\epsilon}{2})qr \leqslant |A_1| \leqslant \epsilon pqr$ and $t \leqslant \frac{2\epsilon}{1+\epsilon}p$. Hence, at least $(1 - \frac{2\epsilon}{1+\epsilon})p$ slices belong to no component. For such a slice G_v, each component of $G_v - S$ is contained in some A_i and thus has at most $(\frac{1+\epsilon}{2})qr$ vertices. That is, $S \cap G_v$ is a $(\frac{1+\epsilon}{2})$-separator of the $q \times r$ grid graph induced by G_v. By Lemma 3, $|S \cap G_v| \geqslant r$. Thus $|S| \geqslant (1 - \frac{2\epsilon}{1+\epsilon})pr \geqslant (\frac{1-\epsilon}{1+\epsilon})qr$. $\qquad\square$

Proof (of Theorem 11). Let $r := k + 1$.

First suppose that $g \leqslant 19$. Let G be the $q \times q \times r$ grid graph where $q \geqslant 2r$. As observed above, G is k-planar and thus (g, k)-planar. Lemma 4 implies that every $\frac{1}{2}$-separator of G has size at least $\frac{1}{3}qr$. Lemma 1 thus implies that G has treewidth at least $\frac{1}{3}qr - 1$, which is at least $\Omega(\sqrt{(g+1)(k+1)n})$, as desired.

Now assume that $g \geqslant 20$. By Lemma 2 there is a 4-regular expander H on $m := \lfloor \frac{g}{4} \rfloor \geqslant 5$ vertices. Thus H has $2m$ edges, H embeds in the orientable surface with $2m$ handles, and thus has Euler genus at most $4m \leqslant g$. We may assume that $q := \sqrt{n/rm}$ is an integer with $q \geqslant 8r$. Let G be obtained from H by replacing each vertex v of H by a copy of the $q \times q \times r$ grid graph with vertex set D_v, and replacing each edge of H by a matching of qr edges, so that $G[D_v \cup D_w]$ is a $2q \times q \times r$ grid, as shown in Fig. 2. Thus G is (g, k)-planar with $q^2rm = n$ vertices.

Let S be a $\frac{1}{2}$-separator in G. Let $A_1, \ldots, A_c$ be the components of $G - S$. Thus $|A_i| \leqslant \frac{1}{2}n$ for $i \in [c]$. Initialise sets $S' := A_1' := \cdots := A_c' := \emptyset$.

For each vertex v of H, if $|S \cap D_v| \geqslant \frac{qr}{14}$ then put $v \in S'$. Otherwise, $|S \cap D_v| < \frac{qr}{14}$. Note that Lemma 4 is applicable with $\epsilon = \frac{13}{15}$ since $q \geqslant 8r > \frac{1}{1-13/15}r$ and $\frac{1-13/15}{1+13/15} = \frac{1}{14}$. Lemma 4 thus implies that $S \cap D_v$ is not a $\frac{13}{15}$-separator. Hence some component of $D_v - S$ has at least $\frac{13}{15}q^2r$ vertices. Since $\frac{13}{15} > \frac{1}{2}$, exactly one component of $D_v - S$ has at least $\frac{13}{15}q^2r$ vertices. This component is a subset of A_i for some $i \in [c]$; add v to A_i'. Thus $S', A_1', \ldots, A_c'$ is a partition of $V(H)$.

We now prove that S' is a $\frac{15}{26}$-separator in H. Suppose that $v \in A_i'$ and $w \in A_j'$ for some edge vw of H. Let D be the vertex set of the $2q \times q \times r$ grid graph induced by $D_v \cup D_w$. Since $v \notin S'$ and $w \notin S'$, we have $|S \cap D_v| < \frac{qr}{14}$ and $|S \cap D_w| < \frac{qr}{14}$. Thus $|S \cap D| < \frac{qr}{7}$. Note that Lemma 4 is applicable with $\epsilon = \frac{3}{4}$ since $q \geqslant 8r > \frac{1}{1-3/4}r$ and $\frac{1-3/4}{1+3/4} = \frac{1}{7}$. Lemma 4 thus implies that $S \cap D$ is not a $\frac{3}{4}$-separator of $G[D]$. Hence some component X of $G[D] - S$ contains at least

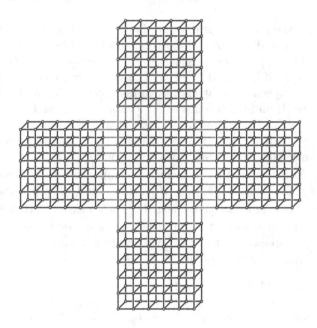

Fig. 2. Construction of G in the proof of Theorem 11.

$\frac{3}{4}|D| = \frac{3}{2}q^2r$ vertices. Each of D_v and D_w can contain at most q^2r vertices in X. Thus D_v and D_w each contain at least $\frac{1}{2}q^2r$ vertices in X. Thus, by construction, v and w are in the same A_i'. That is, there is no edge of H between distinct A_i' and A_j', and each component of $H - S'$ is contained in some A_i'. For each $i \in [c]$, we have $\frac{1}{2}q^2rm \geqslant |A_i| \geqslant \frac{13}{15}q^2r|A_i'|$ implying $|A_i'| \leqslant \frac{15}{26}m$. Therefore S' is a $\frac{15}{26}$-separator in H.

By Lemma 2, $|S'| \geqslant \beta m$ for some constant $\beta > 0$. Thus $|S| \geqslant \frac{qr}{14}|S'| \geqslant \frac{\beta}{14}mqr$. By Lemma 1, G has treewidth at least $\frac{\beta}{14}mqr - 1 = \frac{\beta}{14}\sqrt{mrn} - 1 \geqslant \Omega(\sqrt{g(k+1)n})$, as desired.   □

Note that the proof of Theorem 11 in the case $k = 0$ is very similar to that of Gilbert, Hutchison, and Tarjan [6].

For $gk \geqslant n$ the trivial upper bound of $tw(G) \leqslant n$ is better than that given in Theorem 10. We conclude that the maximum treewidth of (g,k)-planar n-vertex graphs is $\Theta(\min\{n, \sqrt{(g+1)(k+1)n}\})$ for arbitrary g, k, n. This completes the proof of Theorem 2.

5   Drawings with Few Crossings per Edge

This section studies the following natural conjecture: for every surface Σ of Euler genus g, every graph G with m edges has a drawing in Σ with $O(\frac{m}{g+1})$ crossings per edge. This conjecture is trivial at both extremes: with $g = 0$, every graph has a straight-line drawing in the plane (and therefore a drawing in the sphere) with at most m crossings per edge, and with $g = 2m$, every graph has a drawing in the orientable surface with one handle per edge. Moreover, if this conjecture is true, it would provide a simple proof of Theorem 11 in the same manner as the proof of Theorem 8.

Our starting point is the following well-known result of Leighton and Rao [14, Theorem 22, p. 822]:

Theorem 12 (Leighton and Rao [14]). *Let G be a graph with bounded degree and n vertices, mapped one-to-one onto the vertices of an expander graph H. Then the edges of G can be mapped onto paths in H so that each path has length $O(\log n)$ and each edge of H is used by $O(\log n)$ paths.*

It is straightforward to extend this result to regular graphs G of unbounded degree, with the number of paths per edge of H increasing in proportion to the degree. However, there are two difficulties with using it in our application. First, it does not directly handle graphs in which there is considerable variation in degree from vertex to vertex: in such cases we would want the number of paths per edge to be controlled by the average degree in G, but instead it is controlled by the maximum degree. And second, it does not allow us to control separately the sizes of G and H; instead, both must have the same number of vertices. To handle these issues, we do not map the vertices of our input graph G directly to the vertices of an expander H; instead, we keep the vertices of G and the vertices of H disjoint from each other, connecting them by a bipartite graph that balances the degrees, according to the following lemma.

Lemma 5. *Let $d_1, d_2, \ldots, d_n$ be a sequence of positive integers, and let q be any positive integer. Then there exists a bipartite graph with colour classes $\{v_1, \ldots, v_n\}$ and $\{w_1, \ldots, w_q\}$, at most $n + q - 1$ edges, and a labelling of the edges with positive integers, such that*

- *each vertex v_i is incident to a set of edges whose labels sum to d_i, and*
- *each pair of distinct vertices w_i and w_j are incident to sets of edges whose label sums differ by at most 1.*

Proof. Preassign label sums of $\lfloor \sum d_i/q \rfloor$ or $\lceil \sum d_i/q \rceil$ to each vertex w_i so that the resulting values sum to $\sum d_i$. We will construct a bipartite graph and a labelling whose sums match the numbers $d_1, \ldots, d_n$ on one side of the bipartition and whose sums match the preassigned numbers on the other side.

Build this graph and its labelling one edge at a time, starting from a graph with no edges. At each step, let v_i and w_j be the vertices on each side of the bipartition with the smallest indices whose edge labels do not yet sum to the required values, add an edge from v_i to w_j, and label this edge with the largest integer that does not exceed the required sum on either vertex.

Each step completes the sum for at least one vertex. Because the required values on the two sides of the bipartition both sum to $\sum d_i$, the final step completes the sum for two vertices, v_n and w_q. Therefore, the total number of steps, and the total number of edges added to the graph, is at most $n + q - 1$.   □

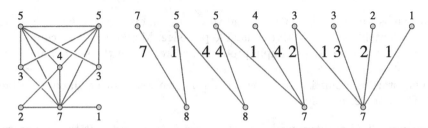

Fig. 3. A graph (left) with degree sequence $7, 5, 5, 4, 3, 3, 2, 1$ and a bipartite graph (right) formed from this degree sequence by Lemma 5. The large numbers are the edge labels of the lemma, and the small numbers along the top and bottom of the bipartite graph give the sums of incident edge labels at each vertex. The top sums match the given degree sequence, while the bottom sums all differ by at most 1.

By combining this load-balancing step with the Leighton-Rao expander-routing scheme, we may obtain a more versatile mapping of our given graph G to a host graph H, with better control over the genus of the surface we obtain from H. This genus will be determined by the *cyclomatic number* of H, where the cyclomatic number of a graph with n vertices and m edges is $m - n + 1$. This number is the dimension of the cycle space of the graph, and the first Betti number of the topological space obtained from the graph by replacing each edge by a line segment.

Lemma 6. *Let G be an arbitrary graph, with m edges, and let Q be a q-vertex bounded-degree expander graph. Then there exists a host graph H, a one-to-one mapping of the vertices of G to a subset of vertices of H, and a mapping of the edges of G to paths in H, with the following properties:*

- *The vertices of H that are not images of vertices in G induce a subgraph isomorphic to Q.*
- *The image of an edge e in G forms a path of length $O(\log q)$ that starts and ends at the image of the endpoints of e, and passes through the image of no other vertex of G.*
- *Each vertex of H that is not an image of a vertex in G is crossed by $O((m \log q)/q)$ paths.*
- *The cyclomatic number of H is $O(q)$.*

Proof. Let the vertices of G be $u_1, \ldots, u_n$. Apply Lemma 5 to the degree sequence of G to form a bipartite graph H with bipartition $\{v_1, \ldots, v_n\}, \{w_1, \ldots, w_q\}$ (Fig. 3). Then add edges between pairs of vertices (w_i, w_j) so that $\{w_1, \ldots, w_q\}$ induces a subgraph isomorphic to Q. In this way, each vertex u_i in G is mapped to a vertex v_i in H so that the mapping is one-to-one and the unmapped vertices form a copy of Q, as required. The cyclomatic number of H equals the cyclomatic number of Q, plus $n + q - 1$ (for the added edges in the bipartite graph), minus n (for the added vertices relative to Q). These two added and subtracted terms cancel, leaving the cyclomatic number of Q plus $q - 1$, which is $O(q)$ as required.

It remains to find paths in H corresponding to the edges in G. Assign each edge $u_i u_j$ of G to a pair of vertices $(w_{i'}, w_{j'})$ adjacent to the images v_i and v_j in H, so that the number of edges of G assigned to each edge between $\{v_1, \ldots, v_n\}$ and $\{w_1, \ldots, w_q\}$ equals the corresponding label. Complete each path by applying Theorem 12 to the copy of Q; this gives paths of length $O(\log q)$ connecting each pair $(w_{i'}, w_{j'})$ obtained in this way. These pairs do not form a bounded-degree graph, but they can be partitioned into $O(m/q)$ bounded-degree graphs, each of which causes each vertex in the copy of Q to be crossed $O(\log q)$ times. Combining these suproblems, each vertex in the copy of Q is crossed by a total of $O((m \log q)/q)$ paths, as required.   □

We are now ready to prove the existence of embeddings with small local crossing number, on surfaces of arbitrary genus.

Proof (of Theorem 3). Given a graph G, to be embedded on a surface with at most g handles and with few crossings per edge, choose q so that the $O(q)$ bound on the cyclomatic number of the graph H in Lemma 6 is at most g, and apply Lemma 6 to find a graph H and a mapping from G to H obeying the conditions of the lemma.

To turn this mapping into the desired embedding of G, replace each vertex of degree d in H by a sphere, punctured by the removal of d unit-radius disks, and form a surface (as a cell complex, not necessarily embedded into three-dimensional space) by replacing each edge xy of H by a unit-radius cylinder connecting boundaries of removed disks on the spheres for vertices x and y. The number of handles on the resulting surface (shown in Fig. 4) equals the cyclomatic number of H, which is at most g.

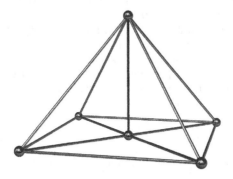

Fig. 4. A topological surface obtained by replacing each vertex of a graph by a punctured sphere, and each edge of the graph by a cylinder connecting two punctures. Image Square_pyramid_pyramid.png by Tom Ruen on Wikimedia commons, made available under a Creative Commons CC-BY-SA 4.0 International license.

Embed each vertex of G as an arbitrarily chosen point on the sphere of the corresponding vertex of H, and each edge of G as a curve through the sequence of spheres and cylinders corresponding to its path in H. Choose this embedding so that no intersection of edge curves occurs within any of the cylinders, and so that every pair of edges that are mapped to curves on the same sphere meet at most once, either at a crossing point or a shared endpoint.

Because the spheres that contain vertices of G only contain curves incident to those vertices, they do not have any crossings. Each edge is mapped to a curve through $O(\log g)$ of the remaining spheres, and can cross at most $O((m \log g)/g)$ other curves within each such sphere. Therefore, the maximum number of crossings per edge is $O((m \log^2 g)/g)$.   □

Acknowledgement. This research was initiated at the Workshop on Graphs and Geometry held at the Bellairs Research Institute in 2015. Vida Dujmović was supported by NSERC. David Eppstein was supported in part by NSF grant CCF-1228639. David Wood was supported by the Australian Research Council.

References

1. Dvořák, Z., Norin, S.: Treewidth of graphs with balanced separations. Electronic preprint arXiv: 1408.3869 (2014)
2. Reed, B.A.: Tree width and tangles: a new connectivity measure and some applications. In: Bailey, R.A. (ed.) Surveys in Combinatorics. London Mathematical Society Lecture Note Series, vol. 241, pp. 87–162. Cambridge University Press, Cambridge (1997)
3. Schaefer, M.: The graph crossing number and its variants: a survey. Electron. J. Combin. DS21 (2014)
4. Grigoriev, A., Bodlaender, H.L.: Algorithms for graphs embeddable with few crossings per edge. Algorithmica **49**(1), 1–11 (2007)
5. Guy, R.K., Jenkyns, T., Schaer, J.: The toroidal crossing number of the complete graph. J. Comb. Theor. **4**, 376–390 (1968)
6. Gilbert, J.R., Hutchinson, J.P., Tarjan, R.E.: A separator theorem for graphs of bounded genus. J. Algorithms **5**(3), 391–407 (1984)
7. Dujmović, V., Morin, P., Wood, D.R.: Layered separators in minor-closed families with applications. Electronic preprint arXiv: 1306.1595 (2013)
8. Shahrokhi, F., Székely, L.A., Sýkora, O., Vrt'o, I.: Drawings of graphs on surfaces with few crossings. Algorithmica **16**(1), 118–131 (1996)
9. Halin, R.: S-functions for graphs. J. Geometry **8**(1–2), 171–186 (1976)
10. Robertson, N., Seymour, P.D.: Graph minors. II. algorithmic aspects of tree-width. J. Algorithms **7**(3), 309–322 (1986)
11. Eppstein, D.: Diameter and treewidth in minor-closed graph families. Algorithmica **27**, 275–291 (2000)
12. Hoory, S., Linial, N., Wigderson, A.: Expander graphs and their applications. Bull. Am. Math. Soc. **43**(4), 439–561 (2006)
13. Grohe, M., Marx, D.: On tree width, bramble size, and expansion. J. Combin. Theory Ser. B **99**(1), 218–228 (2009)
14. Leighton, T., Rao, S.: Multicommodity max-flow min-cut theorems and their use in designing approximation algorithms. J. ACM **46**(6), 787–832 (1999)

Hanani-Tutte for Radial Planarity

Radoslav Fulek[1], Michael Pelsmajer[2], and Marcus Schaefer[3] [(⊠)]

[1] IST Austria, Am Campus 1, Klosterneuburg 3400, Austria
radoslav.fulek@gmail.com
[2] Illinois Institute of Technology, Chicago, IL 60616, USA
pelsmajer@iit.edu
[3] DePaul University, Chicago, IL 60604, USA
mschaefer@cs.depaul.edu

Abstract. A drawing of a graph G is *radial* if the vertices of G are placed on concentric circles $C_1, \ldots, C_k$ with common center c, and edges are drawn *radially*: every edge intersects every circle centered at c at most once. G is *radial planar* if it has a radial embedding, that is, a crossing-free radial drawing. If the vertices of G are ordered or partitioned into ordered levels (as they are for leveled graphs), we require that the assignment of vertices to circles corresponds to the given ordering or leveling.

We show that a graph G is radial planar if G has a radial drawing in which every two edges cross an even number of times; the radial embedding has the same leveling as the radial drawing. In other words, we establish the weak variant of the Hanani-Tutte theorem for radial planarity. This generalizes a result by Pach and Tóth.

1 Introduction

In a *leveled* graph every vertex is assigned a level in $\{1, \ldots, k\}$. We can capture the leveling of the graph visually, by placing the vertices on parallel lines or concentric circles corresponding to the levels of G. To further emphasize the levels, we can require that edges respect the levels in the sense that edges must lie between the levels of their endpoints, and be monotone in the sense that they intersect any line (circle) parallel to (concentric with) the chosen lines (circles) at most once. If we choose lines, we obtain the concept of *level-planarity*; for circles we get *radial (level) planarity*.

Radial planarity was introduced by Bachmaier, Brandenburg and Forster [1] as a generalization of level-planarity [6]. Radial layouts are a popular visualization tool (see [7] for a recent survey); early examples of radial graph layouts can be found in the literature on sociometry [13]. Bachmaier, Brandenburg and Forster [1] showed that radial planarity can be tested, and an embedding can be found, in linear time. Their algorithm is based on a variant of PQ-trees [2] and is rather intricate. It generalizes an earlier linear time algorithm for level-planarity

Radoslav Fulek — The research leading to these results has received funding from the People Programme (Marie Curie Actions) of the European Union's Seventh Framework Programme (FP7/2007-2013) under REA grant agreement no [291734].

© Springer International Publishing Switzerland 2015
E. Di Giacomo and A. Lubiw (Eds.): GD 2015, LNCS 9411, pp. 99–110, 2015.
DOI: 10.1007/978-3-319-27261-0_9

testing by Jünger and Leipert [12]. In this paper, we take the first step toward an alternative algorithm for radial planarity testing via a Hanani-Tutte style characterization.

The classical Hanani-Tutte theorem [5,20] states that a graph is planar if and only if it can be drawn in the plane so that every two independent edges cross an even number of times. A particularly nice algorithmic consequence of this result is that it reduces planarity testing to solving a system of linear equation (of polynomial size) over $\mathbb{Z}_2$, a purely algebraic problem which can be solved in polynomial time.

If we could show that a leveled graph G is radial planar if it has a radial drawing (respecting the leveling) in which every two *independent* edges cross an even number of times, we would have a new, very simple, polynomial-time algorithm for radial planarity. We conjecture that this (strong) Hanani-Tutte characterization of radial planarity is true, and take the first step toward this result: a *weak* Hanani-Tutte theorem. A weak variant of the Hanani-Tutte theorem makes the stronger assumption that *every* two edges cross an even number of times. Often, this leads to stronger conclusions. For example, it is known that if a graph can be drawn in a surface so that every two edges cross evenly, then the graph has an embedding on that surface with the same rotation system, i.e. the cyclic order of ends at each vertex remains the same [3,16].

Our main result, proved in Sect. 3, is the following theorem:

Theorem 1. *Suppose a leveled graph G has a radial drawing in which every two edges cross an even number of times. Then G has a radial embedding with the same rotation system. (All drawings respect the given leveling.)*

Theorem 1 implies a polynomial time algorithm for the radial planarity testing of a leveled graph G if a combinatorial embedding (rotation system) of G is fixed. This algorithm (sketched in Sect. 4) is based on solving a system of linear equations over $\mathbb{Z}_2$, see also [19, Sect. 1.4]. Thus, our algorithm runs in time $O(|V(G)|^{2\omega})$, where $O(n^\omega)$ is the complexity of multiplication of two square $n \times n$ matrices. Since our linear system is sparse, it is also possible to use Wiedemann's randomized algorithm [21], with expected running time $O(n^4 \log n^2)$ in our case.

Remark 1. While we do not know whether the (strong) Hanani-Tutte characterization of radial planarity holds, it can still be used for an algorithmic solution of the radial planarity problem, in the following sense: one can write a polynomial-time algorithm which—given leveled graph G—either returns a radial planar embedding of G, states that G is not radial planar, or stops with "don't know". If the algorithm outputs one of the first two answers, it is correct. If the strong Hanani-Tutte theorem for radial planarity is true, then the third answer will not occur. The details of how this can be done have been explained in a paper by Gutwenger, Mutzel, and Schaefer [11], where such an algorithm was successfully implemented for c-planarity.   ∎

Theorem 1 is a generalization of a weak variant of the Hanani-Tutte theorem for level-planarity[1], first proved by Pach and Tóth [9,14]. The full Hanani-Tutte theorem for level-planarity was established only more recently [9], and it led to a quadratic time level-planarity test. A computational study of Chimani and Zeranski [4] of various algorithms for upward planarity testing (an NP-complete problem related to level-planarity), showed that the algorithm based on the Hanani-Tutte characterization of level-planarity performs very well in practice (it beats all other algorithms in nearly all scenarios).

Hanani-Tutte style characterizations have also been established for partially embedded planar graphs, several classes of simultaneously embedded planar graphs [18], and two-clustered graphs [8]. The family of counterexamples in [8, Sect. 4] shows that a straightforward variant of the Hanani-Tutte theorem for clustered graphs with more than two clusters fails. Gutwenger et al. [11] showed that by using the reduction from [18], this counterexample can be turned into a counterexample for a variant of the Hanani-Tutte theorem for two simultaneously embedded planar graphs [18, Conjecture 6.20]. For higher-genus (compact) surfaces, the weak variant is known to hold in all surfaces [3,17], while the strong variant is known for the projective plane only [15]. It remains an intriguing open problem whether the strong Hanani-Tutte theorem holds for closed surfaces other than the sphere and projective plane.

2   Terminology

For the purposes of this paper, graphs may have multi-edges, but no loops. An *ordered* graph $G = (V, E)$ is a graph whose vertices are equipped with a total order $v_1 < v_2 < \cdots < v_n$. We consider an ordered graph a special case of a *leveled* graph, in which every vertex of G is assigned a *level*, a number in $\{1, \ldots, k\}$ for some k. The leveling of the vertices induces a natural ordering of the vertices.

For convenience we represent radial drawings as drawings on a (standing) cylinder. Intuitively, imagine placing a cylindrically-shaped mirror in the center of a radial drawing as described in the introduction.[2]

The *cylinder* $\mathcal{C}$ is $\mathbb{S}^1 \times I$, where $\mathbb{S}^1$ is a unit circle and I is the unit interval $[0, 1]$. Thus, a *point* on $\mathcal{C}$ is a pair (s, i), where $s \in \mathbb{S}^1$ and $i \in I$. The *projection* of $\mathcal{C}$ to $\mathbb{S}^1$, or I, maps $(s, i) \in \mathcal{C}$ to s, or i. We denote a projection of a point or a subset α of $\mathbb{S}^1 \times I$ to I by $I(\alpha)$. The *winding number* of a closed curve on a cylinder is the number of times the projection to $\mathbb{S}^1$ of the curve winds around $\mathbb{S}^1$, i.e., the number of times the projection passes through an arbitrary point of $\mathbb{S}^1$ in the counter clockwise sense minus the number of times the projection passes through the point in the clockwise sense. A closed curve (or a cycle in a graph) on a cylinder is *essential* if it has an odd winding number.

---

[1] The result is stated for x-monotonicity, the special case of level-planarity in which every level contains a single vertex. As we will see below, this special case is equivalent to the general case.

[2] Search for "cylindrical mirror anamorphoses" on the web for many cool pictures of this transformation.

With this, a *radial drawing* of G is a drawing of G on $\mathcal{C}$ such that the projection to I of every edge is injective (i.e., an edge does not "turn back") and for every pair of vertices $v_i < v_j$ we have $I(v_i) < I(v_j)$. We also speak of an edge being *radial* when it satisfies this condition. In a radial drawing an *upper edge* and *lower edge*, respectively, at v is an edge incident to v for which $\min(I(e)) = I(v)$ and $\max(I(e)) = I(v)$. A vertex v is a *sink* (*source*), if v has no upper (lower) edges. In order to avoid some inconvenient situation we assume that $I(G)$ is contained in the interior of I.

The *rotation* at a vertex in a drawing (on any surface) of a graph is the cyclic order of the ends of edges incident to the vertex in the drawing. The *rotation system* is the set of rotations at all the vertices in the drawing. In the case of radial drawings the *upper* (*lower*) rotation at a vertex v is the linear order of the end pieces of the upper (lower) edges in the rotation at v starting with the direction corresponding to the clockwise orientation of $\mathbb{S}^1$. The rotation at a vertex in a radial drawing is completely determined by its upper and lower rotation. The *rotation system* of a radial drawing is the set of the upper and lower rotations at all the vertices in the drawing.

In what follows we consider drawings of G in the (Euclidean) plane or on a cylinder. Thus, every embedded cycle of G is *separating*, i.e. its complement in the ambient space of G has two components. Also, the complement of any closed curve (possibly with self-crossings) can be *two-colored* so that connected regions each get one color and neighboring regions receive opposite colors.

A drawing of G is *even* if every two edges in the drawing cross an even number of times. After a sufficiently generic continuous deformation of a drawing of G the parity of the number of crossings between a pair of edges changes only when an edge passes through a vertex during the deformation. We call this event an *edge-vertex switch*. In particular, when an edge e passes through a vertex v the parity of the number of crossings between e and every edge incident to v changes.

3   Weak Hanani-Tutte for Radial Drawings

In this section, we prove Theorem 1, the weak Hanani-Tutte theorem for radial planarity. We claim that it will be sufficient to restrict ourselves to the special case in which every level of G contains a single vertex, an *ordered* graph.

Theorem 2. *Suppose the ordered graph G has an even radial drawing. Then G has a radial embedding with the same rotation system, and the winding parity of every cycle remains the same.*[3] *(All drawings respect the ordering.)*

The reduction of Theorem 1 to Theorem 2 is based on the same construction used in [9, Sect. 4.2] to reduce level-planarity to x-monotonicity: Suppose we are given an even radial drawing of a leveled graph G. If any level of G contains

---

[3] The claim about the invariance of the parity of the winding number of every cycle in Theorem 2 is a consequence of the preservation of the rotation system (a proof will be included in the journal version).

more than one vertex, we do the following: if any vertex at that level is a source or a sink, we add a crossing-free edge on the empty side of that vertex. We place the new vertex at a new level, close to the current level we are working on. We now slightly perturb all the vertices of the current level so no two vertices are at the same level (without moving them past any of the new vertices we created). We can do so, while keeping all edges radial, and without introducing any crossings. Since the new vertices we added are at unique levels, we only perform the perturbation on the original levels. Call the resulting ordered graph G'. By Theorem 2, G' has a radial embedding with the same rotation system, and the winding number of every cycle unchanged. We can now move all perturbed vertices back to their original levels, the additional edges we added ensure that this is always possible.

We will make use of the weak Hanani-Tutte theorem for x-monotone graphs due to Pach and Tóth [14], reproved in [9].

Theorem 3 (Pach, Tóth [14]). *Suppose that G can be drawn so that edges are x-monotone and every two edges cross an even number of times. Then there exists an embedding of G, in which the vertices are drawn as in the given drawing of G, the edges are x-monotone, and the rotation system is the same.*

Figure 1 shows an example of a graph for which x-monotonicity and radial planarity differ. A radial embedding of a graph not admitting an x-monotone embedding, must contain an essential cycle.

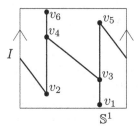 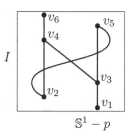

Fig. 1. An instance of an ordered graph that admits a radial embedding but does not admit an x-monotone embedding. The left and the right edge of the rectangle are identified in the left part of the figure.

3.1 Working with Even Radial Drawings

Given a connected graph G with a rotation system, we can define a *facial walk* purely combinatorially by following the edges according to the rotation system (see, for example, [10, Sect. 3.2.6]). A vertex and edge, respectively, is *incident* to a face if it appears on its facial walk. By gluing disks to each facial walk, one obtains a surface with the graph embedded in it. In case the surface is a sphere, each vertex occurs at most once on every facial walk of G if and only if G does not contain a cut-vertex. Any drawing of a graph G on an orientable surface defines a rotation system. For an even drawing of a connected graph G on the plane, the facial walks obtained from the rotation system correspond to a planar embedding. This is the weak Hanani-Tutte theorem [3,16]:

Theorem 4. *Let G be a graph. Suppose that G has an even drawing in the plane. Then G has an embedding in the plane with the same rotation system.*

Let v denote a vertex incident to a face f. Let $e = uv$ and $e' = vw$ denote a pair of consecutive edges on the facial walk of a face f in an embedding of a graph G. (Note that e and e' might be equal.) The edges e and e' define a *wedge* at v in f which is a portion of f in a small topological ϵ-neighborhood of v between e and e'. By Theorem 4, any even drawing of G corresponds to its embedding with the same rotation system. Hence, every consecutive pair of edges $e = uv$ and $e' = vw$ in the rotation at v in an even drawing defines a wedge in the corresponding embedding of G. Thus, a face f in an even drawing of G is given by the set of wedges corresponding to f in the embedding of G. From the set of wedges representing a face f in an even drawing we obtain the facial walk W_f of f, since every two consecutive edges on W_f define a wedge of f. By slightly abusing the notation we will denote the facial walk W_f by f.

We consider a radial embedding of a connected graph G. Let a *maximum* (*minimum*) of a face f in the radial embedding of G be the maximum (minimum) i such that v_i is incident to f. Let a *local maximum* (*minimum*) of a face f be a vertex v incident to f such that $v > u, w$ ($v < u, w$), where u and w are the predecessor and successor of v on the facial walk of f. An *outer face* in a radial embedding is a face containing $0 \times \mathbb{S}^1$ or $1 \times \mathbb{S}^1$. A face is an *inner face* if it is not an outer face.

Let f be a face in G given by an even radial drawing. The *boundary curve* of f is a closed curve traversing the facial walk of f in its close topological neighborhood, and at vertices passing through wedges in f. We naturally extend the notion of the winding number to a face defined as the winding number of its boundary curve. The *two-coloring* of f is the two-coloring of the complement of its boundary curve. A point in the complement of the boundary curve of f is in the *interior* (*exterior*) of f if it receives the same (opposite) color as a wedge in f when we two-color f. Note that in an even drawing all the wedges in a face have the same color. The outer face in an even radial drawing is a face having $0 \times \mathbb{S}^1$ or $1 \times \mathbb{S}^1$ in its interior. In an even radial drawing an outer face containing $0 \times \mathbb{S}^1$ ($1 \times \mathbb{S}^1$) is the *lower* (*upper*) outer face. If G has only one outer face f, f is simultaneously the lower and upper outer face. A face that is not an outer face is an *inner face*. For a closed non-essential curve on the cylinder we define the *interior* (*exterior*) as the union of the connected components in its complement whose color is different (the same) as the color of $0 \times \mathbb{S}^1$ in the two-coloring of its complement.

Let v be a local minimum or maximum of a face f. A wedge ω at v in f in a radial drawing is *convex* (*concave*) if the angle bounding ω in a small topological neighborhood of v is convex and concave. Let C be a cycle in a radial drawing of G. We consider C to be a subgraph of G whose drawing is inherited from that of G. Now, C represents two faces. It is easy to see that C is non-essential if and only if in the two-coloring of (the complement of) C the concave wedge of C at the minimum and maximum of C receive the same color, i.e., are in the same face defined by C. Indeed, the parity of the number of crossings between a "vertical" path in $\mathcal{C}$ joining $0 \times \mathbb{S}^1$ with $1 \times \mathbb{S}^1$, and C equals to the winding number of C by the definition of the winding number of C.

In a radial embedding of a connected graph G we observe that a face f is an outer face if and only if v_n is its maximum, or v_1 is its minimum, and the wedge in f at v_n or v_1 is concave. Thus, either G has two outer faces one of which is lower and one of which is upper, or G has exactly one outer face. In the former, the boundary curve of the outer faces is essential. In the latter G does not contain any essential cycle. The same is true also in even radial drawings.

Lemma 1. *In an even radial drawing of a connected graph G at most one face can have a concave wedge at its maximum or minimum, which necessarily happens only at v_n or v_1. Consequently, either G has two outer faces one of which is lower and one of which is upper, or G has exactly one outer face.*

The next lemma simplifies the type of faces we have to deal with.

Lemma 2. *Let G be a connected graph. Suppose that G has an even radial drawing. Then we can augment the drawing of G by adding edges so that the resulting drawing is still even and radial, every face of G has at most two local minima and two maxima, and each outer face has exactly one local minimum and one local maximum.*

Proofs of previous lemmata will be contained in the journal version of the paper.

3.2 Proof of Theorem 2

A connected component G_1 of G drawn radially is *essential* if it contains an essential cycle.

Given a graph G with a radial drawing, consider the augmentation of G that's guaranteed by Lemma 2. If Theorem 2 is true for the augmented graph, then it is clearly true for G as well. Thus, it suffices to prove Theorem 2 for graphs of the form constructed in Lemma 2. Even more, we can restrict ourselves to connected components of the graph, as the following lemma shows (proof left to the journal version).

Lemma 3. *A counterexample to Theorem 2 with the smallest number of vertices is not disconnected.*

Before we turn to the proof of Theorem 2 we present one more tool which allows us to clear an arbitrary edge in an even radial drawing of crossings while keeping the drawing radial and even. This is a slight extension of redrawing results we have used in previous papers [17, Fig. 3]. A proof will be included in the journal version.

Lemma 4. *Let G denote a graph given by an independently even radial drawing. Let e denote an arbitrary edge of G crossing every other edge in G evenly. We can redraw the edges crossing e inside $I(e) \times \mathbb{S}^1$ so that (i) e is crossing free, (ii) the resulting drawing remains even and radial, (iii) the rotation system and the points representing vertices are the same as in the given drawing, (iv) the parity of winding number remains the same for all cycles.*

Proof (of Theorem 2). For the sake of contradiction let G denote a minimal counterexample with respect to the number N of ordered pairs of vertices (u, v), $u < v$, of G, where u is a source and v is a sink. By Lemma 3, we assume that G is connected and by Lemma 2 we assume that each face of G contains at most two local minima and each outer face at most one local minimum. We construct a radial embedding $\mathcal{D}$ of G inductively as $\mathcal{D}_n$, where every $\mathcal{D}_i$, $1 \le i \le n$, is an even radial drawing of a graph G_i obtained from G by subdividing certain edges in their interior without altering its radial planarity such that $\mathcal{D}_i$ is crossing free in $[0, I(v_i)] \times \mathbb{S}^1$. Since throughout the proof we keep the rotation system of G unchanged (after suppressing the subdividing vertices in G_i), this contradicts the choice of G.

We proceed by constructing compatible cyclic orders $\mathcal{O}_i$ given by the order of appearance of points in $\mathcal{D}_i \cap (I(v_i) \times \mathbb{S}^1)$ along $I(v_i) \times \mathbb{S}^1$. The elements, i.e., points, in $\mathcal{O}_i$'s are denoted by the objects of G_i, i.e., edges and vertices, they belong to. By "compatible" we mean that for two consecutive orders $\mathcal{O}_i$ and $\mathcal{O}_{i+1}$ (after suppressing the subdividing vertices in G_i) the order $\mathcal{O}_{i+1}$ is obtained from $\mathcal{O}_i$ using an auxiliary order $\mathcal{O}_i'$ as follows. We obtain $\mathcal{O}_i'$ from $\mathcal{O}_i$ by replacing v_i with its upper edges. We obtain $\mathcal{O}_{i+1}$ from $\mathcal{O}_i'$ by replacing the lower edges of v_{i+1} appearing consecutively in $\mathcal{O}_i'$ with v_{i+1} if v_{i+1} is not a source. Otherwise, we obtain $\mathcal{O}_{i+1}$ from $\mathcal{O}_i'$ by placing v_{i+1} between two edges bounding the face that contains the concave wedge at v_{i+1}.

In the base case there is nothing to prove. We just put $G_1 := G$. For $i+1 > 1$, we distinguish two cases depending on whether v_{i+1} is a source or not. We work in $\mathcal{D}_i$ from the induction hypothesis. We subdivide in G_i every edge e' whose projection $I(e')$ contains v_i in its interior at a point $p_{e'}$ whose projection $I(p_{e'}) \in (I(v_i), I(v_i)+\epsilon)$, where ϵ is sufficiently small such that $[I(v_i), I(v_i)+\epsilon] \times \mathbb{S}^1$ is free of edge crossings. Let G_{i+1} denote the resulting graph. Clearly, such subdivisions have no effect on the embeddability and do not alter the value N. The order $\mathcal{O}_i'$ is obtained by taking $\mathcal{D}_i \cap ((I(v_i) + \epsilon) \times \mathbb{S}^1)$. There are two cases to distinguish depending on whether v_{i+1} is a source.

The vertex v_{i+1} is not a source. First, if v_{i+1} is not a source let e denote a lower edge at v_{i+1}, see Fig. 2. In $\mathcal{D}_i$ we clear e of crossings by using Lemma 4.

Now, if the lower edges at v_{i+1} do not appear consecutively in $\mathcal{O}_i'$ we necessarily obtain a pair of edges crossing an odd number of times (contradiction). If e is crossing free, every pair of edges cross in $[I(v_i), I(v_{i+1})] \times \mathbb{S}^1$ an even number of times, if and only if $\mathcal{D}_i \cap (I(v_{i+1}) \times \mathbb{S}^1)$ yields the same circular order as $\mathcal{O}_i'$. We continuously deform the drawing in $(I(v_i) - \epsilon, I(v_{i+1}) + \epsilon) \times \mathbb{S}^1$ so that the order of the edges at $I(v_{i+1}) \times \mathbb{S}^1$ is the same as in $\mathcal{O}_i'$ while keeping e crossing free.

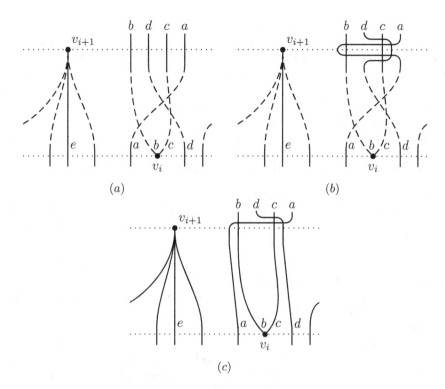

Fig. 2. Case: v_{i+1} is not a source. (a) Drawing of $[I(v_i), I(v_{i+1})] \times \mathbb{S}^1$ (left and right sides are identified), dashed curves may cross. (b) Edges crossing $I(v_{i+1}) \times \mathbb{S}^1$ are deformed in a small neighborhood of $I(v_{i+1}) \times \mathbb{S}^1$ so their cyclic order matches $\mathcal{O}_i$. (c) Edges are redrawn using geodesics.

Thus, in particular we do not perform any edge-vertex switch during the deformation. Now, every pair of edges in $[I(v_i), I(v_{i+1})] \times \mathbb{S}^1$ cross an even number of times. Thus, we can replace the pieces of edges in $[I(v_i), I(v_{i+1})] \times \mathbb{S}^1$ by geodesics connecting the same ends, which yields a desired $\mathcal{D}_{i+1}$. Note that the lower rotation at v_{i+1} is preserved since every pair of edges cross in $[I(v_i), I(v_{i+1})] \times \mathbb{S}^1$ an even number of times.

The vertex v_{i+1} is a source. Let f denote the face in which v_{i+1} has the concave wedge. Let v denote the local maximum of f which is not the maximum of f. If v does not exists it follows that the maximum of f is a cut-vertex, and we let v be this cut-vertex. Let u denote the minimum of f. Let P denote the path in the walk f from u to v not containing v_{i+1}. Let Q denote the path in the walk f from v_{i+1} to v not containing u. In $\mathcal{D}_i$ we clear P of crossings by a repeated application of Lemma 4.

Unlike the previous case, we cannot, in general, deform the edges in $(I(v_i) - \epsilon, I(v_{i+1}) + \epsilon) \times \mathbb{S}^1$ while keeping P crossing free to obtain the order of edges at $I(v_{i+1}) \times \mathbb{S}^1$ compatible with $\mathcal{O}'_i$. This might require an edge-vertex switch with v_{i+1} yielding a pair of edges crossing an odd number of times. Let S denote the cyclic interval between v_{i+1} and P in $(I(v_{i+1}) \times \mathbb{S}^1)$ corresponding to the interior of f determined by the order of end pieces of P and Q in the upper rotation at v. The problem arises if the interior of S is crossed by edges. Thus, we first have to clean S of edges. We proceed by reducing G_{i+1} so that in the reduced instance no edge crosses S.

From now on, an edge e is always an edge such that $I(e)$ contains $I(v_{i+1})$ in its interior. Let G_e denote the connected component of $G_{i+1}[\{v_{i+1}, \ldots v_n\}] \cup \{e\}$ containing e, where $G_{i+1}[\{v_{i+1}, \ldots v_n\}]$ denotes the induced subgraph of G on $\{v_{i+1}, \ldots v_n\}$. An edge e *sprouts* beyond v if $\max(I(G_e)) \geq I(v)$.

The crucial observation is, that after we cleaned P of crossings,

(i) S is crossed only by edges e not sprouting beyond v; and
(ii) the union G_i' of subgraphs G_e, for e's not sprouting beyond v, does not contain an essential cycle. The smaller vertices of e's are *seeds* of G_i'. (We recall that after the subdivisions $I(G_e) \subseteq [I(v_i), 1]$.)

Since the interior of $(I(P) \times \mathbb{S}^1) \setminus P$ is homeomorphic to the plane, (ii) is obvious once we establish (i). To prove (i), consider a closed curve C obtained by concatenating the part of f between v_{i+1} and v (not containing u); the part of P between v and $I(v_{i+1}) \times \mathbb{S}^1$; and the part of $I(v_{i+1}) \times \mathbb{S}^1$ crossing the interior of f, i.e., not crossing P. Claim (i) follows since e crosses C an odd number of times, and hence, the only way for e to sprout beyond v is to have an edge in G_e splitting a wedge in the interior of f which is impossible.

If G_i' is empty, S is not crossed by an edge, and hence, the order of edges in $\mathcal{O}_{i+1}$ is inherited from $\mathcal{O}_i'$. The drawing $\mathcal{D}_{i+1}$ is obtained analogously as in the case when v_{i+1} was not a source. Otherwise, consider a connected component H of G_i' given by the even radial drawing inherited from the drawing $\mathcal{D}_i$ of G_{i+1}. Let $S_H \subseteq \mathbb{S}^1$ denote the smallest connected subset such that the set $\mathcal{C}_H = [I(v_i), I(v_i) + \epsilon] \times S_H$ contains all seeds of H and does not intersect P. Let H' denote the union of H with all G_e for e intersecting $\mathcal{C}_H$. By evenness of the drawing $I(H') \subseteq [I(v_i), \max(I(H))]$, since the curve $(I(v_i) + \epsilon) \times S_H$ and a part of the outer face of H in $[I(v_i) + \epsilon, 1] \times \mathbb{S}^1$ disjoint from P define a (non-essential) cycle having the non-seed vertices of $V(H') \setminus V(H)$ in its interior. Let z be the vertex, for which $I(z) = \max(I(H'))$. Then z has a concave wedge inside a face f' for which $\max(f') > \max(I(H'))$. Indeed, f' is not an outer face by Lemma 1, since f' has a local maximum whose corresponding wedge is concave and different from v_n. Thus, the maximum of f' is different from z which is a local maximum of f'.

We reduce G_{i+1} by removing the edges of H' from it, and identifying the seeds of H' thereby replacing them by a single vertex s_i. In $\mathcal{D}_i$, this corresponds to contracting $\mathcal{C}_H$ to a point while keeping the drawing radial and even. Let G' denote the resulting graph. Note that N necessarily decreased in G'. Indeed, at least the contribution of the source-sink pair (v_{i+1}, z) towards N decreased. Since G_{i+1} was a minimal counterexample, G' has a desired radial embedding. Similarly, let H'' denote the graph obtained from H' by identifying its seeds and denoting the resulting vertex by s_i'. By the weak variant of the Hanani-Tutte theorem for monotone drawings, Theorem 3, we obtain a radial embedding of

H'' without an essential cycle. We combine the embeddings of H'' and G' by identifying s_i and s_i', and uncontracting s_i. We have room to accommodate H'' inside f', since $\max(f') > \max(I(H'))$. ∎

4 Algorithm

Theorem 1 reduces the problem of radial planarity testing with a fixed rotation system to a system of linear equations over $\mathbb{Z}_2$. For planarity testing, systems like this were first constructed by Wu and Tutte [19, Sect. 1.4.2].

Unlike the x-monotone case, two drawings of an edge e with fixed endpoints may not be obtainable from each other by a continuous deformation of e, while keeping e radial: two radial drawings of e may also differ by a number of (Dehn) twists. The system has a variable $x_{e,v}$ for every edge-vertex switch (e,v) such that $I(v) \in I(e)$, and x_e for every edge twist. Since we work in $\mathbb{Z}_2$ orientation of a twist does not matter. We consider a twist of $e = uv$, $u < v$, as being performed very close to v, i.e., the twist is carried out by removing a small portion P_e of e such that we have $I(w) \notin I(P_e)$, for all vertices w, and reconnecting the severed pieces of e by a curve intersecting every edge e', s.t. $I(P_e) \subset I(e')$, exactly once. Observe that with respect to the parity of crossings between edges performing a twist close to v equals performing an edge-vertex switch of e with all the vertices $w < v$ (even those w for which $w < u$). Hence, any twist of e keeping e radial can be simulated by a twist of e very close to v and a set of edge-vertex switches of e with certain vertices w, for which $u < w < v$.

By the previous paragraph a linear system for testing radial planarity with the fixed rotation system looks as follows. Given an arbitrary radial drawing of G we denote by $cr(e,f)$ the parity of the number of crossings between e and f. In the linear system, for each pair of independent edges $(e,f) = (uv, wz)$, where $u < v$, $w < z$, $u < w$, and $w < v$, we have $x_{e,w} + x_{e,z} + x_f = cr(e,f)$ if $z < v$, and $x_{e,w} + x_{f,v} + x_e = cr(e,f)$ if $z > v$. For a pair of edges $(e,f) = (uv, uw)$, where $u < v < w$, we have $x_{f,v} + x_e = cr(e,f)$ and for a pair of edges $(e,f) = (uv, uw)$, where $u > v > w$, we have $x_{f,v} + x_e + x_f = cr(e,f)$. For a pair of edges $(e,f) = (uv, uv)$, where $u < v$, we have $x_e + x_f = cr(e,f)$.

References

1. Bachmaier, C., Brandenburg, F.J., Forster, M.: Radial level planarity testing and embedding in linear time. J. Graph Algorithms Appl. **9**, 2005 (2005)
2. Booth, K.S., Lueker, G.S.: Testing for the consecutive ones property, interval graphs, and graph planarity using PQ-tree algorithms. J. Comput. Syst. Sci. **13**(3), 335–379 (1976)
3. Cairns, G., Nikolayevsky, Y.: Bounds for generalized thrackles. Discrete Comput. Geom. **23**(2), 191–206 (2000)
4. Chimani, M., Zeranski, R.: Upward planarity testing: a computational study. In: Wismath, S., Wolff, A. (eds.) GD 2013. LNCS, vol. 8242, pp. 13–24. Springer, Heidelberg (2013)

5. Chojnacki, C., Hanani, H.: Über wesentlich unplättbare Kurven im dreidimensionalen Raume. Fundamenta Mathematicae **23**, 135–142 (1934)
6. Di Battista, G., Nardelli, E.: Hierarchies and planarity theory. IEEE Trans. Syst. Man Cybern. **18**(6), 1035–1046 (1989)
7. Di Giacomo, E., Didimo, W., Liotta, G.: Spine and radial drawings, chapter 8. In: Roberto, T. (ed.) Handbook of Graph Drawing and Visualization. Discrete Mathematics and Its Applications. Chapman and Hall/CRC, Boca Raton (2013)
8. Fulek, R., Kynčl, J., Malinović, I., Pálvölgyi, D.: Clustered planarity testing revisited. In: Duncan, C., Symvonis, A. (eds.) GD 2014. LNCS, vol. 8871, pp. 428–439. Springer, Heidelberg (2014)
9. Fulek, R., Pelsmajer, M., Schaefer, M., Štefankovič, D.: Hanani-Tutte, monotone drawings, and level-planarity. In: Pach, J. (ed.) Thirty Essays on Geometric Graph Theory, pp. 263–287. Springer, New York (2013)
10. Gross, J.L., Tucker, T.W.: Topological Graph Theory. Dover Publications Inc., Mineola (2001). Reprint of the 1987 original
11. Gutwenger, C., Mutzel, P., Schaefer, M.: Practical experience with Hanani-Tutte for testing *c*-planarity. In: McGeoch, C.C., Meyer, U. (eds.) 2014 Proceedings of the Sixteenth Workshop on Algorithm Engineering and Experiments (ALENEX), pp. 86–97. SIAM, Portland (2014)
12. Jünger, M., Leipert, S.: Level planar embedding in linear time. J. Graph Algorithms Appl. **6**(1), 72–81 (2002)
13. Northway, M.L.: A method for depicting social relationships obtained by sociometric testing. Sociometry **3**(2), 144–150 (1940)
14. Pach, J., Tóth, G.: Monotone drawings of planar graphs. J. Graph Theory **46**(1), 39–47 (2004). Updated version: arXiv:1101.0967
15. Pelsmajer, M.J., Schaefer, M., Stasi, D.: Strong Hanani-Tutte on the projective plane. SIAM J. Discrete Math. **23**(3), 1317–1323 (2009)
16. Pelsmajer, M.J., Schaefer, M., Štefankovič, D.: Removing even crossings. J. Combin. Theor. Ser. B **97**(4), 489–500 (2007)
17. Pelsmajer, M.J., Schaefer, M., Štefankovič, D.: Removing even crossings on surfaces. Eur. J. Comb. **30**(7), 1704–1717 (2009)
18. Schaefer, M.: Toward a theory of planarity: Hanani-Tutte and planarity variants. J. Graph Algortihms Appl. **17**(4), 367–440 (2013)
19. Schaefer, M.: Hanani-Tutte and related results. In: Bárány, I., Böröczky, K.J., Tóth, G.F., Pach, J. (eds.) A Tribute to László Fejes Tóth. Bolyai Society Mathematical Studies, vol. 24, pp. 259–299. Springer, Berlin (2014)
20. Tutte, W.T.: Toward a theory of crossing numbers. J. Comb. Theor. **8**, 45–53 (1970)
21. Wiedemann, D.H.: Solving sparse linear equations over finite fields. IEEE Trans. Inf. Theor. **32**(1), 54–62 (1986)

Experiments

Drawing Planar Cubic 3-Connected Graphs with Few Segments: Algorithms and Experiments

Alexander Igamberdiev[1], Wouter Meulemans[2](✉), and André Schulz[1](✉)

[1] LG Theoretische Informatik, FernUniversität in Hagen, Hagen, Germany
{alexander.igamberdiev,andre.schulz}@fernuni-hagen.de
[2] GiCentre, City University London, London, UK
wouter.meulemans@city.ac.uk

Abstract. A drawing of a graph can be understood as an arrangement of geometric objects. In the most natural setting the arrangement is formed by straight-line segments. Every cubic planar 3-connected graph with n vertices has such a drawing with only $n/2+3$ segments, matching the lower bound. This result is due to Mondal et al. [J. of Comb. Opt., 25], who gave an algorithm for constructing such drawings.

We introduce two new algorithms that also produce drawings with $n/2+3$ segments. One algorithm is based on a sequence of dual edge contractions, the other is based on a recursion of nested cycles. We also show a flaw in the algorithm of Mondal et al. and present a fix for it. We then compare the performance of these three algorithms by measuring angular resolution, edge length and face aspect ratio of the constructed drawings. We observe that the corrected algorithm of Mondal et al. mostly outperforms the other algorithms, especially in terms of angular resolution. However, the new algorithms perform better in terms of edge length and minimal face aspect ratio.

1 Introduction

To assess the quality of network visualizations, many criteria have been investigated, such as crossing minimization, bend minimization and angular resolution (see [9] for an overview). The structural complexity of a graph is often measured in terms of its number of vertices or edges. This, however, does not necessarily correspond to its *cognitive load* (mental effort needed to interpret a drawing). Bends and crossings increase the cognitive load, making it harder to interpret a graph visualization, and should be avoided.

We consider the following measure of *visual complexity* for planar graphs [8]: the number of basic geometric objects that are needed to realize the drawing. For example, if a path in the graph is placed along a line, then we do not need one line segment for each edge in this path; one line segment can represent the entire path. In contrast to bends and crossings, which *increase* the cognitive

---

Funded by the German Research Foundation (DFG) under grant SCHU 2458/4-1.

E. Di Giacomo and A. Lubiw (Eds.): GD 2015, LNCS 9411, pp. 113–124, 2015.
DOI: 10.1007/978-3-319-27261-0_10

load, this definition of visual complexity aims to measure a *reduction* in cognitive load in comparison to the structural complexity. The basic geometric shapes are typically straight-line segments or circular arcs. Upper and lower bounds on the necessary visual complexity of various graph classes are known [2,3,8]. A lower bound for any graph is $N/2$, where N is the number of odd-degree vertices: at least one geometric object must have its endpoint at such a vertex. Computing the optimal visual complexity of line-segment drawings is NP-hard [4].

We consider line-segment drawings for *planar cubic 3-connected graphs*; unless mentioned otherwise, "graph" is used to refer to a graph of this class. Any plane drawing has at least three vertices of the same face on its convex hull: such a vertex is the endpoint of the line segment for each incident edge. Thus, we obtain a lower bound of $n/2+3$ line segments, as n is even. Dujmović gave an algorithm for drawing general planar graphs with low visual complexity [2]. This algorithm will draw a cubic planar graph with $n+2$ line segments. Mondal et al. [7] improve this by giving an algorithm that uses $n/2+4$ segments. Moreover this algorithms places the vertices on a $(n/2+1) \times (n/2+1)$ grid and uses only 6 different slopes. A variant of the algorithm is also suggested, one that does not place the vertices on a grid and uses 7 distinct slopes, but attains a visual complexity of $n/2+3$. The presentation of Mondal et al. contains a flaw, but it can be fixed as discussed in Sect. 4.

To compute a plane drawing matching the lower bound, we are given (or pick) three convex hull vertices; these are referred to as the *suspension vertices*. For all other *internal vertices*, we decide which two incident edges lie on the same line segment, that is, which of the three angles is flat. Hence, this corresponds to a *flat-angle assignment*; we refer to plane drawings that match the lower bound as *flat-angle drawings*. Note that any face in a flat-angle drawing is nonstrictly convex. Aerts and Felsner [1] describe conditions for the stretchability of flat-angle assignments. From a stretchable assignment, a layout can be obtained by solving a system of harmonic (linear) equations with arbitrary edge weights, very similar to the directed version of Tutte's barycentric embedding as presented by Haas et al. [6]. How to efficiently compute stretchable flat-angle assignments remains an open problem.

Contributions. We present two different new $O(n^2)$-time algorithms (Sects. 2 and 3) to construct a plane drawing with $n/2+3$ segments for $n \geq 6$, matching the lower bound. From the constructed drawings, a flat-angle assignment is derived, which is then used to set up a system of harmonic equations [1]. By solving the system using uniform edge weights we can redraw the layouts to (possibly) increase their visual appeal. To the best of our knowledge the new algorithms present novel methods to incrementally build up cubic planar 3-connected graphs by simple and local modifications. These construction sequences might also find applications outside of our applications.

We review the algorithm of Mondal et al. and discuss cases where it might produce degenerate drawings. We then present a fix for these problematic cases. This leaves us with three algorithms that produce drawings of cubic planar 3-connected graphs with low visual complexity; see Fig. 1. We run several exper-

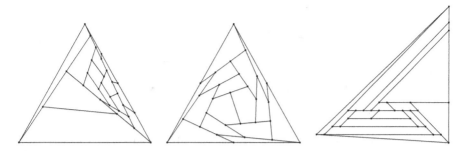

Fig. 1. Result of the various algorithms for the same graph and outer face. (left) Deconstruction algorithm. (middle) Windmill algorithm. (right) Mondal algorithm.

iments and evaluate the performance of these algorithms by measuring geometric features of the produced drawings. In particular, we measure angular resolution, edge length, and face aspect ratio. We use two data sets for our experiments. For the first data set we sample over the set of all cubic planar 3-connected graphs with 24 to 30 vertices. The second data set is given by the set of 146 popular graphs with at most 30 vertices from the Wolfram graph database[1].

2   The Deconstruction Algorithm

For this algorithm we define an operation called *edge insertion*[2]: pick two edges that belong to one face, subdivide both edges and add a new edge between the new degree-2 vertices while preserving planarity. It is folklore that every cubic 3-connected graph can be obtained from K_4 by a sequence of edge insertions (e.g., [5, page 243]). For our purpose we need a slightly stronger version (proven in the full version): any cubic planar 3-connected graph other than K_4 can be constructed by a sequence of edge insertions from the triangular prism, while not adding new edges in a given outer face (though outer-face edges may be subdivided).

An edge whose removal (understood as a reverse edge insertion) maintains planarity, 3-regularity, 3-connectivity, and a chosen outer face is called a *good edge*. We compute a construction order by repeatedly removing a good edge (a good edge always exists as proven in the full version). This procedure always finishes on a triangular prism which has a trivial flat-angle drawing for any outer face. Note that the construction sequence is not necessarily unique. For the later analysis (Sect. 5), we distinguish three different strategies on how to select the removed edge from the set of all good edges in every step:

(R) We select the edge randomly from the set of all good edges.
(S) We select a good edge with minimal sum of the degrees of its incident faces.
(L) Analogous to (S), we select a good edge with maximal sum of degrees.

---

[1] http://reference.wolfram.com/language/ref/GraphData.html.
[2] This is *not* the graph-theoretic notion of edge insertion.

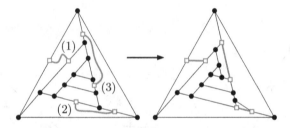

Fig. 2. Edge insertion that connects (1) noncollinear edges of a face, (2) collinear edges separated by one vertex, and (3) collinear edges separated by two or more vertices. In case 2 and 3, we reassign the flat angle at the first and/or last separating vertex.

We remark that there is no basis to suggest that the strategies (S) or (L) might perform particularly well: we study these strategies primarily to have a more structured procedure against which we can compare the randomized strategy.

Once we have obtained the construction order, we can reconstruct the original graph from the triangular prism with a sequence of edge insertions maintaining a flat-angle drawing (see Fig. 2). When inserting edges, we may have different possibilities how to update the flat-angle assignment. Depending on our strategy we may obtain different drawings. If an edge insertion "connects" two edges that are not aligned, we have an obvious way how to add the new edge: we pick a subdivision point on each of the edges and add the new edge as a straight-line segment connecting these points. If the two edges are aligned (part of a common segment ℓ), we need to modify the existing drawing. Let u and v be the new vertices that we introduce and let $s_1, \ldots, s_k$ be the vertices in between u and v on ℓ; see Fig. 3(a). Since the graph after adding e is planar, all segments starting at s_i have to leave ℓ on the same side. We first draw the new edge (u, v) on top of ℓ such that v coincides with s_k. To repair the degeneracy, we tilt the old part of ℓ that was running between u and s_k as done in Fig. 3(b). Here we let s_k "slide" on its segment that was not part of ℓ. For $k \geq 2$ we have also the following alternative how to insert (u, v): we draw a segment parallel to ℓ that runs between the segments starting at s_1 and s_k. We place all endpoints s_i on this new segment without changing any slopes of the old segments. We now take the old vertex s_1 as u, and the old vertex s_k as v as depicted in Fig. 3(c). We refer to the latter strategy as the *alternate insertion operation*.

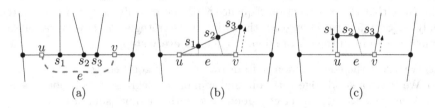

Fig. 3. (a) Inserting edge $e = (u, v)$ with its endpoints on the same side of a face (a). (b) The standard insertion. (c) An alternative strategy.

With three different strategies (R, S and L) and an alternative insertion operation (ALT), we have six variants of the Deconstruction algorithm, referred to as DEC-R, DEC-R-ALT, DEC-S, DEC-S-ALT, DEC-L and DEC-L-ALT.

3 The Windmill Algorithm

The Windmill algorithm computes a flat-angle drawing, working its way inward from the outer face, until all vertices have been processed. It does so recursively, using as parameter a simple cycle C in the graph. It assumes that C is drawn as a nonstrictly convex polygon. Its convex corners correspond to suspension vertices or vertices having an edge outside C; any flat vertex has an edge inside C. Initially, C is the outer face, drawn as an equilateral triangle with the suspension vertices as corners (Fig. 4(a)). Based on the cyclic sequence F of faces along the inside of C, a recursive step for cycle C is done using the first of the cases below:

1. If at most one vertex lies inside C, we draw all chords as line segments. The one vertex (if present) is positioned to lie on a line segment between two of its neighbors. See Fig. 4(b→c,e→f).
2. If a face occurs more than once in F, we draw its paths inside C as line segments and recurse on a subcycle for each path. See Fig. 4(c→d).
3. If two faces share an edge, but are not consecutive in F, we draw three line segments to represent the paths inside C along the two faces and recurse on the two subcycles created. See Fig. 4(a→b).
4. Otherwise, we create a windmill pattern with the sequence of faces along C. We recurse on the cycle inside the windmill. See Fig. 4(d→e).

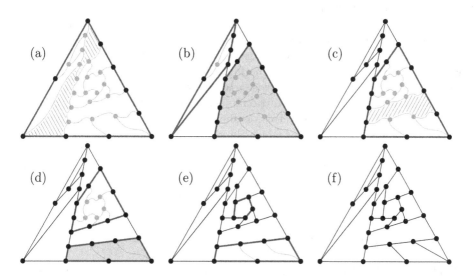

Fig. 4. The Windmill algorithm. Cycles are drawn thick; unshaded cycles are processed in the next step. (a) Initial call. (b→ ... →e) Consecutive states. (f) Final result. (e→f) Two cycles are processed. (a,c) Hashures indicate faces relevant for case 3 and 2.

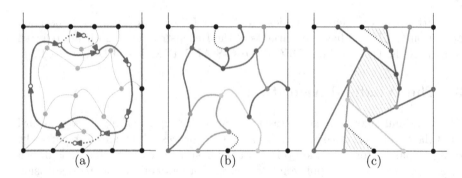

Fig. 5. (a) The dual restricted to F. Shortest cycle is given with solid lines. (b) Decomposing the face boundaries for the windmill structure. (c) Drawing the windmill pattern and the bypassed components as triangles. Three cycles are recursed on (hashures).

There is a subtlety for case 3: the faces must lie on different sides of the polygon for C. Otherwise, the condition on C described above cannot be maintained in a plane drawing. Therefore, we use case 4 to handle such a pattern using additional recursive calls. Below, we sketch how this works.

To construct a windmill, we proceed as follows (see Fig. 5). First, we consider the dual of the given graph, restricted to the vertices that are dual to the faces in F. Ideally, this is a simple cycle. However, two nonconsecutive faces of F that are adjacent (and, because of case 3, lie on the same side of C) cause a chord in this cycle. We find the shortest cycle in this restricted dual to create the windmill (Fig. 5(a)). The face boundaries of the faces on this cycle are decomposed as to provide the basic windmill structure (Fig. 5(b)). This bypasses any components separated by a chord in the dual. These are inserted as triangles at the correct place and recursed on as well, after drawing the windmill pattern (Fig. 5(c)).

Windmills can be created in a clockwise or counterclockwise direction. To decide, we provide two strategies. The first is to always choose the same direction; the other is to alternate clockwise and counterclockwise, depending on the recursion depth. We refer to these variants as WIN and WIN-ALT respectively.

Crucial to the proof of correctness is showing that any cycle C is nonstrictly convex and any vertex on C, for which the third edge is not drawn yet, is either a suspension vertex or its other two edges (part of C) are drawn collinearly: in either case, we need not worry about aligning the undrawn edge with another to obtain minimal visual complexity (since each nonsuspension vertex must have exactly two aligned edges). For full details and proofs, we refer to the full version.

4   The Mondal Algorithm

Mondal et al. [7] describe two linear-time algorithms for drawing cubic planar 3-connected graphs: one results in a grid drawing with $n/2+4$ line segments; the other attains minimal visual complexity but does not produce a grid drawing. Both algorithms introduce the vertices as given by a canonical order.

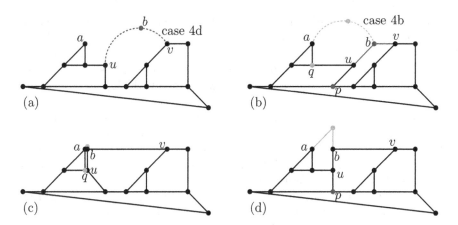

Fig. 6. (a) State before adding b between u and v. (b) After adding b, before adding a vertex between a and b. (c) Rotating about q as described in [7] results in a nonplane drawing. (d) Rotating about p as described here yields a plane result.

We observed that the grid algorithm as described by Mondal et al. [7] is flawed. The example in Fig. 6 illustrates the problem. When adding a chain of vertices from u to v (case 4d in [7], see Fig. 6(a–b)), the vertex at u is "rotated" to give it an incident edge with slope 1. In the next step, we may need to rotate backward to give vertex b an incident edge with slope ∞ (case 4b in [7], see Fig. 6(b–c)). However, the point computed to rotate about is erroneous: it is point q. This causes u to be placed on top of q, resulting in a nonplane drawing.

To resolve this issue, we suggest the following procedure for determining the correct pivot point. For case 4b, we walk downward along the slope 1 edges, until we find a pivot vertex p that has either two slope 0 edges or a downward edge with slope 1 and downward edge with slope ∞. In this case, every vertex w along the path is moved $\ell(w)$ positions to the left, where $\ell(w)$ is the vertical distance between w and pivot vertex p. We refer to this as a *left-rotation*. The correct result for the counter example is given in Fig. 6(d).

Analogously for case 4d, we walk downward along the slope ∞ edges, until we find a pivot vertex p that has either two slope 0 edges or a downward edge with slope 1 and downward edge with slope ∞. In this case, every vertex w along the path is moved $r(w)$ positions to the right, where $r(w)$ is the vertical distance between w and pivot vertex p. We refer to this as a *right-rotation*.

To prove that left- and right-rotations maintain a plane drawing, we must show that for every degree-3 vertex along the path to the pivot vertex, any horizontal edge in the direction of the rotation has sufficient length. This is captured by the invariant below. To simplify notation, we define $r(w)$ and $\ell(w)$ to be 0, if w is not on a path that may be right-rotated or left-rotated respectively.

Invariant 1. *Consider an edge $e = (u, v)$ with slope 0 and let u and v be its left and right vertex respectively. The length of e is at least $1 + r(u) + \ell(v)$.*

Observe that $r(u)$ or $\ell(v)$ is nonzero only in situations where u has been left-rotated or v has been right-rotated. To fully prove this statement is out of scope for this paper. Also, note that this is the invariant for the grid algorithm. For the minimal-complexity algorithm, we must multiply the values of $r(\cdot)$ and $\ell(\cdot)$ by two, and observe that rotations are not performed with slope ∞ edges: their role is taken by slope -1 edges.

Moreover, we observe that the Mondal algorithm achieving minimal visual complexity can be easily adapted to lie fully on a grid and use only six slopes as well. To this end, we need to do only the following: whenever the bottom point is moved to the right, it is moved downwards for an equal distance. This ensures that its incident edge maintains a slope of -1.

Thus, we have two variants of the Mondal algorithms, both on a grid and with only 6 slopes for its edges: one uses $n/2 + 4$ line segments, but draws on a smaller grid than the second algorithm that uses only $n/2 + 3$ line segments. We refer to these as MON-GRID and MON-MIN respectively.

5   Experiments

We have three different algorithms (each with its own variants) to draw planar cubic 3-connected graphs using only $n/2+3$ line segments. The drawings (Fig. 1) are obviously different, but—as the visual complexity is the same—we need criteria to further assess the overall quality. In this section we discuss experimental results comparing the 10 algorithm variants described in the previous sections.

5.1   Graphs

We generated all planar cubic 3-connected graphs with 24, 26, 28 and 30 vertices, using *plantri*[3]. From each batch we sampled 500 graphs uniformly at random, resulting in a total of 2000 graphs. The Wolfram data set shows roughly the same patterns can be observed as for the random data set. As the graphs in this data set are typically smaller, some differences arise.

5.2   Measures

We use the following three measures to quantify the quality of a graph layout.

Angular Resolution. At each internal vertex in the graph, we measure the smallest angle as an indicator of angular resolution. Since one angle is always π, the best angular resolution is $\pi/2$. Angular resolution measures how easily discernible the incident edges are. A high value indicates a good angular resolution.

Edge Length. We measure all edge lengths in the graph, normalized to a percentage of the diagonal of the smallest enclosing axis-aligned square.

---

[3] http://cs.anu.edu.au/~bdm/plantri/.

Though edge lengths should neither be too short nor too long, we in particular look at avoiding long edges[4]: we consider lower values for edge length to be better.

Aspect Ratio. For each face, we measure the aspect ratio of the smallest enclosing (not necessarily axis-aligned) rectangle. To compute this ratio, we divide the length of its shorter side by the length of its longer side, yielding a value between 0 and 1. High values thus indicate a good aspect ratio. This is a simple indicator of fatness, as all faces are convex.

Measuring Procedure. For each graph, we run each algorithm using each possible face of the graph as an outer face. For each measure, we compute both the average value over all elements (vertices, edges, faces) as well as the worst value. The worst value is the minimum value for angular resolution and face aspect ratio, and the maximum value for the edge length. For both the average and worst value, we compute the average over all drawings for a particular graph, i.e., what may be expected for that graph if we had chosen an outer face uniformly at random. Thus, we have six measures in total.

5.3 Algorithm Comparison

Figure 7 shows the measured results for all graphs in the data set, summarized as a box plot. For the DEC algorithms, only the ALT variants are shown, as the results of the other variants are very similar.

Angular Resolution. The MON algorithms clearly perform better than the WIN algorithms, which in turn outperform the DEC algorithms. This was to be expected due to the fixed slopes used in the MON algorithms. We observed that for the Wolfram data set, the angular resolution tends to increase for the WIN and DEC algorithms. However, for the MON algorithms, there in fact seems to be a slight decrease.

Edge Length. The worst-case values show that the MON algorithms perform worst, and the WIN algorithms perform best; average edge length shows that MON is slightly behind the WIN and DEC algorithms. Though statistically significant (later in this section), the differences are only minor. The maximum edge length for the WIN and DEC algorithms is lower due to its placement in an equilateral triangle and the possibility of having additional vertices on all sides of this triangle; the MON algorithms always have a long edge, close to the diagonal of the drawing. For the MON-MIN algorithm, this worst-case edge length is smaller than for MON-GRID. This is caused by our modification which moves one point downward, thereby increasing the grid size.

Face Aspect Ratio. We see that the DEC algorithms are outperformed by the other algorithms in terms of average ratio. MON-MIN outperforms MON-GRID and the WIN algorithms. However, looking at the minimal face aspect ratio of a drawing, we see that DEC outperforms the MON algorithms, and MON-GRID

---

[4] Informal investigation of minimal edge lengths suggested only tiny differences, though MON-GRID was slightly ahead of the other algorithms.

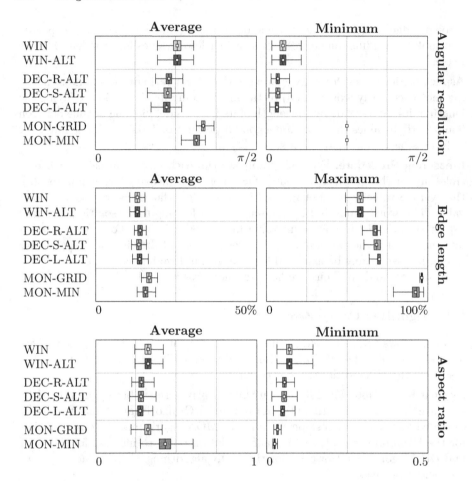

Fig. 7. Box plot of the measured results. For length, lower values indicate better drawings; for the other measurements, higher values indicate better drawings.

is actually slightly ahead of MON-MIN. For the Wolfram data set, containing smaller graphs ($n = 16.5$ on average), we observe that the average face aspect ratio of the MON algorithms decreases: MON-MIN is in line with the DEC algorithms and the MON-GRID algorithm has lost its lead on the WIN algorithms.

We conclude from the above that the WIN algorithms generally outperform DEC algorithms. Between the WIN and MON algorithms, there is no clear agreement between the measures: the MON algorithms perform very well in angular resolution, but worse in edge length; for the face aspect ratio, it depends whether we consider the average or minimum ratio in a drawing.

Statistical Analysis. We further investigate the differences by performing an RM-ANOVA analysis on the measurements with a post-hoc Tukey HSD test to reveal the pairwise differences. The Skewness and Kurtosis of all measurements

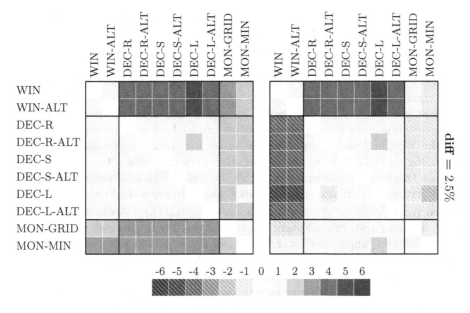

Fig. 8. (Left) Number of "wins" (measures for which the row-algorithm outperforms the column-algorithm), using $p < 0.001$ in a Tukey HSD test with an estimated difference in means of at least 2.5 %. (Right) The number of "wins" minus the number of "losses" in the left table, giving an overall view of relative performance.

are within the range $[-2, 2]$, thus providing evidence for the assumption of the normal distribution for these analyses. The only exception is the minimal angular resolution for the MON algorithms, which are constant at $\pi/4$. These are excluded from the statistical analysis; we consider the MON algorithms to (significantly) outperform the other algorithms due to the high difference in means.

The results of this statistical investigation are summarized in Fig. 8. For this, we require an estimated difference in means of at least 2.5 % of the possible range of values, i.e., a difference of $\pi/80$ for angular resolution, 2.5 % for edge length and 0.025 for face aspect ratio.

We verify that the WIN algorithms clearly outperform the DEC algorithms, in at least four measures (out of six). Between the two variants, there is no difference. As observed above, whether the MON algorithms outperform another algorithm depends highly on the measure. The MON-MIN algorithm "wins" more often than it "loses" compared to any other algorithm. However, the MON-GRID algorithm is outperformed by the WIN algorithms. The DEC algorithms are not distinguishable between themselves. They outperform the MON algorithms for some measures, but are outperformed by the MON algorithms more often.

6   Conclusions

We studied algorithms for drawing cubic planar 3-connected graphs with minimal visual complexity, i.e., with as few line segments as possible. The lower bound is $n/2 + 3$ for a graph with n vertices, and we introduced two new algorithms to match this lower bound. These algorithms may be of independent interest, as a way of constructing planar cubic 3-connected graphs. Moreover, we resolved a flaw in an existing algorithm by Mondal et al. [7].

This leaves us with three algorithms, each with two or more variants. We performed an experiment with two data sets to compare the performance of these algorithms in terms of angular resolution, edge length and face aspect ratio. The Deconstruction algorithm is always outperformed by the Windmill algorithm, but the Windmill algorithm seems to be on par with the Mondal algorithm: depending on the criterion, one or the other performs better. One aspect that was not taken into consideration though, is that the Mondal algorithm comes with a maximum grid size and uses only 6 slopes to draw the line segments.

Future Work. We studied visual complexity for planar cubic 3-connected graphs, which is rather restrictive. Future algorithmic work may aim towards reducing the gap between upper and lower bounds for other graph classes such as triangulations or general planar graphs (see [3]). Moreover, the definition of visual complexity is not limited to line segments, but may include for example the use of circular arcs (see [8]). We may investigate how many vertices are spanned by a line segment—but what is "better" here is not immediately clear. Moreover, we may look into applying the system of harmonic equations to the Mondal layouts.

Furthermore, it would be interesting to investigate whether the definition of visual complexity correlates to an observer's assessment of complexity. In other words, are drawings with minimal visual complexity indeed perceived to be simpler than those with higher visual complexity? Moreover, can we establish a relation between visual complexity and cognitive load? The graph may be visually simpler, but that does not readily imply that it is easier to interpret.

References

1. Aerts, N., Felsner, S.: Straight line triangle representations. In: Wismath, S., Wolff, A. (eds.) GD 2013. LNCS, vol. 8242, pp. 119–130. Springer, Heidelberg (2013)
2. Dujmović, V., Eppstein, D., Suderman, M., Wood, D.: Drawings of planar graphs with few slopes and segments. CGTA **38**, 194–212 (2007)
3. Durocher, S., Mondal, D.: Drawing plane triangulations with few segments. In: Proceedings of 26th CCCG, pp. 40–45 (2014)
4. Durocher, S., Mondal, D., Nishat, R., Whitesides, S.: A note on minimum-segment drawings of planar graphs. J. Graph Algorithm Appl. **17**(3), 301–328 (2013)
5. Grünbaum, B.: Convex Polytopes (Graduate Texts in Math), 2nd edn. Springer-Verlag, New York (2003)
6. Haas, R., Orden, D., Rote, G., Santos, F., Servatius, B., Servatius, H., Souvaine, D.L., Streinu, I., Whiteley, W.: Planar minimally rigid graphs and pseudo-triangulations. Comput. Geom. **31**(1–2), 31–61 (2005)
7. Mondal, D., Nishat, R., Biswas, S., Rahman, S.: Minimum-segment convex drawings of 3-connected cubic plane graphs. J. Comb. Opt. **25**, 460–480 (2013)
8. Schulz, A.: Drawing graphs with few arcs. In: Brandstädt, A., Jansen, K., Reischuk, R. (eds.) WG 2013. LNCS, vol. 8165, pp. 406–417. Springer, Heidelberg (2013)
9. Tamassia, R.: Handbook of Graph Drawing and Visualization. CRC Press, Boca Raton (2013)

The Book Embedding Problem from a SAT-Solving Perspective

Michael A. Bekos$^{(\boxtimes)}$, Michael Kaufmann, and Christian Zielke

Wilhelm-Schickard-Institut Für Informatik, Universität Tübingen,
Tübingen, Germany
{bekos,mk,zielke}@informatik.uni-tuebingen.de

Abstract. In a book embedding, the vertices of a graph are placed on the spine of a book and the edges are assigned to pages, so that edges of the same page do not cross. In this paper, we approach the problem of determining whether a graph can be embedded in a book of a certain number of pages from a different perspective: We propose a simple and quite intuitive SAT formulation, which is robust enough to solve nontrivial instances of the problem in reasonable time. As a byproduct, we show a lower bound of 4 on the page number of 1-planar graphs.

1 Introduction

Embedding graphs in books is a fundamental issue in graph theory that has received considerable attention (see, e.g., [5] for an overview). In a book embedding [26], the vertices of a graph are restricted to a line, referred to as the *spine* of the book, and the edges are drawn at different half-planes delimited by the spine, called *pages* of the book. The task is to find a so-called *linear order* of the vertices along the spine and an assignment of the edges of the graph to the pages of the book, so that no two edges of the same page cross; see Fig. 1b. The *book thickness* or *page number* of a graph is the smallest number of pages that are required by any book embedding of the graph.

Problems on book embeddings are mainly classified into two categories based on whether the graph to be embedded is planar or not. For non-planar graphs, it is known that there exist graphs on n vertices that have book thickness $\Theta(n)$, e.g., the book thickness of the complete graph K_n is $\lceil n/2 \rceil$ [3]. Sublinear book thickness have, e.g., graphs with subquadratic number of edges [24], subquadratic genus [23] or sublinear treewidth [13]. Constant book thickness have, e.g., all minor-closed graphs [6] or the k-trees for fixed k [16]. Another class of nonplanar graphs that was recently proved to have constant book thickness is the class of 1-planar graphs [2].

For planar graphs, a remarkable result is due to Yannakakis, who back in 1986 proved that any planar graph can be embedded in a book with four pages [33]. However, more restricted subclasses of planar graphs allow embeddings in books with fewer pages. Bernhart and Kainen [3] showed that the graphs which can be embedded in single-page books are the outerplanar graphs, while the graphs which can be embedded in books with two pages are the subhamiltonian ones.

© Springer International Publishing Switzerland 2015
E. Di Giacomo and A. Lubiw (Eds.): GD 2015, LNCS 9411, pp. 125–138, 2015.
DOI: 10.1007/978-3-319-27261-0_11

It is known that not all planar graphs are subhamiltonian and the corresponding decision problem whether a maximal planar graph is Hamiltonian (and therefore two-page book embeddable) is NP-complete [32]. However, several subclasses of planar graphs are known to be Hamiltonian or subhamiltonian, see, e.g., [1,11,12,19,21,25].

A well-known non-subhamiltonian graph is the Goldner-Harary one [17]. This graph, however, is a planar 3-tree and hence 3-page book embeddable [18]. To the best of our knowledge, there is no planar graph whose page number is four. In other words, it is not known whether the upper bound of four pages of Yannakakis [33] is tight or not.

Our Contribution. We suggest an alternative approach to the problem of determining whether a graph can be embedded in a book of a certain number of pages. We propose a formulation of the problem as a SAT instance, which can be useful in practice (note that, apart from their independent theoretical interest, book embeddings find applications in several contexts, such as VLSI design, fault-tolerant processing, sorting networks and parallel matrix multiplication, see e.g., [11,20,28,30]). It turns out that our formulation is of a simple nature, quite intuitive and easy-to-implement, but simultaneously robust enough to solve non-trivial instances of the problem in reasonable amount of time (e.g., within 20 min we can test whether a maximal planar graphs with up to 400 vertices is 3-page book embeddable), as we will see in our experimental study.

Note that SAT formulations are not so common in graph drawing. A few notable exceptions are [4,10,15]. In our context, of interest is the work of Biedl et al. [4], who proposed ILP and SAT formulations for several grid-based graph problems. Their general formulation can be extended to solve our problem as well. However, from the authors' experimental evaluation (and we could also confirm it) it follows that their approach is limited to solve relatively small instances within reasonable time, e.g., within 20 min one can cope with graphs whose size in vertices and edges does not exceed 100.

A List of Hypotheses. When we started working on this project, we placed several hypotheses that we wanted to prove or disprove. So, before we proceed with the description of our formulation, we first list and then discuss the most important ones:

H1: There is a (maximal) planar graph whose book thickness is four.

H2: There is a 1-planar graph whose book thickness is (at least) four.

H3: There is a (maximal) planar graph, which cannot be embedded in a book of three pages, if the subgraphs embedded at each page must be acyclic.

H4: There is a maximal planar graph, say G_a, which in any of its book embeddings on three pages has at least one face whose edges are on the same page.

H5: There is a maximal planar graph, say G_c, which in any of its book embeddings on three pages has a face, say f_c^*, whose edges cannot be on the same page.

Summary and Discussion. Clearly, our ultimate goal was to find a planar graph supporting Hypothesis 1. During our extensive practical analysis, we tested several hundreds maximal planar graphs (both randomly created and crafted), but we did not manage to find one supporting Hypothesis 1. We also tested a specific graph with roughly 600 vertices out of the family of planar graphs that Yannakakis proposed to require page number four, but it turned out to be 3-page book embeddable for this particular size.

On the positive side, we proved that the weakest version of Hypothesis 2 holds. In particular, we managed to find a relatively small 1-planar graph whose book thickness is exactly four; see Fig. 1. To the best of our knowledge, this is the first (non-trivial) lower bound on the book thickness of 1-planar graphs.

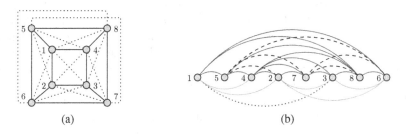

(a)   (b)

Fig. 1. (a) A 1-planar graph, whose underlying planar structure (solid drawn) is the cube graph. (b) A 4-page embedding, in which the fourth page contains a single edge (dotted drawn).

We were surprised that we did not succeed in proving that Hypothesis 3 holds. Note that it is very natural to try to embed a tree-structured subgraph at each of the available pages of the book, if one seeks to prove that indeed all planar graphs can be embedded in books of three pages. For example, Heath [18] who constructively proved that all planar 3-trees fit into books with three pages used exactly this approach: the subgraphs embedded at each of the three pages of the book are acyclic.

Note that we managed to prove a weaker version of Hypothesis 3, according to which the input maximal planar graph has n vertices and cannot be embedded in a book with three pages, so that:(i)the subgraph assigned to each of the three pages is a tree on $n - 1$ vertices and, additionally, (ii) the three vertices, that are not spanned by the three trees are pairwise adjacent forming a face f_o of the graph, say w.l.o.g. its outerface. This negative results implies that it is not always possible to construct a 3-page book embedding based on a Schnyder decomposition into three trees (regardless of the linear order underneath).

From our experimental evaluation (see Sect. 3), we quickly observed that the practical limitation of testing the book-embeddability of maximal planar graphs on three pages with our SAT formulation (that we present in Sect. 2) lies at around 600 to 700 vertices. Larger graphs lead to instance sizes, that excess several gigabytes of random access memory. Hypotheses 4 and 5 in conjunction describe an approach, that could potentially overcome this bottleneck. To see this, assume that the two planar graphs, denoted by G_a and G_c in Hypotheses 4 and 5, exist (note, however, that we did not succeed in finding them). If for each face f_a of G_a, we create a copy of graph G_c and identify each of the vertices of face f_c^* of G_c with one of the vertices of face f_a of G_a, then we will obtain a (drastically larger) planar graph, that is not 3-page book-embeddable. This is because G_a must contain at least one face whose edges are on the same page, while in the same time face f_c^* would require at least one of them not to be at the same page.

2   SAT Formulation

Let $G = (V, E)$, with $V = \{v_1, v_2, \ldots, v_n\}$ and $E = \{e_1, e_2, \ldots, e_m\}$, be a graph for which we seek to decide, whether it can be embedded in a book with $p \geq 2$ pages. Next we describe a logic formula $\mathcal{F}(G, p)$ that will solve this problem by encoding it as a SAT instance. Recall that any SAT problem can be described in *conjunctive normal form* (CNF), which is a conjunction of clauses; each clause being a disjunction of (possibly negated) literals. We will define $\mathcal{F}(G, p)$ by its set of variables and a corresponding set of rules. The rules will ensure the proper assignment of the variables and will be given in propositional logic, which can be converted into CNF clauses straightforwardly [27].

The variables of $\mathcal{F}(G, p)$ model a book embedding of G in a book with p pages, if it exists. Thus, we use variables $\sigma(v_i, v_j)$ for each pair of vertices $v_i, v_j \in V$ to determine, whether vertex v_i is to the left of vertex v_j along the spine. If that is the case, the assignment is $\sigma(v_i, v_j) = \texttt{true}$. In this so-called *relative encoding*, the variables encode a relative order between the vertices. Clearly, asymmetry has to hold for these variables, which is ensured by the following rule:

$$\sigma(v_i, v_j) \leftrightarrow \neg\sigma(v_j, v_i)$$

With this *asymmetry rule*, variables $\sigma(v_i, v_j)$ can be defined only for $i < j$. The other literals, $\sigma(v_i, v_j)$ with $i > j$, can be replaced by $\neg\sigma(v_j, v_i)$. This results in a significantly smaller formula, which is easier to be solved by a SAT solver [31]. The following *transitivity rule* for the σ-relation ensures a proper order of the vertices along the spine:

$$\sigma(v_i, v_j) \wedge \sigma(v_j, v_k) \rightarrow \sigma(v_i, v_k) \quad \forall \text{ pairwise distinct } v_i, v_j, v_k \in V$$

The search space of possible satisfying assignments can be reduced by choosing a particular vertex as the first vertex along the spine and by assuming w.l.o.g. that v_2 is to the left of v_3. These choices can be easily encoded by the *direction rules*:

$$\sigma(v_1, v_i) \; \forall v_i \in V \text{ with } i > 1$$
$$\sigma(v_2, v_3)$$

Note that the direction rules are represented by unit clauses that fix $\sigma(v_1, v_i)$, for $i > 1$ and $\sigma(v_2, v_3)$ to be **true**. Via *unit propagation* (that is, a basic operation performed by all SAT solvers [14]) other constraints may become simpler or even already satisfied.

To encode the assignment of the edges to the pages of the book, we introduce a variable $\phi_q(e_i)$ for every edge $e_i \in E$ and every possible page $1 \leq q \leq p$. Thereby $\phi_q(e_i) = \text{true}$ means that the edge e_i is assigned to the q-th page of the book. Every edge has to be assigned to one page, which is ensured by the *page assignment rule*:

$$\phi_1(e_i) \vee \phi_2(e_i) \vee \ldots \vee \phi_p(e_i) \quad \forall e_i \in E$$

We can again reduce the search space by the *fixed page assignment rule*, that fixes a single edge on a particular page, e.g., edge $e_1 \in E$ to page 1:

$$\phi_1(e_1)$$

To forbid crossings among edges of the same page, we first introduce a variable $\chi(e_i, e_j)$ for each pair of edges $e_i, e_j \in E$, which describes whether e_i and e_j are assigned to the same page. $e_i, e_j \in E$ are assigned to the same page, if and only if, they are both assigned to one of the available pages, which is ensured by the *same page rule*:

$$((\phi_1(e_i) \wedge \phi_1(e_j)) \vee \ldots \vee (\phi_k(e_i) \wedge \phi_k(e_j))) \rightarrow \chi(e_i, e_j) \quad \forall e_i, e_j \in E$$

To ensure planarity, it is enough to ensure that two edges which are assigned to the same page do not cross. So, if $(v_i, v_j), (v_k, v_\ell) \in E$ are two edges of G, such that vertices v_i, v_j, v_k and v_ℓ are pairwise different, this can be ensured by the following *planarity rule*:

$$\chi((v_i, v_j), (v_k, v_\ell)) \rightarrow$$
$$\neg(\sigma(v_i, v_k) \wedge \sigma(v_k, v_j) \wedge \sigma(v_j, v_\ell)) \wedge \neg(\sigma(v_i, v_\ell) \wedge \sigma(v_\ell, v_j) \wedge \sigma(v_j, v_k))$$
$$\wedge \neg(\sigma(v_j, v_k) \wedge \sigma(v_k, v_i) \wedge \sigma(v_i, v_\ell)) \quad \wedge \neg(\sigma(v_j, v_\ell) \wedge \sigma(v_\ell, v_i) \wedge \sigma(v_i, v_k))$$
$$\wedge \neg(\sigma(v_k, v_i) \wedge \sigma(v_i, v_\ell) \wedge \sigma(v_\ell, v_j)) \quad \wedge \neg(\sigma(v_k, v_j) \wedge \sigma(v_j, v_\ell) \wedge \sigma(v_\ell, v_i))$$
$$\wedge \neg(\sigma(v_\ell, v_i) \wedge \sigma(v_i, v_k) \wedge \sigma(v_k, v_j)) \quad \wedge \neg(\sigma(v_\ell, v_j) \wedge \sigma(v_j, v_k) \wedge \sigma(v_k, v_i))$$

Theorem 1. *Let $G = (V, E)$ be a graph and $p \in \mathbb{N}$. Then, G admits a book embedding on p pages, if and only if, $\mathcal{F}(G, p)$ is satisfiable. In addition, $\mathcal{F}(G, p)$ has $O(n^2 + m^2 + pm)$ variables and $O(n^3 + m^2)$ clauses.*

Proof. The number of σ-, χ- and ϕ-variables are $O(n^2)$, $O(m^2)$ and $O(pm)$, respectively, which implies that $\mathcal{F}(G, p)$ has $O(n^2 + m^2 + pm)$ variables. The number of clauses of $\mathcal{F}(G, p)$ is dominated by the number of transitivity, same-page and planarity rules, which yield in total $O(n^3 + m^2)$ clauses. So, to prove this theorem, it remains to show that: (i) a book embedding on p pages yields a satisfying assignment of $\mathcal{F}(G, p)$ and (ii) a satisfying assignment of $\mathcal{F}(G, p)$ yields a book embedding on p pages.

(i) *From an embedding to an assignment:* Assume that G has a book embedding $\mathcal{E}(G, p)$ on p pages. From $\mathcal{E}(G, p)$, we obtain an order of the vertices along the spine and an assignment of the edges to the pages. We define an assignment $(\hat{\sigma}, \hat{\phi}, \hat{\chi})$ to the σ-, ϕ- and χ-variables of $\mathcal{F}(G, p)$ consistent with the intended meaning of the variables as follows. (a) $\hat{\sigma}(v_i, v_j) = \mathtt{true}$, if and only if v_i is before v_j along the spine, (b) $\hat{\phi}_q(e_i) = \mathtt{true}$, if and only if e_i is assigned to the q-th page, (c) $\hat{\chi}(e_i, e_j) = \mathtt{true}$, if and only if e_i and e_j are assigned to the same page. To prove that assignment $(\hat{\sigma}, \hat{\phi}, \hat{\chi})$ satisfies $\mathcal{F}(G, p)$, we consider all rules of $\mathcal{F}(G, p)$:

- The *transitivity* and *asymmetry rules* are satisfied by $(\hat{\sigma}, \hat{\phi}, \hat{\chi})$, since $\hat{\sigma}$ is a complete order over the vertices of G (by definition of the assignment).
- The *direction rules* are also satisfied, since we can assume w.l.o.g. that in $\mathcal{E}(G, p)$ vertex $v_1 \in V$ is the first vertex along the spine and v_2 is to the left of v_3. Note that if this is not the case, then we can circularly-shift the vertices of G along the spine and potentially mirror $\mathcal{E}(G, p)$ and obtain an equivalent embedding which has the aforementioned properties; see e.g., [33].
- The *page assignment rule* is trivially satisfied by the definition of the assignment and the fact that $\mathcal{E}(G, p)$ was given.
- The *fixed page assignment rule* can be satisfied as well, since we can assume w.l.o.g. that the first page of $\mathcal{E}(G, p)$ is the page where edge $e_1 \in E$ is assigned to.
- The *same page rule* is trivially satisfied due to the definition of the assignment.
- It remains to show that all *planarity rules* are satisfied. For the sake of contradiction, assume that the assignment $(\hat{\sigma}, \hat{\phi}, \hat{\chi})$ violates a planarity rule for some pair of edges (v_i, v_j) and (v_k, v_ℓ). We know that $\hat{\chi}((v_i, v_j), (v_k, v_\ell)) = \mathtt{true}$ and further $\hat{\sigma}(v_i, v_k) = \hat{\sigma}(v_k, v_j) = \hat{\sigma}(v_j, v_\ell) = \mathtt{true}$. Hence, in $\mathcal{E}(G, p)$ we have that v_k is between v_i and v_j, while v_ℓ is not between v_i and v_j. Thus, (v_i, v_j) and (v_k, v_ℓ) are on the same page and cross in $\mathcal{E}(G, p)$, which is a clear contradiction.

(ii) *From an assignment to an embedding:* Let $(\hat{\sigma}, \hat{\phi}, \hat{\chi})$ be a satisfying assignment to $\mathcal{F}(G, p)$. Let $\xi : V \mapsto \{1, \ldots, n\}$ be a function which maps each vertex $v \in V$ to a position along the spine. Based on $(\hat{\sigma}, \hat{\phi}, \hat{\chi})$, map ξ can be defined as follows:

$$\xi(v_i) = 1 + |\{v_j : \sigma(v_j, v_i) = \mathtt{true}, 1 \leq j \leq n, \; j \neq i\}|$$

Since $(\hat{\sigma}, \hat{\phi}, \hat{\chi})$ satisfies the *asymmetry* and *transitivity rules*, it follows that all positions assigned to the vertices of G are pairwise different. Therefore, a proper global ordering is obtained. Since by the *page assignment rule*, every edge of G is assigned to at least one page, we only have to show that each page is crossing-free. Assume for the sake of contradiction that (v_i, v_j) and (v_k, v_ℓ) are two edges of G that are assigned to the same page and cross. In this case, one of the following relationships must hold:

$$\min\{\xi(v_i), \xi(v_j)\} < \min\{\xi(v_k), \xi(v_\ell)\} < \max\{\xi(v_i), \xi(v_j)\} < \max\{\xi(v_k), \xi(v_\ell)\}$$

$$\min\{\xi(v_k), \xi(v_\ell)\} < \min\{\xi(v_i), \xi(v_j)\} < \max\{\xi(v_k), \xi(v_\ell)\} < \max\{\xi(v_i), \xi(v_j)\}$$

However, neither is possible, since both configurations do not comply with the *planarity rule* of $(\hat{\sigma}, \hat{\phi}, \hat{\chi})$. Therefore, each page is crossing-free, as desired. $\square$

So far, we have described a SAT formulation that tests, whether a given graph $G = (V, E)$ admits an embedding in a book with $p \geq 2$ pages. Of course, this formulation can be extended with additional variables and rules. In the following, we will introduce three different extensions, which encode Hypotheses 3, 4 and 5.

2.1   A First Variant to check Hypothesis 3

In this subsection, we present a SAT formulation to check Hypothesis 3. Recall that, we seek to check whether a maximal planar graph G can be embedded in $p = 3$ pages, so that the subgraph assigned to each of the three pages is an acyclic graph. In the following, we will extend formula $\mathcal{F}(G, 3)$ with new variables and rules to encode the additional requirement. We denote by $\mathcal{F}_f(G, 3)$ the resulting formula.

Let $\mathcal{N}(v)$ be the set of vertices adjacent to $v \in V$. For every edge $(v_i, v_j) \in E$, variable $\pi_q(v_i, v_j)$ describes, whether vertex v_i is the *parent* of vertex v_j in the forest of page $q \in \{1, 2, 3\}$. Variable $\pi_q(v_j, v_i)$ is defined symmetrically. We ensure, that exactly one of the two variables is `true` when (v_i, v_j) is assigned to page q, and both of the variables are `false`, when (v_i, v_j) is not assigned to page q, by the *parent rules*:

$$\phi_q((v_i, v_j)) \rightarrow (\pi_q(v_i, v_j) \wedge \neg\pi_q(v_j, v_i)) \vee (\neg\pi_q(v_i, v_j) \wedge \pi_q(v_j, v_i))$$
$$\neg\phi_q((v_i, v_j)) \rightarrow (\neg\pi_q(v_i, v_j) \wedge \neg\pi_q(v_j, v_i))$$

We also have to ensure, that every vertex $v_i \in V$ has at most one parent vertex in the forest of page q, which can be done via the *single parent rule*:

$$(\neg\pi_q(v_k, v_i) \vee \neg\pi_q(v_\ell, v_i)), \ \forall v_k, v_\ell \in \mathcal{N}(v_i) : v_k \neq v_\ell$$

To ensure acyclicity, we use variables $\alpha_q(v_i, v_j)$ that describe, whether vertex v_i is an *ancestor* of v_j in the forest of page q. We know that whenever for an edge $(v_i, v_j) \in E$ vertex v_i is the parent of vertex v_j on page q, that v_i is the ancestor for v_j on that page as well, which results in the *parent ancestor rule*:

$$\pi_q(v_i, v_j) \rightarrow \alpha_q(v_i, v_j)$$

Clearly, *transitivity* as well as *antisymmetry* has to hold for the ancestor relationship:

$$(\alpha_q(v_i, v_j) \wedge \alpha_q(v_j, v_k)) \rightarrow \alpha_q(v_i, v_k) \ \forall \text{ pairwise distinct } v_i, v_j, v_k \in V$$
$$\alpha_q(v_i, v_j) \rightarrow \neg\alpha_q(v_j, v_i) \ \forall \text{ pairwise distinct } v_i, v_j \in V$$

Theorem 2. *Let $G = (V, E)$ be a (maximal) planar graph. Then, G admits a book embedding on three pages, so that the subgraph assigned to each of the three pages is a forest, if and only if $\mathcal{F}_f(G, 3)$ is satisfiable.*

Proof. We use the same technique as in the proof of Theorem 1. So, consider an embedding $\mathcal{E}(G, 3)$ in three pages yield by our formulation. We claim that the subgraphs embedded at each page are acyclic. For contradiction, assume that there is a cycle C_q at page q. If we direct each edge of C_q from the child to the parent vertex, then all edges of C_q must have the same orientation, that is, either clockwise or counterclockwise along C_q (otherwise, there is a vertex of C_q that has two parents, deviating the single parent rule). The transitivity of the ancestor relationship implies that the antisymmetry property is deviated along C_q, which is a contradiction. Hence, the subgraphs embedded at each page of $\mathcal{E}(G,3)$ are indeed acyclic. Following similar arguments as in the second part of the proof of Theorem 1, we can easily prove that a satisfying assignment of $\mathcal{F}_f(G,3)$ yields a book embedding on 3 pages, in which the subgraph assigned to each page is a forest, which completes the proof.   □

Note that $\mathcal{F}_f(G,3)$ has asymptotically the same number of variables and clauses as $\mathcal{F}(G,3)$. Our formulation can be easily adjusted to check whether the subgraph assigned to each page is a tree. In this scenario, we employ an additional variable $\rho_q(v_i)$ for each vertex $v_i \in V$ that describes whether v_i is the *root* of the tree of page $q \in \{1,2,3\}$. Vertex v_i is the root of the tree of page q if and only if it has no parent and (at least) one child at page q, which can be ensured via the following *root rule*:

$$(\bigwedge_{v_j \in \mathcal{N}(v_i)} \neg \pi_q(v_j, v_i)) \wedge (\bigvee_{v_k \in \mathcal{N}(v_i)} \pi_q(v_i, v_k)) \leftrightarrow \rho_q(v_i)$$

We ensure that there are not two or more roots on the same page q via the *single root rule*:

$$(\neg \rho_q(v_i) \vee \neg \rho_q(v_j)), \ \forall v_i, v_j \in V; i \neq j$$

2.2   A Second Variant to Check Hypothesis 4

Assume that $G_a = (V_a, E_a)$ is a maximal planar graph, that is embeddable in a book with 3 pages. Let $\Delta(G_a) = \{f_1, f_2, \dots, f_{2|V_a|-4}\}$ be the set of faces of G_a. In the following, we describe an extension to the formula $\mathcal{F}(G_a, 3)$ that forbids the so-called *unicolored* faces, that is, faces whose edges are assigned to the same page of the book. We denote the resulting formula by $\mathcal{F}_a(G_a, 3)$.

In comparison to our previous approaches, we are not searching for a single book embedding, for which an additional property holds, but rather we have to test whether all possible book embeddings have this property. We will use the *same page variables* already present in $\mathcal{F}(G_a, 3)$ to ensure this property via the *forbid unicolored face rule*:

$$(\neg \chi(e_i, e_j) \vee \neg \chi(e_i, e_k)) \ \forall f = \{e_i, e_j, e_k\} \in \Delta(G_a)$$

Theorem 3. $\mathcal{F}_a(G_a, 3)$ *is unsatisfiable, if and only if, for every possible book embedding* $\mathcal{E}(G_a, 3)$ *there exists a unicolored face* $f_i \in \Delta(G_a)$, $i = 1, \dots, 2|V_a|-4$.

Proof. Directly follows from the validity of $\mathcal{F}(G_a, 3)$.   □

2.3   A Third Variant to Check Hypothesis 5

Assume that $G_c = (V_c, E_c)$ is a maximal planar graph, that is embeddable in a book with 3 pages and let $\Delta(G_c)$ be the set of faces of G_c. To check whether a particular face $f^* = \{e_i, e_j, e_k\} \in \Delta(G_c)$ cannot be unicolored in any possible book embedding of G_c, we again use the already present *same page variables* of $\mathcal{F}(G_c, 3)$:

$$(\chi(e_i, e_j) \wedge \chi(e_i, e_k)), f^* = \{e_i, e_j, e_k\} \in \Delta(G_c)$$

The *force unicolored face rule* yields a new formula, which we denote by $\mathcal{F}_c(G_c; f^*, 3)$. By the following theorem, it follows that in order to check Hypothesis 5 one has to check $2|V_c| - 4$, different formulas; one for each face of G_c.

Theorem 4. $\mathcal{F}_c(G_c; f^*, 3)$ *is unsatisfiable, if and only, if there exists no book embedding of G_c with f^* being unicolored.*

Proof. Directly follows from the validity of $\mathcal{F}(G_c, 3)$. $\square$

3   Experiments

In this section, we present an experimental evaluation of our SAT formulation. We ran our experiments on a Linux machine with four cores at $2, 5$ GHz and 8 GB of RAM. The implementation that creates the SAT instances was done in Java. For solving the SAT instances, we used the *SApperloT* solver [22]. This solver is as fast as the well-known *minisat* [14] solver for smaller graphs, but it considerably outperforms minisat for increasing instance sizes. The runtime we report consists of both, the time to create the instance and the time to solve it. Note that the time to create the SAT instance for small graphs is neglectable. For large graphs, however, that step can take a few minutes.

Established Benchmark Sets. Since the Rome and the North graphs are popular test sets for planar and nearly planar graphs, we also used them as test sets for our experiment (cf. http://www.graphdrawing.org). The *Rome graphs* are 11534 graphs; 3281 of them are planar and 8253 are non-planar. Their average density is 0.069, where the *density* of a graph $G = (V, E)$ is $2|E|/(|V|(|V|-1))$. The number of vertices of the Rome graphs range from 10 to 110. The corresponding number of edges range from 9 to 158.

It is eye-catching, that all planar Rome graphs are 2-page book embeddable (see Table 1). The non-planar ones are 3-page embeddable. But since the Rome graphs are very sparse this result was more or less expected. Note that 99 % of the planar Rome graphs (that is, 3248 out of 3282) are solved within 2 s. For the non-planar Rome graphs, the same ratio (that is, 8169 out of 8253) is achieved after 6.25 s.

As a second benchmark set, we used the *North graphs* which are 1277 graphs; 854 of them are planar and 423 are non-planar. The number of vertices of these graphs range from 10 to 100. The corresponding number of edges range from 9 to 241. Their average density is 0.13. Again, all planar graphs were 2-page book embeddable. The runtime to compute the corresponding embedding for the vast majority of the planar North graphs was rather small. In particular, 97.5 % of them (that is, 833 out of 854 graphs) were solved within 3 s, with the maximum runtime being 9.4 s.

Table 1. Overview of the results for the established benchmark sets.

planar			nonplanar				
Graph class	#	$p = 2$	#	$p = 3$	$p = 4$	$p = 5$	see below
Rome	3281	3281	8253	8253	0	0	
North	854	854	423	329	25	8	61

The non-planar North graphs show the practical limitations of our formulation. In particular, we could determine the page number of only 344 out of 423 graphs within the time limit of 1200 s. Finding a 3-page book embedding is much faster than proving that such an embedding does not exist (see Fig. 2b). For the remaining 79 graphs, we increased the timeout to 3 h and we managed to get at least some partial results: (i) for 18 graphs we computed their exact page number, (ii) 27 graphs fit in four pages but we were not able to determine whether they could fit in three pages, (iii) 32 graphs did not fit in three pages (and 6 out of them did not even fit in four pages), but we did not manage to determine their page number. Nevertheless, all non-planar North graphs could fit into 8 pages and since the focus of the paper is mostly on planar and 1-planar graphs, we did not further investigated the book embeddability of these graphs.

Crafted Graphs. To prove Hypothesis 1, we also crafted several maximal planar graphs with at least 500 vertices each, which we tested for 3-page book embeddability. To avoid testing Hamiltonian graphs, we adopted a two-step approach that was inspired by the graph class that Yannakakis proposed as candidate to require four pages. In the first step, we chose a triangulated planar (not necessarily non-Hamiltonian) graph as the base for the second step. In the second step, we augmented the base graph by specific operations to make it non-Hamiltonian (and therefore not 2-page embeddable). Examples of these operations are: (i) stellate a face f, that is, introduce a new vertex and connect it to all vertices of f, (ii) replace a triangular face by an octahedron, (iii) embed a non-Hamiltonian planar graph G_f to a face f by identifying the vertices of f with the vertices of a particular face of G_f.

We observed that these operations most of the times yield non-Hamiltonian planar graphs. Note that they do not generate the whole class of non-Hamiltonian planar graphs and not even a uniformly-distributed random subset. The graphs, that we crafted and tested with this approach, were all maximal planar with at least 500 and at most 700 vertices. The runtime to check each instance ranged from few hours to a couple of days.

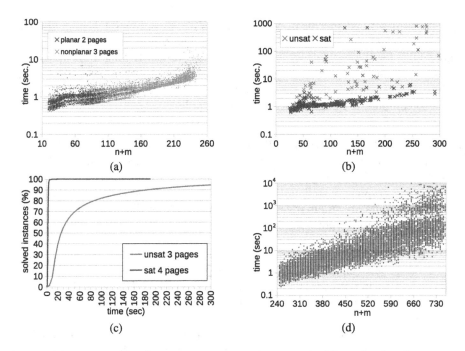

Fig. 2. (a) Rome graphs: Runtime to compute 2-page embeddings for the planar ones (green) and 3-page embeddings for the non-planar ones (red). (b) Non-planar North graphs: Time needed to compute 3-page embeddings (green) and to prove that no 3-page embedding exist (red). (c) The runtime for maximal 1-planar graphs with 25 vertices. The red curve shows the runtime to prove that no 3-page embedding exist; the green curve shows the runtime to compute 4-page embeddings. (d) The runtime to compute 4-page embeddings for randomized maximal 1-planar graphs.

1-Planar Graphs. To check Hypothesis 2 for more than four pages, we initially generated all 2,098,675 planar triconnected quadrangulations with 25 vertices and minimum degree three using plantri [8]. By augmenting every face with two crossing edges, the generated quadrangulations yield optimal 1-planar graphs (recall that a 1-planar graph on n vertices is said to be *optimal*, if it has exactly $4n - 8$ edges, which is the maximum possible [7]). Our experiments showed that all tested optimal 1-planar graphs required four pages. The runtime distribution is shown in Fig. 2c. Computing a 4-page embedding was always fast: For 99.06 % of the graphs the solver found a solution within 4.7 s. The maximum runtime for a single instance was 186 s. Proving that no 3-page embedding existed was harder. In less than 5 min., 94.4 % of the instances could be solved. However, for very few instances this could take up to two hours.

To obtain a better understanding of the connection between the runtime of our approach and the size of the graphs, we randomly created 8312 optimal 1-planar graphs of different sizes varying from 50 to 155 vertices. Starting from the cube graph (see Fig. 1), we iteratively applied at random one of the two operations described in [29] in order to generate all optimal 1-planar graphs, until we reached the desired size of the graph. The runtime to compute 4-page embeddings for these graphs is shown in Fig. 2d. For nearly all graphs up to 100 vertices a 4-page embedding could be computed within two minutes. However, with increasing vertex-count, the amount of graphs that could take up to several hours to compute a 4-page embedding rises rapidly.

Randomized Planar Graphs. To check Hypotheses 3-5, we generated a large set of random maximal planar graphs as follows. We applied Delaunay triangulation on a set of randomly created points within a triangular region. To avoid Hamiltonian graphs, we stellated every face of the produced Delaunay triangulations. Our results are summarized in the following. Note that none of the tested graphs corroborate Hypotheses 3-5.

- For Hypothesis 3, we tested 15,040 maximal planar graphs of varying number of vertices between 25 and 80. We were able to solve 70.78 % of the instances (that is, 10,646 out of 15,040) within 3 min. and 76.37 % (that is, 11,487 out of 15,040) of the instances within 20 min., which was the time limit of the computation.
- For Hypothesis 4, we tested 7,174 maximal planar graphs of varying number of vertices between 59 and 125. We were able to solve 92.75 % of the instances (that is, 6621 out of 7174) within 10 min. and 99 % of the instances within an hour. The maximum time that was needed to solve a single instance was 5 h and 6 min.
- For Hypothesis 5, we managed to test 277,284 maximal planar graphs of varying number n of vertices between 59 and 95. Every single instance, each containing $2n - 4$ different SAT formulas, required only few seconds to be tested.

4   Conclusions and Discussion

In this paper, we approached the problem of determining whether a graph can be embedded in a book of a given number of pages from a SAT-solving perspective. By elaborating natural hypotheses for planar and 1-planar graphs, we gained valuable insights to the problem. We now tend to believe that the following refined hypotheses hold:

H6: All planar graphs admit 3-page book embeddings, in which additionally the subgraphs assigned to each page are trees.
H7: Optimal 1-planar graphs have book thickness four.
H8: The book thickness of general 1-planar graphs is not more than five.

Our experimental evaluation showed that our approach can be useful, since it can cope with non-trivial instances in reasonable amount of time. However, around the optimal solution, where the problem switches from unsatisfiable to satisfiable, we observed the well-known *phase transitional behavior* for SAT problems [9]. So, in this context, we mention two more major open problems. The first one is to refine our SAT approach to support denser graphs. The second one is to extend it to other grid-based problems. For example, with our formulation it could be possible to solve larger instances for several of the problems studied, e.g., by Biedl et al. [4].

References

1. Bekos, M., Gronemann, M., Raftopoulou, C.N.: Two-page book embeddings of 4-planar graphs. In: STACS, vol. 25, pp. 137–148. LIPIcs, Schloss Dagstuhl (2014)
2. Bekos, M.A., Bruckdorfer, T., Kaufmann, M., Raftopoulou, C.: 1-planar graphs have constant book thickness. In: Bansal, N., Finocchi, I. (eds.) ESA 2015. LNCS, vol. 9294, pp. 130–141. Springer, Heidelberg (2015)
3. Bernhart, F., Kainen, P.: The book thickness of a graph. Comb. Theory **27**(3), 320–331 (1979)
4. Biedl, T., Bläsius, T., Niedermann, B., Nöllenburg, M., Prutkin, R., Rutter, I.: Using ILP/SAT to determine pathwidth, visibility representations, and other grid-based graph drawings. In: Wismath, S., Wolff, A. (eds.) GD 2013. LNCS, vol. 8242, pp. 460–471. Springer, Heidelberg (2013)
5. Bilski, T.: Embedding graphs in books: a survey. IEEE Proc. Comput. Digit. Tech. **139**(2), 134–138 (1992)
6. Blankenship, R.: Book embeddings of graphs. Ph.D. thesis, Louisiana State University (2003)
7. Bodendiek, R., Schumacher, H., Wagner, K.: Über 1-optimale graphen. Math. Nachr. **117**(1), 323–339 (1984)
8. Brinkmann, G., Greenberg, S., Greenhill, C.S., McKay, B.D., Thomas, R., Wollan, P.: Generation of simple quadrangulations of the sphere. Discrete Math. **305**(1–3), 33–54 (2005)
9. Cheeseman, P., Kanefsky, B., Taylor, W.M.: Where the really hard problems are. In: Mylopoulos, J., Reiter, R. (eds.) AI, pp. 331–340. Morgan Kaufmann (1991)
10. Chimani, M., Zeranski, R.: Upward planarity testing via SAT. In: Didimo, W., Patrignani, M. (eds.) GD 2012. LNCS, vol. 7704, pp. 248–259. Springer, Heidelberg (2013)
11. Chung, F.R.K., Leighton, F.T., Rosenberg, A.L.: Embedding graphs in books: a layout problem with applications to VLSI design. SIAM J. Algebraic Discrete Method **8**(1), 33–58 (1987)
12. Cornuéjols, G., Naddef, D., Pulleyblank, W.: Halin graphs and the travelling salesman problem. Math. Programm. **26**(3), 287–294 (1983)
13. Dujmović, V., Wood, D.: Graph treewidth and geometric thickness parameters. Discrete Comput. Geom. **37**(4), 641–670 (2007)
14. Eén, N., Sörensson, N.: An extensible SAT-solver. In: Giunchiglia, E., Tacchella, A. (eds.) SAT 2003. LNCS, vol. 2919, pp. 502–518. Springer, Heidelberg (2004)
15. Gange, G., Stuckey, P.J., Marriott, K.: Optimal k-level planarization and crossing minimization. In: Brandes, U., Cornelsen, S. (eds.) GD 2010. LNCS, vol. 6502, pp. 238–249. Springer, Heidelberg (2011)
16. Ganley, J.L., Heath, L.S.: The pagenumber of k-trees is $O(k)$. Discrete Appl. Math. **109**(3), 215–221 (2001)

17. Goldner, A., Harary, F.: Note on a smallest nonhamiltonian maximal planar graph. Bull. Malays. Math. Sci. Soc. **1**(6), 41–42 (1975)

18. Heath, L.: Embedding planar graphs in seven pages. In: FOCS, pp. 74–83. IEEE Computer Society (1984)

19. Heath, L.: Algorithms for embedding graphs in books. Ph.D. thesis, University of N. Carolina (1985)

20. Heath, L.S., Leighton, F.T., Rosenberg, A.L.: Comparing queues and stacks as machines for laying out graphs. SIAM J. Discrete Math. **3**(5), 398–412 (1992)

21. Kainen, P.C., Overbay, S.: Extension of a theorem of Whitney. Appl. Math. Lett. **20**(7), 835–837 (2007)

22. Kottler, S.: Description of the SApperloT, SArTagnan and MoUsSaka solvers for the SAT-competition 2011 (2011)

23. Malitz, S.: Genus g graphs have pagenumber $O(\sqrt{q})$. J. Algorithms **17**(1), 85–109 (1994)

24. Malitz, S.: Graphs with e edges have pagenumber $O(\sqrt{E})$. J. Algorithms **17**(1), 71–84 (1994)

25. Nishizeki, T., Chiba, N.: Hamiltonian cycles. In: Planar Graphs: Theory and Algorithms, chap. 10, pp. 171–184. Dover Books on Mathematics, Courier Dover Publications (2008)

26. Ollmann, T.: On the book thicknesses of various graphs. In: Hoffman, F., Levow, R., Thomas, R. (eds.) Southeastern Conference on Combinatorics, Graph Theory and Computing. Congressus Numerantium, vol. VIII, p. 459 (1973)

27. Plaisted, D.A., Greenbaum, S.: A structure-preserving clause form translation. J. Symbolic Comput. **2**(3), 293–304 (1986)

28. Rosenberg, A.L.: The Diogenes approach to testable fault-tolerant arrays of processors. IEEE Trans. Comput. **C–32**(10), 902–910 (1983)

29. Suzuki, Y.: Optimal 1-planar graphs which triangulate other surfaces. Discrete Math. **310**(1), 6–11 (2010)

30. Tarjan, R.: Sorting using networks of queues and stacks. J. ACM **19**(2), 341–346 (1972)

31. Velev, M.N., Gao, P.: Efficient SAT techniques for relative encoding of permutations with constraints. In: Nicholson, A., Li, X. (eds.) AI 2009. LNCS, vol. 5866, pp. 517–527. Springer, Heidelberg (2009)

32. Wigderson, A.: The complexity of the Hamiltonian circuit problem for maximal planar graphs. Technical report TR-298, EECS Department, Princeton University (1982)

33. Yannakakis, M.: Embedding planar graphs in four pages. J. Comput. Syst. Sci. **C–38**(1), 36–67 (1989)

Size- and Port-Aware Horizontal Node Coordinate Assignment

Ulf Rüegg[(✉)], Christoph Daniel Schulze, John Julian Carstens,
and Reinhard von Hanxleden

Department of Computer Science, Christian-Albrechts-Universität zu Kiel,
Kiel, Germany
{uru,cds,jjc,rvh}@informatik.uni-kiel.de

Abstract. The approach by Sugiyama et al. is widely used to automatically draw directed graphs. One of its steps is to assign horizontal coordinates to nodes. Brandes and Koepf presented a method that proved to work well in practice. We extend this method to make it possible to draw diagrams with nodes that have considerably different sizes and with edges that have fixed attachment points on a node's perimeter (ports). Our extensions integrate seamlessly with the original method and preserve the linear execution time.

1 Introduction

The layer-based approach to graph layout as introduced by Sugiyama et al. [6] is a well-established methodology to automatically draw directed graphs in the plane. It is defined as a pipeline of three subsequent phases: *node layering* distributes the nodes into subsequent layers such that edges only point from lower to higher layers; *crossing minimization* orders the nodes in each layer such that the number of edge crossings is minimized; finally x-coordinate assignment (or *node placement*) determines x coordinates for nodes. In practice, an initial *cycle breaking* phase as well as a final *edge routing* phase are often added to support cyclic graphs and non-simple edge routing styles.

In the area of model-driven engineering (MDE), graphical languages are often used to model complex software systems. For instance, tools such as *LabVIEW* (National Instruments), *EHANDBOOK* (ETAS), and *Ptolemy* (UC Berkeley) allow to model systems using *data flow diagrams* and make use of automatic layout algorithms to arrange nodes and edges. In such diagrams edges are usually routed in an orthogonal fashion and connect to nodes through dedicated attachment points on a node's boundary (so-called *ports*). Also, nodes have considerably different sizes, see Fig. 1 for examples.

All of these characteristics pose challenges for automatic graph drawing algorithms that are rarely addressed by existing solutions. Previous work by Schulze et al. [5] introduced methods that extend the layer-based approach to support the special requirements of data flow diagrams, focusing on crossing minimization and edge routing. In this paper, we focus on node placement.

© Springer International Publishing Switzerland 2015
E. Di Giacomo and A. Lubiw (Eds.): GD 2015, LNCS 9411, pp. 139–150, 2015.
DOI: 10.1007/978-3-319-27261-0_12

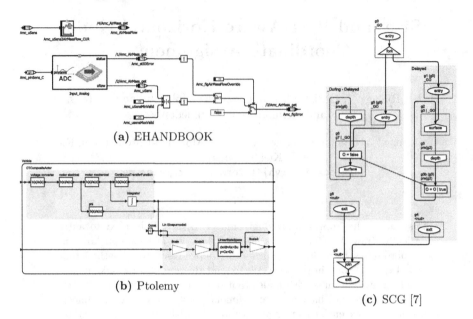

(a) EHANDBOOK

(b) Ptolemy

(c) SCG [7]

Fig. 1. Exemplary drawings using the methods presented here. They would not be drawable solely with the existing algorithm [1] since they contain ports and nodes of considerably different sizes.

While we refer to Healy and Nikolov [3] for a general overview of existing node placement approaches, it is worth noting that most of them try to a certain extent to reduce the number of edge bend points. For one thing, the approach introduced by Sander [4] ensures that long edges are always drawn straight, but uses a barycenter-like balanced placement for all other edges. Once a node has more than one outgoing edge, this usually results in two bend points per edge. For another thing, the approach introduced by Brandes and Koepf [1], extending ideas of Buchheim et al. [2], tries to draw as many edges straight as possible.

Contributions. Brandes and Koepf assume that all nodes have the same size and do not take ports into account; thus their algorithm straightens at most one outgoing edge per node. In this paper, we extend the approach by Brandes and Koepf to remove these restrictions and take the opportunity to place nodes such that more than one outgoing edge per node can be drawn straight. This leads to drawings as seen in Fig. 1. Throughout the paper we will assume that the node placement algorithm cannot change the size of nodes and the position of ports.

Outline. Following the usual conventions, we start by introducing the required terminology in the next section. Sect. 3 then gives an overview of the algorithm by Brandes and Koepf before Sect. 4 introduces our extensions. We evaluate our algorithm in Sect. 5 and close with a conclusion and future work in Sect. 6.

2 Preliminaries

Let $G = (V, P, \pi, E)$ denote a directed graph with ports, where V is a set of nodes and P a set of ports, i.e. attachment points on a node's boundary. $\pi : P \mapsto V$ assigns each port to a node. $E \subseteq P \times P$ is a set of directed edges connecting the ports.

During the first steps of the layer-based approach cyclic graphs are made acyclic, a *layering* is calculated, and an ordering is determined for each layer. A layering $\mathcal{L}$ is an ordered partition of V into non-empty *layers* $L_1, \ldots, L_{|\mathcal{L}|}$ and $\mathcal{L}(v) \rightarrow \{1, \ldots, |\mathcal{L}|\}$ maps each node $v \in V$ to the index of its respective layer. Since all edges must point in the same direction, $\mathcal{L}(\pi(p)) < \mathcal{L}(\pi(q))$ must hold for all edges $(p, q) \in E$. An edge (p, q) is *short* if $\mathcal{L}(\pi(q)) - \mathcal{L}(\pi(p)) = 1$; it is *long* otherwise. A layering is *proper* if all edges are short. Note that a layering can be made proper by splitting long edges and introducing *dummy nodes*. We refer to the short edges of a proper layering as *edge segments*. That is, an original edge can be represented by one or more edge segments. Each layer $L_i \in \mathcal{L}$ is an ordered tuple of nodes $(v_1^i, \ldots, v_n^i)$, where $n = |L_i|$. The position of a node v_j^i in layer i is $pos(v_j^i) = j$ and the predecessor of a node v_j^i with $j > 1$ is $pred(v_j^i) = v_{j-1}^i$. This gives a properly layered, directed, acyclic graph with ports (LDAGP) $G' = (V', P', \pi', E', \mathcal{L})$. The set of nodes now includes a set of dummy nodes $\mathcal{D}$ such that $V' = V \cup \mathcal{D}$. For each dummy node two ports are introduced and edges are added and reconnected accordingly.

Finally, let $width : V' \mapsto \mathbb{R}$ assign a width to each node. Throughout this paper, we assume that for an edge $(p, q) \in E$, p is on the lower boundary of $\pi(p)$ and q is on the upper boundary of $\pi(q)$ to prevent edges from crossing nodes. Let $x_p : P' \mapsto \mathbb{R}$ assign positions to ports relative to the leftmost point on their respective boundary.

3 The Original Algorithm

In this section we give a brief summary of the original algorithm of Brandes and Koepf. For further details we refer to the paper itself [1]. The basic idea of the algorithm is to traverse a given graph in different directions to calculate four extremal layouts and combine them into a balanced final layout. The algorithm is divided into the following steps: (1) During *Vertical Alignment* nodes are combined into so-called *blocks*. Different directions may result in different blocks. Edges between the nodes in a block will be drawn straight. (2) *Horizontal Compaction* moves the calculated blocks as close to each other as possible and assigns explicit x coordinates to nodes. Depending on the direction nodes are either compacted leftwards or rightwards. (3) *Balancing* combines the four extremal layouts resulting from the previous two steps to a final drawing.

A direction is a combination of traversing the layers of $\mathcal{L}$ either downwards or upwards and traversing the nodes in each layer either rightwards or leftwards. For brevity, we will limit our explanations and examples throughout this paper to the combination of downwards and rightwards. The other three combinations are easy to infer.

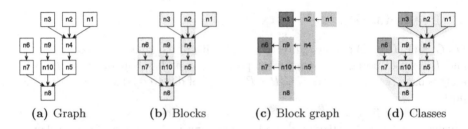

(a) Graph   (b) Blocks   (c) Block graph   (d) Classes

Fig. 2. Gray boxes in (b) show the calculated alignments (blocks) for the graph (a). In (c) the block graph is depicted with the two sinks in darker gray, (d) shows the corresponding classes.

During the alignment step nodes are aligned with their median neighbor in the preceding layer. Consecutively aligned nodes are referred to as a block, see Fig. 2b for an illustration. Let $\mathcal{B}$ denote the set of blocks of an LDAGP G', where each block $b \in \mathcal{B}$ is represented as an ordered tuple of edges $(e_0, \ldots, e_n)$. For compaction, an auxiliary *block graph* is constructed as seen in Fig. 2c. Blocks are the nodes in the block graph and are connected by an edge if two nodes of different blocks are consecutive in their layer. Within the block graph, blocks are divided into *classes*. A class is defined by a unique sink that is reachable by all of the class's nodes. Positions subject to a global separation value δ_g are then assigned to blocks using a longest path layering within each class, which recursively assigns positions relative to the class's sink. If two adjacent blocks are part of the same class, their relative positions can be determined immediately. If they belong to different classes, the blocks impose a minimum required separation between the involved classes. This separation is remembered and applied after all blocks have been placed.

As mentioned earlier, the original approach does not cater for varying node sizes and ports. For one thing, ports reveal two problems that are illustrated in Fig. 3a. First, in the depicted graph no edge is drawn straight even though all nodes of the blocks B1 and B2 are neatly left-aligned. Second, node n1 has two ports both of which would allow the connected edge to be drawn straight. Yet, n1 and n4 are part of different blocks that will be separated during the compaction step. In addition, different node sizes increase the two aforementioned problems and render the global separation value δ_g impractical. δ_g would have to be larger than the widest node of the graph to avoid overlapping nodes, possibly leaving a lot of whitespace. Figure 3b and c show two drawings that would be more desirable using a local separation δ_l. Furthermore, in conjunction with orthogonally drawn edges, as opposed to general polylines, the balancing step often yields undesirable bendpoints (see Fig. 3d). For this reason we consider the balancing step to be optional and, if discarded, choose the final layout out of the four possible candidates based on the smallest width.

During the rest of this paper, we will keep our explanations and pseudo code as close as possible to the style and notation of the original paper. There, the following data records are used for a node $v \in V$: $root[v]$ denotes the root node

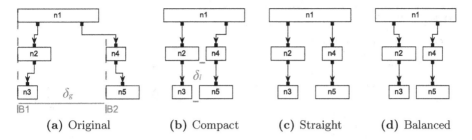

(a) Original   **(b)** Compact   **(c)** Straight   **(d)** Balanced

Fig. 3. Illustration of the additional challenges for the node placement phase imposed by ports. In (a) a global separation value δ_g is used to space blocks B1 and B2, and ports are neglected. (b) shows a compact drawing where ports are considered and the same blocks can flow into one another using a local separation δ_l. (d) shows the result of executing a balancing step (as it is part of the original algorithm [1]) in conjunction with orthogonally drawn edges that, as opposed to (c), yields more edge bends.

of v's block; $align[v]$ maps to the next node within v's block in the current iteration direction and represents a cyclically linked list; $sink[v]$ stores the sink of the class v belongs to; $shift[v]$ holds the distance by which the class of v should be moved during compaction.

4   Size- and Port-Aware Node Coordinate Assignment

In this section we present our extensions. First, we add a step that we call *inner shift*. It calculates offsets for nodes within a block to account for ports and simultaneously determines the width of the blocks, which is required to calculate the size of a layout. Second, we extend the compaction phase to consider node sizes when calculating explicit x-coordinates. Third, we modify the objective such that more straight edges are, to a certain extent, favored over achieving the most compact layout possible. All additions integrate seamlessly with the original algorithm and preserve its linear execution time.

While the first two modifications share the objectives of the original algorithm, the third one considers straight edges to be more important than compactness. Since different diagram types demand different aesthetics, the third change is optional in our algorithm.

Node Size and Port Support. The original algorithm assigns the same x-coordinate to all nodes within a block. This automatically yields straight edges if all nodes have the same size and the same attachment points for edges. Here we extend this in two ways. First, blocks have a width that depends on the sizes of the block's nodes. Second, each node has an *inner shift*, which is an offset relative to a block's left border. The inner shift is used to properly deal with ports.

Given a set of blocks $\mathcal{B}$ calculated by the vertical alignment method of the original algorithm, we execute Algorithm 1. For each block, it iterates through the block's edges (p, q), considers p to be fixed and determines an offset value for

---

Algorithm 1. inner_shift

Input: LDAGP with blocks $\mathcal{B}$
Output: innerShift[v] (inner-block offset of node v),
  blockSize[b] (size of block b)
1   **function** inner_shift()
2   innerShift[v] $\leftarrow$ 0 $\forall v \in V$
3   **for** $b \in \mathcal{B}$ **do**
4   left $\leftarrow$ 0; right $\leftarrow$ 0
5   **for** $(p,q) \in b$ **do**
6         $s \leftarrow$ innerShift[$\pi(p)$] + $x_p(p) - x_p(q)$
7   innerShift[$\pi(q)$] $\leftarrow s$
8   left $\leftarrow min$(left, s)
9   right $\leftarrow max$(right, s+width($\pi(q)$))
10   **for** all nodes v in block b **do**
11   innerShift[v] $-=$ left
12   blockSize[b] $\leftarrow$ right $-$ left

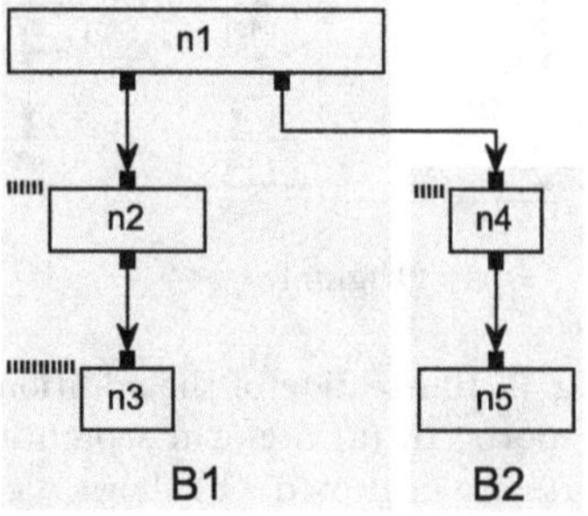

Fig. 4. Illustration of the inner shift. Nodes n2, n3, and n4 have an inner shift value different from zero. It defines the offset within the node's block and is depicted by the dashed lines. Also, both blocks B1 and B2 have an extent.

$\pi(q)$ such that (p,q) can be drawn straight. Additionally, the maximum extent of the nodes to either side of the starting node's leftmost coordinate is recorded. Using these values, the size of each block is calculated and all inner shift values are shifted to be relative to the leftmost coordinate of any of the block's nodes. The block size is used to determine the width of each extremal layout. Figure 4 illustrates the effect of the inner shift.

Given an inner shift for the nodes of each block, the horizontal compaction technique is applied with the alterations seen in lines 10 and 13 of Algorithm 2. Contrary to the original method, the inner shift and the width of the nodes are considered while iterating through the block. Note that we consider the individual width of every node and do not use the overall width of a block. This allows blocks to "flow" into each other, as seen in Fig. 3b.

Moreover, the inner shift of a node and its size have to be considered during the final balancing step, which is easy to incorporate into the original algorithm.

Improving Straightness. A wider node can allow for more than one edge to be drawn straight. The original algorithm did not have to address this since nodes were considered to be uniform. We solve this as follows. Remember that our extended compaction step as shown in Algorithm 2 compacts blocks and classes as much as possible. This implies that for a given iteration direction only such edges are possible candidates for additional straightening where one of the involved blocks was moved "too far" (for instance node n4 in Fig. 3b). In other words, we have to prevent the blocks of such edges to be compacted too far in order to get more straight edges.

The procedure we apply can be seen in Fig. 5. In (a) everything is compacted as much as possible. In (b) a threshold value thresh is used to prevent node n5 from moving further to the left, resulting in a straight edge.

Algorithm 2. place_block

Input: v (root node of a block)
1 **function** place_block(v)
2   **if** x[v] undefined **then**
3   x[v] ← 0; initial ← true; w ← v
4   **repeat**
5   **if** pos[w] > 0 **then**
6   n ← pred[w]; u ← root[n]
7   place_block(u)
8   **if** sink[v] = v **then** sink[v] ← sink[u]
9   **if** sink[v] ≠ sink[u] **then**
10   s_c ← x[v] + innerShift[w] − x[u] − innerShift[n] − width[n] − δ_l
11   shift[sink[u]] ← min(shift[sink[u]], s_c)
12   **else**
13   s_b ← x[u] + innerShift[n] + width[n] − innerShift[w] + δ_l
14   **if** initial **then** x[v] ← s_b **else** x[v] ← max(x[v], s_b)
15   initial ← false
16   w ← align[w]
17   **until** w = v

A threshold value can, however, only be determined if the connected block is already placed. Consider Fig. 5a and the iteration direction down and right. The algorithm starts by placing block B2, but has to place block B1 before it can finish B2. Now, when placing node n4, no threshold can be calculated because node n3 has not been placed yet. In such a case we delay the straightening of the outgoing edge of n4 until all blocks have been placed. A queue is used to store such edges. Imagine a further node n4' connected to n3 and located between n4 and n2. Just as n4 it will be delayed. To give both edges a chance to be straightened later, it is important to post-process n4 prior to n4'. Using a queue allows to do exactly this. When all blocks have been placed we fetch edge by edge from the queue and check for the involved block how far it can be moved without exceeding the threshold or overlapping other nodes. This way the edge becomes either straight or shortens as much as possible (see Fig. 5c for a final result). Algorithm 3 shows the modification of the place_block function. A threshold value is calculated and used as an additional bound for a new block position s_b. We only list code that is added to Algorithm 2 and prefix code lines by a fractional number, e.g. 2.5 to denote an addition between lines 2 and 3.

As it is now, thresholds are only calculated for edges incident to either the root or the last node of a block. Note that blocks can consist of a single node in which case this node is both the root and the last node of the block. To pick an edge in line 4 of Algorithm 3 we use the first edge incident to w that connects to an already placed node. There are three points by which this procedure might be improved in the future: (1) check whether edges between nodes that are neither root nor last in a block can be drawn straight, i.e. calculate thresholds for those nodes as well (2) when multiple edges incident to a block are candidates to be drawn straight, choose the one that allows the most compact layout, i.e. the one with the smallest difference of node and threshold value, and (3) be more intelligent in picking an edge instead of just using the first one that is

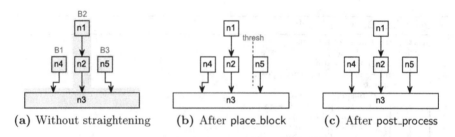

(a) Without straightening   (b) After place_block   (c) After post_process

Fig. 5. Illustration of the procedure to straighten additional edges during the compaction step. A threshold value is used to prevent n5 in (b) from compacting "too far" in order to get an additional straight edge.

encountered. Nevertheless, the described method removes bends on edges that, to a human, are obvious candidates for straightening. For instance, nodes that are connected by a single edge to a larger node.

Execution Time. For an LDAGP G', the original algorithm runs in time linear to the number of nodes and edge segments, $O(|V'| + |E'|)$. Algorithm 1 is linear in the number of edge segments that are involved in blocks. Algorithm 2 only adds constant time operations to the procedure of the original algorithm. Algorithm 3 additionally calculates the threshold value which influences which edge will be picked later. To pick an edge, for every node the incident edge segments are touched at most once. Adding elements to and removing them from a queue can be done in constant time and the post processing step is bounded by the number of nodes and edge segments. Therefore, the overall execution time remains linear in the number of nodes and edge segments.

5   Evaluation

All drawings seen in Fig. 1 were created using the methods of Schulze et al. [5] in combination with our extensions. The methods are implemented in the KLay Layered algorithm and the drawings are created using the KLighD framework, both of which are part of the KIELER open source project.[1]

Recall that we present two contributions here: (1) Supporting varying node sizes and ports, which allows us to draw more diagram types in the first place. (2) The possibility to further increase the number of straight edges if desired. To measure the performance of our second contribution, we need to know how many edges can be drawn straight theoretically for a given graph without violating any overlap and separation constraints. To obtain such numbers we formulated an optimization problem and solved it using CPLEX. In 38 out of 9729 layout executions the solver did not finish within our set time limit of one hour. Nevertheless, we use the reported results since they are always equal to or better than

---

[1] http://www.rtsys.informatik.uni-kiel.de/en/research/kieler

---

Algorithm 3. place_block with straightening

Input: v: root node of a block
function place_block(v)
   | 2.5: thresh $\leftarrow -\infty$
   | 7.5, 15.5 (as **else** of **if** in line 5): thresh $\leftarrow$ calculate_threshold(v, u, thresh)
   | 13: $s_b \leftarrow max($thresh, $x[u] + $innerShift$[n] + $width$[n] - $innerShift$[w] + \delta_l)$

Input: v (root node of a block); w (current node); ot (current threshold value); Q (queue)
1 **function** calculate_threshold(v, w, ot)
2 | thresh $\leftarrow ot$
3 | **if** $v = w$ **then**
4 | $(p, q) \leftarrow$ pick incoming edge of w
5 | **if** block of root$[\pi(p)]$ placed **then**
6 | thresh $\leftarrow x[$root$[\pi(p)]] + $innerShift$[\pi(p)] + x_p(p) - $innerShift$[\pi(q)] - x_p(q)$
7 | **else if** w has incoming edges **then**
8 | enqueue w to Q

9 | **if** thresh $= -\infty$ and align$[w] = v$ **then**
10 | symmetric to before, this time picking an outgoing edge

11 | **return** thresh

12 // the following method is called after all blocks have been placed
13 **function** post_process()
14 | **while** Q not empty **do**
15 | $w \leftarrow$ dequeue from Q
16 | $(p, q) \leftarrow$ previously picked edge for w (line 4)
17 | $t_1 \leftarrow x[$root$[\pi(p)]] + $innerShift$[\pi(p)] + x_p(p)$
            $- x[$root$[\pi(q)] - $innerShift$[\pi(q)] - x_p(q)$
18 | $t_2 \leftarrow$ minimum distance between block of w and its neighbors
19 | $t \leftarrow$ **if** $abs(t_1) < abs(t_2)$ **then** t_1 **else** t_2
20 | move all nodes v in w's block by t

---

the results of the constructive methods discussed above and thus provide reasonable bounds. As a second metric we use the width of a drawing. It was already noted by Brandes and Koepf that straightening edges might hamper reducing the width of a drawing [1]. A drawing with minimum width can be achieved by placing all nodes of a layers as close to each other as possible and centering every layer in the drawing. Now, given an optimum number of straight edges and a minimum width for a certain graph, we can compare the performance of our extensions, once without straightening (BK) and once with straightening (BKS). It did not make much sense to try and compare our algorithm to the plain original or to other node placement algorithms as they either do not support ports and variable node sizes or do not try to maximize the number of straight edges, or both.

We use four different diagram types for the evaluation: (1) randomly generated graphs with same-sized nodes, (2) data flow diagrams shipping with the academic Ptolemy project[2], (3) data flow diagrams from the commercial interactive model browsing solution EHANDBOOK[3], and (4) SCGs, which are specialized control flow graphs for sequentially constructive programs [7]. The Ptolemy and

---

[2] http://ptolemy.eecs.berkeley.edu/
[3] http://www.etas.com/de/products/ehandbook.php

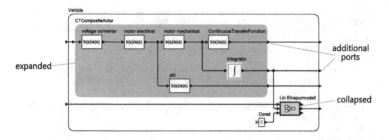

Fig. 6. Same diagram as in Fig. 1b, however, now with the bottom right node collapsed. Note how it is not possible anymore, as opposed to Fig. 1b, to draw the edge of the Const node straight.

EHANDBOOK diagrams are meant to be navigated using an expand/collapse mechanism. Figure 6 shows a diagram with both an expanded hierarchical node and a collapsed one. Scenarios where more than one edge can be drawn straight are more likely in the presence of expanded nodes as they are wider. We therefore fully expanded existing diagrams for our evaluation and then extracted each hierarchical level into a separate diagram. KLay supports hierarchical graphs by introducing additional ports where an edge crosses a hierarchy boundary, see for instance Fig. 6. The layout is then performed in a bottom-up fashion and additional ports are considered to be dummy nodes. After the evaluations we realized that the aforementioned extraction of subdiagrams kept several edges from being drawn straight since additional ports were fixed at disadvantageous positions. We believe the results could be better than reported.

Table 1 summarizes the characteristics of each type of diagram and Fig. 7 shows a scatter plot for each one of them. It can be seen that for diagrams with same-sized nodes BK finds optimal or near-optimal solutions. The other three plots indicate that while BK's overall performance is still very good, there are diagrams for which the number of straight edges can be improved. This is due to variable node sizes. BKS performs better here. The overall number of straight edges increases as well as the number of diagrams for which an optimum solution is found. For SCGs BKS produces more straight edges for almost every diagram. The average width of the tested diagrams on the other hand does not increase notably, which implies that for the tested graphs the additionally straightened edges did not negatively affect the width.

Execution Time. We measured the execution time of BK and BKS using randomly generated graphs with 40 different node counts between 10 and 1000, 1.5 edges per node, and node widths varying between 20 and 100. For each graph size, we generated 10 random graphs and ran the algorithm 10 times, using the average execution time as result. The tests were executed using a 64 bit JVM on a laptop with an Intel i7 2 GHz CPU and 8 GB memory.

For graphs with up to 100 nodes both strategies finish in under 2.5ms and require about 62ms for graphs with 1000 nodes. The average difference between BK and BKS is below 1ms. Therefore, both strategies are fast enough to be used in interactive modeling and browsing tools.

Table 1. Summary of the evaluation data. For each diagram type the number of diagrams d is listed alongside the average number of nodes $\bar{n}$ and edges $\bar{m}$ per diagram. IE is the percentage increase for BKS compared to BK in the overall sum of all diagrams' straight edges. IS indicates the increase of the average diagram size. By size we mean the width of top-down drawings and the height of left-right drawings. ID represents the number of diagrams for which BKS found more straight edges than BK. OBK and $OBKS$ represent the number of diagrams for which BK and BKS found the optimum number of straight edges.

Type	d	$\bar{n}$	$\bar{m}$	$IE(\%)$	$IS(\%)$	$ID(\%)$	$OBK(\%)$	$OBKS(\%)$
Random	106	29.5	46.5	0.1	0.0	4.7	58.5	60.4
SCGs	107	134.4	268.7	3.5	0.0	96.3	2.8	47.7
EHANDBOOK	97	21.6	24.1	3.7	2.1	18.6	58.8	66.0
Ptolemy	1140	10.6	13.7	2.3	0.2	15.4	74.6	87.0

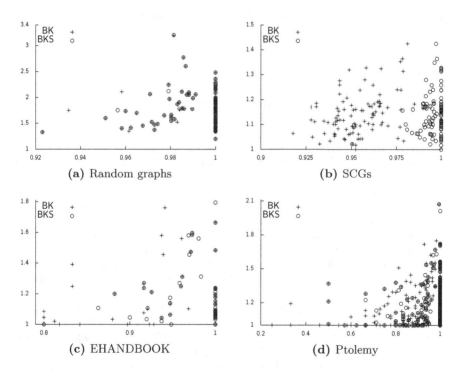

(a) Random graphs   (b) SCGs

(c) EHANDBOOK   (d) Ptolemy

Fig. 7. A scatter plot for each diagram type. The performance in terms of straight edges (x-axis) is plotted against the diagram size (y-axis). The diagram size is either the width or the height of the diagram depending on the layout direction. Each data point represents the performance of BK (or BKS) for a given graph instance relative to the optimum performance for that graph. Thus, the closer a data point is to the bottom right corner of the coordinate system (or 1.0) the better.

6   Final Remarks

We presented extensions to the node placement algorithm presented by Brandes and Koepf [1] to support different node sizes and ports. These extensions make the algorithm usable for a wider range of diagram types, including data flow diagrams. We evaluated our extensions on randomly generated diagrams as well as on three sets of real-world diagrams and found that the results often were near the optimum in terms of straight edges and compactness does not suffer. Performance-wise, the algorithm fares well enough to be used in interactive applications.

For certain graphs, straightening edges may still lead to less compact diagrams. Our intuition is that drawing very few edges in a given diagram non-straight would often lead to a more compact layout. Future work could go into confirming or refuting this intuition and developing methods to find such edges.

Acknowledgements. This work was supported by the German Research Foundation under the project *Compact Graph Drawing with Port Constraints* (ComDraPor, DFG HA 4407/8-1).

References

1. Brandes, U., Köpf, B.: Fast and simple horizontal coordinate assignment. In: Mutzel, P., Jünger, M., Leipert, S. (eds.) GD 2001. LNCS, vol. 2265, p. 31. Springer, Heidelberg (2002)
2. Buchheim, C., Jünger, M., Leipert, S.: A fast layout algorithm for k-level graphs. In: Marks, J. (ed.) GD 2000. LNCS, vol. 1984, pp. 229–240. Springer, Heidelberg (2001)
3. Healy, P., Nikolov, N.S.: Hierarchical drawing algorithms. In: Tamassia, R. (ed.) Handbook of Graph Drawing and Visualization, pp. 409–453. CRC Press, Boca Raton (2013)
4. Sander, G.: A fast heuristic for hierarchical Manhattan layout. In: Brandenburg, F.J. (ed.) GD 1995. LNCS, vol. 1027, pp. 447–458. Springer, Heidelberg (1996)
5. Schulze, C.D., Spönemann, M., von Hanxleden, R.: Drawing layered graphs with port constraints. J. Vis. Lang. Comput. Spec. Issue Diagr. Aesthet. Layout **25**(2), 89–106 (2014)
6. Sugiyama, K., Tagawa, S., Toda, M.: Methods for visual understanding of hierarchical system structures. IEEE Trans. Syst. Man Cybern. **11**(2), 109–125 (1981)
7. von Hanxleden, R., Mendler, M., Aguado, J., Duderstadt, B., Fuhrmann, I., Motika, C., Mercer, S., O'Brien, O., Roop, P.: Sequentially Constructive Concurrency–a conservative extension of the synchronous model of computation. ACM Trans. Embed. Comput. Syst. Spec. Issue Appl. Concurrency Syst. Des. **13**(4s), 144:1–144:26 (2014)

Area, Bends, Crossings

Small-Area Orthogonal Drawings
of 3-Connected Graphs

Therese Biedl[1] and Jens M. Schmidt[2]([✉])

[1] David R. Cheriton School of Computer Science,
University of Waterloo, Waterloo, Canada
[2] Institute of Mathematics, TU Ilmenau, Ilmenau, Germany
jens.schmidt@tu-ilmenau.de

Abstract. It is well-known that every graph with maximum degree 4 has an orthogonal drawing with area at most $\frac{49}{64}n^2 + O(n) \approx 0.76n^2$. In this paper, we show that if the graph is 3-connected, then the area can be reduced even further to $\frac{9}{16}n^2 + O(n) \approx 0.56n^2$. The drawing uses the 3-canonical order for (not necessarily planar) 3-connected graphs, which is a special Mondshein sequence and can hence be computed in linear time. To our knowledge, this is the first application of a Mondshein sequence in graph drawing.

1 Introduction

An orthogonal drawing of a graph $G = (V, E)$ is an assignment of vertices to *points* and edges to *polygonal lines* connecting their endpoints such that all edge-segments are horizontal or vertical. Edges are allowed to intersect, but only in single points that are not bends of the polygonal lines. Such an orthogonal drawing can exist only if every vertex has degree at most 4; we call such a graph a *4-graph*. It is easy to see that every 4-graph has an orthogonal drawing with area $O(n^2)$, and this is asymptotically optimal [17].

For planar 2-connected graphs, several authors showed independently [10,15] how to achieve area $n \times n$, and this is optimal [16]. We measure the drawing-size as follows. Assume (as we do throughout the paper) that all vertices and bends are at points with integral coordinates. If H rows and W columns of the integer grid intersect the drawing, then we say that the drawing occupies a $W \times H$-*grid* with *width* W, *height* H, *half-perimeter* $H + W$ and *area* $H \cdot W$. Some papers count as width/height the width/height of the smallest enclosing axis-aligned box. This is one unit less than with our measure.

For arbitrary graphs (i.e., graphs that are not necessarily planar), improved bounds on the area of orthogonal drawings were developed much later, decreasing from $4n^2$ [11] to n^2 [1] to $0.76n^2$ [9]. (In all these statements, we omit lower-order terms for ease of notation.)

T. Biedl—Supported by NSERC.

E. Di Giacomo and A. Lubiw (Eds.): GD 2015, LNCS 9411, pp. 153–165, 2015.
DOI: 10.1007/978-3-319-27261-0_13

Our Results: In this paper, we decrease the area-bound for orthogonal drawings further to $0.56n^2 + O(n)$ under the assumption that the graph is 3-connected. The approach is similar to the one by Papakostas and Tollis [9]: add vertices to the drawing in a specific order, and pair some of these vertices so that in each pair one vertex re-uses a row or column that was used by the other. The main difference in our paper is that 3-connectivity allows the use of a different, stronger, vertex order.

It has been known for a long time that any *planar* 3-connected graph has a so-called canonical order [7], which is useful for planar graph drawing algorithms. It was mentioned that such a canonical order also exists in non-planar graphs (e.g. in [4, Remark on p.113]), but it was not clear how to find it efficiently, and it has to our knowledge not been used for graph drawing algorithms. Recently, the second author studied the so-called Mondshein sequence, which is an edge partition of a 3-connected graph with special properties [8], and showed that it can be computed in linear time [13]. A Mondshein sequence is the appropriate generalization of the canonical order to (not necessarily planar) 3-connected graphs [13] and is most naturally defined by ear decompositions. However, in order to highlight its relation to canonical orders, we define a Mondshein sequence here as a special vertex partition and call it a 3-canonical order.

We use this 3-canonical order to add vertices to the orthogonal drawing. This almost immediately lowers the resulting area, because vertices with one incoming edge can only occur in chains. We then mimic the pairing-technique of Papakostas and Tollis, and pair groups of the 3-canonical order in such a way that even more rows and columns can be saved, resulting in a half-perimeter of $\frac{3}{2}n + O(1)$ and the area-bound follows.

No previous algorithms were known that achieve smaller area for 3-connected 4-graphs than for 2-connected 4-graphs. For *planar* graphs, the orthogonal drawing algorithm by Kant [7] draws 3-connected planar 4-graphs with area $(\frac{2}{3}n)^2 + O(n)$ [14], while the best-possible area for planar 2-connected graphs is n^2 [16].

2   Preliminaries

Let $G = (V, E)$ be a graph with $n = |V|$ vertices and $m = |E|$ edges. The *degree* of a vertex v is the number of incident edges. In this paper, all graphs are assumed to be *4-graphs*, i.e., all vertex degrees are at most 4. A graph is called *4-regular* if every vertex has degree exactly 4; such a graph has $m = 2n$ edges.

A graph G is called *connected* if, for any two vertices u, v, there is a path in G connecting u and v. It is called 3-connected if $n > 3$ and, for any two vertices u, v, the graph $G - \{u, v\}$ is connected.

A *loop* is an edge (v, v) that connects an endpoint with itself. A *multi-edge* is an edge (u, v) for which another copy of edge (u, v) exists. When not otherwise stated, the graph G that we want to draw is *simple*, i.e., it has neither loops nor multi-edges. While modifying G, we will sometimes temporarily add a *double edge*, i.e., an edge for which exactly one other copy exists (we refer always to the added edge as double edge, the copy is not a double edge).

2.1   The 3-Canonical Order

Definition 1. Let G be a 3-connected graph. A *3-canonical order* (or *Mondshein sequence*) is a partition of V into groups $V = V_1 \cup \cdots \cup V_k$ such that

- $V_1 = \{v_1, v_2\}$, where (v_1, v_2) is an edge.
- $V_k = \{v_n\}$, where (v_1, v_n) is an edge.
- For any $1 < i < k$, one of the following holds:
    - $V_i = \{z\}$, where z has at least two predecessors and at least one successor.
    - $V_i = \{z_1, \ldots, z_\ell\}$ for some $\ell \geq 2$, where
        - $z_1, \ldots, z_\ell$ is an induced path in G (i.e. edges $z_1 - z_2 - \cdots - z_\ell$ exist, and there are no edges (z_i, z_j) with $i < j - 1$),
        - z_1 and z_ℓ have exactly one predecessor each, and these predecessors are different,
        - z_j for $1 < j < \ell$ has no predecessor,
        - $z_j \in V_i$ for $1 \leq j \leq \ell$ has at least one successor.

Here, a *predecessor* [*successor*] of a vertex in V_i is a neighbor that occurs in a group V_h with $h < i$ [$h > i$]. See Fig. 1 for a 3-canonical order.

We call a vertex group V_i a *singleton* if $|V_i| = 1$, and a *chain* if $|V_i| \geq 2$ and $i \geq 2$. We distinguish chains further into *short chains* with $|V_i| = 2$ and *long chains* with $|V_i| \geq 3$. A 3-canonical order imposes a natural orientation on the edges of the graph from lower-indexed groups to higher-indexed groups and, for edges within a chain, from one (arbitrary) end of the path to the other. This implies in-degree $indeg(v) \geq 2$ for any singleton, $indeg(v) = 2$ for exactly one vertex of each chain, and $indeg(v) = 1$ for all other vertices of a chain.

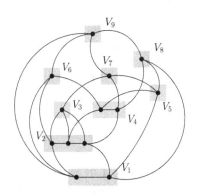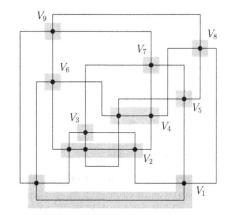

Fig. 1. A 4-regular 3-connected graph with a 3-canonical order, and the drawing created with our algorithm. For illustrative purposes, we show the drawing exactly as created, even though many more grid lines and bends could be saved with straightforward compaction steps. V_2 is a long chain, V_4 is a short chain, V_5 is a 2-2-singleton, V_3, V_6, V_7 and V_8 are 3-1-singletons.

Numerous related methods of ordering vertices of 3-connected graphs exist, e.g. *(2,1)-sequences* [8], *non-separating ear decompositions* [2,13], and, limited to planar graphs, *canonical orders for maximal planar graphs* [6], *canonical orders for 3-connected planar graphs* [7] and *orderly spanning trees* [3]. A Mondshein sequence (i.e. a 3-canonical order) of a 3-connected graph implies all these orders, up to minor subtleties.

The most efficient way known to compute a Mondshein sequence (proving in particular that one exists) uses *non-separating ear decompositions* [2,13]. This is a partition of the edges into *ears* $P_1 \cup \cdots \cup P_k = E$ such that P_1 is an induced cycle, P_i for $i > 1$ is a non-empty induced path that intersects $P_1 \cup \cdots \cup P_{i-1}$ in exactly its endpoints, and $G - \bigcup_{j=1}^{i} P_j$ is connected for every $i < k$. Such a non-separating ear decomposition exists for any 3-connected graph [2], and we can even fix two edges $v_1 v_2$ and $v_2 v_n$ and require that $v_1 v_2$ is in the cycle P_1 and that v_n is the only vertex in P_k; hence, P_k will be a singleton.

Further, such a non-separating ear decomposition can be computed in linear time [12,13]. The sets of newly added vertices for each P_i will be the vertex groups of a 3-canonical order (additionally, P_1 is split into the groups $V_1 := \{v_1, v_2\}$ and $V_2 := V(P_1) - \{v_1, v_2\}$). Although vertices in V_i may have arbitrarily many incident edges in a non-separating ear decomposition, we can easily get rid of these extra edges by a simple short-cutting routine in linear time (see Lemmas 8 and 12 in [12]). This gives a 3-canonical order. Thus, a linear-time algorithm for computing a 3-canonical order follows immediately from the one for non-separating ear decompositions.

2.2   Making 3-Connected 4-Graphs 4-Regular

It will greatly simplify the description of the algorithm if we only give it for 4-regular graphs. Thus, we want to modify a 3-connected 4-graph G such that the resulting graph G' is 4-regular, draw G', and then delete added edges to obtain a drawing of G. However, we must maintain a simple graph since the existence of 3-canonical orders depends on simplicity. This turns out to be impossible (e.g. for the graph obtained from the octahedron by subdividing two distinct edges with a new vertex and joining the new vertices by an edge), but allowing one double edge is sufficient.

Lemma 2. *Let G be a simple 3-connected 4-graph with $n \geq 5$. Then we can add edges to G' such that the resulting graph G' is 3-connected, 4-regular, and has at most one double edge.*

Proof. Since G is 3-connected, any vertex has degree 3 or 4. If there are four or more vertices of degree 3, then they cannot be mutually adjacent (otherwise $G = K_4$, which contradicts $n \geq 5$). Hence, we can add an edge between two non-adjacent vertices of degree 3; this maintains simplicity and 3-connectivity.

We repeat until only two vertices of degree 3 are left (recall that the number of vertices of odd degree is even). Now an edge between these two vertices is added, even if one existed already; this edge is the only one that may become a double edge. The resulting graph is 4-regular and satisfies all conditions.   □

3   Creating Orthogonal Drawings

From now on, let G be a 3-connected 4-regular graph that has no loops and at most one double edge. Compute a 3-canonical order $V = V_1 \cup \cdots \cup V_k$ of G with $V_k = \{v_n\}$, choosing $v_1 v_n$ to be the double edge if there is one. Let x^{short} and x^{long} be the number of short and long chains. Let $x^{j-\ell}$ be the number of vertices with in-degree j and out-degree ℓ. Since G is 4-regular, we must have $j + \ell = 4$. A j-ℓ-*singleton* is a vertex z that constitutes a singleton group V_i for $1 < i \leq k$ and that has in-degree j and out-degree ℓ.

Observation 3. *Let G be a 4-regular graph with a 3-canonical order. Then*

1. $x^{0-4} = x^{4-0} = 1$
2. $x^{1-3} = x^{3-1}$
3. *Every chain V_i contributes one to x^{2-2} and $|V_i| - 1$ to x^{1-3}.*

Proof. (1) holds, since every vertex that is different from v_n has a successor, and every vertex that is different from v_1 has an incoming edge from either a predecessor or within its chain. For (2), observe that $2n = m = \sum_v indeg(v) = x^{1-3} + 2x^{2-2} + 3x^{3-1} + 4x^{4-0}$ and $n = x^{0-4} + x^{1-3} + x^{2-2} + x^{3-1} + x^{4-0}$, and rearrange. For (3), say $V_i = \{z_1, \ldots, z_\ell\}$ is directed from z_1 to z_ℓ. Then $indeg(z_\ell) = 2$ and $indeg(z_j) = 1$ for all $j < \ell$. $\qquad\square$

3.1   A Simple Algorithm

As in many previous orthogonal drawing papers [1,7,9], the idea is to draw the graph G_i induced by $V_1 \cup \cdots \cup V_i$ in such a way that all *unfinished edges* (edges with one end in G_i and the other in $G - G_i$) end in a column that is unused above the point where the drawing ends.

Embedding the First Two Vertices: If (v_1, v_n) is a single edge, then v_1 and v_2 are embedded exactly as in [1]: refer to Fig. 2. If (v_1, v_n) is a double edge, then it was added only for the purpose of making the graph 4-regular and need not be drawn. In that case, we omit one of the outgoing edges of v_1 that has a bend.

Embedding a Singleton: If V_i is a singleton $\{z\}$, we embed z exactly as in [1]: refer to Fig. 2. For $indeg(z) \in \{2, 3\}$, this adds one new row and $outdeg(z) - 1 = 3 - indeg(z)$ many new columns. For $indeg(z) = 4$, $z = v_n$; if (v_1, v_n) is a double edge, we omit the edge having two bends.

Embedding Chains: Let V_i be a chain, say $V_i = \{z_1, \ldots, z_\ell\}$ with $\ell \geq 2$. For chains, our algorithm is substantially different from [1]. Only z_1 and z_ℓ have predecessors. We place the chain-vertices on a new horizontal row above the previous drawing, between the edges from the predecessors; see Fig. 3. We add new columns as needed to have space for new vertices and outgoing edges without using columns that are in use for other unfinished edges. We also use a second new row if the chain is a long chain.

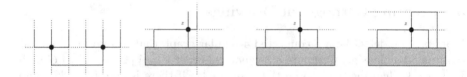

Fig. 2. Embedding the first two vertices, and a singleton with in-degree $2, 3, 4$. Newly added grid-lines are dotted.

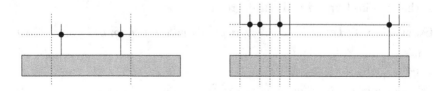

Fig. 3. Embedding short and long chains.

Observation 4. *The increase in the half-perimeter is as follows:*

- *For the first and last vertex-group:* $O(1)$
- *For a 3-1-singleton:* $+1$ *(we add one row)*
- *For a 2-2-singleton:* $+2$ *(we add one row and one column)*
- *For a short chain:* $+3$ *(we add one row and two columns)*
- *For a long chain V_i:* $+2|V_i|$ *(we add two rows and $2|V_i| - 2$ columns)*

Corollary 5. *The half-perimeter is at most $\frac{3}{2}n + \frac{1}{2}x^{2\text{-}2} - x^{short} + O(1)$.*

Proof. From Observation 4 and using Observation 3.3 the half-perimeter is at most $x^{3\text{-}1} + 2x^{2\text{-}2} + 2x^{1\text{-}3} - x^{short} + O(1)$. By Observation 3.2 this is at most $\frac{3}{2}x^{3\text{-}1} + 2x^{2\text{-}2} + \frac{3}{2}x^{1\text{-}3} - x^{short} + O(1)$, which gives the result. □

Theorem 6. *Every simple 3-connected 4-graph has an orthogonal drawing of area at most $\frac{25}{36}n^2 + O(n) \approx 0.69n^2$.*

Proof. First, make the graph 4-regular, compute the 3-canonical order, and consider the number $x^{2\text{-}2}$ of 2-2-vertices.

1. If $x^{2\text{-}2} \leq n/3$, apply the above algorithm. By Corollary 5, the half-perimeter is at most $\frac{3}{2}n + \frac{1}{6}n + O(1) \leq \frac{5}{3}n + O(1)$.
2. If $x^{2\text{-}2} \geq n/3$, apply the algorithm from [9]. They state their area bound as $0.76n^2 + O(1)$, but one can observe (see [9, Theorem 3.1, ll.2–5]) that their half-perimeter is at most $2n - \frac{1}{2}(x^{1\text{-}3} + x^{2\text{-}2}) + O(1)$, since they pair at least $x^{1\text{-}3} + x^{2\text{-}2}$ vertices. Using Observation 3.2 and ignoring $O(1)$ terms, we have $x^{1\text{-}3} + x^{2\text{-}2} = \frac{1}{2}x^{1\text{-}3} + x^{2\text{-}2} + \frac{1}{2}x^{3\text{-}1} = \frac{1}{2}n + \frac{1}{2}x^{2\text{-}2}$. Hence, the half-perimeter of their algorithm is at most $\frac{7}{4}n - \frac{1}{4}x^{2\text{-}2} + O(1) \leq (\frac{7}{4} - \frac{1}{12})n + O(1) = \frac{5}{3}n + O(1)$.

In both cases, we get a drawing with half-perimeter $\frac{5}{3}n + O(1)$. The area of it is maximal if the two sides are equally large and thus at most $(\frac{5}{6}n + O(1))^2$. □

3.2 Improvement via Pairing

We already know a bound of $\frac{3}{2}n+\frac{1}{2}x^{2-2}-x^{\text{short}}+O(1)$ on the half-perimeter. This section improves this further to half-perimeter $\frac{3}{2}n + O(1)$. The idea is strongly inspired by the pairing technique of Papakostas and Tollis [9]. They created pairs of vertices with special properties such that at least $\frac{1}{2}(x^{2-2} + x^{1-3})$ such pairs must exist. For each pair, they can save at least one grid-line, compared to the $2n + O(1)$ grid-lines created with [1].

Our approach is similar, but instead of pairing vertices, we pair groups of the canonical order by scanning them in backward order as follows:

1. Initialize $i := k - 1$. (We ignore the last group, which is a 4-0-singleton.)
2. While V_i is a 3-1-singleton and $i > 2$, set $i := i - 1$.
3. If $i = 2$, break. Else, V_i is a chain or a 2-2-singleton and we choose the partner of V_i as follows: Initialize $j := i-1$. While V_j is a 3-1-singleton whose successor is not in V_i, set $j := j - 1$. Now, pair V_i with V_j. Observe that such a V_j with $j \geq 2$ always exists, since $i > 2$ and V_2 is not a 3-1-singleton.
4. Update $i := j - 1$ and repeat from Step (2) onwards.

In the small example in Fig. 1, the 2-2-singleton V_5 gets paired with the short chain V_4, and all other groups are not paired.

Observe that, with the possible exception of V_2, every chain is paired and every 2-2-vertex is in a paired group (either as 2-2-singleton or as part of a chain). Hence there are at least $\frac{1}{2}(x^{2-2} - 1)$ pairs. The key observation is the following:

Lemma 7. *Let V_i, V_j be two vertex groups that are paired. Then there exists a method of drawing V_i and V_j (without affecting the layout of any other vertices) such that the increase to rows and columns is at most $2|V_i \cup V_j| - 1$.*

We defer the (lengthy) proof of Lemma 7 to the next section, and study here first its consequences. We can draw V_1 and V_k using $O(1)$ grid-lines. We can draw V_2 using $2|V_2| = 2x_{V_2}^{2-2} + 2x_{V_2}^{1-3}$ new grid-lines, where $x_W^{\ell-k}$ denotes the number of vertices of in-degree ℓ and out-degree k in vertex set W. We can draw any unpaired 3-1-singleton using one new grid-line. Finally, we can draw each pair using $2|V_i \cup V_j|-1 = 2x_{V_i \cup V_j}^{2-2} + 2x_{V_i \cup V_j}^{1-3} - 1$ new grid-lines. This covers all vertices, since all 2-2-singletons and all chains belong to pairs or are V_2, and since there are no 1-3-singletons.

Putting it all together and using Observation 3.2, the number of grid-lines hence is $2x^{1-3} + 2x^{2-2} + x^{3-1} - \#\text{pairs} + O(1) \leq 2x^{1-3} + \frac{3}{2}x^{2-2} + x^{3-1} + O(1) = \frac{3}{2}n + O(1)$ as desired. Since a drawing with half-perimeter $\frac{3}{2}n$ has area at most $\left(\frac{3}{4}n\right)^2 = \frac{9}{16}n^2$, we can conclude:

Theorem 8. *Every simple 3-connected 4-graph has an orthogonal drawing of area at most $\frac{9}{16}n^2 + O(n) \approx 0.56n^2$.*

We briefly discuss the run-time. The 3-canonical order can be found in linear time. Most steps of the drawing algorithm work in constant time per vertex, hence $O(n)$ time total. One difficulty is that to place a group we must know the relative order of the columns of the edges from the predecessors. As discussed extensively in [1], we can do this either by storing columns as a balanced binary search tree (which uses $O(\log n)$ time per vertex-addition), or using the data structure by Dietz and Sleator [5] which allows to find the order in $O(1)$ time per vertex-addition. Thus, the worst-case run-time to find the drawing is $O(n)$.

4   Proof of Lemma 7

Recall that we must show that two paired vertex groups V_i and V_j, with $j < i$, can be embedded such that we use at most $2|V_i| + 2|V_j| - 1$ new grid-lines. The proof of this is a massive case analysis, depending on which type of group V_i and V_j are, and whether there are edges between them or not.[1] We first observe some properties of pairs.

Observation 9. *By choice of the pairing, the following holds:*

1. *For any pair (V_i, V_j) such that $j < i$, V_i is either a 2-2-singleton or a chain.*
2. *If V_i is paired with V_j such that $j < i$, then all predecessors of V_i are in V_j or occurred in a group before V_j.*

The following notation will cut down the number of cases a bit. We say that groups V_i and V_j are *adjacent* if there is an edge from a vertex in one to a vertex in the other group. If two paired groups V_i, V_j are not adjacent, then by Observation 9.2 all predecessors of V_i occur before group V_j. We hence can safely draw V_i first, and then draw V_j, thereby effectively exchanging the roles of V_i and V_j in the pair. Now, we distinguish five cases:

1. At least one of V_i and V_j is a short chain. Say V_i is the short chain, the other case is similar. Recall that the standard layout for a short chain uses 3 new grid-lines, but $x_{V_i}^{2\text{-}2} + x_{V_i}^{1\text{-}3} = 2$. So the layout of a short chain automatically saves one grid-line. We do not change the algorithm at all in this case; laying out V_i and V_j exactly as before results in at most $2x_{V_i \cup V_j}^{2\text{-}2} + 2x_{V_i \cup V_j}^{1\text{-}3} - 1$ new grid-lines. (This is what happens in the example of Fig. 1.)

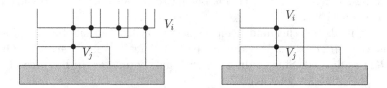

Fig. 4. Reusing the column freed by a 3-1-singleton with a later chain or singleton. In this and the following figures, the re-used grid-line is dotted and red.

---

[1] The constructions we give have been designed as to keep the description simple; often even more grid-lines could be saved by doing more complicated constructions.

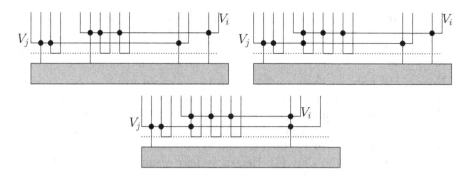

Fig. 5. Sharing the extra row between two long chains when there are 0, 1 or 2 predecessors of V_i in V_j.

2. One of V_i and V_j is a 3-1-singleton. By Observation 9, the 3-1-singleton must be V_j. By the pairing algorithm, the unique outgoing edge of the 3-1-singleton must lead to V_i. Draw V_j as before. We can then draw V_i such that it re-uses one of the columns that were freed by V_j; see Fig. 4.

3. V_i and V_j are both long chains. In this case, both V_i and V_j can use the same extra row for the "detours" that their *middle* vertices (by which we mean vertices that are neither the first nor the last vertex of the chain) use. Since we can freely choose into which columns these middle vertices are placed, we can ensure that none of these "detours" overlap and, hence, one row suffices for both chains. This holds even if one or both of the predecessors of V_i are in V_j, as these are distinct and the two corresponding incoming edges of V_i extend the edges that were already drawn for V_j; see Fig. 5.

4. None of the previous cases applies and V_j is a 2-2-singleton. By Observation 9.1 and since Case (1) does not apply, V_i is either a 2-2-singleton or a long chain. There are two columns reserved for edges from predecessors of V_j. Since predecessors of V_i are distinct, at most one of them can be the 2-2-singleton in V_j. Thus, there also is at least one column reserved for an edge from a predecessor of V_i not in V_j. We call these three or four columns the *predecessor-columns*. We have three sub-cases depending on the relative location of these columns:

   (a) The leftmost predecessor-column leads to V_j. In this case, we save a column almost exactly as in [9]. Place V_j as before, in the right one of its predecessor-columns. This leaves the leftmost predecessor-column free to be reused. Now no matter whether V_i is a 2-2-singleton or a long chain, or whether V_i is adjacent to V_j or not, we can re-use this leftmost column for one outgoing edge of V_i with a suitable placement; see Fig. 6.

   (b) The rightmost predecessor-column leads to V_j. This case is symmetric to the previous one.

   (c) The leftmost and rightmost predecessor-columns lead to V_i. This implies that V_i has two predecessors not in V_j. Hence, V_i cannot be adjacent to V_j. If V_i is a 2-2-singleton, then (as discussed earlier) we can exchange the roles of V_i and V_j, which brings us to Case 4(a). If V_i is a long chain, then place V_j in the standard fashion. We then place the long chain V_i such that the "detours" of its middle vertices re-use the row of V_j. See Fig. 7.

5. None of the previous cases applies. Then V_j is a chain, say $V_j = \{z_1, \ldots, z_\ell\}$, and $\ell \geq 3$ since Case (1) does not apply. We assume the naming is such that the predecessor column of z_1 is left of the predecessor column of z_ℓ.

   Since we are not in a previous case, V_i must be a 2-2-singleton, say z. If V_i is not adjacent to V_j, then we can again exchange the roles of V_i and V_j, which brings us to Case (4). Hence, we may assume that there are edges between V_j and V_i. We distinguish the following sub-cases depending on how many such edges there are and whether their ends are middle vertices.

   (a) z has exactly one neighbor in V_j, and it is either z_1 or z_ℓ. We rearrange $V_i \cup V_j$ into two different chains. Let z be adjacent to z_1 (the other case is symmetric). Then $\{z, z_1\}$ forms one chain and $\{z_2, \ldots, z_\ell\}$ forms another. Embed these two chains as usual. Since $\{z, z_1\}$ forms a short chain, this saves one grid-line; see Fig. 8(left).

   (b) z has exactly one neighbor in V_j, and it is z_h for some $1 < h < \ell$. Embed the chain V_j as usual, but omit the new column next to z_h. For embedding z, we place a new row *below* the rows for the chain. Using this new row, we can connect the bottom outgoing edge of z_h to the horizontal incoming edge of z; see Fig. 8(right).

   (c) z has two neighbors in V_j, and both of them are middle vertices z_g, z_h for $1 < g < h < \ell$. Embed the chain V_j as usual, but omit the new columns next to z_g and z_h. Place a new row *between* the two rows for the chain and use it to connect the two bottom outgoing edges of z_g and z_h to place z, re-using the row for the detours to place the bottom outgoing edge of z. This uses an extra column for z, but saved two columns at z_g and z_h, so overall one grid-line has been saved; see Fig. 9(top left).

   (d) z is adjacent to z_1 and z_2 (the case of adjacency to $z_{\ell-1}$ and z_ℓ is symmetric). Embed $z_2, \ldots, z_\ell$ as usual for a chain, then place z_1 below z_2. The horizontally outgoing edge of z_2 intersects one outgoing edge of z_1. Put z at this place to save a row and a column; see Fig. 9(top right).

   (e) z is adjacent to z_1 and z_h with $h > 2$ (the case of adjacency to z_ℓ and z_h with $h < \ell - 1$ is symmetric). Draw the chain V_j with the modification that z_h is *below* z_{h-1}, but still all middle vertices use the same extra row for their downward outgoing edges. This uses 3 rows, but now z can be placed using the two left outgoing edges of z_1 and z_h, saving a row for z and a column for the left outgoing edge of z_h; see Fig. 9(bottom), both for $h < \ell$ and $h = \ell$.

This ends the proof of Lemma 7 and hence shows Theorem 8.

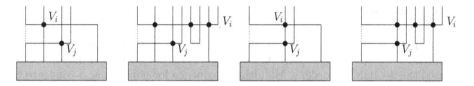

Fig. 6. Reusing the predecessor-column freed by a 2-2-singleton V_j in Case 4(a). Left two pictures: V_i is not adjacent to V_j. Right two: V_i is adjacent to V_j.

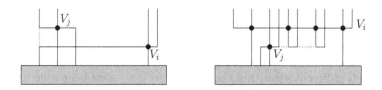

Fig. 7. If the predecessor-columns of V_j are between the ones of V_i, then we can either revert to Case 4(a) or the long chain V_i can re-use the row of V_j.

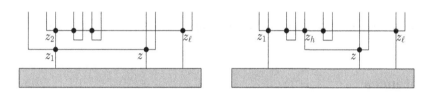

Fig. 8. V_j is a long chain, V_i is a 2-2-singleton with one predecessor in V_j.

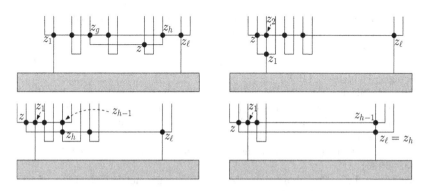

Fig. 9. V_j is a long chain, V_i is a 2-2-singleton, and there are exactly two edges between them. (Top) Cases 5(c) and (d). (Bottom) Case 5(e).

5 Conclusion

In this paper, we gave an algorithm to create an orthogonal drawing of a 3-connected 4-graph that has area at most $\frac{9}{16}n^2 + O(n) \approx 0.56n^2$. As a main tool, we used the 3-canonical order / Mondshein sequence for non-planar 3-connected graphs, whose existence was long known but only recently efficient algorithms for it were found. To our knowledge, this is the first application of the 3-canonical order on non-planar graphs in graph-drawing. Among the many remaining open problems are the following:

- Can we draw 2-connected 4-graphs with area less than $0.76n^2$? A natural approach would be to draw each 3-connected component with area $0.56n^2$ and to merge them suitably, but there are many cases depending on how the cut-vertices and virtual edges are drawn, and so this is far from trivial.
- Can we draw 3-connected 4-graphs with $(2 - \varepsilon)n$ bends, for some $\varepsilon > 0$? With an entirely different algorithm (not given here), we have been able to prove a bound of $2n - x^{2-2} + O(1)$ bends, so an improved bound seems likely.
- Our algorithm was strongly inspired by the one of Kant [7] for 3-connected planar graphs. Are there other graph drawing algorithms for planar 3-connected graphs that can be transferred to non-planar 3-connected graphs by using the 3-canonical order?

Acknowledgments. We wish to thank the anonymous reviewers for their constructive comments.

References

1. Biedl, T.C., Kant, G.: A better heuristic for orthogonal graph drawings. Comput. Geom. **9**(3), 159–180 (1998)
2. Cheriyan, J., Maheshwari, S.N.: Finding nonseparating induced cycles and independent spanning trees in 3-connected graphs. J. Algorithms **9**(4), 507–537 (1988)
3. Chiang, Y.-T., Lin, C.-C., Lu, H.-I.: Orderly spanning trees with applications. SIAM J. Comput. **34**(4), 924–945 (2005)
4. de Fraysseix, H., de Mendez, P.O.: Regular orientations, arboricity and augmentation. In: Tamassia, R., Tollis, I.G. (eds.) GD 1994. LNCS, vol. 894, pp. 111–118. Springer, Heidelberg (1995)
5. Dietz, P., Sleator, D.: Two algorithms for maintaining order in a list. In: 19th Annual ACM Symposium on Theory of Computing, pp. 365–372 (1987)
6. de Fraysseix, H., Pollack, R., Pach, J.: How to draw a planar graph on a grid. Combinatorica **10**, 41–51 (1990)
7. Kant, G.: Drawing planar graphs using the canonical ordering. Algorithmica **16**(1), 4–32 (1996)
8. Mondshein, L.F.: Combinatorial ordering and the geometric embedding of graphs. Ph.D. thesis, M.I.T. Lincoln Laboratory / Harvard University (1971)
9. Papakostas, A., Tollis, I.G.: Algorithms for area-efficient orthogonal drawings. Comput. Geom. **9**(1–2), 83–110 (1998)

10. Rosenstiehl, P., Tarjan, R.E.: Rectilinear planar layouts and bipolar orientation of planar graphs. Discrete Comput. Geom. **1**, 343–353 (1986)
11. Schäffter, M.: Drawing graphs on rectangular grids. Discrete Appl. Math. **63**, 75–89 (1995)
12. Schmidt, J.M.: The Mondshein sequence (2013). http://arxiv.org/pdf/1311.0750.pdf
13. Schmidt, J.M.: The Mondshein sequence. In: Esparza, J., Fraigniaud, P., Husfeldt, T., Koutsoupias, E. (eds.) ICALP 2014. LNCS, vol. 8572, pp. 967–978. Springer, Heidelberg (2014)
14. Biedl, T.: Optimal orthogonal drawings of triconnected plane graphs. In: McCune, W., Padmanabhan, R. (eds.) Automated Deduction in Equational Logic and Cubic Curves. LNCS, vol. 1095, pp. 333–344. Springer, Heidelberg (1996)
15. Tamassia, R., Tollis, I.: A unified approach to visibility representations of planar graphs. Discrete Comput. Geom. **1**, 321–341 (1986)
16. Tamassia, R., Tollis, I.G., Vitter, J.S.: Lower bounds for planar orthogonal drawings of graphs. Inf. Process. Lett. **39**, 35–40 (1991)
17. Valiant, L.G.: Universality considerations in VLSI circuits. IEEE Trans. Comput. **C–30**(2), 135–140 (1981)

Simultaneous Embeddings with Few Bends and Crossings

Fabrizio Frati[1]([✉]), Michael Hoffmann[2], and Vincent Kusters[2]

[1] Dipartimento di Ingegneria, University Roma Tre, Rome, Italy
frati@dia.uniroma3.it
[2] Department of Computer Science, ETH Zürich, Zürich, Switzerland
{hoffmann,vincent.kusters}@inf.ethz.ch

Abstract. A *simultaneous embedding with fixed edges* (SEFE) of two planar graphs R and B is a pair of plane drawings of R and B that coincide when restricted to their common vertices and edges. We show that whenever R and B admit a SEFE, they also admit a SEFE in which every edge is a polygonal curve with few bends and every pair of edges has few crossings. Specifically: (1) if R and B are trees then one bend per edge and four crossings per edge pair suffice, (2) if R is a planar graph and B is a tree then six bends per edge and eight crossings per edge pair suffice, and (3) if R and B are planar graphs then six bends per edge and sixteen crossings per edge pair suffice. This improves on results by Grilli et al. (GD'14), who prove that nine bends per edge suffice, and by Chan et al. (GD'14), who prove that twenty-four crossings per edge pair suffice.

1 Introduction

Let $R = (V_R, E_R)$ and $B = (V_B, E_B)$ be two planar graphs sharing a *common graph* $C = (V_R \cap V_B, E_R \cap E_B)$. The vertices and edges of C are *common*, while the other vertices and edges are *exclusive*. We refer to the edges of R, B, and C as the *red, blue,* and *black edges*, respectively. A *simultaneous embedding* of R and B is a pair of plane drawings of R and B, respectively, that agree on the common vertices (see Fig. 1a–b).

Simultaneous graph embeddings have been a central topic of investigation for the graph drawing community in the last decade, because of their applicability to the visualization of dynamic graphs and of multiple graphs on the same vertex set [6,11], and because of the depth and breadth of the theory they have been found to be related to.

Brass et al. [6] initiated the research on this topic by investigating *simultaneous geometric embeddings* (or SGEs), which are simultaneous embeddings where

F. Frati is partially supported by MIUR project AMANDA, prot. 2012C4E3KT_001.
M. Hoffmann and V. Kusters are partially supported by the ESF EUROCORES programme EuroGIGA, CRP GraDR and the Swiss National Science Foundation, Project 20GG21-134306.

E. Di Giacomo and A. Lubiw (Eds.): GD 2015, LNCS 9411, pp. 166–179, 2015.
DOI: 10.1007/978-3-319-27261-0_14

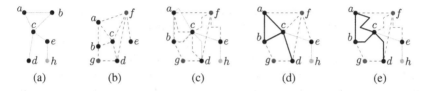

Fig. 1. (a-b) R and B with $V_C = \{a, b, c, d, e\}$ and $E_C = \{(a, b), (b, c), (a, c), (c, d)\}$. (c) Simultaneous embedding of R and B. (d) SGE of R and B. (e) SEFE of R and B.

all edges are represented by straight-line segments (see Fig. 1d). This setting proved to be fairly restrictive: there exist two trees [16] and even a tree and a path [2] with no SGE. Furthermore, the problem of deciding whether two graphs admit an SGE is NP-hard [12].

Two relaxations of SGE have been considered in the literature in which edges are not forced to be straight-line segments. In the first setting, we look for a simultaneous embedding of two given planar graphs R and B in which every edge is drawn as a polygonal curve with few bends. Di Giacomo and Liotta [9] proved that two bends per edge always suffice. If R and B are trees, then one bend per edge is sufficient [10]. Note that black edges may be represented by different curves in each drawing. The variant in which the edges of R and B might only cross at right angles has also been considered [3]. In the second setting, we look for a *simultaneous embedding with fixed edges* (or SEFE) of R and B: a simultaneous embedding in which every common edge is represented by the same simple curve in the plane (see Fig. 1e). In other words, a SEFE is a drawing of the *union graph* $(V_R \cup V_B, E_R \cup E_B)$ that determines a plane drawing of R (of B) when restricted to the vertices and edges of R (resp. of B). While not every two planar graphs admit a SEFE, this setting is less restrictive than SGE: for example, every tree and every planar graph admit a SEFE [13]. Determining the complexity of deciding whether two given graphs admit a SEFE is a major open problem in the field of graph drawing. Polynomial-time testing algorithms are known in many restricted cases, such as when the common graph C is biconnected [1] or when C is a set of disjoint cycles [5]. We refer to an excellent survey by Bläsius et al. [4] for many other results.

In this paper we present algorithms to construct SEFEs in which edges are represented by polygonal curves. For the purpose of guaranteeing the readability of the representation, we aim at minimizing two natural aesthetic criteria: the number of bends per edge and the number of crossings per edge pair. Both criteria have been recently considered in relation to the construction of a SEFE. Namely, Grilli et al. [17] proved that every combinatorial SEFE can be realized as a SEFE with at most nine bends per edge, a bound which improves to three bends per edge when the common graph is biconnected. Further, Chan et al. [7] proved that if R and B admit a SEFE, then they admit a SEFE in which every red-blue edge pair crosses at most twenty-four times.

Contribution. We improve on the results of Grilli et al. [17] and of Chan et al. [7] by proving the following results. (1) Any two trees admit a SEFE with one bend per edge; thus, every two edges cross at most four times. The number of bends is the best possible, since two trees exist with no SGE [16]. (2) Any planar graph and any tree admit a SEFE with six bends per edge in which every two edges cross at most eight times. (3) Any two planar graphs that admit a SEFE also admit a SEFE with six bends per edge in which every two edges cross at most sixteen times. In all cases, the common edges are straight-line segments. Because of space limits, we present the result for trees and we just sketch the ideas for the other results. For the full version of the paper see [14].

2   Preliminaries

A plane drawing of a (multi)graph G determines a circular ordering of the edges incident to each vertex of G; the set of these orderings is called a *rotation system*. Two plane drawings of G are *equivalent* if they have the same rotation system, the same containment relationship between cycles, and the same outer face (the second condition is redundant if G is connected). A *planar embedding* is an equivalence class of plane drawings. Analogously, a SEFE of two planar graphs R and B determines a circular ordering of the edges incident to each vertex (comprising edges incident to both R and B); the set of these orderings is the *rotation system* of the SEFE. Two SEFEs of R and B are *equivalent* if they have the same rotation system and if their restriction to the vertices and edges of R (of B) determines two equivalent plane drawings of R (resp. of B). Finally, a *combinatorial* SEFE $\mathcal{E}$ for two planar graphs R and B is an equivalence class of SEFEs; we denote by $\mathcal{E}|_R$ (by $\mathcal{E}|_B$) the planar embedding of R (resp. of B) obtained by restricting $\mathcal{E}$ to the vertices and edges of R (resp. of B).

A *subdivision* of a multigraph G is a graph G' obtained by replacing edges of G with paths, whose internal vertices are called *subdivision vertices*. If G' is a subdivision of G, the operation of *flattening* subdivision vertices in G' returns G. The *contraction* of an edge (u, v) in a multigraph G leads to a multigraph G' by replacing (u, v) with a vertex w incident to all the edges u and v are incident to in G; k parallel edges (u, v) in G lead to $k - 1$ self-loops incident to w in G' (the contracted edge is not in G'). If G has a planar embedding $\mathcal{E}_G$, then G' *inherits* a planar embedding $\mathcal{E}_{G'}$ as follows. Let $a_1, \ldots, a_k, v$ and $b_1, \ldots, b_\ell, u$ be the clockwise orders of the neighbors of u and v in $\mathcal{E}_G$, respectively. Then the clockwise order of the neighbors of w is $a_1, \ldots, a_k, b_1, \ldots, b_\ell$. The contraction of a connected graph is the contraction of all its edges.

The straight-line segment between points p and q is denoted by $\overline{pq}$. The *angle* of $\overline{pq}$ is the angle between the ray from p in positive x-direction and the ray from p through $\overline{pq}$. A polygon P is *strictly-convex* if at every vertex the interior angle is $< \pi$; also, P is *star-shaped* if there exists a point p^* such that $\overline{pp^*} \subset P$, for every vertex p of P; the *kernel* of P is the set of all such points p^*.

A *1-page book embedding* (1PBE) is a plane drawing where all vertices are placed on an oriented line ℓ called *spine* and all edges are curves in the halfplane to the left of ℓ. A *2-page book embedding* (2PBE) is a plane drawing where all vertices are placed on a spine ℓ and each edge is a curve in one of the two halfplanes delimited by ℓ.

3   Two Trees

In this section we describe an algorithm that computes a SEFE of any two trees R and B with one bend per edge. Let C be the common graph of R and B.

We outline our algorithm. In Step 1, we compute a combinatorial SEFE of R and B for every vertex the incident black edges are consecutive in the circular order of incident edges. In Step 2, we contract each component of C, obtaining trees R' from R and B' from B. In Step 3, we independently augment R' and B' to Hamiltonian planar graphs, so as to satisfy topological constraints that are necessary for the subsequent drawing algorithms. In Step 4, we use the Hamiltonian augmentations to construct a simultaneous embedding of R' and B' with one bend per edge, similarly to an algorithm of Erten and Kobourov [10]. Finally, in Step 5, we expand the components of C by modifying the simultaneous embedding of R' and B' in a neighborhood of each vertex to make room for the components of C. We now describe these steps in detail.

Step 1: Combinatorial SEFE. Fix the clockwise order of edges incident to each vertex as follows: all black edges in any order, then all red edges in any order, and then all blue edges in any order (each sequence might be empty). As any rotation system for a tree determines a planar embedding, this results in a combinatorial SEFE $\mathcal{E}$ of R and B (Fig. 2a). We may assume that every component S of C is incident to at least one red and one blue edge: If S is not incident to any, say, blue edge, then $B = S = C$, since B is connected, and any plane straight-line drawing of R is a SEFE of R and B.

(a)   (b)

Fig. 2. (a) A connected component S of C, together with its incident exclusive edges. (b) Vertex v resulting from the contraction of S.

For every component S of C we pick two incident edges $r(S)$ and $b(S)$ as follows. In any SEFE equivalent to $\mathcal{E}$ let γ be a simple closed curve surrounding S and close enough to it so that γ has no crossing in its interior. Note that γ intersects all the exclusive edges incident to S in some clockwise order in which all the exclusive edges incident to a single vertex of S appear consecutively. Let $r(S)$ be any red edge not preceded by a red edge in this order and let $b(S)$ be the first blue edge after $r(S)$. We define a total ordering ϱ_S of the vertices of S, as the order in which their exclusive edges intersect γ (a curve is added incident to every vertex of S with no incident exclusive edge for this purpose), where the first vertex of ϱ_S is the endvertex of $r(S)$.

Lemma 1. *The straight-line drawing of S obtained by placing its vertices on a strictly-convex curve λ in the order defined by ϱ_S is plane.*

Proof. For every vertex v of S, shrink γ along an exclusive edge incident to v so that γ passes through v and still every edge of S lies in its interior. Eventually γ passes through all the vertices of S in the order ϱ_S. The planarity of the drawing of S implies that there are no two edges whose endvertices alternate along γ. Then placing the vertices of S on λ in the order ϱ_S leads to a plane straight-line drawing of S.   □

Step 2: Contractions. Contract each component S of C to a single vertex v. The resulting trees $R' = (V'_R, E'_R)$ and $B' = (V'_B, E'_B)$ have planar embeddings $\mathcal{E}_{R'}$ and $\mathcal{E}_{B'}$ inherited from $\mathcal{E}_R$ and $\mathcal{E}_B$, respectively. Vertex v is common to R' and B'; let $r(v)$ and $b(v)$ be the edges corresponding to $r(S)$ and $b(S)$ after the contraction. See Fig. 2b.

Step 3: Hamiltonian Augmentations. We describe this step for R' only; the treatment of B' is analogous and independent. The goal is to find a vertex order corresponding to a 1PBE of R'. All edges between consecutive vertices along the spine ℓ, as well as the edge between the first and last vertex along ℓ, can be added to a 1PBE while maintaining planarity: hence the 1PBE is essentially a Hamiltonian augmentation of R'. For Step 5 we need to place $r(v)$, for each common vertex v, as in the following.

Lemma 2. *There is a 1PBE for R' equivalent to $\mathcal{E}_{R'}$ such that, for every common vertex v, the spine passes through v right before $r(v)$ in clockwise order around v.*

Proof. We construct the embedding recursively. For each exclusive vertex v, let $r(v)$ be an arbitrary edge incident to v. Arbitrarily choose a vertex s as the root of R' and place s on ℓ. Place the other endpoint of $r(s)$ after s on ℓ and all remaining neighbors of s, if any, in between in the order given by $\mathcal{E}_{R'}$. Then process every child v of s (and the subtree below v) recursively as follows (and ensure that all subtrees stay in pairwise disjoint parts of the spine, for instance, by assigning a specific region to each).

(a)   (b)

Fig. 3. Embedding the children of v if (a) $p \neq v'$ or (b) $p = v'$. Parts of the embedding already constructed are in the shaded regions.

Note that both v and the parent p of v are already embedded. By symmetry we can assume that p lies before v on the spine. Let v' be the endvertex of $r(v)$ different from v. If $p \neq v'$, we place the other endvertex of $r(v)$ right before v. Both if $p \neq v'$ (see Fig. 3a) and if $p = v'$ (see Fig. 3b), we place the other children of v, if any, according to $\mathcal{E}_{R'}$, in the parts of the spine between p and v', and after v. If v is not a leaf, then all its children are processed recursively in the same fashion. It is easily checked that the resulting embedding is a 1PBE that satisfies the stated properties.   □

Step 4: Simultaneous Embedding. We now construct a simultaneous embedding of R' and B'. Let σ_v be the order of the edges around a vertex v obtained by sweeping a ray clockwise around v, starting in direction of the negative x-axis.

Lemma 3. *For every $\varepsilon > 0$, R' and B' admit a simultaneous embedding with one bend per edge in which:*

– *all edges of $E_{R'}$ ($E_{B'}$) incident to each vertex v in V'_R (resp. V'_B) leave v within an angle of $[-\varepsilon; +\varepsilon]$ with respect to the positive y-direction (resp. x-direction);*
– *the drawing restricted to R' (to B') is equivalent to $\mathcal{E}_{R'}$ (resp. to $\mathcal{E}_{B'}$); and*
– *for every common vertex v, the first red (blue) edge in σ_v is $r(v)$ (resp. $b(v)$).*

Proof. Our algorithm is very similar to algorithms due to Brass et al. [6] and Erten and Kobourov [10]. These algorithms, however, do not guarantee the construction of a simultaneous embedding in which the order of the edges incident to each vertex is as stated in the lemma. This order is essential for the upcoming expansion step.

We assign x-coordinates $1, \ldots, |V_{R'}|$ (y-coordinates $|V_{B'}|, \ldots, 1$) to the vertices of R' (resp. of B') according to the order in which they occur on the spine in the 1PBE of R' (resp. of B') computed in Lemma 2. This determines the placement of every common vertex. Set any not-yet-assigned coordinate to 0.

We now draw the edges of R' (the construction for B' is symmetric). The idea is to realize the 1PBE of R' with its vertices placed as above and its edges drawn as x-monotone polygonal curves with one bend. We proceed as follows. The 1PBE of R' defines a partial order of the edges corresponding to the way they nest. For example, denoting the vertices by their order along the spine, edge $(3, 4)$ preceeds $(3, 5)$ and $(2, 5)$, while $(1, 2)$ and $(6, 7)$ are incomparable. We draw the edges of R' in any linearization of this partial order. Suppose we have drawn some edges and let (u, v) be the next edge to be drawn. Assume w.l.o.g. that the x-coordinate of u is smaller than the one of v. For some $\varepsilon_{uv} > 0$, consider the ray ϱ_u emanating from u with an angle of $\pi/2 - \varepsilon_{uv}$ (with respect to the positive x-axis). Similarly, let ϱ_v be the ray emanating from v with an angle of $\pi/2 + \varepsilon_{uv}$. We choose $\varepsilon_{uv} < \varepsilon$ sufficiently small so that:

(1) no vertex in $V_{R'} \setminus \{u\}$ lies in the region to the left of the underlying (oriented) line of ϱ_u and to the right of the vertical line through u;
(2) no vertex in $V_{R'} \setminus \{v\}$ lies in the region to the right of the underlying (oriented) line of ϱ_v and to the left of the vertical line through v; and
(3) neither ϱ_u nor ϱ_v intersects any previously drawn edge.

As no two vertices of R' have the same x-coordinate, we can choose ε_{uv} as claimed. The corresponding rays ϱ_u and ϱ_v intersect in some point: this is where we place the bend-point of (u, v). The resulting drawing is equivalent to the 1PBE of R' and therefore to $\mathcal{E}_{R'}$. The remaining claimed properties are preserved from the 1PBE.   □

Step 5: Expansion. We now expand the components of C in the drawing produced by Lemma 3 one by one in any order. Let Γ be the current drawing, v be a vertex corresponding to a not-yet-expanded component S of C, and p be the point on which v is placed in Γ. Note that the red and blue edges incident to v may be incident to different vertices in S. Let $\sigma_v = (e_1, \ldots, e_\ell)$, where $e_1, \ldots, e_k$ are red and $e_{k+1}, \ldots, e_\ell$ are blue. By Lemma 3, $r(v) = e_1$ and $b(v) = e_{k+1}$. Each edge incident to v is drawn as a polygonal curve with one bend. Let b_i be the bend-point of e_i. The plan is to delete p and segments $\overline{pb_i}$ in Γ to obtain Γ'. Then draw S in Γ' inside a small disk around p and draw segments from S to $b_1, \ldots, b_\ell$. See Fig. 4. For an $\varepsilon \geq 0$, let D_ε be the disk with radius ε centered at p. Let Γ_R (Γ'_R) be the restriction of Γ (resp. Γ') to the red and black edges. We state the following propositions only for the red graph; the propositions for the blue graph are analogous. By continuity, v can be moved around slightly in Γ_R while maintaining a plane drawing for the red graph. This implies the following.

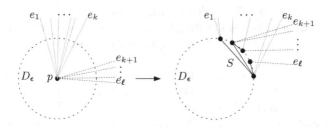

Fig. 4. Expanding a component S in a small disk D_ε around p.

Proposition 1. *There exists a $\delta_R > 0$ with the following property. For every drawing Γ_R^* obtained from Γ'_R by drawing S in D_{δ_R}, the red segments from S to $b_1, \ldots, b_k$ do not cross any segment already present in Γ'_R.*

Proposition 2. *There exists an $\varepsilon_R > 0$ with the following property. Let $q_1, \ldots, q_k$ be any k (not necessarily distinct) points in this clockwise order on the upper semicircle of D_{ε_R}. Then the segments $\overline{q_1 b_1}, \ldots, \overline{q_k b_k}$ do not intersect except at common endpoints.*

Proof. The angles of $\overline{pb_1}, \ldots, \overline{pb_k}$ are distinct and strictly decreasing, by Lemma 3 and by the way $e_1, \ldots, e_k$ are labeled. We claim that ε_R can be chosen sufficiently small so that the angles of $\overline{q_1 b_1}, \ldots, \overline{q_k b_k}$ are also distinct and strictly decreasing. For a certain ε, let $I_i(\varepsilon)$ be the interval of all angles α such that the ray with angle α from b_i intersects D_ε. Since the angles of $\overline{pb_1}, \ldots, \overline{pb_k}$ are distinct, it follows that the intervals $I_1(0), \ldots, I_k(0)$ are disjoint. By continuity, there exists an $\varepsilon_R > 0$ for which $I_1(\varepsilon_R), \ldots, I_k(\varepsilon_R)$ are also disjoint, and the claim follows for this ε_R. Finally, two segments $\overline{q_i b_i}$ and $\overline{q_j b_j}$ with $i < j$ and $q_i \neq q_j$ can intersect only if the angle of $\overline{q_i b_i}$ is smaller than the angle of $\overline{q_j b_j}$, which does not happen by the claim. $\qquad\square$

Lemma 4. *There exists an $\varepsilon > 0$ with the following property. We can expand S to obtain a simultaneous embedding Γ^* from Γ' by drawing the vertices of S on the boundary of D_ε, the edges of S as straight-line segments, and by connecting S to $b_1, \ldots, b_\ell$ with straight-line segments.*

Proof. Let δ_R, δ_B, ε_R, and ε_B be the constants given by Propositions 1 and 2 and their analogous formulations for B. Let $\varepsilon := \min\{\delta_R, \delta_B, \varepsilon_R, \varepsilon_B\}$. Place the vertices of S as distinct points on the boundary of the upper-right quadrant of D_ε in the order ϱ_S. By Lemma 1, this placement determines a straight-line plane drawing of S. Draw straight-line segments from the vertices of S to $b_1, \ldots, b_\ell$, thus completing the drawing of the exclusive edges incident to S. We prove that the red segments incident to S do not cross any red or black edge; the proof for the blue segments is analogous. By Proposition 1, the red segments incident to S do not cross the red and black segments not incident to S. Also, they do not cross the edges of S, which are internal to D_ε. Further, Proposition 2 ensures that these segments do not cross each other. Namely, the linear order of the vertices of S defined by the sequence of red edges $e_1, \ldots, e_k$ is a subsequence of ϱ_S, given that the embedding $\mathcal{E}_{R'}$ of R' is the one inherited from $\mathcal{E}_R$, given that Lemma 3 produces a drawing of R' respecting $\mathcal{E}_{R'}$ and in which $e_1 = r(v)$, and given that the endvertex of $r(S)$ in S is the first vertex of ϱ_S.   □

Theorem 1. *Let R and B be two trees. There exists a SEFE of R and B in which every exclusive edge is a polygonal curve with one bend, every common edge is a straight-line segment, and every two exclusive edges cross at most four times.*

Proof. By Lemma 3, R' and B' admit a simultaneous embedding with one bend per edge. By repeated applications of Lemma 4, the simultaneous embedding of R' and B' can be turned into a SEFE of R and B in which every exclusive edge has one bend and every common edge is a straight-line segment. Finally, any two exclusive edges cross at most four times, given that each of them consists of two straight-line segments.   □

4   A Planar Graph and a Tree (sketch)

In this section we sketch an algorithm that computes a SEFE of any planar graph R and any tree B in which every edge of R has at most six bends and every edge of B has one bend. The common graph C of R and B is a forest, as it is a subgraph of B. The algorithm is similar to the one for trees (Sect. 3), however, it encounters some of the complications one needs to handle when dealing with pairs of planar graphs (Sect. 5). A detailed description of the algorithm and a proof of its correctness can be found in [14]. The algorithm consists of several steps.

Step 1: Antennas. We modify R and B as follows. Each red edge (u, v), with u and v in C, is replaced by a path (u, u_e, v_e, v), whose first and third edge are black, and whose second edge is red; also, each red edge (u, v), with u in

C and v not in C, is replaced by a path (u, u_e, v), with (u, u_e) black and with (u_e, v) red. Denote by R' and B' the resulting planar graph and tree. While this modification costs two extra bends per edge of R in the final SEFE of R and B (which will be obtained from a SEFE of R' and B' by removing vertices and edges not in R and B), it establishes the property that, for every exclusive edge e of R', every common endvertex of e is incident to e, to a common edge, and to no other edge.

Step 2: Combinatorial SEFE. We construct a combinatorial SEFE $\mathcal{E}'$ of R' and B' such that at each vertex v of the common graph C' of R' and B', all the edges of C' are consecutive in the circular order of the edges incident to v. While this is done similarly to the case of tree-tree pairs, the existence of this combinatorial SEFE here is possible only because of the antennas introduced in Step 1. Edges $r(S)$ and $b(S)$, ordering ϱ_S, and planar embeddings $\mathcal{E}_{R'}$ and $\mathcal{E}_{B'}$ are defined as in Sect. 3.

Step 3: Contractions. We contract each component of C' to a vertex in R' and in B', determining a planar multigraph R'' (with loops) and a tree B'', respectively. Graphs R'' and B'' inherit planar embeddings $\mathcal{E}_{R''}$ and $\mathcal{E}_{B''}$ from $\mathcal{E}_{R'}$ and $\mathcal{E}_{B'}$, respectively. Let $r(v)$ and $b(v)$ be the edges corresponding to $r(S)$ and $b(S)$.

Step 4: Hamiltonian Augmentations. A Hamiltonian augmentation of B'' is computed by Lemma 2. A Hamiltonian augmentation of R'' might not exist, thus we subdivide some edges of R'' and then augment the subdivided R'' into a graph R''' containing a Hamiltonian cycle $\mathcal{C}$ (which we assume to be oriented counter-clockwise), none of whose edges is part of an original edge of R''.

The augmentation does not alter the embedding of R'', that is, it produces a planar embedding $\mathcal{E}_{R'''}$ that contains a subdivision of $\mathcal{E}_{R''}$; also, for each common vertex v of R'' and B'', the edge of $\mathcal{C}$ entering v is right before $r(v)$ in the clockwise order of edges incident to v in $\mathcal{E}_{R'''}$. Each edge e of R'' either is also an edge of R''' (as it has not been subdivided) or corresponds to a path with three edges in R''' (as it has been subdivided twice). In the former case, e is to the left of $\mathcal{C}$; in the latter case, the path corresponding to e starts to the left of $\mathcal{C}$, then moves to its right, and then ends again to its left.

The augmentation can be computed as follows: let T be a spanning tree of R''; draw a closed curve γ in $\mathcal{E}_{R''}$ around T crossing twice every edge of R'' not in T; replace such crossings with subdivision vertices for the edges of R'' and insert dummy vertices on γ; modify γ in a neighborhood of each vertex v of R'', so that γ passes through v.

Step 5: Simultaneous Embedding. In order to construct a simultaneous embedding of R'' and B'', we would like to use known algorithms that embed planar graph pairs simultaneously with two bends per edge [8,9,19]. However, the existence of self-loops in R'' prevents us from doing that. Thus, we modify those algorithms to prove that a simultaneous embedding of R'' and B'' exists in which every edge of R'' (of B'') is a polygonal curve with at most four bends (resp. with one bend) and every two edges cross at most eight times. Further, the planar embeddings $\mathcal{E}_{R''}$ and $\mathcal{E}_{B''}$ of R'' and B'' are respected, the first red (blue) edge in σ_v is $r(v)$ (resp. $b(v)$), where σ_v is defined as in Sect. 3, and all the edges of R'' leave their incident vertices within an angle of $[-\varepsilon; +\varepsilon]$ with respect to the positive y-direction (resp. x-direction).

The embedding algorithm is similar to the one in Lemma 3. First, the vertices of R''' (of B'') are assigned increasing x-coordinates (decreasing y-coordinates), according to their order along the Hamiltonian cycle in R''' (in B''). The edges of B'' and the not subdivided edges of R'' are drawn as steep 1-bend curves. Every other edge e of R'' is a path with three edges in R'''; drawing each of these edges as a steep 1-bend curve would result in e having five bends (one per edge of R''' composing e, plus two corresponding to the subdivision vertices for e); one bend is saved by placing all the subdivision vertices for the edges of R'' on a strictly-convex curve, so that the edges of R''' between them can be drawn as straight-line segments rather than as 1-bend curves.

Step 6: Expansion. Expand the components of C' in the simultaneous embedding of R'' and B'', as in Sect. 3; this results in a SEFE of R' and B'. Remove vertices and edges not in R and B, obtaining a SEFE of R and B. We get the following.

Theorem 2. *Let R be a planar graph and let B be a tree. There exists a SEFE of R and B in which every exclusive edge of R is a polygonal curve with at most six bends, every exclusive edge of B is a polygonal curve with one bend, every common edge is a straight-line segment, and every two exclusive edges cross at most eight times.*

5   Two Planar Graphs (sketch)

In this section we sketch an algorithm that computes a SEFE of any two planar graphs R and B in which every edge has at most six bends. A detailed description of the algorithm and a proof of its correctness can be found in [14].

We assume that a combinatorial SEFE $\mathcal{E}$ of R and B is given, that no exclusive vertex or edge lies in the outer face of C in $\mathcal{E}$, and that R and B are connected. We make the first assumption since determining the existence of such a SEFE is a problem of unknown complexity [4]; the last two assumptions can be met after an initial augmentation. We introduce antennas, as in Sect. 4, turning R and B into planar graphs R' and B' with a common graph C'; however, here the modification is performed for both graphs. This costs two extra bends per edge in the final SEFE of R and B; however, it establishes the property that, for every exclusive edge e, every common endvertex of e is incident to e, to a common edge, and to no other edge. A combinatorial SEFE $\mathcal{E}'$ of R' and B' is derived from $\mathcal{E}$ by drawing the antennas as "very small" curves on top of the edges they partially replace. Let $\mathcal{E}_{C'}$ be the restriction of $\mathcal{E}'$ to C'.

We now construct a SEFE of R' and B'. Similarly to Sects. 3 and 4, we would like to *contract* each component S of C', construct a *simultaneous embedding* of the resulting graphs, and finally *expand* the components of C'. However, S is not a tree here, but rather a planar graph containing other components of C' in its internal faces. Hence, the *contraction – simultaneous embedding – expansion* process does not happen just once, but rather we proceed from the outside to the inside of C' iteratively, each time applying that process to draw certain subgraphs of R' and B', until R' and B' have been entirely drawn. We now sketch how this is done.

We start by representing the cycle δ^* delimiting the outer face of C' in $\mathcal{E}'$ as a strictly-convex polygon Δ^*. Next, assume that a SEFE Γ'' of two subgraphs R'' of R' and B'' of B' has been constructed. Let C'' be the common graph of R'' and B'' and let $\mathcal{E}_{R''}$, $\mathcal{E}_{B''}$, and $\mathcal{E}_{C''}$ be the planar embeddings of R'', B'', and C'' in $\mathcal{E}'$, respectively. Assume that the following properties hold for Γ''.

- (BENDS AND CROSSINGS): every edge is a polygonal curve with at most four bends, every common edge is a straight-line segment, and every two exclusive edges cross at most sixteen times;
- (EMBEDDING): the restrictions of Γ'' to R'', B'', and C'' are equivalent to $\mathcal{E}_{R''}$, $\mathcal{E}_{B''}$, and $\mathcal{E}_{C''}$, respectively; and
- (POLYGONS): each not-yet-drawn vertex or edge of R' or B' lies in $\mathcal{E}'$ inside a simple cycle δ_f in C'' which is represented in Γ'' by a star-shaped empty polygon Δ_f; further, if an edge exists in C' that lies inside δ_f in $\mathcal{E}'$ and that belongs to the same 2-connected component of C' as δ_f, then Δ_f is a strictly-convex polygon.

These properties are initially met with $R'' = B'' = C'' = \delta^*$ and with $\Gamma'' = \Delta^*$. It remains to describe how to insert in Γ'' vertices and edges of R' and B' that are not yet in Γ'', while maintaining these properties. Since R' and B' are finite graphs, this will eventually lead to a SEFE of R' and B'. We distinguish two cases.

In Case 1, a 2-connected component S_f of C' exists such that: (i) the outer face of S_f in $\mathcal{E}_{C'}$ is delimited by a simple cycle δ_f belonging to C'' and containing no vertex or edge of C'' in its interior in $\mathcal{E}_{C''}$; and (ii) S_f contains edges inside δ_f in $\mathcal{E}_{C'}$, hence by property POLYGONS, δ_f is a strictly-convex polygon Δ_f in Γ''. As observed in [18], a straight-line plane drawing Γ_f of S_f exists in which the outer face of S_f is delimited by Δ_f and every internal face is delimited by a star-shaped polygon. Plugging Γ_f in Γ'' maintains properties BENDS AND CROSSINGS, EMBEDDING, and POLYGONS.

In Case 2, let δ_f be a simple cycle belonging to C'', containing no vertex or edge of C'' in its interior in $\mathcal{E}_{C'}$, and containing a not-yet-drawn vertex or edge in its interior in $\mathcal{E}'$. By property POLYGONS, δ_f is a star-shaped polygon Δ_f in Γ''. Since Case 1 does not apply, δ_f delimits a face f of $\mathcal{E}_{C'}$ in its interior (possibly with other cycles of C'). Let $C'(f)$ be the subgraph of C' composed of the vertices and edges incident to f in $\mathcal{E}_{C'}$. Also, let $R'(f)$ $(B'(f))$ be the subgraph of R' (of B') composed of $C'(f)$ and of the red (blue) vertices and edges lying in f in $\mathcal{E}'$; these are the graphs we draw while maintaining properties BENDS AND CROSSINGS, EMBEDDING, and POLYGONS. This proof is the most involved part of the paper.

We give an algorithm that draws $R'(f)$ and $B'(f)$ in four steps, with the approach of Sects. 3 and 4: (Step 1) contract each component of $C'(f)$, obtaining planar multigraphs $R''(f)$ and $B''(f)$; (Step 2) independently compute Hamiltonian augmentations of $R''(f)$ and $B''(f)$; (Step 3) construct a simultaneous embedding $\Gamma''(f)$ of $R''(f)$ and $B''(f)$, relying on their Hamiltonian augmentations; and (Step 4) expand each component of $C'(f)$ in $\Gamma''(f)$, obtaining a SEFE $\Gamma'(f)$ of $R'(f)$ and $B'(f)$.

Differently from the previous sections, a simultaneous embedding has to be constructed for two planar multigraphs; this is not a big issue though, other than for the number of bends of $B'(f)$ in $\Gamma'(f)$. What is a major complication is that, in order to extend the SEFE Γ'' of R'' and B'' by plugging $\Gamma'(f)$ into it, we need to ensure that Γ'' and $\Gamma'(f)$ coincide along the part they share, which is polygon Δ_f. That is, the SEFE $\Gamma'(f)$ of $R'(f)$ and $B'(f)$ we construct has to coincide with Δ_f when restricted to δ_f.

The impact of this constraint on the *contraction – simultaneous embedding – expansion* process is as follows. The contraction and Hamiltonian augmentation steps stay unchanged. Denote by u^* the vertex of $R''(f)$ and $B''(f)$ to which the 2-connected component S^* of $C'(f)$ containing δ_f has been contracted. The simultaneous embedding step is also very similar to the one in Sect. 4, except that it ensures that u^* and its adjacent bends are in certain geometric positions. The expansion step changes heavily. Namely: (i) we expand the components $S \neq S^*$ of $C'(f)$ in Γ''_f in the usual way; (ii) we define a region $\mathcal{H}^*$ inside the kernel of Δ_f; (iii) we construct a drawing Γ^* of S^* such that δ_f is represented as Δ_f and all the other vertices and edges of S^* are inside Δ_f but outside $\mathcal{H}^*$; we rotate and scale Γ''_f and place it in $\mathcal{H}^*$; and we finally connect Γ^* with Γ''_f via straight-line segments, thus obtaining Γ'_f. We then plug Γ'_f in Γ'', so that they coincide along Δ_f, obtaining a drawing satisfying Properties BENDS AND CROSSINGS, EMBEDDING, and POLYGONS. We get the following.

Theorem 3. *Let R and B be two planar graphs. If there exists a SEFE of R and B, then there also exists a SEFE of R and B in which every edge is a polygonal curve with at most six bends, every common edge is a straight-line segment, and every two exclusive edges cross at most sixteen times.*

6   Conclusions

In this paper we proved upper bounds for the number of bends per edge and the number of crossings per edge pair required to realize a SEFE with polygonal curves as edges.

While the bound on the number of bends per edge we presented for tree-tree pairs is tight, there is room for improvement for pairs of planar graphs, as the best known lower bound [6] only states that one bend per edge might be needed. We suspect that our upper bound could be improved by designing an algorithm that constructs a simultaneous embedding of two planar multigraphs with less than four bends per edge. A related interesting problem is to determine how many bends per edge are needed to construct a simultaneous embedding of pairs of (simple) planar graphs. The best known upper bound is two [8,9,19] and the best known lower bound is one [15].

As a final research direction, we mention the problem of constructing SEFEs of pairs of planar graphs in polynomial area, while matching our bounds for the number of bends per edge and crossings per pair of edges.

Acknowledgments. This research initiated at the Workshop on Geometry and Graphs, held at the Bellairs Research Institute in Barbados in March 2015. The authors thank the other participants for a stimulating atmosphere. Frati also wishes to thank Anna Lubiw and Marcus Schaefer for insightful ideas they shared during the research for [7].

References

1. Angelini, P., Di Battista, G., Frati, F., Patrignani, M., Rutter, I.: Testing the simultaneous embeddability of two graphs whose intersection is a biconnected or a connected graph. J. Discr. Algorithms **14**, 150–172 (2012)
2. Angelini, P., Geyer, M., Kaufmann, M., Neuwirth, D.: On a tree and a path with no geometric simultaneous embedding. J. Graph Algorithms Appl. **16**(1), 37–83 (2012)
3. Bekos, M.A., van Dijk, T.C., Kindermann, P., Wolff, A.: Simultaneous drawing of planar graphs with right-angle crossings and few bends. In: Rahman, M.S., Tomita, E. (eds.) WALCOM 2015. LNCS, vol. 8973, pp. 222–233. Springer, Heidelberg (2015)
4. Bläsius, T., Kobourov, S.G., Rutter, I.: Simultaneous embeddings of planar graphs. In: Tamassia, R. (ed.) Handbook of Graph Drawing and Visualization, Discrete Mathematics and Its Applications, chapter 11, pp. 349–382. Chapman and Hall/CRC (2013)
5. Bläsius, T., Rutter, I.: Disconnectivity and relative positions in simultaneous embeddings. Comp. Geom. **48**(6), 459–478 (2015)
6. Brass, P., Cenek, E., Duncan, C.A., Efrat, A., Erten, C., Ismailescu, D.P., Kobourov, S.G., Lubiw, A., Mitchell, J.S.: On simultaneous planar graph embeddings. Comput. Geom. Theory Appl. **36**(2), 117–130 (2007)
7. Chan, T.M., Frati, F., Gutwenger, C., Lubiw, A., Mutzel, P., Schaefer, M.: Drawing partially embedded and simultaneously planar graphs. In: Duncan, C., Symvonis, A. (eds.) GD 2014. LNCS, vol. 8871, pp. 25–39. Springer, Heidelberg (2014)
8. Di Giacomo, E., Didimo, W., Liotta, G., Wismath, S.K.: Curve-constrained drawings of planar graphs. Comput. Geom. Theory Appl. **30**(1), 1–23 (2005)
9. Di Giacomo, E., Liotta, G.: Simultaneous embedding of outerplanar graphs, paths, and cycles. Int. J. Comput. Geom. Appl. **17**(2), 139–160 (2007)
10. Erten, C., Kobourov, S.G.: Simultaneous embedding of planar graphs with few bends. J. Graph Algorithms Appl. **9**(3), 347–364 (2005)
11. Erten, C., Kobourov, S.G., Le, V., Navabi, A.: Simultaneous graph drawing: layout algorithms and visualization schemes. J. Graph Algorithms Appl. **9**(1), 165–182 (2005)
12. Estrella-Balderrama, A., Gassner, E., Jünger, M., Percan, M., Schaefer, M., Schulz, M.: Simultaneous geometric graph embeddings. In: Hong, S.-H., Nishizeki, T., Quan, W. (eds.) GD 2007. LNCS, vol. 4875, pp. 280–290. Springer, Heidelberg (2008)

13. Frati, F.: Embedding graphs simultaneously with fixed edges. In: Kaufmann, M., Wagner, D. (eds.) GD 2006. LNCS, vol. 4372, pp. 108–113. Springer, Heidelberg (2007)

14. Frati, F., Hoffmann, M., Kusters, V.: Simultaneous embeddings with few bends and crossings (2015). CoRR abs/1508.07921

15. Frati, F., Kaufmann, M., Kobourov, S.G.: Constrained simultaneous and near-simultaneous embeddings. J. Graph Algorithms Appl. **13**(3), 447–465 (2009)

16. Geyer, M., Kaufmann, M., Vrt'o, I.: Two trees which are self-intersecting when drawn simultaneously. Discrete Math. **309**(7), 1909–1916 (2009)

17. Grilli, L., Hong, S.-H., Kratochvíl, J., Rutter, I.: Drawing simultaneously embedded graphs with few bends. In: Duncan, C., Symvonis, A. (eds.) GD 2014. LNCS, vol. 8871, pp. 40–51. Springer, Heidelberg (2014)

18. Hong, S., Nagamochi, H.: An algorithm for constructing star-shaped drawings of plane graphs. Comput. Geom. Theory Appl. **43**(2), 191–206 (2010)

19. Kammer, F.: Simultaneous embedding with two bends per edge in polynomial area. In: Arge, L., Freivalds, R. (eds.) SWAT 2006. LNCS, vol. 4059, pp. 255–267. Springer, Heidelberg (2006)

Rook-Drawing for Plane Graphs

David Auber[1,2], Nicolas Bonichon[1,2], Paul Dorbec[1,2],
and Claire Pennarun[1,2]([✉])

[1] University of Bordeaux, LaBRI, UMR 5800, 33400 Talence, France
{david.auber,nicolas.bonichon,paul.dorbec,claire.pennarun}@labri.fr
[2] CNRS, LaBRI, UMR 5800, 33400 Talence, France

Abstract. Motivated by visualization of large graphs, we introduce a
new type of graph drawing called "rook-drawing". A *rook-drawing* of a
graph G is obtained by placing the n nodes of G on the intersections
of a regular grid, such that each row and column of the grid supports
exactly one node. This paper focuses on rook-drawings of planar graphs.
We first give a linear algorithm to compute a planar straight-line rook-
drawing for outerplanar graphs. We then characterize the maximal planar
graphs admitting a planar straight-line rook-drawing, which are unique
for a given order. Finally, we give a linear time algorithm to compute a
polyline planar rook-drawing for plane graphs with at most $n - 3$ bent
edges.

1 Introduction

Nowadays, large and dynamic graphs are widely used in the context of Big Data,
and their visualization is a classical tool for their analysis. On the one hand, when
representing dynamic graphs, it is necessary to handle easily the addition or
deletion of nodes or edges. On the other hand, when using hierarchical views, the
ability to aggregate or de-aggregate sets of nodes is required [1,8]. When doing
such operations, it is important to preserve the mental map of the graph [3], as
well as to compute the changes in the representation efficiently, both in order to
guarantee a smooth use.

In the following, we define a particular type of graph drawing on a grid, that
we call *rook-drawing*. In a rook-drawing, we require that the nodes of the graph
lie on the intersections of a $(n - 1) \times (n - 1)$ regular grid, in such a way that
each row and column hosts exactly one node. Then, the addition or deletion
of a node impacts only the row and column it lies on, without interfering with
other nodes or other parts of the drawing. In particular, dealing with aggregated
data consists in stretching the grid to create enough room for the new appearing
nodes (see Fig. 1). These operations clearly preserve orthogonal ordering, which
is the first type of mental map defined in [12]. Observe that this technique of

This work has been carried out as part of the "REQUEST" project (PIAO18062-
645401) supported by the French "Investissement d'Avenir" Program (Big Data -
Cloud Computing topic) and has been supported by ANR grant JCJC EGOS ANR-
12-JS02-002-01.

© Springer International Publishing Switzerland 2015
E. Di Giacomo and A. Lubiw (Eds.): GD 2015, LNCS 9411, pp. 180–191, 2015.
DOI: 10.1007/978-3-319-27261-0_15

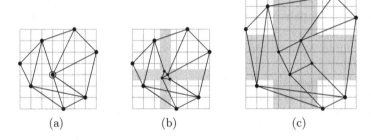

Fig. 1. Expansion of an aggregated node in a rook-drawing of a non-planar graph.

having exactly one node per row and per column is also used by Kornaropoulos et al. [10,11], who represented edges with overlapping orthogonal polylines.

We here explore the existence of rook-drawings for planar graphs. The first question that comes to mind is: *Does every planar graph admit a* planar straight-line rook-drawing, *i.e. a rook-drawing in which each edge is represented by a segment and no two edges cross?* De Fraysseix et al. showed that every planar graph admits a straight-line drawing on an $(n-2) \times (2n-4)$ grid [9]. Schnyder improved this result by proving the existence of such a drawing on an $(n-2) \times (n-2)$-grid [14]. But in such drawings, some columns and rows contain several nodes and some others may be empty. Upward-rightward drawings presented by Di Giacomo et al. [7] are grid drawings in which all nodes have different y-coordinates, but empty rows and several nodes in a same column are allowed.

Contrasting with these results, we show in Sect. 4 that almost every maximal planar graph admits no planar rook-drawing. Yet, every outerplanar graph admits a planar straight-line rook-drawing computable in linear time, as shown in Sect. 3. We then consider *polyline planar rook-drawings*, in which edges are drawn as polylines with bends placed on grid intersections. We show in Sect. 5) that every planar graph admits a polyline planar rook-drawing. Moreover, this drawing can be computed in linear time, each edge is bent at most once and the total number of bends is at most $n-3$.

2   Definitions

A *drawing* of a graph G is a mapping of the nodes of G to points of the plane and of the edges of G to curves between their endpoints. The drawing is *straight-line* if the edges are mapped to line segments. It is *polyline* if the edges are series of line segments. A *grid drawing* is a drawing in which the nodes are mapped to intersections of a regular grid. In such a drawing, we use positive coordinates $(x(u), y(u))$ for each node u. A $k \times l$-grid is a grid of width k and height l. Recall that a *rook-drawing* of a graph with n vertices is a $(n-1) \times (n-1)$-grid drawing, i.e. the functions x and y are bijections from the set of vertices to $\{1, \ldots, n\}$. For simplicity, throughout this paper, the term "rook-drawing" denotes a *straight-line rook-drawing* unless otherwise precised.

A *planar graph* is a graph admitting a planar drawing, i.e. a drawing on the plane in which no pair of edges crosses. Such a drawing can be characterized by the collection of circular permutations of incident edges around each node, called *embedding*. A connected planar graph together with an embedding is called a *plane graph*.

In a plane graph, the edges partition the plane into regions called *faces*. A *rooted plane graph* is a plane graph in which one face (called *outer face*) and one node (called *root*) lying on this face are distinguished. The nodes lying on the outer face are called *outer nodes*, all other nodes are *inner nodes*. Similarly, *outer edges* are edges belonging to the outer face, the other edges are called *inner edges*. An *outerplane graph* is a rooted plane graph in which every node is on the outer face. A *maximal plane graph* is a plane graph with maximal number of edges, implying that every face is a triangle if there are at least three nodes.

A *tree* is a rooted plane graph without cycles. In a tree, a node u is a *descendant* of a node v (or v is an *ancestor* of u) if v is on the path from the root to u. Moreover, if v is connected to u, we say that v is the *parent* of u (and u is a *child* of v). Two nodes are said *unrelated* if one is neither ancestor nor descendant of the other. A *leaf* of a tree is a node of the tree without descendants. The *depth* of a tree is the length of the longest path from a leaf to the root in the tree. For a tree T a node u, the *subtree* of u, denoted $T(u)$, is the tree induced on u and all of its descendants.

The clockwise *preorder* of a tree T is a list of the nodes of T in the order of a clockwise depth-first search algorithm on T. The clockwise *postorder* of a tree T is a list of the nodes of T in the order of their last visit in a clockwise depth-first search algorithm of T. Counterclockwise preorder and postorder are defined similarly.

3   Planar Rook-Drawing for Outerplane Graphs

In this section, we prove the following theorem:

Theorem 1. *Every outerplane graph admits a planar rook-drawing. This drawing can be computed in linear time.*

To prove Theorem 1, we use a partition of the edges of outerplane graphs introduced by Bonichon et al. [4]:

Theorem 2 ([4]). *Let G be an outerplane graph rooted in r. There exists a unique partition of the edges of G into two sets T and S such that:*

– *T is a tree rooted in r*
– *edges of S join a node u to the first node after u in the counterclockwise postorder of T.*

Such a partition can be computed in linear time.

Denote by $y(v)$ the index of v in a counterclockwise postorder of T. We consider an orientation of the edges of T and S such that all edges of T are oriented towards the root r and the edges (uv) of S are oriented from u to v if $y(u) > y(v)$. If G is maximal, then S is a tree rooted in w with $y(w) = 0$ that does not contain the root r of G.

The tree T can be computed by Algorithm 1 due to Bonichon et al. [4]. A call Traversal$(G, \varnothing, r)$ returns the tree T of G rooted in r, the second parameter stands for the current set of edges of the tree during the execution.

Algorithm 1. Traversal(G, T, u)

---

begin
$\quad C \leftarrow \{(u, v) \in G \mid v \notin T\}$
$\quad T \leftarrow T \cup C$
$\quad$**for** *all edges* $(u, v) \in C$ *taken in the clockwise order around* u **do**
$\quad\quad \mid \quad T \leftarrow$ Traversal(H, T, v)
$\quad$return T

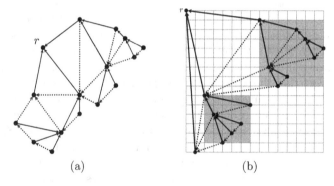

(a)   (b)

Fig. 2. (a) The decomposition of an outerplane graph G rooted in r into T (solid edges) and S (dotted edges). (b) A rook-drawing of G showing the induction process of the proof for Lemma 1.

For each node v of the outerplane graph G, we denote $x(v)$ its index in counterclockwise preorder of T. Recall that $y(v)$ is its index in counterclockwise postorder of T.

Lemma 1. *Placing each node v of G at coordinates $(x(v), y(v))$ produces a planar rook-drawing of G.*

Proof. By construction, the drawing $\mathcal{D}(G)$ obtained is a rook-drawing. It remains to show that this drawing is planar.

For v a node of G, let T_v be the subtree of T rooted in v. Let $G(v)$ be the subgraph of G induced by the nodes of T_v. Let $\mathcal{D}(G(v))$ be the drawing induced by the edges and nodes of $G(v)$. The *left branch* of v in T_v denotes the path between v and the first leaf found in a counterclockwise postorder of T_v.

Let u and v be two nodes of G. The following observations are direct consequences of the definition of the x- and y-coordinates:

(i) If u is before v in counterclockwise preorder of T (i.e. $x(u) < x(v)$) and they are unrelated, then $y(u) < y(v)$ and for each descendant w of u in T, $x(w) < x(v)$ and $y(w) < y(v)$.

(ii) If u is parent of v in T, then $x(u) < x(v)$ and $y(u) > y(v)$.

(iii) Let (uv) be an edge of S with $y(u) > y(v)$. Then v is before u in counterclockwise preorder of T (i.e. $x(v) < x(u)$) and as they are unrelated, v is also before u in counterclockwise postorder of T (i.e. $y(v) < y(u)$). Thus the edges of S are going down and to the left.

(iv) The coordinates of the nodes of the left branch of v are x-increasing and y-decreasing.

We now want to prove by induction the following proposition : $\mathcal{D}(G(u))$ is planar and drawn in the subgrid $[x(u), x(u) + |T_u| - 1] \times [y(u) - |T_u| + 1, y(u)]$.

When T_u is reduced to a single node, the proposition clearly holds.

Now assume the proposition holds for nodes having a subtree of depth at most k. Let u be a node with a subtree T_u of depth $k + 1$. Denote by $u_1, ..., u_m$ the children of u in clockwise order. Their subtrees in T are denoted $T_{u_1}, ..., T_{u_m}$. By induction hypothesis, the subtrees $T_{u_1}, ..., T_{u_m}$ are placed in disjoint areas (see Fig. 2). Then $\mathcal{D}(G(u_i))$ and $\mathcal{D}(G(u_j))$ with $i \neq j$ do not intersect. Thus T_u is planarly drawn in the sub-grid $[x(u), x(u) + |T_u| - 1] \times [y(u) - |T_u| + 1, y(u)]$.

We now prove that the edges of S joining nodes belonging to different subtrees do not create any crossing in $\mathcal{D}(G(u))$. Let v and w be nodes from different subtrees linked by an edge of S, and such that $x(w) < x(v)$. Recall that by definition of S, w is the first node unrelated to v with $y(w) < y(v)$. So v and w are in consecutive trees, say T_{u_i} and $T_{u_{i+1}}$ and $w = u_{i+1}$. Thus all edges of S joining T_i to T_{i+1} have u_{i+1} as an end: edges of S join nodes of the left branch of u_i to u_{i+1}. Then by remarks (iii) and (iv), the edges of S can not cross each other or edges of the tree T.

Thus $\mathcal{D}(G(u))$ is planar. This concludes the proof.   □

Remark that as Andrews [2] showed that a strictly convex drawing of a cycle of n nodes with integer coordinates requires area $\Omega(n^3)$, whereas a rook-drawing requires area $\Omega(n^2)$, our algorithm can not produce strictly convex drawings for outerplane graphs for large n.

Also note that the existence of n nodes both in rook position and in general position (i.e. such that no three nodes are colinear [13]) would imply an algorithm for generating a rook drawing of outerplane graphs (from [6]). We do not know how to prove whether such a configuration exists. Remark though that the algorithm in [6] is of complexity $n \log^3(n)$, while the algorithm presented here is linear.

4   Existence of a Planar Rook-Drawing

We define the *tower plane graph* T_n of order $n \geq 3$ as the plane join graph $K_2 + P_{n-2}$ (i.e. a complete graph K_2 and a path on $n - 2$ nodes P_{n-2} together with all the edges joining nodes from K_2 to nodes of P_{n-2}) drawn in such a way that the nodes of K_2 are on the outer face (see Fig. 3 for a drawing of T_6).

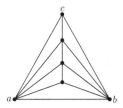

Fig. 3. The tower plane graph T_6.

Theorem 3. *There exists a unique maximal plane graph on $n \geq 3$ nodes admitting a planar rook-drawing, namely the tower plane graph T_n.*

Proof. Suppose we have a planar rook-drawing of a maximal plane graph G. We prove that G is the tower plane graph T_n.

Let a, b, c be the three outer nodes of G. To maintain planarity, the inner nodes are placed at coordinates inside the area defined by the edges (ab), (bc) and (ca). Thus the outer nodes must occupy altogether the four borders of the grid, and one of them has to be placed in a corner. Without loss of generality, assume that a occupies the bottom-left corner.

Consider the positions of the two other outer nodes of G. Suppose one of them is in the top-right corner (without loss of generality, say b). If the third node c is placed below the edge (ab) (see Fig. 4a), then the second column on the left can not contain a node: the coordinates $(k, 2)$ are outside the area delimited by the edges (ab), (bc) and (ca) for all $k > 2$. The point $(2, 2)$ is covered by (ab) and the point $(2, 1)$ can not contain a node because a is already on the first row. If c is above (ab), then for similar reasons the column left to b can not contain a node. Thus b is not in a corner. Without loss of generality, assume b is on the top row and c on the rightmost column of the grid.

Now consider the positions of the inner nodes of G. Let α be the angle between the column containing b and the edge (bc) and β be the angle between the row containing c and the edge (bc) (see Fig. 4b). Consider the row just below b: the angle between the edge (ab) and the column containing b is less or equal to $45°$ thus no nodes can be placed at the left of b on the row below it. No node can be placed on the same column as b either. No node can be placed at the right of the intersection between the edge (bc) and the row below b. Thus for the row under b to contain a node we must have $\alpha \geq 45°$. With similar arguments, for the column on the left of c to contain a node, we must have $\beta \geq 45°$. We have thus $\alpha = \beta = 45°$. Thus c is the node placed on the row below b and b is placed on the column left to c and $x(b) = y(c) = n - 1$. Finally, the inner nodes must be placed on coordinates (i, i) for $2 \leq i \leq n - 2$, i.e. along a diagonal of the grid (see Fig. 4c).

Now the positions of the nodes are determined and there is only one way to complete the drawing into a maximal plane graph, forming the graph $\mathcal{T}_n$.   □

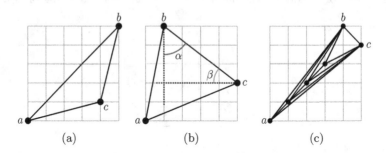

(a)   (b)   (c)

Fig. 4. (a) and (b) Illustrations of the proof of Theorem 3. (c) A planar rook-drawing of $\mathcal{T}_6$.

5   Polyline Rook-Drawing for Planar Graphs

As we proved that some plane graphs do not admit a planar rook-drawing with straight lines, we now relax the straight-line constraint and look at planar polyline rook-drawings. We first recall the definition of Schnyder woods.

5.1   Properties of Schnyder Woods

Definition 1 (Schnyder [14]). *A Schnyder wood of a maximal plane graph G is a partition of the inner edges of G into three directed trees T_0, T_1, T_2 with the following properties:*

- *each tree T_i is rooted on a distinct outer node v_i;*
- *the edges of each tree are directed toward the root;*
- *each inner node u of G has one parent in each T_i, denoted $P_i(u)$;*
- *in counterclockwise order around each inner node, the outgoing edges are in T_0 then T_1 then T_2;*
- *each ingoing edge belonging to the tree T_i is placed after the outgoing edge in $T_{i+1 \bmod 3}$ and before the outgoing edge in $T_{i-1 \bmod 3}$ in counterclockwise order around an inner node.*

The orientation of edges around an inner node is shown in Fig. 5, where T_0 is drawn solid, T_1 is dotted and T_2 is dotted-dashed. Throughout the paper, we call a 0-edge (respectively 1-edge, 2-edge) an edge belonging to the tree T_0 (resp. T_1, T_2).

Two properties of Schnyder woods follow.

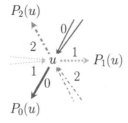

Fig. 5. Orientation around an inner node u in a Schnyder wood.

Proposition 1 (Bonichon et al. [5]). *If u is a descendant of v in T_i, then u is unrelated to v in T_j, $j \neq i$.*

Proposition 2. *If u is the parent of v in T_i, then u is before v in counterclockwise preorder of T_{i-1} and after v in counterclockwise preorder of T_{i+1}.*

Proof. Without loss of generality, assume $i = 2$. In this proof, an i-path denotes a directed path in the tree T_i. Recall that v_i denotes the root of the tree T_i. Let u be the parent of v in T_2 (see Fig. 6).

Suppose that u is after v in the counterclockwise preorder of T_1. By orientation around the node v, the 1-path from v to v_1 has to cross the 2-path from u to v_2. Let t be the intersection of these two paths. Then t is an ancestor of v in T_2 (it is an ancestor of u and thus of v). But t is also an ancestor of v in T_1 because it is on the 1-path from v_1 to v. Though this contradicts Proposition 1. So u is before v in the counterclockwise preorder of T_1, as claimed.

A similar argument proves that u is after v in the counterclockwise preorder of T_0. $\square$

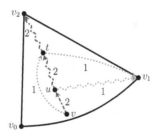

Fig. 6. Illustration of the proof of Proposition 2

5.2 Polyline Rook-Drawing Algorithm

We here describe an algorithm to produce a planar polyline rook-drawing of a maximal plane graph of order n. The algorithm is inspired by an algorithm for polyline drawings proposed by Bonichon et al. [5]. The original algorithm was designed to minimize the grid size and thus many rows and columns support several nodes. This new algorithm shares with the former the edge bending strategy, but the node placement is different.

Theorem 4. *Every maximal plane graph G with n nodes admits a polyline planar rook-drawing $\mathcal{D}(G)$, which can be computed with Algorithm 2 in linear time. This drawing has $n - 3$ bends.*

Algorithm 2. Planar polyline rook-drawing for a maximal plane graph G

$(T_0, T_1, T_2) \leftarrow$ Schnyder wood of G
add the oriented edge $(v_1 v_0)$ to T_0
add the oriented edge $(v_2 v_0)$ to T_0
add the oriented edge $(v_2 v_1)$ to T_1
column order $C \leftarrow$ clockwise preorder of T_0
row order $R \leftarrow$ clockwise postorder of T_1
for u *node of G* **do**
$\quad \mid \quad (x(u), y(u)) = (C(u), R(u))$
draw all T_2 edges with straight lines
for $e = (u, P_0(u))$ *edge of T_0* **do**
$\quad \mid \quad$ **if** $x(u) = x(P_0(u)) + 1$ **then** draw e with a straight line **else** Bend e at
$\quad \mid \quad (x(u), y(P_0(u)) + 1)$
for $e = (u, P_1(u))$ *edge of T_1* **do**
$\quad \mid \quad v \leftarrow ll_0(u).$
$\quad \mid \quad$ bend e at $(x(v), y(u))$

In Algorithm 2 and later, $ll_0(u)$ denotes the last leaf found in a clockwise preorder of u in T_0. An example of the result of Algorithm 2 on a maximal plane graph is presented in Fig. 7b.

We first make the following observations on the placement of nodes after applying Algorithm 2:

- Since the nodes are placed according to their position in a preorder and a postorder, each row and column contains exactly one node. Thus $\mathcal{D}(G)$ is a rook-drawing.
- When u is a leaf of T_0, then $ll_0(u) = u$ and this is the only case when the edge from u to $P_1(u)$ is drawn straight.

Number of Bends. Let k be the number of leaves in T_0. By construction, T_0 contains $n - 1$ edges, T_1 contains $n - 2$ edges and T_2 contains $n - 3$ edges. The edges of T_0 are all bent, except one for each non-leaf node in T_0. Thus $n - 1 - (n - k)$ 0-edges are bent. The edges of T_1 are all bent, except k. Thus $n - 2 - k$ 1-edges are bent. Finally, the edges of T_2 are never bent. Thus, there are exactly $n - 3$ bends in the drawing of G.

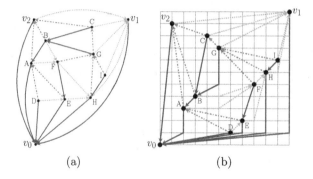

Fig. 7. (a) Schnyder wood of a maximal plane graph G. (b) Result of our polyline algorithm on G.

Planarity. Most of the proofs for the planarity of the drawing are placed in the appendix. We describe in the following some structural properties of the drawing with Lemmas 2, 3 and 4.

Lemma 2. *In $\mathcal{D}(G)$, for each inner node u:*

- $x(P_0(u)) < x(u)$ *and* $y(P_0(u)) < y(u)$: $P_0(u)$ *is left and below* u.
- $x(P_1(u)) > x(u)$ *and* $y(P_1(u)) > y(u)$: $P_1(u)$ *is right and above* u.
- $x(P_2(u)) < x(u)$ *and* $y(P_2(u)) > y(u)$: $P_2(u)$ *is left and above* u.

From Lemma 2 and the coordinates of bends chosen for the edges in Algorithm 2, we observe that the configuration around an inner node follow the scheme illustrated in Fig. 8.

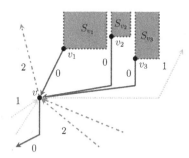

Fig. 8. Edges orientation around an inner node v. S_{v_i} denotes the area in which the subtree of v_i in T_0 is drawn.

This drawing gives a good intuition of why the edges within T_0 do not cross. The detailed proof is not given here, but is based on the following lemmas.

Lemma 3. *For every inner node u, every node v such that $x(P_0(u)) < x(v) < x(u)$ is a descendant of $P_0(u)$ in T_0.*

Proof. This is a direct consequence of the fact that the x-coordinates are given by the clockwise preorder of T_0.   □

Lemma 4. *For every inner node u, every node v such that $x(u) < x(v) < x(P_1(u))$ (resp. $x(P_2(u)) < x(v) < x(u)$) is either a descendant of u (resp. $P_2(u)$) in T_0 or $y(v) < y(u)$ (resp. $y(v) < y(P_2(u))$) in $\mathcal{D}(G)$.*

The final step is to explicitly state that the edges drawn do not cross. The proofs are not given due to space limitation. The idea is the following: we first show that edges inside each tree T_0, T_1 and T_2 do not cross. Then we prove that edges from different trees do not cross.

6   Conclusion

In this paper, we observed that all maximal planar graphs but the tower graphs admit no planar straight-line rook-drawing. On the other hand we showed that every outerplane graph admits a planar straight-line rook-drawing. A natural question is: are there usual classes of plane graphs that all admit a planar straight-line rook-drawing? A plane graph that has a triangular outer face and admits a planar straight-line rook-drawing is necessarily a subgraph of the tower plane graph we described earlier. However, if we consider plane graphs with an outer face with at least 4 vertices, it seems that many of them should admit such a drawing. Then, plane graphs that do not contain non-facial triangles, as, for instance, quadrangulations or 4-connected triangulations with outer face of degree at least 4, are possibly good candidates for admitting a planar rook drawing.

We also showed that every plane graph admits a planar polyline rook-drawing with at most $n - 3$ bent edges. Even if this number of bends is reasonable, one could ask if a linear number of bends is needed for allowing a planar rook-drawing of any planar graph.

Another interesting question would be to consider *relaxed rook-drawing* in which each row and column contains at most one node (and no longer exactly one node). Clearly every plane graph admits a planar relaxed rook-drawing: it suffices to consider a straight-line planar drawing of the plane graph and add a tiny perturbation to nodes sharing some coordinates. This naive approach produces drawings with a huge number of empty columns and rows, which is not suitable in practice. Hence the good question would be: does every plane graph admits a planar relaxed rook-drawing with a small (i.e. linear or sub-linear) number of empty rows and columns? There are no evidence yet that even a constant number of empty rows and columns would not suffice.

References

1. Abello, J., Van Ham, F., Krishnan, N.: ASK-GraphView: a large scale graph visualization system. IEEE Trans. Vis. Comput. Graph. **12**(5), 669–676 (2006)
2. Andrews, G.E.: A lower bound for the volume of strictly convex bodies with many boundary lattice points. Trans. Am. Math. Soc. **106**, 270–279 (1963)

3. Archambault, D., Purchase, H.C.: Mental map preservation helps user orientation in dynamic graphs. In: Didimo, W., Patrignani, M. (eds.) GD 2012. LNCS, vol. 7704, pp. 475–486. Springer, Heidelberg (2013)
4. Bonichon, N., Gavoille, C., Hanusse, N.: Canonical decomposition of outerplanar maps and application to enumeration, coding, and generation. J. Graph Algorithms Appl. **9**(2), 185–204 (2005)
5. Bonichon, N., Le Saëc, B., Mosbah, M.: Optimal area algorithm for planar polyline drawings. In: Kučera, L. (ed.) WG 2002. LNCS, vol. 2573, pp. 35–46. Springer, Heidelberg (2002)
6. Bose, P.: On embedding an outerplanar graph in a point set. Comput. Geom. **23**(3), 303–312 (2002)
7. Di Giacomo, E., Didimo, W., Kaufmann, M., Liotta, G., Montecchiani, F.: Upward-rightward planar drawings. In: The 5th International Conference on Information, Intelligence, Systems and Applications, IISA 2014, pp. 145–150. IEEE (2014)
8. Eades, P., Feng, Q.-W.: Multilevel visualization of clustered graphs. In: North, S.C. (ed.) GD 1996. LNCS, vol. 1190, pp. 101–112. Springer, Heidelberg (1997)
9. de Fraysseix, H., Pach, J., Pollack, R.: Small sets supporting fary embeddings of planar graphs. In: Proceedings of the twentieth annual ACM symposium on Theory of computing, pp. 426–433. ACM (1988)
10. Kornaropoulos, E.M., Tollis, I.G.: Overloaded orthogonal drawings. In: Speckmann, B. (ed.) GD 2011. LNCS, vol. 7034, pp. 242–253. Springer, Heidelberg (2011)
11. Kornaropoulos, E.M., Tollis, I.G.: DAGView: an approach for visualizing large graphs. In: Didimo, W., Patrignani, M. (eds.) GD 2012. LNCS, vol. 7704, pp. 499–510. Springer, Heidelberg (2013)
12. Misue, K., Eades, P., Lai, W., Sugiyama, K.: Layout adjustment and the mental map. J. Visual Lang. Comput. **6**(2), 183–210 (1995)
13. Pach, J., Gritzmann, P., Mohar, B., Pollack, R.: Embedding a planar triangulation with vertices at specified points. Am. Math. Monthly **98**, 165–166 (1991)
14. Schnyder, W.: Embedding planar graphs on the grid. Symp. Discrete Algorithms **90**, 138–148 (1990)

On Minimizing Crossings in Storyline Visualizations

Irina Kostitsyna[1], Martin Nöllenburg[2], Valentin Polishchuk[3], André Schulz[4], and Darren Strash[5]([✉])

[1] Department of Mathematics and Computer Science, TU Eindhoven,
Eindhoven, The Netherlands
i.kostitsyna@tue.nl

[2] Algorithms and Complexity Group, TU Wien, Vienna, Austria
noellenburg@ac.tuwien.ac.at

[3] Communications and Transport Systems, ITN, Linköping University,
Linköping, Sweden
valentin.polishchuk@liu.se

[4] LG Theoretische Informatik, FernUniversität in Hagen, Hagen, Germany
andre.schulz@fernuni-hagen.de

[5] Institute of Theoretical Informatics, Karlsruhe Institute of Technology,
Karlsruhe, Germany
strash@kit.edu

Abstract. In a storyline visualization, we visualize a collection of interacting characters (e.g., in a movie, play, etc.) by x-monotone curves that converge for each interaction, and diverge otherwise. Given a storyline with n characters, we show tight lower and upper bounds on the number of crossings required in any storyline visualization for a restricted case. In particular, we show that if (1) each meeting consists of exactly two characters and (2) the meetings can be modeled as a tree, then we can always find a storyline visualization with $O(n \log n)$ crossings. Furthermore, we show that there exist storylines in this restricted case that require $\Omega(n \log n)$ crossings. Lastly, we show that, in the general case, minimizing the number of crossings in a storyline visualization is fixed-parameter tractable, when parameterized on the number of characters k. Our algorithm runs in time $O(k!^2 k \log k + k!^2 m)$, where m is the number of meetings.

1 Introduction

Ever since an xkcd comic[1] featured storyline visualizations of various popular films, storyline visualizations have increasingly gained popularity as an area of research in the information visualization community (although the precursors of this kind of visualization may date back to Minard's 1861 visualization of Napoleon's Russian campaign of 1812). Informally, a storyline consists of characters (e.g., in a movie, play, etc.) who meet at certain times during a story.

---

[1] http://xkcd.com/657.

E. Di Giacomo and A. Lubiw (Eds.): GD 2015, LNCS 9411, pp. 192–198, 2015.
DOI: 10.1007/978-3-319-27261-0_16

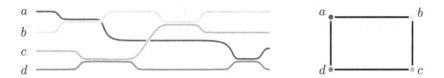

Fig. 1. Left: a storyline visualization with characters a, b, c, d. Right: the event graph.

In a storyline visualization, each character is represented as an x-monotone curve. When characters meet (e.g., appear together in a scene, or interact), their representative curves should be grouped close together vertically, and otherwise their curves should be separate (see Fig. 1, left). We assume that every character can only be in one meeting group at every point in time. One of the main goals for producing readable storyline visualizations is to minimize the number of crossings between character curves. Most previous results for constructing storyline visualizations are practical, implementing drawing routines that rely on heuristics or genetic algorithms [5,10]. However, there are only few theoretical results for storyline visualizations. Storyline visualization is tightly related to layered graph drawing [9], where layers correspond to meeting times in the storyline, and a permutation of all character curves needs to be computed for each time point. Minimizing crossings in a storyline visualization is also related to bounding the ratio of (proper) crossings to touchings for families of monotone curves [6].

Our Results. While previous results focus on drawing storyline visualizations in practice using heuristics [5,10], here we investigate the minimum number of crossings required in any storyline visualization. First, we investigate storyline visualizations in a restricted case. We show that if (1) each meeting consists of exactly two characters and (2) the meetings can be modeled as a tree, then we can always find a storyline visualization with $O(n \log n)$ crossings, where n is the number of characters. Furthermore, we show that there exist storylines in this restricted case that require $\Omega(n \log n)$ crossings. Lastly, we show that, in the general case, minimizing the number of crossings in a storyline visualization is fixed-parameter tractable, when parameterized on the number of characters k. Our algorithm runs in time $O(k!^2 k \log k + k!^2 m)$, where m is the number of meetings.

Problem Formulation. In the *storyline problem*, we are given a storyline $S = (C, \mathcal{T}, \mathcal{E})$, that is defined by set of characters $C = \{1, \ldots, n\}$, that meet during closed time intervals $\mathcal{T} \subset \{[s, t] \mid s, t \in \mathbb{N}, s \leq t\}$. We call a meeting an *event*, and denote the set of events as $\mathcal{E} \subset 2^C \times \mathcal{T}$, where each event $E_i = (C_i, [s_i, t_i]) \in \mathcal{E}$ (with $1 \leq i \leq m$) is defined by a subset $C_i \subseteq C$ of characters that meet for the entire time interval $[s_i, t_i] \in \mathcal{T}$ (naturally, a character cannot participate in two overlapping events). The goal then is to produce a 2D drawing of S, called a *storyline visualization*, where the x-axis represents time, and characters are drawn as x-monotone curves placed in some vertical order for each point in time. During each event $E_i = (C_i, [s_i, t_i])$, curves representing characters in C_i

should be grouped within some small vertical distance δ_{group} of each other, and otherwise the characters should be separated by some larger vertical distance $\delta_{\text{separate}} > \delta_{\text{group}}$.

2   Pairwise Single-Meeting Storylines

We focus on a simplified version of the storyline problem, where each event consists of exactly two characters, and these characters meet exactly once in $\mathcal{E}$. For this simplified version, we can represent our events as a graph where every vertex is a character, and every edge is a meeting of the corresponding characters. We call this graph an *event graph* (Fig. 1, right).

2.1 $O(n \log n)$ Crossings for Tree Event Graphs

Let our event graph be a tree T with n nodes. Then we show that we can always draw a storyline visualization with $O(n \log n)$ crossings. Our result relies on decomposing T into disjoint subtrees that are drawn in disjoint axis-aligned rectangles. We reach this bound by using the *heavy path decomposition* technique [8].

Definition 1 (heavy path decomposition [8]). Let T be a rooted tree. For each internal node v in T, we choose a child w with the largest subtree among all of v's children. We call the edge (v, w) a *heavy edge*, and the edges to v's other children *light edges*. We call a maximal path of heavy edges a *heavy path*, and the decomposition of T into heavy paths and light edges a heavy path decomposition.

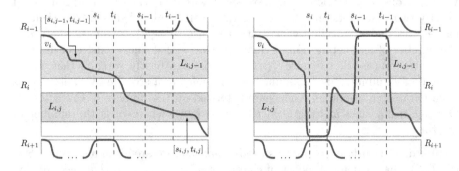

Fig. 2. The curve for v_i before (left) and after (right) introducing detours.

We first arbitrarily root T, and compute its heavy path decomposition. Note that any root-leaf path of the event graph T contains at most $\lceil \log n \rceil$ light edges [8]. Let P be the heavy path beginning at the root of T. We denote the node on P at depth i in T by v_i. For each v_i, with l_i light children, we first lay out each light subtree $L_{i,j}$ for $1 \leq j \leq l_i$. We then order these layouts vertically in increasing order of meeting start time between v_i and the root $r_{i,j}$ of $L_{i,j}$, separating each layout by vertical distance δ_{separate}. We denote the rectangle containing all layouts $L_{i,j}$ by R_i (see Fig. 2). Then, we draw a single x-monotone curve from the top left to the bottom right of R_i, passing through the layout of each $L_{i,j}$, meeting the curve for each root $r_{i,j}$ at time $s_{i,j}$, and leaving at time $t_{i,j}$, for each event $(\{v_i, r_{i,j}\}, [s_{i,j}, t_{i,j}])$.

Now for each v_i, we have a layout of v_i and its light subtrees in a rectangle R_i. We now show how to draw events between characters that are adjacent via a heavy edge in P. We first place all R_i vertically in order along the path P (from R_1 to $R_{|P|}$), separated by distance δ_{group}. We must have the curves meet for each event $(\{v_i, v_{i+1}\}, [s_i, t_i])$. We show how to introduce detours so that the curve v_i joins curve v_{i+1} at time s_i. Let n_i be the number of curves in the light subtrees of v_i. Before time s_i, curve v_i has intersected some number γ of the curves from its light subtrees, and has $n_i - \gamma$ curves still to intersect. Just before time s_i, we divert the curve so that it intersects the remaining $n_i - \gamma$ curves and reaches the bottom of rectangle R_i to meet with v_{i+1} at time s_i. Then at time t_i, we return the curve back to between curves γ and $\gamma + 1$ and allow the curve to continue as before, passing through the remaining $n_i - \gamma$ curves. For each v_i we must also introduce a similar detour to the top of its rectangle R_i so that it can meet the curve of v_{i-1} at time s_{i-1}; see Fig. 2 (right).

We introduce at most two such detours for each rectangle R_i, and therefore increase the number of crossings of each curve v_i by a constant factor of at most five. Therefore, the total number of crossings $N(T)$ in our drawing of T satisfies $N(T) \leq \sum_{i=1}^{|P|} \sum_{j=1}^{l_i} N(L_{i,j}) + 5n$, with base case $N((\{v\}, \emptyset)) = 0$. Since all $L_{i,j}$ are disjoint, each iteration of the recurrence contributes at most $O(n)$ crossings. Further, since there are $O(\log n)$ light edges on the simple path from the root to any leaf in the heavy path decomposition [8], the recurrence reaches the base case after $O(\log n)$ iterations. Therefore, the recurrence solves to $N(T) = O(n \log n)$ crossings.

Theorem 1. *Any pairwise single-meeting storyline with a tree event graph has a storyline visualization with $O(n \log n)$ crossings.*

2.2 A Lower Bound

Consider some storyline visualization $\mathcal{V}$ with an event graph G with n nodes and m edges. Let π_0 be the ordering of the characters along a vertical line $t = 0$ in $\mathcal{V}$. Assign labels $[1, \dots, n]$ to the characters according to π_0. Then permutation π_0 defines an embedding of G on the line $t = 0$. As time progresses and character curves intersect, the corresponding vertices in the embedding of G are swapped, see Fig. 3.

For every edge $e = (i, j) \in G$ define its cost $c_t(e)$ to be the number of characters between i and j on the vertical line at any given time t.

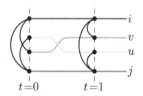

Fig. 3. Event graph on line $t = 0$ before and on line $t = 1$ after swapping u and v.

Then initially $c_0(e) = |i - j| - 1$. So before i and j can meet, their curves must cross at least $|i - j| - 1$ curves that were initially between them, which may be 0.

When two character curves cross, their corresponding vertices u and v swap in the embedding of G on the vertical line. Notice that, after the swap, the costs of edges incident to u or v change by ± 1, and there is no change for non-incident edges. Thus, the crossing changes the cost of at most $\deg(u) + \deg(v)$ edges in G.

Let $C_0 = \sum c_0(e)$ be the total initial cost of the edges of G embedded on the line $t = 0$. Then C_0 is the number of decrements in edge costs needed before all edges would have had cost 0 at some moment in time. Every crossing of character curves u and v in $\mathcal{V}$ decreases this cost by at most $\deg(u) + \deg(v)$. Therefore, there are at least $\frac{\min_{\pi_0} C_0}{2\Delta}$ crossings in any storyline visualization $\mathcal{V}$ with an event graph G, where Δ is the maximum degree of G. Notice that $\min_{\pi_0} C_0 = L^* - m$, where L^* is the total edge length in the *optimal linear ordering* of graph G (the numbering of its vertices that minimizes the sum of differences of numbers over the graph's edges; see [1] and [3, Problem GT42]).

Theorem 2. *Any storyline visualization with an event graph G requires $\Omega(\frac{L^* - m}{2\Delta})$ crossings, where L^* is the total edge length of the optimal linear ordering of G, and Δ is the maximum degree of G.*

Corollary 1. *There exists a pairwise single-meeting storyline with a tree event graph whose storyline visualization requires $\Omega(n \log n)$ crossings.*

Proof. Let G be a full binary tree. Chung [2] showed that for any assignment of unique labels $[1, \ldots, n]$ to vertices of a full binary tree, the sum of label differences $|i - j|$ over all edges $(i, j) \in G$ is $\Omega(n \log n)$ (see also [7]). Therefore, there will be $\Omega(\frac{\Omega(n \log n) - n + 1}{2 \times 3}) = \Omega(n \log n)$ crossings. □

3   An FPT Algorithm for the Storyline Problem

We now consider general storylines, where any number of characters may participate in an event, and we have no restrictions on the event (hyper-)graph structure. The general storyline problem is NP-complete, by a straightforward reduction from BIPARTITE CROSSING NUMBER [4]. However, in real-world storylines, there may be only a few characters of interest and these characters participate frequently in events. We therefore are interested in a parameterized algorithm to better capture the complexity in this scenario. Let $k = |C|$ be the number of characters in a storyline, and let $m = |\mathcal{E}|$ be the number of events. We show that the storyline problem is *fixed-parameter tractable* when parameterized on k. A problem is said to be fixed-parameter tractable if it can be solved in time $f(k)m^{O(1)}$, where f is some function of k that is independent of m.

Theorem 3. *For storylines with k characters and m events, we can solve the storyline problem in time $O(k!^2 k \log k + k!^2 m)$.*

Proof. We show how to reduce the storyline problem to finding shortest path in a graph. For each time interval $[s_i, t_i]$ in the storyline we take its start time s_i and create a vertex for each of the $O(k!)$ possible vertical orderings of the curves that satisfy the event groupings at s_i. We denote the vertices for time s_i by $v_{i,j}$, where $1 \leq j \leq k!$, and say these vertices are on *level i*.

Denote the minimum number of crossings to transform one ordering $v_{i,j}$ at level i to ordering $v_{i+1,l}$ at level $i + 1$ by $I(v_{i,j}, v_{i+1,l})$. For all levels, we connect each vertex $v_{i,j}$ to each vertex $v_{i+1,l}$ by a directed edge with weight $I(v_{i,j}, v_{i+1,l})$. We then create source and terminal vertices s and t and connect them with edges of weight 0 to vertices on levels 1 and m, respectively. Then the weight of a shortest path from s to t is the minimum number of crossings in any embedding, and this path specifies the vertical orderings of the curves at each time step s_i.

We now compute the number of crossings to transform between vertical orderings. First note that we can compute the minimum number of swaps between two vertical orderings of size k in time $O(k \log k)$ by counting inversions with merge sort. Thus, we can precompute the weights between all pairs of orderings in time $O(k!^2 k \log k)$, and assign edge weights when building the graph at a cost of $O(k!^2)$ per level.

Now a minimum-weight path from s to t fully specifies a storyline visualization. We can lay out each curve by the vertical ordering specified by each vertex on the path with its time step, swapping curve order between time steps. Then during each event we group the curves together, otherwise we separate them.

In total there are m levels, each with $O(k!)$ vertices and $O(k!^2)$ edges. Thus, there are $O(k!m)$ vertices and $O(k!^2 m)$ edges. We can compute a shortest path from s to t in time linear in the number of vertices and edges, by dynamic programming: For each level i, we compute the minimum weight for each vertex v by iterating over all incoming edges from vertices on level $i-1$ and choosing the one that minimizes the total weight to v. Thus we can compute a shortest path from s to t in time $O(k!^2 m)$. Including the time to precompute edge weights, we get total time $O(k!^2 k \log k) + O(k!^2 m) = O(k!^2 k \log k + k!^2 m)$.   □

Acknowledgments. We thank the anonymous referees for their helpful comments. This research was initiated at the 2nd International Workshop on Drawing Algorithms for Networks in Changing Environments (DANCE 2015) in Langbroek, the Netherlands, supported by the Netherlands Organisation for Scientific Research (NWO) under project no. 639.023.208. IK is supported in part by the NWO under project no. 639.023.208. VP is supported by grant 2014-03476 from the Sweden's innovation agency VINNOVA.

References

1. Adolphson, D., Hu, T.C.: Optimal linear ordering. SIAM J. Appl. Math. **25**(3), 403–423 (1973)
2. Chung, F.R.K.: A conjectured minimum valuation tree (I. Cahit). SIAM Rev. **20**(3), 601–604 (1978)

3. Garey, M.R., Johnson, D.S.: Computers and Intractability: A Guide to the Theory of NP-Completeness. W. H. Freeman & Co., New York (1979)
4. Garey, M.R., Johnson, D.S.: Crossing number is NP-complete. SIAM J. Alg. Disc. Meth. 4(3), 312–316 (1983)
5. Muelder, C., Crnovrsanin, T., Sallaberry, A., Ma, K.-L.: Egocentric storylines for visual analysis of large dynamic graphs. In: IEEE Big Data'13, pp. 56–62 (2013)
6. Pach, J., Rubin, N., Tardos, G.: On the Richter-Thomassen conjecture about pairwise intersecting closed curves. In: Discrete Algorithms (SODA'15), pp. 1506–1516 (2015)
7. Šeĭdvasser, M.A.: The optimal numbering of the vertices of a tree. Diskret. Analiz 17, 56–74 (1970)
8. Sleator, D.D., Tarjan, R.E.: A data structure for dynamic trees. J. Comput. System Sci. 26(3), 362–391 (1983)
9. Sugiyama, K., Tagawa, S., Toda, M.: Methods for visual understanding of hierarchical system structures. IEEE Trans. Syst. Man Cybernet. 11(2), 109–125 (1981)
10. Tanahashi, Y., Ma, K.-L.: Design considerations for optimizing storyline visualizations. IEEE Trans. Vis. Comput. Graph. 18(12), 2679–2688 (2012)

Maximizing the Degree of (Geometric) Thickness-t Regular Graphs

Christian A. Duncan[✉]

Department of Mathematics and Computer Science,
Quinnipiac University, Hamden, CT, USA
christian.duncan@acm.org

Abstract. In this paper, we show that there exist $(6t-1)$-regular graphs with thickness t, by constructing such an example graph. Since all graphs of thickness t must have at least one node with degree less than $6t$, this construction is optimal. We also show, by construction, that there exist $5t$-regular graphs with geometric thickness at most t. Our construction for the latter builds off of a relationship between geometric thickness and the Cartesian product of two graphs.

1 Introduction

A straight-line drawing Γ of a graph $G = (V, E)$ is a mapping of vertices to points in the plane and edges to curves between the endpoints. A drawing is planar if and only if the edges only intersect at the endpoints. For convenience, we often refer to the vertices of a given graph G as $V(G)$ and the edges as $E(G)$. The *order* of a graph, $|V|$, is the number of vertices in the graph. The *thickness* $\theta(G)$ of a graph G is the minimum number of planar subgraphs whose union forms G. The edges of these subgraphs form a partitioning of $E(G)$. For convenience, we identify each partition with a unique color.

The *geometric thickness* of G, $\bar{\theta}(G)$, is the smallest integer t such that there is a *straight-line drawing* $\Gamma(G)$ whose edges can be colored with t colors such that no two edges with the same color intersect, except at the endpoints. That is, each coloring (layer) is a planar drawing. We refer to such a drawing, with associated coloring, as a *t-layered planar drawing*.

When discussing a drawing $\Gamma(G)$, it often helps to describe the grid size $h \times w$ of the drawing. We take the convention that the vertices (points) all lie on integer coordinates. In addition, to count the width and height we use the number of grid points in the smallest axis aligned bounding box of the drawing. This convention means that a single vertex has dimension 1, as opposed to 0. See, for example, Fig. 2a, which has height 5 and width 4.

Generalized by Tutte [8], the thickness problem began as an exploration of the biplanarity of a graph, where biplanar refers to having thickness two [1,4,7]. Numerous research articles have also explored geometric thickness and its relationship to thickness, e.g. [2,5]. Since the research in graph thickness is too large to summarize adequately in this technical note, we refer the interested reader to a survey by Mutzel *et al.* [6].

© Springer International Publishing Switzerland 2015
E. Di Giacomo and A. Lubiw (Eds.): GD 2015, LNCS 9411, pp. 199–204, 2015.
DOI: 10.1007/978-3-319-27261-0_17

In [3], Durocher *et al.* explore the relationship between the colorability and minimum degree of thickness-t graphs and pose several questions. We investigate the question of finding the largest k for a given t, such that there exist k-regular graphs of (geometric) thickness t, as this provides a lower bound on the minimum degree. We provide a construction that shows that there exist $(6t-1)$-regular graphs of thickness t, which is optimal since every graph of thickness t must have at least one node of degree less than $6t$. This observation can be seen by recalling that the average degree of any planar graph is less than 6. We also show that there exist 5t-regular graphs of geometric thickness at most t, but do not claim that 5t is optimal.

2 $(6t-1)$-Regular Thickness-t Graphs

To prove that there exist $(6t-1)$-regular thickness-t graphs, we start by creating a graph $\mathcal{G}$ with $48(t-1)$ vertices of degree 6 and 48 vertices of degree 5 and a larger graph $\mathcal{G}_C$ composed of several disjoint copies of $\mathcal{G}$. We then create t layers of $\mathcal{G}_C$ using the same vertex set but in different permutations to ensure that every vertex has degree 5 in exactly one layer, leading to the following theorem.

Theorem 1. *For any $t \in \mathbb{Z}^+$, there exist $(6t-1)$-regular graphs with thickness t.*

Proof. For $t = 1$, it is well-known that there exist 5-regular planar graphs. (See for example, Fig. 2d). Therefore, we assume that $t > 1$.

Our construction starts with a base graph $\mathcal{G}$ shown in Fig. 1a. The graph consists of an outer collection of 24 vertices, a nested sequence of $16(t-1)$ triangles, and an inner collection of 24 vertices, constructed in the same manner as the outer collection. The outer collection of vertices is formed by taking three wheel graphs of order 6 and connecting each vertex of the outermost triangle (with vertices labelled $1-3$) to two adjacent outer vertices of one of the wheels. The wheel graphs are then connected to each other via two additional nodes to ensure that every one of the outer vertices has degree 5. The inner collection of vertices are similarly constructed by connecting to the innermost nested triangle. Observe that this ensures that every vertex in the nested triangle subgraph has degree 6. Consequently, $\mathcal{G}$ is a planar graph of order $48t$ containing exactly 48 vertices of degree 5 and $48(t-1)$ vertices of degree 6.

Let $\mathcal{G}_C$ be the graph of order $48tC$ formed by the union of $C = 48t$ disjoint copies of $\mathcal{G}$.[1] We shall form our $(6t-1)$-regular graph $G = (V, E)$ of thickness t by creating t layers of $\mathcal{G}_C$ where each layer uses the same vertex set, in a different permutation. That is, V consists of $48tC$ vertices, and E is the union of t sets, $E_0, E_1, \ldots, E_{t-1}$, with each set representing a different layer (color) of the graph. For any $v \in V$ and $0 \leq i < t$, we refer to $\pi_i(v)$ as the permuted vertex in $\mathcal{G}_C$ of the i-th layer. For any two vertices $v, w \in V$, the edge (v, w) is in E_i if and only if $(\pi_i(v), \pi_i(w)) \in E(\mathcal{G}_C)$, the permuted vertices have an edge in $\mathcal{G}_C$. See Fig. 1b for an illustration using a simpler base graph, two copies of a 4-cycle.

---

[1] Although this means there are $(48t)^2$ vertices in $\mathcal{G}_C$, it is more convenient to refer to C distinctly for now.

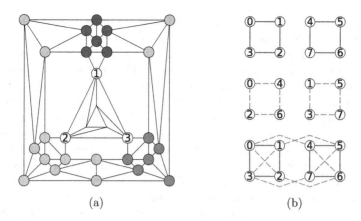

(a)   (b)

Fig. 1. (a) The base graph $\mathcal{G}$ consisting of 24 outer shaded vertices, the nested triangle graph of $48(t-1)$ vertices, and another 24 inner vertices (not shown), symmetric to the outer vertices. (b) An illustration of merging two layers using different permutations. The top graph of two 4-cycles represents an example base graph. The middle graph is the same graph but with the vertices permuted and a different color (and pattern) shown for the edges. The bottom graph is the resulting 4-regular thickness-two graph.

In our mapping of these vertices of $\mathcal{G}_C$, we must ensure two conditions:

1. Every vertex gets mapped to a degree-5 vertex *exactly once*.
2. If $(\pi_i(v), \pi_i(w)) \in E(\mathcal{G}_C)$ then $(\pi_j(v), \pi_j(w)) \notin E(\mathcal{G}_C)$ for all $j \neq i$. That is, we do not have any duplicate edges.

If we guarantee these two conditions, then we know that the degree of every vertex in G is exactly $6t - 1$ and given the construction we know that the graph has thickness t, completing the proof.

We refer to a vertex in $\mathcal{G}_C$ by the notation $\rho_{a,\ell,c}$ where $0 \leq a < 48$, $0 \leq \ell < t$, $0 \leq c < C$. We conceptually partition the $16(t-1)$ nested triangles of $\mathcal{G}$ into $t-1$ groups of 16 triangles, called *levels*. In addition, we refer to level 0 as the group formed by the outer and inner (degree-5) vertices. The index ℓ corresponds to vertices within level ℓ. The index c represents one of the C disjoint copies of $\mathcal{G}$. Consequently, if two vertices share an edge, they must have the same c-index. For any given level ℓ and copy c, there are exactly 48 vertices of $\mathcal{G}_C$. The specific ordering of these 48 vertices does not particularly matter so long as there is no edge connecting $\rho_{a,\ell,c}$ and $\rho_{a,\ell',c}$. That is, two vertices with the same indices a and c cannot share an edge. By ensuring that the innermost three vertices of one level do not share an a-index with the outermost three vertices of the next level, we can easily construct such an ordering.

For convenience, we label the vertices of G in the same manner such that $\pi_0(v_{a,\ell,c}) = \rho_{a,\ell,c}$. We define the permutations as follows:

$$\pi_i(v_{a,\ell,c}) = \rho_{a,(\ell+i) \mod t,(c+ai) \mod C}, \text{ for } 0 \leq i < t. \tag{1}$$

We now show that our two conditions hold. For condition 1, note that the only degree-5 vertices are those with $\ell = 0$. Since i ranges from 0 to $t - 1$, the permutation guarantees that $\ell + i \equiv 0 \mod t$ for exactly one value of i.

Suppose now that condition 2 does not hold. That is, there exist two vertices $v = v_{a,\ell,c}$ and $w = v_{a',\ell',c'}$ in $V(G)$ and two layers $i \neq j$ such that $(\pi_i(v), \pi_i(w)) \in E(\mathcal{G}_C)$ and $(\pi_j(v), \pi_j(w)) \in E(\mathcal{G}_C)$. Assume, without loss of generality, that $i < j$. Since two vertices with the same a-index do not share an edge, we know that $a \neq a'$. In addition, since the c copies are disjoint in $\mathcal{G}_C$, we know that $c + ai \equiv c' + a'i \mod C$ (or else there would be no edge in the i-th permutation). Similarly, we know that $c + aj \equiv c' + a'j \mod C$. Subtracting the two values, we see that $a(j - i) \equiv a'(j - i) \mod C$. However, since $0 \leq a, a' < 48$ and $0 < j - i < t$ and because we chose to use $C = 48t$ copies, we know that $0 \leq a(j - i), a'(j - i) < C$. Thus, since $a \neq a'$, the equivalence only holds when $j = i$, a contradiction. □

3   5t-Regular Geometric Thickness-t Graphs

Although we have shown an optimal example for the thickness problem, when we look at the same problem with the restriction that the graph have *geometric thickness* t, the optimal solution appears more challenging. Before we prove our main theorem for this section, we first discuss a relationship between the *Cartesian product* of two graphs, $G_1 \square G_2$, and geometric thickness.

Lemma 1. *The Cartesian product of two graphs, $G = G_1 \square G_2$, has geometric thickness at most $t_1 + t_2$ where t_1 and t_2 are the geometric thicknesses of G_1 and G_2, respectively. Furthermore, if G_1 (resp., G_2) can be drawn on an integer grid of dimension $h_1 \times w_1$ (resp., $h_2 \times w_2$) such that each row and column contains at most one vertex, then G can be drawn on an integer grid of dimension $h_1 h_2 \times w_1 w_2$ such that each row and column contains at most one vertex.*

Proof. Let G_1 and G_2 be two graphs with geometric thicknesses t_1 and t_2 respectively. In addition, for $i \in \{1, 2\}$, let Γ_i be a t_i-layered planar drawing of G_i. We treat the t_1 colors used to partition the edges of Γ_1 as distinct from the t_2 colors of Γ_2. Note, if the drawings do not have the property that each row and column contains at most one vertex (that is, if two vertices share the same y-coordinate or x-coordinate), we can slightly rotate the drawing so that this property holds, although the grid dimension would increase significantly. Therefore, we assume that the drawings are given on an integer grid of dimension $h_i \times w_i$, for $i \in \{1, 2\}$ such that no two vertices share the same x-coordinate or y-coordinate.

The construction, illustrated in Fig. 2a–c, begins by (non-uniformly) scaling Γ_1 vertically so that it has height $h_1 h_2$ and (non-uniformly) scaling Γ_2 horizontally so that it has width $w_1 w_2$. Call the resulting drawings Γ_1' and Γ_2', respectively. We place $|V(G_1)|$ copies of Γ_2', such that the leftmost vertex of each Γ_2' lines up with one of the vertices of Γ_1'. We color the edges of the $|V(G_1)|$ copies of Γ_2' using the t_2 colors. We similarly create the $|V(G_2)|$ copies of Γ_1' using the t_1 colors, aligning the lowest vertex of each Γ_1' with each vertex of

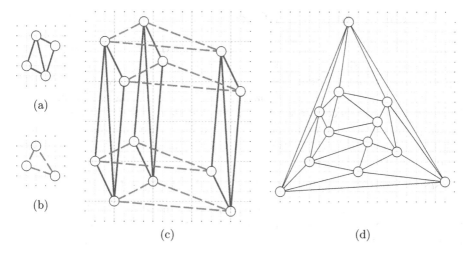

Fig. 2. (a) A planar drawing Γ_1 of an example graph G_1. (b) A planar drawing Γ_2 of an example graph G_2. (c) A 2-layered planar drawing Γ of $G = G_1 \square G_2$ using the scaled drawings of Γ_1 and Γ_2. (d) An 18×18 drawing Γ of a 5-regular planar graph G of order 12 with each vertex having its own row and column.

the lowest copy of Γ_2'. Let (v_1, v_2) be a vertex in $G_1 \square G_2$ with $v_1 \in V(G_1)$ and $v_2 \in V(G_2)$. It is easy to verify that (v_1, v_2) has a corresponding vertex in the merged drawing. Thus, we have a drawing Γ of $G = G_1 \square G_2$ using $t_1 + t_2$ colors.

We now argue that no two edges in the same layer (with same color) cross. Suppose for the sake of contradiction that two edges with color c cross. First, assume that c is one of the t_1 colors from Γ_1. Since Γ_1' is a t_1-layered planar drawing, we know that the two crossing edges cannot be from the same copy of Γ_1'. However, the origins of the Γ_1' copies are separated by at least w_1 units, because they are placed at vertices of Γ_2', which have been scaled horizontally by a factor of w_1, and because no two vertices in Γ_2' share the same column. Since w_1 is the width of Γ_1', no two copies of Γ_1' can intersect each other. So, c cannot be one of the t_1 colors. Similarly, c cannot be one of the t_2 colors. Since these are the only colors used, we have a contradiction, and the drawing is a $(t_1 + t_2)$-layered planar drawing. □

Theorem 2. *For any $t \in \mathbb{Z}^+$, there exist 5t-regular graphs with geometric thickness at most t.*

Proof. Our construction for such a graph starts with any 5-regular graph G. To use the integer grid property of Lemma 1, we create a drawing Γ of G such that every vertex has a unique column and row; see Fig. 2d.

We now simply compute $\mathbb{G} = G \square G \square \cdots \square G$, applying the Cartesian product $t - 1$ times. This results in a 5t-regular graph and by Lemma 1 we know that $\mathbb{G}$ has thickness at most t. Using the example drawing from Fig. 2d, we observe that the resulting drawing has area $18^t \times 18^t$ and that $\mathbb{G}$ has order 12^t. □

4 Conclusion and Open Questions

Ideally, to create a thickness-t regular graph from $\mathcal{G}$, we would simply create t permutations of the vertex order with each permutation corresponding to a separate layer to ensure that every vertex is assigned a degree-5 role exactly once and that two vertices never share an edge in more than one permutation layer. Although this is certainly plausible, we did not see a simple description for such a permutation that would not result in a large case analysis. In addition, our aim was not to construct the smallest order of such a regular graph but simply to maximize the degree.

Question 1. What is the smallest $(6t - 1)$-regular graph of thickness t?

Our solution for geometric thickness t is very generalized and has plenty of room to add extra edges. It seems possible that one might be able to create regular graphs with larger degree while still maintaining the same thickness.

Question 2. What is the largest k such that there exists a k-regular graph of geometric thickness t? Or is $5t$ optimal?

In Lemma 1, we were careful to state that our Cartesian product produced graphs with geometric thickness *at most* $t_1 + t_2$, because we did not prove that the geometric thickness was not less than this value. This is clearly possible as the Cartesian product of two line segments yields a planar graph (a 4-cycle). Nonetheless, $\mathbb{G}$ has geometric thickness *exactly* t for $t < 7$.

Question 3. Does the graph $\mathbb{G}$ have geometric thickness *exactly* t for all $t \in \mathbb{Z}^+$?

We thank the anonymous reviewers for their helpful comments and suggestions.

References

1. Battle, J., Harary, F., Kodama, Y.: Every planar graph with nine points has a nonplanar complement. Bull. Amer. Math. Soc. **68**(6), 569–571 (1962). http://dx.doi.org/10.1090/S0002-9904-1962-10850-7
2. Dillencourt, M.B., Eppstein, D., Hirschberg, D.S.: Geometric thickness of complete graphs. J. Graph Algorithms Appl. **4**(3), 5–17 (2000). http://dx.doi.org/10.7155/jgaa.00023
3. Durocher, S., Gethner, E., Mondal, D.: Thickness and colorability of geometric graphs. In: Brandstädt, A., Jansen, K., Reischuk, R. (eds.) WG 2013. LNCS, vol. 8165, pp. 237–248. Springer, Heidelberg (2013)
4. Harary, F.: Research problem. Bull. Amer. Math. Soc. **67**, 542 (1961)
5. Kainen, P.: Thickness and coarseness of graphs. Abh. aus dem Mathematischen Semin. der Univ. Hamburg **39**(1), 88–95 (1973). http://dx.doi.org/10.1007/BF02992822
6. Mutzel, P., Odenthal, T., Scharbrodt, M.: The thickness of graphs: a survey. Graphs Comb. **14**(1), 59–73 (1998). http://dx.doi.org/10.1007/PL00007219
7. Tutte, W.T.: The non-biplanar character of the complete 9-graph. Can. Math. Bull. **6**(3), 319–330 (1963)
8. Tutte, W.T.: The thickness of a graph. Indagationes Math. **25**, 567–577 (1963)

Intersection Representations

On the Zarankiewicz Problem for Intersection Hypergraphs

Nabil H. Mustafa[1](✉) and János Pach[2]

[1] Laboratoire d'Informatique Gaspard-Monge, Equipe A3SI, ESIEE Paris,
Université Paris-Est, Paris, France
mustafan@esiee.fr
[2] EPFL, Lausanne and Rényi Institute, Budapest, Hungary
pach@cims.nyu.edu

Abstract. Let d and t be fixed positive integers, and let $K_{t,\ldots,t}^d$ denote the complete d-partite hypergraph with t vertices in each of its parts, whose hyperedges are the d-tuples of the vertex set with precisely one element from each part. According to a fundamental theorem of extremal hypergraph theory, due to Erdős [7], the number of hyperedges of a d-uniform hypergraph on n vertices that does not contain $K_{t,\ldots,t}^d$ as a subhypergraph, is $n^{d-\frac{1}{t^{d-1}}}$. This bound is not far from being optimal.

We address the same problem restricted to *intersection hypergraphs* of $(d-1)$-dimensional *simplices* in $\mathbb{R}^d$. Given an n-element set $\mathcal{S}$ of such simplices, let $\mathcal{H}^d(\mathcal{S})$ denote the d-uniform hypergraph whose vertices are the elements of $\mathcal{S}$, and a d-tuple is a hyperedge if and only if the corresponding simplices have a point in common. We prove that if $\mathcal{H}^d(\mathcal{S})$ does not contain $K_{t,\ldots,t}^d$ as a subhypergraph, then its number of edges is $O(n)$ if $d = 2$, and $O(n^{d-1+\epsilon})$ for any $\epsilon > 0$ if $d \geq 3$. This is almost a factor of n better than Erdős's above bound. Our result is tight, apart from the error term ϵ in the exponent.

In particular, for $d = 2$, we obtain a theorem of Fox and Pach [8], which states that every $K_{t,t}$-free intersection graph of n *segments* in the plane has $O(n)$ edges. The original proof was based on a separator theorem that does not generalize to higher dimensions. The new proof works in any dimension and is simpler: it uses *size-sensitive cuttings*, a variant of random sampling. We demonstrate the flexibility of this technique by extending the proof of the planar version of the theorem to intersection graphs of x-monotone curves.

1 Introduction

Let $\mathcal{H}$ be a d-uniform hypergraph on n vertices. One of the fundamental questions of extremal graph and hypergraph theory goes back to Turán and Zarankiewicz: What is the largest number $\mathrm{ex}^d(n, K)$ of hyperedges (or, in short, edges) that $\mathcal{H}$

N.H. Mustafa—Supported by the grant ANR SAGA (JCJC-14-CE25-0016-01).
J. Pach—Supported by Swiss National Science Foundation Grants 200020-144531 and 200021-137574.

© Springer International Publishing Switzerland 2015
E. Di Giacomo and A. Lubiw (Eds.): GD 2015, LNCS 9411, pp. 207–216, 2015.
DOI: 10.1007/978-3-319-27261-0_18

can have if it contains no subhypergraph isomorphic to fixed d-uniform hypergraph K. (See Bollobás [2]). In the most applicable special case, $K = K^d(t, \ldots, t)$ is the complete d-partite hypergraph on the vertex set $V_1, \ldots, V_d$ with $|V_1| = \ldots = |V_d| = t$, consisting of all d-tuples that contain one point from each V_i. For graphs $(d = 2)$, it was proved by Erdős (1938) and Kővári-Sós-Turán [14] that

$$\mathrm{ex}^2(n, K^2_{t,t}) \leq n^{2-1/t}.$$

The order of magnitude of this estimate is known to be best possible only for $t = 2$ and 3 (Reiman [18]; Brown [3]). The constructions for which equality is attained are algebraic.

Erdős (1964) generalized the above statement to d-uniform hypergraphs for all $d \geq 2$. A hypergraph $\mathcal{H}$ is said to be K-*free* if it contains no copy of K as a (not necessarily induced) subhypergraph.

Theorem A ([7]). *The maximum number of hyperedges that a d-uniform, $K^d_{t,\ldots,t}$-free hypergraph of n vertices can have satisfies*

$$\mathrm{ex}^d(n, K^d_{t,\ldots,t}) \leq n^{d - \frac{1}{t^{d-1}}}.$$

It was further shown in [7] that this bound cannot be substantially improved. There exists an absolute constant $C > 0$ (independent of n, t, d) such that

$$\mathrm{ex}^d(n, K^d_{t,\ldots,t}) \geq n^{d - \frac{C}{t^{d-1}}}.$$

In particular, for every $\epsilon > 0$, there exist $K^d_{t,\ldots,t}$-free d-uniform hypergraphs with $t \approx (C/\epsilon)^{1/(d-1)}$, having at least $n^{d-\epsilon}$ edges. The construction uses the probabilistic method.

In certain geometric scenarios, better bounds are known. For instance, consider a bipartite graph with $2n$ vertices that correspond to n distinct points and n distinct lines in the plane, and a vertex representing a point p is connected to a vertex representing a line ℓ if and only if p is incident to ℓ (i.e., $p \in \ell$). Obviously, this graph is $K^2_{2,2}$-free (in short, $K_{2,2}$-free). Therefore, by the above result, it has at most $O(n^{3/2})$ edges. On the other hand, according to a celebrated theorem of Szemerédi and Trotter (1983), the number of edges is at most $O(n^{4/3})$, and the order of magnitude of this bound cannot be improved. In [15], this result was generalized to incidence graphs between points and more complicated curves in the plane.

Given a set $\mathcal{S}$ of geometric objects, their *intersection graph* $\mathcal{H}(\mathcal{S})$ is defined as a graph on the vertex set $\mathcal{S}$, in which two vertices are joined by an edge if and only if the corresponding elements of $\mathcal{S}$ have a point in common. To better understand the possible intersection patterns of edges of a *geometric graph*, that is, of a graph drawn in the plane with possibly crossing straight-line edges, Pach and Sharir [15] initiated the investigation of the following problem. What is the maximum number of edges in a $K_{t,t}$-free intersection graph of n segments in the plane? The Kővári-Sós-Turán theorem (Theorem A for $d = 2$) immediately implies the upper bound $n^{2-1/t}$. Pach and Sharir managed to improve this bound to $O(n)$ for $t = 2$ and to $O(n \log n)$ for any larger (but fixed) value of t. They conjectured that here $O(n \log n)$ can be replaced by $O(n)$ for every t, which was proved by Fox and Pach [8]. They later extended their proof to *string graphs*, that is, to intersection graphs of arbitrary continuous arcs in the plane [9,10]. Some weaker results were established by Radoičić and Tóth [17] by the "discharging method".

The aim of the present note is to generalize the above results to *d-uniform intersection hypergraphs of (d-1)-dimensional simplices* in $\mathbb{R}^d$, for any $d \geq 2$. The arguments used in the above papers are based on planar separator theorems that do not seem to allow higher dimensional extensions applicable to our problem.

Given an n-element set $\mathcal{S}$ of $(d-1)$-dimensional simplices in general position in $\mathbb{R}^d$, let $\mathcal{H}^d(\mathcal{S})$ denote the d-uniform hypergraph on the vertex set $\mathcal{S}$, consisting of all unordered d-tuples of elements $\{S_1, \ldots, S_d\} \subset \mathcal{S}$ with $S_1 \cap \ldots \cap S_d \neq \emptyset$. We prove the following theorem, providing an upper bound on the number of hyperedges of a $K^d_{t,\ldots,t}$-free intersection hypergraph $\mathcal{H}^d(\mathcal{S})$ of $(d-1)$-dimensional simplices. This bound is almost a factor of n better than what we obtain using the abstract combinatorial bound of Erdős (Theorem A), and it does not depend strongly on t.

Theorem 1. *Let $d, t \geq 2$ be integers, let $\mathcal{S}$ be an n-element set of $(d-1)$-dimensional simplices in $\mathbb{R}^d$, and let $\mathcal{H}^d(\mathcal{S})$ denote its d-uniform intersection hypergraph.*

If $\mathcal{H}^d(\mathcal{S})$ is $K^d_{t,\ldots,t}$-free, then its number of edges is $O(n^{d-1+\epsilon})$ for any $\epsilon > 0$. For $d = 2$, the number of edges is at most $O(n)$.

To see that this bound is nearly optimal for every d, fix a hyperplane h in $\mathbb{R}^d$ with normal vector v, and pick $d - 1$ sets $P_1, \ldots, P_{d-1}$, each consisting of $\frac{n-1}{d-1}$ parallel $(d-2)$-dimensional planes in h, with the property that any $d - 1$ members of $P_1 \cup \ldots \cup P_{d-1}$ that belong to different P_is have a point in common. For each i ($1 \leq i \leq d - 1$), replace every plane $p_i \in P_i$ by a hyperplane h_i parallel to u such that $h_i \cap h = p_i$. Clearly, the d-uniform intersection hypergraph of these n hyperplanes, including h, is $K^d_{2,\ldots,2}$-free, and its number of edges is $(\frac{n-1}{d-1})^{d-1} = \Omega(n^{d-1})$. In each hyperplane, we can take a large $(d-1)$-dimensional simplex so that the intersection pattern of these simplices is precisely the same as the intersection pattern (i.e., the d-uniform intersection hypergraph) of the underlying hyperplanes.

The proof of Theorem 1 is based on a partitioning scheme, which was first formulated by Pellegrini [16]. Given an n-element set $\mathcal{S}$ of $(d-1)$-dimensional simplices (or other geometric objects) in $\mathbb{R}^d$, let m denote the number of hyperedges in their d-uniform intersection hypergraph, that is, the number of d-tuples of elements of $\mathcal{S}$ having a point in common. For a parameter $r \leq n$, a $\frac{1}{r}$-*cutting* with respect to $\mathcal{S}$ is a partition of $\mathbb{R}^d$ into simplices such that the interior of every simplex intersects at most n/r elements of $\mathcal{S}$. The *size* of a $\frac{1}{r}$-cutting is the number of simplices it consists of. (See Matoušek [13].)

Theorem B ([6,16]). *Let $d \geq 2$ be an integer, let $\mathcal{S}$ be an n-element set of $(d-1)$-dimensional simplices in $\mathbb{R}^d$, and let m denote the number of d-tuples of simplices in $\mathcal{S}$ having a point in common.*

Then, for any $\epsilon > 0$ and any $r \leq n$, there is a $\frac{1}{r}$-cutting with respect to $\mathcal{S}$ of size at most

$$C_2 \cdot (r + \frac{mr^2}{n^2}) \quad \text{if } d = 2, \text{ and}$$

$$C_{d,\epsilon} \cdot (r^{d-1+\epsilon} + \frac{mr^d}{n^d}) \quad \text{if } d \geq 3.$$

Here C_2 is an absolute constant and $C_{d,\epsilon}$ depends only on d and ϵ.

To construct a $\frac{1}{r}$-cutting, we have to take a *random sample* $\mathcal{R} \subseteq \mathcal{S}$, where each element of $\mathcal{S}$ is selected with probability r/n. It can be shown that, for every $k > 0$, the expected value of the total number of k-dimensional faces of all cells of the cell decomposition induced by the elements of $\mathcal{R}$ is $O(r^{d-1})$, while the expected number of vertices (0-dimensional faces) is $m(r/n)^d$. This cell decomposition can be further subdivided to obtain a partition of $\mathbb{R}^d$ into simplices that meet the requirements. The expected number of elements of $\mathcal{S}$ that intersect a given cell is at most n/r.

Cuttings have been successfully used before, e.g., for an alternative proof of the Szemerédi-Trotter theorem [5]. In our case, the use of this technique is somewhat unintuitive, as the size of the cuttings we construct depends on the number of intersecting d-tuples, that is, on the parameter we want to bound. This sets up an unusual recurrence relation, where the required parameter appears on both sides, but nonetheless, whose solution implies Theorem 1.

The use of cuttings is versatile. In particular, we show how this technique can be applied to establish the generalization of Theorem 1 to x-monotone curves with a bounded number of pairwise intersections.

Theorem 2. *Let $\mathcal{S} = \{S_1, \ldots, S_n\}$ be a set of n x-monotone curves in $\mathbb{R}^2$, and where every pair of curves intersect at most a constant number of times. Let $\mathcal{H}(\mathcal{S})$ denote its intersection graph, and $t \geq 2$ be an integer. If $\mathcal{H}(\mathcal{S})$ is $K_{t,t}$-free, then its number of edges is $O(n)$.*

Fox and Pach [9,10] managed to prove the same result for all continuous curves in the plane, using separator theorems. The weakness of their method is that it is inherently planar. As we will see, the weakness of the cutting technique is that the cell decomposition defined by the randomly selected objects needs to be further refined. To make sure that the number of cells remains under control in this step, we have to bound the number of d-wise intersection points between the objects.

2   Proof of Theorem 1

For any element $S_k \in \mathcal{S}$, let $\text{supp}S_k$ denote the supporting hyperplane of S_k. By slightly perturbing the arrangements, if necessary, we can assume without loss of generality that the supporting hyperplanes of the elements of $\mathcal{S}$ are in *general position*, that is,

(a) no $d - j + 1$ of them have a j-dimensional intersection ($0 \leq j \leq d - 1$), and
(b) the intersection of any d of them is empty or a point that lies in the relative interior of these d elements.

Theorem 1 is an immediate corollary of the following lemma.

Lemma 1. *Let $d, t \geq 2$ be fixed integers. Let $\mathcal{S}$ be an n-element set of $(d-1)$-dimensional simplices in general position in $\mathbb{R}^d$. Assume that their d-uniform intersection hypergraph $\mathcal{H}^d(\mathcal{S})$ has m edges and is $K^d_{t,\ldots,t}$-free.*

If, for suitable constants $C \geq 1$ and u, there exists a $\frac{1}{r}$-cutting of size at most $C(r^u + \frac{mr^d}{n^d})$ with respect to $\mathcal{S}$, consisting of full-dimensional simplices, then

$$m \leq C' \cdot n^u,$$

where C' is another constant (depending on $d, t, C,$ and u).

Proof. For some value of the parameter r to be specified later, construct a $\frac{1}{r}$-cutting $\{\Delta_1, \ldots, \Delta_k\}$ with respect to $\mathcal{S}$, where $k \leq C(r^u + mr^d/n^d)$. Using our assumption that the elements of $\mathcal{S}$ are in general position, we can suppose that all cells Δ_i are full-dimensional.

For every i ($1 \leq i \leq k$), let $\mathcal{S}_i^{\text{int}} \subseteq \mathcal{S}$ denote the set of all elements in $\mathcal{S}$ that intersect the interior of Δ_i. As $\{\Delta_1, \ldots, \Delta_k\}$ is a cutting with respect to $\mathcal{S}$, we have $|\mathcal{S}_i^{\text{int}}| \leq n/r$. Let $\mathcal{S}_i^{\text{bd}} \subseteq \mathcal{S}$ be the set of all elements $S_k \in \mathcal{S}$ such that the supporting hyperplane $\text{supp}S_k$ of S_k contains a j-dimensional face of Δ_i for some j ($0 \leq j \leq d - 1$). Set $\mathcal{S}_i = \mathcal{S}_i^{\text{int}} \cup \mathcal{S}_i^{\text{bd}}$.

Using the general position assumption, we obtain that every j-dimensional face F of Δ_i is contained in the supporting hyperplanes of at most $d - j \leq d$ elements of $\mathcal{S}$ ($0 \leq j \leq d - 1$). The total number of proper faces of Δ_i of all dimensions is smaller than 2^{d+1}, and each is contained in at most d elements of $\mathcal{S}$. Therefore, we have

$$|\mathcal{S}_i| < |\mathcal{S}_i^{\text{int}}| + |\mathcal{S}_i^{\text{bd}}| \leq n/r + d2^{d+1}.$$

Fix an intersection point $q = S_1 \cap \ldots \cap S_d$, and let $\mathcal{S}(q) = \{S_1, \ldots, S_d\} \subseteq \mathcal{S}$. Then either

1. q lies in the interior of some Δ_i, in which case $\mathcal{S}(q) \subseteq \mathcal{S}_i^{\text{int}} \subseteq \mathcal{S}_i$, or
2. q lies at a vertex (0-dimensional face) F or in the interior of a j-dimensional face F of some Δ_i, where $1 \leq j \leq d - 1$. Take any $S_k \in \mathcal{S}(q)$. If $F \subset \text{supp}S_k$, then $S_k \in \mathcal{S}_i^{\text{bd}} \subseteq \mathcal{S}_i$. If $F \not\subset \text{supp}S_k$, then S_k intersects the interior of Δ_i, and since q lies in the relative interior of S_k, we have that $S_k \in \mathcal{S}_i^{\text{int}} \subseteq \mathcal{S}_i$.

In both cases, $\mathcal{S}(q) \subseteq \mathcal{S}_i$.

This means that if for each i we bound the number of d-wise intersection points between the elements of $\mathcal{S}_i$, and we add up these numbers, we obtain an upper bound on $\mathcal{H}^d(\mathcal{S})$, the number of d-wise intersection points between the elements of $\mathcal{S}$. Within each $\mathcal{S}_i$, we apply the abstract hypergraph-theoretic bound of Erdős (Theorem A) to conclude that $\mathcal{H}^d(\mathcal{S}_i)$ has at most $|\mathcal{S}_i|^{d-1/t^{d-1}}$ edges. Hence,

$$m \le \sum_{i=1}^{k} |\mathcal{S}_i|^{d-1/t^{d-1}}.$$

As $|\mathcal{S}_i| \le n/r + d2^{d+1}$, substituting the bound on k, we get

$$m \le C \cdot (r^u + \frac{mr^d}{n^d}) \cdot (d2^{d+1} + \frac{n}{r})^{d-1/t^{d-1}}$$

$$\le 2^d C \cdot (r^u + \frac{mr^d}{n^d}) \cdot (\frac{n}{r})^{d-1/t^{d-1}},$$

provided that $\frac{n}{r} \ge d2^{d+1}$. Setting $r = \frac{n}{C_0}$, where $C_0 = (2^{d+1}C)^{t^{d-1}}$, we obtain

$$m \le 2^d C \left(\frac{n^u}{C_0^u} + \frac{m}{C_0^d} \right) C_0^{d-1/t^{d-1}}$$

$$m \le \frac{C_0^{d-u}}{2} n^u + \frac{m}{2},$$

which implies that $m \le C_0^{d-u} n^u$, as required.   □

Now Theorem 1 follows from Theorem B, as one can choose $u = 1$ if $d = 2$ and $u = d - 1 + \epsilon$ if $d \ge 3$. This completes the proof.

Remark 1. Every d-uniform hypergraph $\mathcal{H}$ has a d-partite subhypergraph $\mathcal{H}'$ that has at least $\frac{d!}{d^d}$ times as many hyperedges as $\mathcal{H}$. Therefore, if K is d-partite, the maximum number of hyperedges that a d-partite K-free hypergraph on n vertices can have is within a factor of $\frac{d!}{d^d}$ from the same quantity over all K-free hypergraphs on n vertices. If instead of abstract hypergraphs, we restrict our attention to intersection graphs or hypergraphs of geometric objects, the order of magnitudes of the two functions may substantially differ. Given two sets of segments $\mathcal{S}$ and $\mathcal{T}$ in the plane with $|\mathcal{S}| = |\mathcal{T}| = n$, let $\mathcal{B}(\mathcal{S}, \mathcal{T})$ denote their *bipartite* intersection graph, in which the vertices representing $\mathcal{S}$ and $\mathcal{T}$ form two independent sets, and a vertex representing a segment in $\mathcal{S}$ is joined to a vertex representing a segment in $\mathcal{T}$ are joined by an edge if and only if they intersect. It was shown in [11] that any $K_{2,2}$-free bipartite intersection graph of n vertices has $O(n^{4/3})$ edges and that this bound is tight. In fact, this result generalizes the Szemerédi-Trotter theorem mentioned above. On the other hand, if we assume that the (non-bipartite) intersection graph associated with the set $\mathcal{S} \cup \mathcal{T}$, which contains the bipartite graph $\mathcal{B}(\mathcal{S}, \mathcal{T})$, is also $K_{2,2}$-free, then Theorem 1 implies that the number of edges drops to linear in n. In the examples where $\mathcal{B}(\mathcal{S}, \mathcal{T})$ has a superlinear number of edges, there must be many intersecting pairs of segments in $\mathcal{S}$ or in $\mathcal{T}$.

Remark 2. The key assumption in Lemma 1 is that there exists a $\frac{1}{r}$-cutting, whose size is sensitive to the number of intersecting d-tuples of objects.

Under these circumstances, in terms of the smallest size of a $\frac{1}{r}$-cutting, one can give an upper bound on the number of edges of $K^d_{t,\ldots,t}$-free intersection hypergraphs with n vertices. For $d = 3$, we know some stronger bounds on the size of vertical decompositions of space induced by a set of triangles [6,19], which imply the existence of $\frac{1}{r}$-cuttings of size $O(r^2\alpha(r) + \frac{mr^3}{n^3})$. Thus, in this case, we can deduce from Lemma 1 that every 3-uniform $K^3_{t,t,t}$-free intersection hypergraph of n triangles has $O(n^2\alpha(n))$ edges. It is an interesting open problem to establish nearly tight bounds on the maximum number of edges that a d-uniform $K^d_{t,\ldots,t}$-free intersection hypergraph induced by n semialgebraic sets in $\mathbb{R}^d$ can have. Lemma 1 does not apply in this case, because in the best currently known constructions of cuttings for semialgebraic sets, the exponent u is larger than d [1].

3   Proof of Theorem 2

Given a set $\mathcal{S}$ of curves in $\mathbb{R}^2$, we will assume that no three curves pass through a common point, and that no two intersection points have the same y-coordinate.

The proof of Theorem 2 follows from an appropriate modification of cuttings for x-monotone curves. Define a *cell* to be a closed set in $\mathbb{R}^2$ that is homeomorphic to a disk. A cell c is *induced by* $\mathcal{S}$ if its boundary is composed of subcurves of elements of $\mathcal{S}$, and (possibly) line segments. The size of a cell c is the number of its boundary curves and segments. For a cell c and a set $\mathcal{R} \subseteq \mathcal{S}$, define $\mathcal{R}^{\mathrm{int}}_c$ to be the curves in $\mathcal{R}$ intersecting the interior of c. A *decomposition* $\mathcal{T}$ of a set $\mathcal{R} \subseteq \mathcal{S}$ is a set of interior-disjoint cells induced by $\mathcal{R}$, and covering $\mathbb{R}^2$ such that $\mathcal{R}^{\mathrm{int}}_c = \emptyset$ for each cell $c \in \mathcal{T}$. The size of a decomposition is its number of cells. Each cell c will be associated with a unique subset $\mathcal{S}^{\mathrm{bd}}_c \subseteq \mathcal{S}$. Given $\mathcal{S}$, the set of decompositions of every $\mathcal{R} \subseteq \mathcal{S}$ is called a *canonical decomposition scheme* if it satisfies the following two properties (see [13, Section 6.5] for details):

1. for every cell c in all decompositions, the size of $\mathcal{S}^{\mathrm{bd}}_c$ can be bounded from above by a constant, and
2. a cell c belongs to the decomposition of $\mathcal{R}$ if and only if $\mathcal{S}^{\mathrm{bd}}_c \subseteq \mathcal{R}$ and $\mathcal{R} \cap \mathcal{S}^{\mathrm{int}}_c = \emptyset$.

A $(1/r)$-cutting Π for $\mathcal{S}$ is a partition of $\mathbb{R}^2$ into interior-disjoint cells of bounded size such that the interior of each cell $c \in \Pi$ is intersected by at most n/r curves of $\mathcal{S}$. Given a canonical decomposition scheme for a set of objects, the existence of small-sized cuttings follows from

Theorem C ([6]). *Let $\mathcal{S}$ be a set of n objects in $\mathbb{R}^2$, and let $r \leq n$ be a parameter. Assume that there exists a $(1/a)$-cutting for any $\mathcal{S}' \subseteq \mathcal{S}$ and any $a > 0$, of size $O(a^C)$ where C is some constant. Then there exists a $(1/r)$-cutting for $\mathcal{S}$ of size $O(\tau(r))$, where $\tau(r)$ is the expected number of cells in the canonical decomposition of a random subset of $\mathcal{S}$ where each element of $\mathcal{S}$ is picked with probability r/n.*

From here, using standard methods, one can deduce the following theorem, whose proof is sketched here, for completeness.

Theorem 3 (Cuttings for x-monotone curves). *Let S be an n-element set of x-monotone curves in $\mathbb{R}^2$, such that every pair of curves intersect at most a constant number of times. Let m be the number of pairs of intersecting curves in S. Then there exists a $(1/r)$-cutting for S of size $O(r + mr^2/n^2)$.*

Proof. The canonical decomposition of any $S' \subseteq S$ will be the *vertical decomposition* of S'. Recall that the vertical decomposition of S' is constructed by extending, from each endpoint of a curve of S' as well as from each intersection point, a vertical segment above and below until it hits another curve of S'. It was verified in [4] that this decomposition is a canonical decomposition scheme.

The vertical decomposition of any $S' \subseteq S$ has size $O(|S'| + I_{S'})$, where $I_{S'}$ is the number of intersection points of the curves in S'. Let $m_{S'}$ be the number of edges in the intersection graph of S'. As each pair of curves can intersect at most a constant number of times, we have $I_{S'} = O(m_{S'})$. Thus, the vertical decomposition of S' has size $O(|S'| + m_{S'})$.

Let $\mathcal{R} \subseteq S$ be a set formed by picking each curve of S with probability $p = r/n$. The expected number of edges in the intersection graph of $\mathcal{R}$ is $\mathbf{E}[m_{\mathcal{R}}] = mp^2$. Therefore, the expected size of the canonical decomposition of $\mathcal{R}$ is $O(|S|p + mp^2) = O(r + mr^2/n^2)$.

It remains to show that for any $a > 0$ and for any $S' \subseteq S$, there exists a $(1/a)$-cutting for S' of size $O(a^C)$. Fix a set S', and let $\mathcal{U}$ be the set of all cells present in the vertical decomposition of any subset S'' of S'. Construct the following set-system on S':

$$\Phi(S') = \{S' \cap \text{interior}(U) \mid U \in \mathcal{U}\}$$

Each boundary vertex of a cell in $\mathcal{U}$ is an (i) intersection point of two curves in S', or (ii) the endpoint of a curve in S', or (iii) the intersection point of a curve in S' and a vertical line passing through either an endpoint or an intersection point of two curves in S'. Since every pair of curves intersect at most a constant number of times, the total number of such vertices is $O(|S'|^3)$. Each cell in the vertical decomposition of any subset of S' can be uniquely identified by the sequence of its boundary vertices. As each cell has at most 4 boundary vertices, we get that $|\mathcal{U}| = O(|S'|^{12})$. Therefore, $|\Phi(S')| = O(|S'|^{12})$, which implies that the VC-dimension of $\Phi(S')$ is bounded by a constant. By the ϵ-net theorem of Haussler and Welzl [12], there exists a subset $\mathcal{R}' \subseteq S'$ of size $O(a \log a)$, such that $S'' \cap \mathcal{R}' \neq \emptyset$ for any $S'' \in \Phi(S')$ of size at least $|S'|/a$. The required $(1/a)$-cutting is the vertical decomposition of $\mathcal{R}'$, as by property 2 of canonical decompositions, for any cell c in the vertical decomposition, we have $|S_c^{\text{int}}| < |S'|/a$. Finally, note that the size of the vertical decomposition of $\mathcal{R}'$ is $O(a^2 \log^2 a)$. $\qquad\square$

Now are in a position to complete the proof of Theorem 2. Construct a $(1/r)$-cutting Π for S, of size $C_3 \cdot (r + mr^2/n^2)$, where C_3 is a constant. For each cell $c \in \Pi$, let $S_c = S_c^{\text{bd}} \cup S_c^{\text{int}}$. Note that $|S_c| \leq O(1) + n/r$ for all c. For a point $p \in S_i \cap S_j$, there are three possibilities:

1. p lies in the interior of a cell $c \in \Pi$. Then $S_i, S_j \in \mathcal{S}_c^{\text{int}}$.
2. p lies in the interior of the boundary of a cell c. By the assumption that every pair of curves of $\mathcal{S}$ intersect in at most a constant number of points and no three pass through a common point, $S_i \in \mathcal{S}_c^{\text{bd}}$ and $S_j \in \mathcal{S}_c^{\text{int}}$ (or vice versa).
3. p lies at a vertex of c. The total number of such intersection points is bounded by the number of vertices of Π, $O(r + mr^2/n^2)$.

This implies that

$$m \leq O(r + \frac{mr^2}{n^2}) + \sum_{c \in \Pi} |\mathcal{S}_c|^{2-1/t}$$

As in proof of Theorem 1, set $r = n/C$ for a sufficiently large constant C to get $m = O(n)$.

Remark. Theorem 2 implies that if the intersection graph of n constant-degree algebraic curves in the plane is $K_{t,t}$-free, then it has $O(n)$ edges.

References

1. Agarwal, P.K., Matousek, J., Sharir, M.: On range searching with semialgebraic sets. II. SIAM J. Comput. **42**(6), 2039–2062 (2013)
2. Bollobás, B.: Modern Graph Theory. Springer (1998)
3. Brown, W.G.: On graphs that do not contain a Thomsen graph. Canad. Math. Bull. **9**, 281285 (1966)
4. Chazelle, B., Edelsbrunner, H., Guibas, L.J., Sharir, M., Snoeyink, J.: Computing a face in an arrangement of line segments and related problems. SIAM J. Comput. **22**(6), 1286–1302 (1993)
5. Clarkson, K., Edelsbrunner, H., Guibas, L., Sharir, M., Welzl, E.: Combinatorial complexity bounds for arrangements of curves and spheres. Discrete Comput. Geom. **5**, 99–160 (1990)
6. de Berg, M., Schwarzkopf, O.: Cuttings and applications. Int. J. Comput. Geom. Appl. **5**(4), 343–355 (1995)
7. Erdös, P.: On extremal problems of graphs and generalized graphs. Israel J. Math **2**, 183–190 (1964)
8. Fox, J., Pach, J.: Separator theorems and Turán-type results for planar intersection graphs. Adv. Math. **219**(3), 1070–1080 (2008)
9. Fox, J., Pach, J.: A separator theorem for string graphs and its applications. Comb. Probab. Comput. **19**(3), 371–390 (2010)
10. Fox, J., Pach, J.: Applications of a new separator theorem for string graphs. Comb., Probab. Comput. **23**, 66–74 (2014)
11. Fox, J., Pach, J., Sheffer, A., Suk, A., Zahl, J.: A semi-algebraic version of Zarankiewicz's problem. ArXiv e-prints (2014)
12. Haussler, D., Welzl, E.: Epsilon-nets and simplex range queries. Discrete Comput. Geom. **2**, 127–151 (1987)
13. Matoušek, J.: Lectures in Discrete Geometry. Springer, New York (2002)
14. Kővári, T., Sós, V.T., Turán, P.: On a problem of K. Zarankiewicz. Colloquium Math. **3**, 50–57 (1954)
15. Pach, J., Sharir, M.: On planar intersection graphs with forbidden subgraphs. J. Graph Theory **59**(3), 205–214 (2008)

16. Pellegrini, M.: On counting pairs of intersecting segments and off-line triangle range searching. Algorithmica **17**(4), 380–398 (1997)
17. Radoičić, R., Tóth, G.: The discharging method in combinatorial geometry and the Pach-Sharir conjecture. In: Goodman, J.E., Pach, J., Pollack, J. (eds.) Proceedings of the Joint Summer Research Conference on Discrete and Computational Geometry, vol. 453, pp. 319–342. Contemporary Mathematics, AMS (2008)
18. Reiman, I.: Uber ein problem von K. Zarankiewicz. Acta Mathematica Academiae Scientiarum Hungarica **9**, 269–273 (1958)
19. Tagansky, B.: A new technique for analyzing substructures in arrangements of piecewise linear surfaces. Discrete Comput. Geom. **16**(4), 455–479 (1996)

Intersection-Link Representations of Graphs

Patrizio Angelini[1], Giordano Da Lozzo[2], Giuseppe Di Battista[2],
Fabrizio Frati[2]([⊠]), Maurizio Patrignani[2], and Ignaz Rutter[3]

[1] Tübingen University, Tübingen, Germany
angelini@informatik.uni-tuebingen.de
[2] Roma Tre University, Rome, Italy
{dalozzo,gdb,frati,patrigna}@dia.uniroma3.it
[3] Karlsruhe Institute of Technology, Karlsruhe, Germany
rutter@kit.edu

Abstract. We consider drawings of graphs that contain dense subgraphs. We introduce *intersection-link representations* for such graphs,
in which each vertex u is represented by a geometric object $R(u)$ and in
which each edge (u, v) is represented by the intersection between $R(u)$
and $R(v)$ if it belongs to a dense subgraph or by a curve connecting the
boundaries of $R(u)$ and $R(v)$ otherwise. We study a notion of planarity,
called CLIQUE PLANARITY, for intersection-link representations of graphs
in which the dense subgraphs are cliques.

1 Introduction

In several applications there is the need to represent graphs that are globally
sparse but contain dense subgraphs. As an example, a social network is often
composed of communities, whose members are closely interlinked, connected by
a network of relationships that are much less dense. The visualization of such
networks poses challenges that are attracting the study of several researchers
(see, e.g., [6,11]). One frequent approach is to rely on clustering techniques to
collapse dense subgraphs and then represent only the links between clusters.
However, this has the drawback of hiding part of the graph structure. Another
approach that has been explored is the use of hybrid drawing standards, where
different conventions are used to represent the dense and the sparse portions
of the graph: In the drawing standard introduced in [4,12] each dense part is
represented by an adjacency matrix while two adjacent dense parts are connected
by a curve.

In this paper we study *intersection-link representations*, which are hybrid
representations where in the dense parts of the graph the edges are represented
by the intersection of geometric objects (*intersection* representation) and in the
sparse parts the edges are represented by curves (*link* representation). More
formally and more specifically, we introduce the following problem. Suppose that

Angelini was partially supported by DFG grant Ka812/17-1. Da Lozzo, Di Battista,
Frati, and Patrignani were partially supported by MIUR project AMANDA, prot.
2012C4E3KT_001. Rutter was partially supported by DFG grant WA 654/21-1.

E. Di Giacomo and A. Lubiw (Eds.): GD 2015, LNCS 9411, pp. 217–230, 2015.
DOI: 10.1007/978-3-319-27261-0_19

a pair (G, S) is given where G is a graph and S is a set of cliques that partition the vertex set of G. In an *intersection-link* representation, vertices are geometric objects that are translates of the same rectangle. Consider an edge (u, v) and let $R(u)$ and $R(v)$ be the rectangles representing u and v, respectively. If (u, v) is part of a clique (*intersection-edge*) we represent it by drawing $R(u)$ and $R(v)$ so that they intersect, else (*link-edge*) we represent it by a curve connecting $R(u)$ and $R(v)$; see Fig. 1.

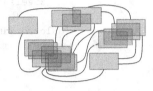

We study the CLIQUE PLANARITY problem that asks to test whether a pair (G, S) has an intersection-link representation such that link-edges do not cross each other and do not traverse any rectangle. The main challenge of the problem lies in the interplay between the geometric constraints imposed by the rectangle arrangements and the topological constraints imposed by the link edges.

Fig. 1. Intersection-link representation of a graph with five cliques.

Several problems are related to CLIQUE PLANARITY; here we mention two notable ones. The problem of recognizing intersection graphs of translates of the same rectangle is $\mathcal{NP}$-complete [7]. Note that this does not imply $\mathcal{NP}$-hardness for our problem, since cliques always have such a representation. *Map graphs* allow to represent graphs containing large cliques in a readable way; they are contact graphs of internally-disjoint connected regions of the plane, where the contact can be even a single point. The recognition of map graphs has been studied in [8,15]. One can argue that there are graphs that admit a clique-planar representation, while not admitting any representation as a map graph, and vice versa.

We now describe our contribution. Our study encountered several interesting and at a first glance unrelated theoretical problems. In more detail, our results are as follows.

- In Sect. 3 we show that CLIQUE PLANARITY is $\mathcal{NP}$-complete even if S contains just one clique with more than one vertex. This result is established by observing a relationship between CLIQUE PLANARITY and a natural constrained version of the CLUSTERED PLANARITY problem, in which we ask whether a path (rather than a tree as in the usual CLUSTERED PLANARITY problem) can be added to each cluster to make it connected while preserving clustered planarity; we prove this problem to be $\mathcal{NP}$-complete, a result which might be interesting in its own right.
- In Sect. 4, we show how to decide CLIQUE PLANARITY in linear time in the case in which each clique has a prescribed geometric representation, via a reduction to the problem of testing planarity for a graph with a given partial representation.
- In Sect. 5, we concentrate on instances of CLIQUE PLANARITY composed of two cliques. While we are unable to settle the complexity of this case, we show that the problem becomes equivalent to an interesting variant of the 2-PAGE BOOK EMBEDDING problem, in which the graph is bipartite and the vertex ordering in the book embedding has to respect the vertex partition of the

graph. This problem is in our opinion worthy of future research efforts. For now, we use this equivalence to establish a polynomial-time algorithm for the case in which the link-edges are assigned to the pages of the book embedding.
– In Sect. 6, we study a Sugiyama-style problem where the cliques are arranged on levels according to a hierarchy. In this practical setting we show that CLIQUE PLANARITY is solvable in polynomial time. This is achieved via a reduction to the T-LEVEL PLANARITY problem [3].

Conclusions and open problems are presented in Sect. 7. Because of space limitations, complete proofs are deferred to the full version of the paper [2].

2 Intersection-Link Model

Let G be a graph and S be a set of cliques inducing a partition of the vertex set of G. In an *intersection-link representation* of (G, S): (i) each vertex u is a translate $R(u)$ of an axis-aligned rectangle $\mathcal{R}$; (ii) $R(u)$ and $R(v)$ intersect if and only if edge (u, v) is an intersection-edge; and (iii) each link-edge (u, v) is a curve connecting the boundaries of $R(u)$ and $R(v)$. To avoid degenerate intersections we assume that no two rectangles have their sides on the same line. The CLIQUE PLANARITY problem asks whether an intersection-link representation of a pair (G, S) exists such that no two curves intersect and no curve intersects the interior of a rectangle. Such a representation is called *clique-planar*. A pair (G, S) is *clique-planar* if it admits a clique-planar representation. Let Γ be an intersection-link representation of $(K_n, \{K_n\})$. We have the following.

Lemma 1. *Traversing the outer boundary B of Γ clockwise, the sequence of encountered rectangles is a subset of $R(u_1), R(u_2), \ldots, R(u_n), R(u_{n-1}), \ldots, R(u_2)$, for some permutation $u_1, \ldots, u_n$ of the vertices of K_n.*

Proof sketch: The statement follows from the following claims: (a) the sequence of encountered rectangles is not of the form $\ldots, R(u), \ldots, R(v), \ldots, R(u), \ldots, R(v)$, for any $u, v \in K_n$; (b) every maximal portion of B belonging to a single rectangle $R(u)$ contains (at least) one corner of $R(u)$; (c) if two adjacent corners of the same rectangle $R(u)$ both belong to B, then the entire side of $R(u)$ between them belongs to B; and (d) any rectangle $R(u)$ does not define three distinct maximal portions of B.   □

The following lemma allows us to focus, without loss of generality, on special clique-planar representations, which we call *canonical*.

Lemma 2. *Let (G, S) admit a clique-planar representation Γ. There exists a clique-planar representation Γ' of (G, S) such that: (i) each vertex is represented by an axis-aligned unit square and (ii) for each clique $s \in S$, all the squares representing vertices in s have their upper-left corners along a common line with slope 1.*

Proof sketch: Initialize $\Gamma' = \Gamma$. Scale Γ' so that each rectangle has both sides of length larger than 2. For any clique $s \in S$, traverse the boundary of the rectangle arrangement representing s in Γ clockwise. By Lemma 1, the circular sequence of encountered rectangles is of the form $R(u_1), \ldots, R(u_{|s|}), R(u_{|s|-1}), \ldots, R(u_2)$, for some permutation $u_1, \ldots, u_{|s|}$ of the vertices of s. Place pairwise-intersecting unit squares $Q(u_1), \ldots, Q(u_{|s|})$ representing $u_1, \ldots, u_{|s|}$ in the interior of $R(u_1)$, as required by the lemma. Remove $R(u_1), \ldots, R(u_{|s|})$ and reroute the curves representing link-edges from the border of the rectangle arrangement to the suitable ending squares. This can be done without introducing any crossings, because the circular sequence of the squares encountered when traversing the boundary of the square arrangement clockwise is $Q(u_1), \ldots, Q(u_{|s|}), Q(u_{|s|-1}), \ldots, Q(u_2)$. □

3 Hardness Results on Clique Planarity

In this section we prove that the CLIQUE PLANARITY problem is not solvable in polynomial time, unless $\mathcal{P}=\mathcal{NP}$. In fact, we have the following.

Theorem 1. *It is $\mathcal{NP}$-complete to decide whether a pair (G, S) is clique-planar, even if S contains just one clique with more than one vertex.*

We prove Theorem 1 by showing a reduction from a constrained clustered planarity problem, which we prove to be $\mathcal{NP}$-complete, to the CLIQUE PLANARITY problem.

A *clustered graph* (G, T) is a pair where G is a graph and T is a rooted tree whose leaves are the vertices of G; the internal nodes of T distinct from the root correspond to subsets of vertices of G, called *clusters*. A clustered graph is *flat* if every cluster is a child of the root. A *c-planar drawing* of (G, T) is a planar drawing of G, together with a representation of each cluster μ as a simple region R_μ of the plane such that: (i) every region R_μ contains all and only the vertices in μ; (ii) every two regions R_μ and R_ν are either disjoint or one contains the other; and (iii) every edge intersects the boundary of each region R_μ at most once. A graph is *c-planar* if it admits a c-planar drawing. The CLUSTERED PLANARITY problem asks whether a given clustered graph is c-planar. Polynomial-time algorithms for testing c-planarity are known only in special cases, most notably, for *c-connected* clustered graphs, in which each cluster induces a connected graph [9,10]. A clustered graph is c-planar if and only if a set of edges can be added to it so that the resulting graph is c-planar and c-connected [10]. Any such set of edges is a *saturator*, and the subset of a saturator composed of the edges between vertices of the same cluster μ defines a *saturator for μ*. A saturator is *linear* if the saturator for each cluster is a path. The CLUSTERED PLANARITY WITH LINEAR SATURATORS (CPLS) problem asks whether a flat clustered graph such that each cluster induces an independent set of vertices admits a linear saturator.

Lemma 3. *Let (G, T) be an instance of CPLS with $G = (V, E)$ and let $E^\star \subseteq \binom{V}{2} \setminus E$ be such that in $G^\star = (V, E \cup E^\star)$ every cluster induces a path. Then $E^\star$ is a linear saturator for (G, T) if and only if $G^\star$ is planar.*

The following lemma connects the problem CLIQUE PLANARITY with CPLS.

Lemma 4. *Given an instance (G, T) of the CPLS problem, an equivalent instance (G', S) of the CLIQUE PLANARITY problem can be constructed in linear time.*

Proof sketch: Initialize $G' = G$. Then, for each cluster μ of (G, T), add a clique s_μ on the vertex set of μ to (G', S). Clearly, (G', S) can be constructed in linear time. We now prove the equivalence between (G, T) and (G', S).

If (G, T) admits a linear saturator E^*, then there exists a c-planar drawing Γ^* of (G^*, T), where G^* is obtained by adding E^* to G. We construct a clique-planar representation of (G', S) as follows. For each cluster μ, replace the interior of the region representing μ in Γ^* with a set of $|s_\mu|$ pairwise-intersecting axis-aligned unit squares, where the order of such squares is the same as the one of the corresponding vertices in the linear saturator for μ; complete the drawing of each link-edge (u, v) of G' with curves from the squares representing u and v to the boundaries of the regions representing the clusters containing u and v. The correspondence between the order of the squares and the order of the vertices in E^* guarantees the absence of crossings.

If (G', S) has a clique-planar representation Γ, which we can assume to be canonical by Lemma 2, then we define a set E^* as follows. For each $s \in S$, we add to E^* the edges of path $(u_1, \ldots, u_k)$ on the vertex set of s, where the unit squares $R(u_1), \ldots, R(u_k)$ are in this order in their arrangement in Γ. We claim that E^* is a linear saturator for (G, T). By Lemma 3, it suffices to show that $G + E^*$ admits a planar drawing: Starting from Γ, we place each vertex v at the center of $R(v)$, replace $R(v)$ with straight-line segments from v to the endpoints of its incident link-edges, and draw the edges of E^* as straight-line segments. This does not produce crossings. In particular, any two straight-line segments not in E^* incident to vertices u and v in the same clique are separated by the line through the intersection points of the boundaries of $R(u)$ and $R(v)$, as in Fig. 2(a). Further, any segment (u, v) in E^* and any segment not in E^* incident to a vertex w in the same clique as u and v are separated by the line through the intersection points of the boundaries of $R(u)$ and $R(w)$ or of $R(v)$ and $R(w)$, as in Fig. 2(b).   □

Next, we prove that the CPLS problem is $\mathcal{NP}$-complete.

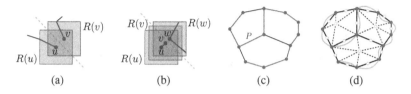

(a)   (b)   (c)   (d)

Fig. 2. (a)–(b) Construction of a linear saturator from a clique-planar representation. (c) A biconnected planar graph G with a Hamiltonian path P. (d) The clustered graph (G', T) obtained from G and the linear saturator for (G', T) corresponding to P.

Theorem 2. *The* CPLS *problem is* $\mathcal{NP}$*-complete, even for instances in which just one cluster contains more than one vertex.*

Proof sketch: The problem clearly lies in $\mathcal{NP}$. We give a polynomial-time reduction from the HAMILTONIAN PATH problem in biconnected planar graphs [14]. Given an instance G of HAMILTONIAN PATH, we construct an instance (G',T) of CPLS as follows. Assume G has an associated planar embedding. Initialize $G' = G$, as in Fig. 2(c). Add a vertex v_f inside each face f and connect it to all the vertices incident to f (this results in a triangulated planar graph G'); then subdivide with a vertex each edge of G' which is also in G, as in Fig. 2(d). Finally, add a cluster μ to T containing all the vertices of G' which are also in G and, for each of the remaining vertices, add to T a cluster containing only that vertex. Now, any Hamiltonian path $P = (v_1, \ldots, v_n)$ in G can be drawn in G' without crossings by letting each edge (v_i, v_{i+1}) lie in one of the two faces of G' incident to the dummy vertex for edge (v_i, v_{i+1}). It follows from Lemma 3 that the edge set of P is a linear saturator for (G',T). Conversely, any linear saturator for (G',T) defines a path on the vertex set of μ, which is a Hamiltonian path in G. □

4   Clique-Planarity with Given Vertex Representations

In this section we show how to test CLIQUE PLANARITY in linear time for instances (G, S) with given vertex representations. That is, a clique-planar representation Γ' of (G', S) is given, where G' is obtained from G by removing its link-edges, and the goal is to test whether the link-edges of (G, S) can be drawn in Γ' to obtain a clique-planar representation of (G, S). We start with a linear-time preprocessing in which we verify that every vertex of G incident to a link-edge is represented in Γ' by a rectangle incident to the outer boundary of the clique it belongs to. If the test fails, the instance is negative. Otherwise, we proceed as follows.

We show a reduction to the PARTIAL EMBEDDING PLANARITY problem [1], which asks whether a planar drawing of a graph H exists extending a given drawing $\mathcal{H}'$ of a subgraph H' of H. First, we define a connected component H'_s of H' corresponding to a clique $s \in S$ and its drawing $\mathcal{H}'_s$. We remark that H'_s is a *cactus graph*, that is a connected graph that admits a planar embedding in which all the edges are incident to the outer face. Denote by B the boundary of the representation of s in Γ' (see Fig. 3(a)). If s has one or two vertices, then H'_s

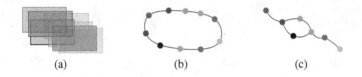

(a)   (b)   (c)

Fig. 3. (a) An intersection-link representation Γ of $(K_7, \{s = K_7\})$. (b) A simple cycle with a vertex for each maximal portion of the boundary of Γ belonging to a single rectangle. (c) Planar drawing $\mathcal{H}'_s$ of graph H'_s corresponding to Γ.

is a vertex or an edge, respectively (and $\mathcal{H}'_s$ is any drawing of H'_s). Otherwise, initialize H'_s to a simple cycle containing a vertex for each maximal portion of B belonging to a single rectangle (see Fig. 3(b)). Let $\mathcal{H}'_s$ be any planar drawing of H'_s with a suitable orientation. Each rectangle in Γ' may correspond to two vertices of H'_s, but no more than two by Lemma 1. Insert an edge in H'_s between every two vertices representing the same rectangle and draw it in the interior of $\mathcal{H}'_s$. By Lemma 1, these edges do not alter the planarity of $\mathcal{H}'_s$. Contract the inserted edges in H'_s and $\mathcal{H}'_s$ (see Fig. 3(c)). This completes the construction of H'_s, together with its planar drawing $\mathcal{H}'_s$. Graph H' is the union of graphs H'_s, over all the cliques $s \in S$; the drawings $\mathcal{H}'_s$ of H'_s are in the outer face of each other in $\mathcal{H}'$. Note that, because of the preprocessing, the endvertices of each link-edge of G are vertices of H'; then we define H as the graph obtained from H' by adding, for each link-edge (u, v) of G, an edge between the vertices of H' corresponding to u and v. We have the following:

Lemma 5. *There exists a planar drawing of H extending $\mathcal{H}'$ if and only if there exists a clique-planar representation of (G, S) coinciding with Γ' when restricted to (G', S).*

Proof sketch: Let $\mathcal{H}$ be a planar drawing of H extending $\mathcal{H}'$. We construct a clique-planar representation of (G, S) as follows: (i) for each $s \in S$, enclose $\mathcal{H}'_s$ with a closed polyline P_s; (ii) scale $\mathcal{H}$ so that the bounding box of the representation of s in Γ' fits inside P_s; (iii) replace the interior of P_s with a copy of the representation of s in Γ'; and (iv) reroute the link-edges from P_s to the suitable rectangles; this creates no crossing, because vertices along the outer face of $\mathcal{H}'_s$ are in the same order as the corresponding rectangles along the boundary of the representation of s in Γ'.

Let Γ be a clique-planar representation of (G, S). We construct a planar drawing of H extending $\mathcal{H}'$ as follows: (i) for each $s \in S$, enclose the representation of s in Γ by a closed polyline P_s; (ii) replace the interior of P_s with a scaled copy of $\mathcal{H}'_s$; and (iii) reroute the curves representing link-edges from P_s to the suitable endvertices (as in the previous direction, this can be done without introducing crossings). $\qquad \square$

We get the main theorem of this section.

Theorem 3. Clique Planarity *can be decided in linear time for a pair (G, S) if the rectangle representing each vertex of G is given as part of the input.*

Proof. First, we check whether, for each $s \in S$, all the rectangles representing vertices in s are pairwise intersecting. This can be done in $O(|s|)$ time by computing the maximum x- and y-coordinates x_M and y_M among all bottom-left corners, the minimum x- and y-coordinates x_m and y_m among all top-right corners, and by checking whether $x_M < x_m$ and $y_M < y_m$. The described reduction to Partial Embedding Planarity can be performed in linear time by traversing the boundary B of each clique $s \in S$; namely, as a consequence of Lemma 1, B has linear complexity. Contracting an edge requires merging the adjacency lists of its endvertices; this can be done in constant time since these vertices have constant degree, again by Lemma 1. Finally, the PARTIAL EMBEDDING PLANARITY problem can be solved in linear time [1]. □

5 Testing Clique Planarity for Graphs Composed of Two Cliques

In this section we study the CLIQUE PLANARITY problem for pairs (G, S) such that $|S| = 2$. Observe that, if $|S| = 1$, then the CLIQUE PLANARITY problem is trivial, since in this case G is a clique with no link-edge, hence a clique-planar representation of (G, S) can be easily constructed. The case in which $|S| = 2$ is already surprisingly non-trivial. Indeed, we could not determine the computational complexity of CLIQUE PLANARITY in this case. However, we establish the equivalence between our problem and a book embedding problem whose study might be interesting in its own; by means of this equivalence we show a polynomial-time algorithm for a special version of the CLIQUE PLANARITY problem. This book embedding problem is defined as follows.

A 2-*page book embedding* is a plane drawing of a graph where the vertices are cyclically arranged along a closed curve ℓ, called the *spine*, and each edge is drawn in one of the two regions of the plane delimited by ℓ. The 2-PAGE BOOK EMBEDDING problem asks whether a 2-page book embedding exists for a given graph. This problem is $\mathcal{NP}$-complete [16]. Now consider a bipartite graph $G(V_1 \cup V_2, E)$. A *bipartite 2-page book embedding* of G is a 2-page book embedding such that all vertices in V_1 occur consecutively along the spine (and all vertices in V_2 occur consecutively, as well). Finally, we define a *bipartite 2-page book embedding with spine crossings* (B2PBESC), as a bipartite 2-page book embedding in which edges are not restricted to lie in one of the two regions delimited by ℓ, but each of them might cross ℓ once; these crossings are only allowed to happen in the two portions of ℓ delimited by a vertex of V_1 and a vertex of V_2. We call the corresponding embedding problem BIPARTITE 2-PAGE BOOK EMBEDDING WITH SPINE CROSSINGS (B2PBESC).

We now prove that B2PBESC is equivalent to CLIQUE PLANARITY for instances (G, S) such that $|S| = 2$. Consider any instance $(G', \{s_1, s_2\})$ of CLIQUE PLANARITY. We define an instance $G(V_1 \cup V_2, E)$ of B2PBESC so that V_i is the vertex set of s_i, for $i = 1, 2$, and E is the set of link-edges of G'. Conversely, given an instance $G(V_1 \cup V_2, E)$ of B2PBESC, an instance $(G', \{s_1, s_2\})$ of CLIQUE PLANARITY can be constructed in which s_i is a clique on V_i, for $i = 1, 2$, and the set of link-edges of G' coincides with E. Since link-edges only connect vertices of different cliques and since each edge of E only connects a vertex of V_1 to one of V_2, each mapping generates a valid instance for the other problem. Also, these mappings define a bijection, hence the following lemma establishes the equivalence between the two problems.

Lemma 6. $(G', \{s_1, s_2\})$ *is clique-planar if and only if* $G(V_1 \cup V_2, E)$ *admits a* B2PBESC.

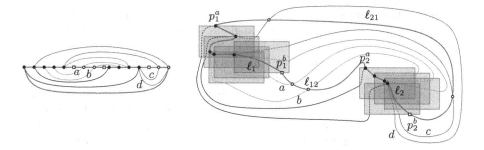

Fig. 4. Constructing a B2PBESC from a clique-planar representation.

Proof sketch: Starting from a B2PBESC $\mathcal{B}$ of G, draw disjoint closed curves λ_1 and λ_2 respectively containing the vertices in V_1 and V_2. Scale $\mathcal{B}$ so that a square of size 2×2 fits in the interiors of λ_1 and λ_2. Replace such interiors with canonical representations of s_1 and s_2 in which the order of the unit squares corresponds to the order of the vertices along ℓ. Each link-edge of (G', S) is then composed of three curves: one is the portion of the corresponding edge of G outside λ_1 and λ_2, and two are curves from the boundaries of λ_1 and λ_2 to the suitable ending squares. This can be done without crossings as the unit squares are in the same order as the vertices along ℓ.

Starting from a clique-planar representation of $(G', \{s_1, s_2\})$, construct a B2PBESC $\mathcal{B}$ of G as follows. Refer to Fig. 4. By Lemma 1, the rectangles representing vertices $u_1, \ldots, u_k \in V_1$ appear along the boundary B_1 of s_1 in an order which is a subsequence of $R(u_1), \ldots, R(u_k), \ldots, R(u_2)$. Draw a curve ℓ_1 between two points $p_1^a \in R(u_1) \cap B_1$ and $p_1^b \in R(u_k) \cap B_1$ entering $R(u_1), \ldots, R(u_k)$ in this order. Place u_i at the point where ℓ_1 enters $R(u_i)$. Define ℓ_2, B_2, p_2^a and p_2^b, and draw the vertices of V_2 analogously. Add to $\mathcal{B}$ curves ℓ_{12} and ℓ_{21}, not intersecting each other, not intersecting the same link-edge, each intersecting any link edge at most once, and connecting p_1^a to p_2^b, and p_2^a to p_1^b, respectively. Let $\ell = \ell_1 \cup \ell_2 \cup \ell_{12} \cup \ell_{21}$. The correspondence between the vertex ordering along ℓ and the order of the rectangles along B_1 and B_2 allows us to extend the link-edges from B_1 and B_2 to the suitable vertices of G inside them. The vertices of V_1 (of V_2) are consecutive along ℓ, since they lie on ℓ_1 (on ℓ_2); also, each edge $e \in E$ crosses ℓ at most once, either on ℓ_{12} or on ℓ_{21}. Hence, $\mathcal{B}$ is a B2PBESC of G.   □

We now consider a variant of the CLIQUE PLANARITY problem for two cliques in which each clique is associated with a 2-partition of the link-edges incident to it, and the goal is to construct a clique-planar representation in which the link-edges in different sets of the partition exit the clique on "different sides". This constraint corresponds to the variant of the 2-page book embedding problem, called PARTITIONED 2-PAGE BOOK EMBEDDING problem, in which the vertices are allowed to be arbitrarily permuted along the spine, while the edges are pre-assigned to the pages of the book [13].

Let $(G, S = \{s_1, s_2\})$ be an instance of CLIQUE PLANARITY and let $\{E_i^a, E_i^b\}$ be a partition of the link-edges incident to s_i, with $i \in \{1, 2\}$. Consider an intersection-link representation Γ_i of s_i with outer boundary B_i, let p_i be the bottom-left corner of the leftmost rectangle in Γ_i, and let q_i be the upper-right corner of the rightmost rectangle in Γ_i. Let B_i^a (B_i^b) be the part of B_i from p_i to q_i (from q_i to p_i) in clockwise direction; this is the *top side* (the *bottom side*) of Γ_i. We aim to construct a clique-planar representation of (G, S) in which all the link-edges in E_i^a (resp. in E_i^b) intersect the arrangement of rectangles representing s_i on its top side (resp. bottom side). We call the problem of determining whether such a representation exists 2-PARTITIONED CLIQUE PLANARITY. We prove that 2-PARTITIONED CLIQUE PLANARITY can be solved in quadratic time. The algorithm is based on a reduction to equivalent instances of SIMULTANEOUS EMBEDDING WITH FIXED EDGES (SEFE) that can be decided in quadratic time. Given two graphs G_1 and G_2 on the same vertex set V, the SEFE problem asks to find planar drawings of G_1 and G_2 that coincide on V and on the common edges of G_1 and G_2.

Lemma 7. *Let* $(G, \{s_1, s_2\})$ *and* $\{E_1^a, E_1^b, E_2^a, E_2^b\}$ *be an instance of* 2-PARTITIONED CLIQUE PLANARITY. *An equivalent instance* $\langle G_1, G_2 \rangle$ *of* SEFE *such that* $G_1 = (V, E_1)$ *and* $G_2 = (V, E_2)$ *are 2-connected and such that the common graph* $G_\cap = (V, E_1 \cap E_2)$ *is connected can be constructed in linear time.*

Proof sketch: By Lemma 6, we can describe $(G, \{s_1, s_2\})$ by its equivalent instance $G(V_1 \cup V_2, E)$ of B2PBESC, where V_i is the vertex set of s_i, for $i = 1, 2$, and E is the set of link-edges of $(G, \{s_1, s_2\})$; partition $\{E_i^a, E_i^b\}$ translates to constraints on the side of the spine ℓ each of these edges has to be incident to. Namely, for each $u_i \in V_1$, all the edges in E_1^a (in E_1^b) incident to u_i have to exit u_i from the internal (resp. external) side of ℓ; and analogously for the edges of E_2^a and E_2^b. Hence, the edges in $E_1^a \cap E_2^a$ (in $E_1^b \cap E_2^b$) lie in the internal (resp. external) side of ℓ, while the other edges cross ℓ.

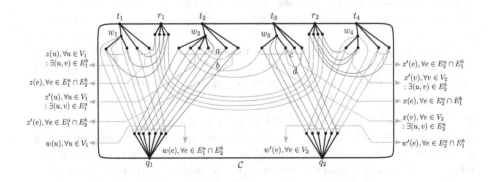

Fig. 5. The SEFE $\langle \Gamma_1, \Gamma_2 \rangle$ of $\langle G_1, G_2 \rangle$ corresponding to Γ from Fig. 4.

We sketch the construction of $\langle G_1, G_2 \rangle$. Refer to Fig. 5. The common graph $G_\cap$ contains a cycle $\mathcal{C} = (t_1, r_1, t_2, t_3, r_2, t_4, q_2, q_1)$ and trees $Q_1, Q_2, R_1, R_2,$

$T_1, \ldots, T_4$ rooted at q_1, q_2, r_1, r_2, $t_1, \ldots, t_4$, respectively. The leaves of these trees are associated to the vertices of G and to the edges crossing ℓ, as described in Fig. 5. Thus, the circular order of the leaves of these trees corresponds to the order in which the vertices of G and the spine-crossings of the edges of $E_2^a \cap E_1^b$ and $E_1^a \cap E_2^b$ appear along ℓ. In particular, $w_1, \ldots, w_4$ enforce the consecutivity of the vertices of V_1 (of V_2) along ℓ.

Some edges of G_1 and G_2 are used to enforce the coherence of such ordering in all stars and trees. Some other edges of G_1 and G_2 represent the edges of G, possibly subdivided if they cross ℓ; in particular, (portions of) edges of G that have to lie on the internal side of ℓ are edges in G_1 between the leaves of T_2 and T_3, while (portions of) edges of G that have to lie on the external side of ℓ are edges in G_2 between the leaves of R_2 and T_4. Thus, (portion of) edges on the same side of ℓ are represented by edges in the same graph, either G_1 or G_2, so that they are not allowed to cross in the SEFE as well as in the B2PBESC. This realizes the equivalence between the SEFE and the B2PBESC and completes the sketch of the proof. □

Theorem 4. 2-PARTITIONED CLIQUE PLANARITY *can be solved in quadratic time for instances* (G, S) *in which* $|S| = 2$.

Proof. Apply Lemma 7 to construct in linear time an instance $\langle G_1, G_2 \rangle$ of SEFE that is equivalent to (G, S) such that G_1 and G_2 are biconnected and their intersection graph $G_\cap$ is connected. The statement follows from the fact that instances of SEFE with this property can be solved in quadratic time [5]. □

6 Clique Planarity with Given Hierarchy

In this section we study a version of the CLIQUE PLANARITY problem in which the cliques are given together with a hierarchical relationship among them. Namely, let (G, S) be an instance of CLIQUE PLANARITY and let $\psi : S \to \{1, \ldots, k\}$, with $k \leq |S|$, be an assignment of the cliques in S to k levels such that, for each link-edge (u, v) of G connecting a vertex u of a clique s' to a vertex v of a clique s'', we have $\psi(s') \neq \psi(s'')$; an instance is *proper* if $\psi(s') = \psi(s'') \pm 1$ for each link-edge.

We aim to construct canonical clique-planar representations of (G, S) such that (Property 1) for each clique $s \in S$, the top side of the bounding box of the representation of s lies on line $y = 2\psi(s)$, while the bottom side lies above line $y = 2\psi(s) - 2$, and (Property 2) each link-edge (u, v), with $u \in s'$, $v \in s''$, $\psi(s') < \psi(s'')$, is drawn as a y-monotone curve from the top side of $R(u)$ to the bottom side of $R(v)$. We call the problem of testing whether such a representation exists LEVEL CLIQUE PLANARITY.

We show how to test level clique planarity in quadratic time for proper instances via a linear-time reduction to equivalent proper instances of T-LEVEL PLANARITY [3].

A $\mathcal{T}$-*level graph* $(V, E, \gamma, \mathcal{T})$ consists of (i) a graph $G = (V, E)$, (ii) a function $\gamma : V \to \{1, ..., k\}$ such that $\gamma(u) \neq \gamma(v)$ for each $(u, v) \in E$, where the set

$V_i = \{v \mid \gamma(v) = i\}$ is the i-th *level* of G, and (iii) a set $\mathcal{T} = \{T_1, \ldots, T_k\}$ of rooted trees such that the leaves of T_i are the vertices in V_i. A $\mathcal{T}$-*level planar drawing* of $(V, E, \gamma, \mathcal{T})$ is a planar drawing of G where the edges are y-monotone curves and the vertices in V_i are placed along line $y = i$, denoted by ℓ_i, according to an order *compatible* with T_i, that is, for each internal node μ of T_i, the leaves of the subtree of T_i rooted at μ are consecutive along ℓ_i. T-LEVEL PLANARITY asks to test whether a $\mathcal{T}$-level graph is $\mathcal{T}$-level planar.

Lemma 8. *Given a proper instance of* LEVEL CLIQUE PLANARITY, *an equivalent proper instance of* T-LEVEL PLANARITY *can be constructed in linear time.*

Proof sketch: Given an instance $(G(V, E), S, \psi)$, an instance $(V, E', \gamma, \mathcal{T})$ of T-LEVEL PLANARITY can be constructed as follows. Their vertex sets coincide and E' coincides with the set of link-edges in E. For each vertex v in a clique $s \in S$ we have $\gamma(v) = \psi(s)$. Finally, for $i = 1, \ldots, k$, where k is the number of levels in (G, S, ψ), tree $T_i \in \mathcal{T}$ has root r_i, a child w_s of r_i for each $s \in S$, and the vertices of s as children of w_s.

Suppose that $(V, E', \gamma, \mathcal{T})$ admits a $\mathcal{T}$-level planar drawing Γ. Construct a clique-planar representation satisfying Properties 1 and 2 as follows. Construct a canonical representation of each clique $s \in S$ with the top side of the bounding box on $y = 2\psi(s)$; cliques on the same level are side-by-side. The order along $y = 2i$ of the cliques $s \in S$ with $\psi(s) = i$ and the order of the rectangles in each of these cliques is dictated by the order of the vertices in V_i along ℓ_i. Finally, each edge $(u, v) \in E'$ consists of three straight-line segments: two segments connect rectangles $R(u)$ and $R(v)$ to points p_u and p_v on the bounding boxes of their cliques and a segment connects p_u and p_v.

Suppose that $(G(V, E), S, \psi)$ admits a clique-planar representations satisfying Properties 1 and 2. We construct a $\mathcal{T}$-level planar drawing of $(V, E', \gamma, \mathcal{T})$ as follows. Place each vertex $v \in V_i$ at the intersection between the left side of $R(v)$ and the line $\ell_i : y = 2i - 1$. Each edge (u, v) is a curve composed of three parts: The middle part coincides with the drawing of the link-edge (u, v) outside $R(u)$ and $R(v)$, while the other parts connect points on their boundaries to u and v. The ordering of the vertices of V_i along ℓ_i is compatible with T_i since ℓ_i intersects all the rectangles of each clique s with $\psi(s) = i$, and since rectangles in different cliques are disjoint. Thus, vertices in the same clique, and hence children of the same node of T_i, are consecutive along ℓ_i.

The construction can be performed in linear time, thus proving the lemma. $\qquad \square$

Theorem 5. LEVEL CLIQUE PLANARITY *is solvable in quadratic time for proper instances and in quartic time for general instances.*

Proof. Any instance (G, S, ψ) of LEVEL CLIQUE PLANARITY can be made proper by introducing dummy cliques composed of single vertices to split link-edges spanning more than one level. This does not alter the level clique planarity of the instance and might introduce a quadratic number of vertices. Lemma 8 constructs in linear time an equivalent proper instance of T-LEVEL PLANARITY. The statement follows since T-LEVEL PLANARITY can be solved in quadratic time [3] for proper instances. $\qquad \square$

7   Conclusions and Open Problems

We initiated the study of hybrid representations of graphs in which vertices are geometric objects and edges are either represented by intersections (if part of dense subgraphs) or by curves (otherwise). Several intriguing questions arise from our research. (1) How about considering families of dense graphs richer than cliques? Other natural families of dense graphs could be considered, say interval graphs, complete bipartite graphs, or triangle-free graphs. (2) How about using different geometric objects for representing vertices? Even simple objects like equilateral triangles or unit circles seem to pose great challenges, as they give rise to arrangements with a complex combinatorial structure. For example, we have no counterpart of Lemma 1 in those cases. (3) What is the complexity of the bipartite 2-page book embedding problem? We remark that, in the version in which spine crossings are allowed, this problem is equivalent to the clique planarity problem for instances with two cliques.

References

1. Angelin, P., Di Battista, G., Frati, F., Jelinek, V., Kratochvíl, J., Patrignani, M., Rutter, I.: Testing planarity of partially embedded graphs. ACM Trans. Algorithms **11**(4), 32:1–32:42 (2015). doi:10.1145/2629341
2. Angelini, P., Da Lozzo, G., Di Battista, G., Frati, F., Patrignani, M., Rutter, I.: Intersection-link representations of graphs. CoRR abs/1508.07557 (2015). http://arxiv.org/abs/1508.07557
3. Angelini, P., Da Lozzo, G., Di Battista, G., Frati, F., Roselli, V.: The importance of being proper (in clustered-level planarity and T-level planarity). Theor. Comp. Sci. **571**, 1–9 (2015)
4. Batagelj, V., Brandenburg, F., Didimo, W., Liotta, G., Palladino, P., Patrignani, M.: Visual analysis of large graphs using (x, y)-clustering and hybrid visualizations. IEEE Trans. Vis. Comput. Graph. **17**(11), 1587–1598 (2011)
5. Bläsius, T., Rutter, I.: Simultaneous PQ-ordering with applications to constrained embedding problems. In: Khanna, S. (ed.) SODA 2013, pp. 1030–1043. SIAM (2013)
6. Brandes, U., Raab, J., Wagner, D.: Exploratory network visualization: Simultaneous display of actor status and connections. J. Soc. Struct. 2 (2001)
7. Breu, H.: Algorithmic Aspects of Constrained Unit Disk Graphs. Ph.D. thesis, The University of British Columbia, Canada (1996)
8. Chen, Z., Grigni, M., Papadimitriou, C.H.: Map graphs. J. ACM **49**(2), 127–138 (2002)
9. Cortese, P.F., Di Battista, G., Frati, F., Patrignani, M., Pizzonia, M.: C-planarity of c-connected clustered graphs. J. Graph Algorithms Appl. **12**(2), 225–262 (2008)
10. Feng, Q.-W., Cohen, R.F., Eades, P.: Planarity for clustered graphs. In: Spirakis, P.G. (ed.) ESA 1995. LNCS, vol. 979, pp. 213–226. Springer, Heidelberg (1995)
11. Heer, J., Boyd, D.: Vizster: Visualizing online social networks. In: Stasko, J.T., Ward, M.O. (eds.) InfoVis 2005, 23–25 October 2005, Minneapolis, USA, p. 5. IEEE Computer Society (2005)
12. Henry, N., Fekete, J., McGuffin, M.J.: Nodetrix: a hybrid visualization of social networks. IEEE Trans. Vis. Comput. Graph. **13**(6), 1302–1309 (2007)

13. Hong, S.-H., Nagamochi, H.: Simpler algorithms for testing two-page book embedding of partitioned graphs. In: Cai, Z., Zelikovsky, A., Bourgeois, A. (eds.) COCOON 2014. LNCS, vol. 8591, pp. 477–488. Springer, Heidelberg (2014)
14. Irzhavsky, P.: Information System on Graph Classes and their Inclusions (ISGCI). http://graphclasses.org/classes/refs1600.html#ref_1660
15. Thorup, M.: Map graphs in polynomial time. In: FOCS 1998, pp. 396–405. IEEE (1998)
16. Wigderson, A.: The complexity of the Hamiltonian circuit problem for maximal planar graphs. EECS Department Report 298, Princeton University (1982)

Combinatorial Properties of Triangle-Free Rectangle Arrangements and the Squarability Problem

Jonathan Klawitter[1,2], Martin Nöllenburg[3]([✉]), and Torsten Ueckerdt[2]

[1] Institut für Theoretische Informatik, Karlsruhe Institute of Technology,
Karlsruhe, Germany
[2] Institut für Algebra und Geometrie, Karlsruhe Institute of Technology,
Karlsruhe, Germany
torsten.ueckerdt@kit.edu
[3] Algorithms and Complexity Group, TU Wien, Vienna, Austria
noellenburg@ac.tuwien.ac.at

Abstract. We consider arrangements of axis-aligned rectangles in the plane. A geometric arrangement specifies the coordinates of all rectangles, while a combinatorial arrangement specifies only the respective intersection type in which each pair of rectangles intersects. First, we investigate combinatorial contact arrangements, i.e., arrangements of interior-disjoint rectangles, with a triangle-free intersection graph. We show that such rectangle arrangements are in bijection with the 4-orientations of an underlying planar multigraph and prove that there is a corresponding geometric rectangle contact arrangement. Using this, we give a new proof that every triangle-free planar graph is the contact graph of such an arrangement. Secondly, we introduce the question whether a given rectangle arrangement has a combinatorially equivalent square arrangement. In addition to some necessary conditions and counterexamples, we show that rectangle arrangements pierced by a horizontal line are squarable under certain sufficient conditions.

1 Introduction

We consider arrangements of axis-aligned rectangles and squares in the plane. Besides *geometric rectangle arrangements*, in which all rectangles are given with coordinates, we are also interested in *combinatorial rectangle arrangements*, i.e., equivalence classes of combinatorially equivalent arrangements. Our contribution is two-fold.

First we consider maximal (with a maximal number of contacts) combinatorial rectangle contact arrangements, in which no three rectangles share a point. For rectangle arrangements this is equivalent to the contact graph being *triangle-free*, unlike, e.g., for triangle contact arrangements. We prove a series of analogues to the well-known maximal combinatorial triangle contact arrangements and to Schnyder realizers. The contact graph G of a maximal triangle contact arrangement is a maximal planar graph. A *3-orientation* is an orientation of the edges

© Springer International Publishing Switzerland 2015
E. Di Giacomo and A. Lubiw (Eds.): GD 2015, LNCS 9411, pp. 231–244, 2015.
DOI: 10.1007/978-3-319-27261-0_20

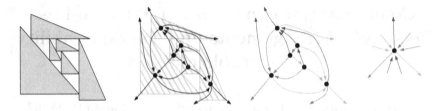

Fig. 1. Left to right: maximal combinatorial contact arrangement with axis-aligned triangles, no three sharing a point. 3-orientation of G'. Schnyder realizer of G'. Local coloring rules for Schnyder realizer (Color figure online).

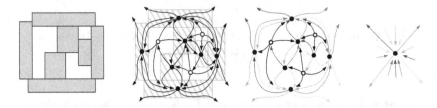

Fig. 2. Left to right: maximal combinatorial contact arrangement with axis-aligned rectangles, no three sharing a point. 4-orientation of underlying graph. Corner-edge-labeling of underlying graph. Local coloring rules for corner-edge-labeling (Color figure online).

of a graph G', obtained from G by adding six edges (two at each outer vertex), in which every vertex has exactly three outgoing edges. Each outer vertex has two outgoing edges that end in the outer face without having an endpoint there. A *Schnyder realizer* [10,11] is a 3-orientation of G' together with a coloring of its edges with colors $0, 1, 2$ such that every vertex has exactly one outgoing edge in each color and incoming edges are colored in the color of the "opposite" outgoing edge. The three outgoing edges represent the three corners of a triangle and the color specifies the corner, see Fig. 1. De Fraysseix *et al.* [3] proved that the maximal combinatorial triangle contact arrangements of G are in bijection with the 3-orientations of G' and the Schnyder realizers of G'. Schnyder proved that for every maximal planar graph G, G' admits a Schnyder realizer and hence G is a triangle contact graph.

In this paper we prove an analogous result, which, roughly speaking, is the following. We consider maximal triangle-free combinatorial rectangle contact arrangements. The corresponding contact graph G is planar with all faces of length 4 or 5. We define an underlying plane multigraph $\bar{G}$, whose vertex set also includes a vertex for each inner face of the contact graph, and define 4-*orientations* of $\bar{G}$. Here, every vertex has exactly four outgoing edges, where each outer vertex has two edges ending in the outer face. For a 4-orientation we introduce *corner-edge-labelings* of $\bar{G}$, which are, similar to Schnyder realizers, colorings of the outgoing edges at vertices of $\bar{G}$ corresponding to rectangles with colors $0, 1, 2, 3$ satisfying certain local rules. Each outgoing edge represents a corner of a rectangle and the color specifies which corner it is, see Fig. 2. We then prove that the combinatorial contact arrangements of G are in bijection with the 4-orientations of $\bar{G}$ and the corner-edge-labelings of $\bar{G}$.

Thomassen [12] proved that rectangle contact graphs are precisely the graphs admitting a planar embedding in which no triangle contains a vertex in its interior. We also prove here that for every maximal triangle-free planar graph G, $\bar{G}$ admits a 4-orientation, obtaining a new proof that G is a rectangle contact graph.

Our second result is concerned with the question whether a given geometric rectangle arrangement can be transformed into a combinatorially equivalent square arrangement. The similar question whether a pseudocircle arrangement can be transformed into a combinatorially equivalent circle arrangement has recently been studied by Kang and Müller [6], who showed that the problem is NP-hard. We say that a rectangle arrangement can be *squared* (or is *squarable*) if an equivalent square arrangement exists. Obviously, squares are a very restricted class of rectangles and not every rectangle arrangement can be squared. The natural open question is to characterize the squarable rectangle arrangements and to answer the complexity status of the corresponding decision problem. As a first step towards solving these questions, we show, on the one hand, some general necessary conditions and, on the other hand, sufficient conditions implying that certain subclasses of rectangle arrangements are always squarable.

Related Work. Intersection graphs and contact graphs of axis-aligned rectangles or squares in the plane are a popular, almost classic, topic in discrete mathematics and theoretical computer science with lots of applications in computational geometry, graph drawing and VLSI chip design. Most of the research for rectangle intersection graphs concerns their recognition [14], colorability [1] or the design of efficient algorithms such as for finding maximum cliques [5]. On the other hand, rectangle contact graphs are mainly investigated for their combinatorial and structural properties. Almost all the research here concerns edge-maximal 3-connected rectangle contact graphs, so called *rectangular duals*. These can be characterized by the absence of separating triangles [9,13] and the corresponding representations by touching rectangles can be seen as dissections of a rectangle into rectangles. Combinatorially equivalent dissections are in bijection with regular edge labelings [7] and transversal structures [4]. The question whether a rectangular dual has a rectangle dissection in which all rectangles are squares has been investigated by Felsner [2].

2   Preliminaries

In this paper a *rectangle* is an axis-aligned rectangle in the plane, i.e., the cross product $[x_1, x_2] \times [y_1, y_2]$ of two bounded closed intervals. A *geometric rectangle arrangement* is a finite set $\mathcal{R}$ of rectangles; it is a *contact arrangement* if any two rectangles have disjoint interiors. In a contact arrangement, any two non-disjoint rectangles R_1, R_2 have one of the two contact types *side contact* and *corner contact*, see Fig. 3 (left); we exclude the degenerate case of two rectangles

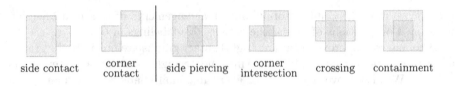

Fig. 3. Contact types (left) and intersection types (right) of rectangles.

sharing only one point. If $\mathcal{R}$ is not a contact arrangement, four intersection types are possible: *side piercing, corner intersection, crossing,* and *containment,* see Fig. 3 (right). Note that side contact and corner contact are degenerate cases of side piercing and corner intersection, whereas crossing and containment have no analogues in contact arrangements. If no two rectangles form a crossing, we say that $\mathcal{R}$ is *cross-free.* Moreover, in each type (except containment) it is further distinguished which sides of the rectangles touch or intersect.

Two rectangle arrangements $\mathcal{R}_1$ and $\mathcal{R}_2$ are *combinatorially equivalent* if $\mathcal{R}_1$ can be continuously deformed into $\mathcal{R}_2$ such that every intermediate state is a rectangle arrangement with the same intersection or contact type for every pair of rectangles. An equivalence class of combinatorially equivalent arrangements is called a *combinatorial rectangle arrangement.* So while a geometric arrangement specifies the coordinates of all rectangles, think of a combinatorial arrangement as specifying only the way in which any two rectangles touch or intersect. In particular, a combinatorial rectangle arrangement is defined by **(1)** for each rectangle R and each side of R the counterclockwise order of all intersecting (touching) rectangle edges, labeled by their rectangle R' and the respective side of R' (top, bottom, left, right), **(2)** for containments the respective component of the arrangement, in which a rectangle is contained.

In the *intersection graph* of a rectangle arrangement there is one vertex for each rectangle and two vertices are adjacent if and only if the corresponding rectangles intersect. As combinatorially equivalent arrangements have the same intersection graph, combinatorial arrangements themselves have a well-defined intersection graph. For rectangle contact arrangements (combinatorial or geometric) the intersection graph is also called the *contact graph.* Note that such contact graphs are planar, as we excluded the case of four rectangles meeting in a corner.

3   Statement of Results

3.1   Maximal Triangle-Free Planar Graphs and Rectangle Contact Arrangements

We consider so-called *MTP-graphs,* that is, (M)aximal (T)riangle-free (P)lane graphs with a quadrangular outer face. Note that each face in such an MTP-graph is a 4-cycle or 5-cycle, and that every plane triangle-free graph is an induced subgraph of some MTP-graph. Given an MTP-graph G a rectangle contact arrangement of G is one whose contact graph is G, where the embedding

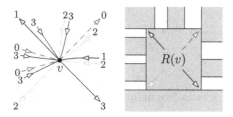

Fig. 4. Local color patterns in corner-edge-labelings of an MTP-graph at a vertex v, together with the corresponding part in a rectangle contact arrangement (Color figure online).

inherited from the arrangement is the given embedding of G, and where each outer rectangle has two corners in the unbounded region[1]. We define the closure, 4-orientations and corner-edge-labelings:

The *closure* $\bar{G}$ of G is derived from G by replacing each edge of G with a pair of parallel edges, called an *edge pair*, and adding into each inner face f of G a new vertex, also denoted by f, connected by an edge, called a *loose edge*, to each vertex incident to that face. At each outer vertex we add two loose edges pointing into the outer face, although we do not add a vertex for the outer face. Note that $\bar{G}$ inherits a unique plane embedding with each inner face being a triangle or a 2-gon.

A 4-orientation of $\bar{G}$ is an orientation of the edges and half-edges of $\bar{G}$ such that every vertex has outdegree exactly 4. An edge pair is called *uni-directed* if it is oriented consistently and *bi-directed* otherwise.

A corner-edge-labeling of $\bar{G}$ is a 4-orientation of $\bar{G}$ together with a coloring of the outgoing edges of $\bar{G}$ at each vertex of G with colors $0, 1, 2, 3$ (see Fig. 4) such that

(i) around each vertex v of G we have outgoing edges in color $0, 1, 2, 3$ in this counterclockwise order and

(ii) in the wedge, called *incoming wedge*, at v counterclockwise between the outgoing edges of color i and $i+1$ there are some (possibly none) incoming edges colored $i + 2$ or $i + 3$, $i = 0, 1, 2, 3$, all indices modulo 4.

In a corner-edge-labeling the four outgoing edges at a vertex of $\bar{G}$ corresponding to a face of G are not colored. Further we remark that (i) implies that uni-directed pairs are colored i and $i - 1$, while (ii) implies that bi-directed pairs are colored i and $i + 2$, for some $i \in \{0, 1, 2, 3\}$, where all indices are considered modulo 4. The following theorem is proved in Sect. 4.

Theorem 1. *Let G be an MTP-graph, then each of the following are in bijection:*

- *the combinatorial rectangle contact arrangements of G*
- *the corner-edge-labelings of $\bar{G}$*
- *the 4-orientations of $\bar{G}$.*

---

[1] Other configurations of the outer four rectangles can be easily derived from this.

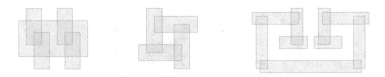

Fig. 5. Three cross-free unsquarable rectangle arrangements.

Using the bijection between 4-orientations of $\bar{G}$ and combinatorial rectangle contact arrangements of G given in Theorem 1, we can give a new proof that every MTP-graph G is a rectangle contact graph, which is the statement of the next theorem; its proof is given in the full paper [8] and sketched in Sect. 5.

Theorem 2. *For every MTP-graph G, $\bar{G}$ has a 4-orientation and it can be computed in linear time. In particular, G has a rectangle contact arrangement.*

We remark that our technique in the proof of Theorem 1 constructs from a given 4-orientation of $\bar{G}$ in linear time a geometric rectangle contact arrangement of G in the $2n \times 2n$ square grid, where n is the number of vertices in G. Thus also the rectangle contact arrangement in Theorem 2 can be computed in linear time and uses only a linear-size grid.

3.2 Squarability and Line-Pierced Rectangle Arrangements

In the squarability problem, we are given a rectangle arrangement $\mathcal{R}$ and want to decide whether $\mathcal{R}$ can be squared. The first observation is that there are obvious obstructions to the squarability of a rectangle arrangement. If any two rectangles in $\mathcal{R}$ are crossing (see Fig. 3) then there are obviously no two combinatorially equivalent squares.

But even if we restrict ourselves to cross-free rectangle arrangements, we can find unsquarable configurations. One such arrangement is depicted in Fig. 5 (left). To get an unsquarable arrangement with a triangle-free intersection graph, we can use the fact that two side-piercing rectangles translate immediately into a smaller-than relation for the corresponding squares: the side length of the square to pierce into the side of another square needs to be strictly smaller. Hence any rectangle arrangement that contains a cycle of side-piercing rectangles cannot be squarable, see Fig. 5 (middle). Moreover, we may even create a counterexample of a rectangle arrangement whose intersection graph is a path and that causes a geometrically infeasible configuration for squares, see Fig. 5 (right).

Proposition 1. *Some cross-free rectangle arrangements are unsquarable, even if the intersection graph is a path.*

Therefore we focus on a non-trivial subclass of rectangle arrangements that we call line-pierced. A rectangle arrangement $\mathcal{R}$ is *line-pierced* if there exists a horizontal line ℓ such that $\ell \cap R \neq \emptyset$ for all $R \in \mathcal{R}$. The line-piercing strongly restricts the possible vertical positions of the rectangles in $\mathcal{R}$, which lets us prove two sufficient conditions for squarability in the following theorem.

Theorem 3. *Let $\mathcal{R}$ be a cross-free, line-pierced rectangle arrangement.*

– *If $\mathcal{R}$ is triangle-free, then $\mathcal{R}$ is squarable.*
– *If $\mathcal{R}$ has only corner intersections, then $\mathcal{R}$ is squarable, even using line-pierced unit squares.*

On the other hand, cross-free, line-pierced rectangle arrangements in general may have forbidden cycles or other geometric obstructions to squarability. We give two examples in Sect. 6, together with a sketch of the proof of Theorem 3.

4 Bijections Between 4-Orientations, Corner-Edge-Labelings and Rectangle Contact Arrangements – Proof of Theorem 1

Throughout this section let $G = (V, E)$ be a fixed MTP-graph and $\bar{G}$ be its closure. By definition, every corner-edge-labeling of $\bar{G}$ induces a 4-orientation of $\bar{G}$. We prove Theorem 1, i.e., that combinatorial rectangle contact arrangements of G, 4-orientations of $\bar{G}$ and corner-edge-labelings of $\bar{G}$ are in bijection, in three steps:

– Every rectangle contact arrangement of G induces a 4-orientation of $\bar{G}$. (Lemma 1)
– Every 4-orientation of $\bar{G}$ induces a corner-edge-labeling of $\bar{G}$. (Lemma 3)
– Every corner-edge-labeling of $\bar{G}$ induces a rectangle contact arrangement of G. (Lemma 4)

Omitted proofs are provided in the full version of this paper [8].

4.1 From Rectangle Arrangements to 4-Orientations

Lemma 1. *Every rectangle contact arrangement of G induces a 4-orientation of $\bar{G}$.*

The proof idea is already given in Fig. 2: For every rectangle draw an outgoing edge through each of the four corners and for every inner face draw an outgoing edge through each of the four extremal sides.

We continue with a crucial property of 4-orientations. For a simple cycle C of G, consider the corresponding cycle $\bar{C}$ of edge pairs in $\bar{G}$. The *interior* of $\bar{C}$ is the bounded component of $\mathbb{R}^2$ incident to all vertices in C after the removal of all vertices and edges of $\bar{C}$. In a fixed 4-orientation of $\bar{G}$ a directed edge $e = (u, v)$ *points inside* C if $u \in V(C)$ and e lies in the interior of $\bar{C}$, i.e., either v lies in the interior of C, or e is a chord of $\bar{C}$ in the interior of $\bar{C}$.

Lemma 2. *For every 4-orientation of $\bar{G}$ and every simple cycle C of G the number of edges pointing inside C is exactly $|V(C)| - 4$.*

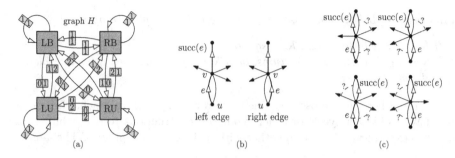

Fig. 6. (a) The graph H. **L**, **R**, **U**, **B** stands for left edge, right edge, uni-directed and bi-directed edge pair, respectively. The number of outgoing edges in the left and right wedge are shown on the left and right of the corresponding arrow. (b) Illustration of the definition of $\mathrm{succ}(e)$. (c) Summarizing the 16 possible cases for e and $\mathrm{succ}(e)$. Edges connected by a dashed arc may or may not coincide.

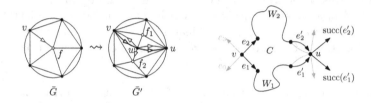

Fig. 7. Left: Stacking a new vertex w into a 5-face f of G. The orientation of edges on the boundary of f, as well as outgoing edges at f, f_1, f_2 is omitted. The directed edge (v, w) and its successor (w, u) are highlighted. Right: Illustration of the proof of the Claim in the proof of Lemma 3.

4.2   From 4-Orientations to Corner-Edge-Labelings

Next we shall show how a 4-orientation of $\bar{G}$ can be augmented (by choosing colors for the edges) into a corner-edge-labeling. Fix a 4-orientation. If e is a directed edge in an edge pair, then e is called a *left edge*, respectively *right edge*, when the 2-gon enclosed by the edge pair lies on the right, respectively on the left, when going along e in its direction. Thus, a uni-directed edge pair consists of one left edge and one right edge, while a bi-directed edge pair either consists of two left edges (clockwise oriented 2-gon) or two right edges (counterclockwise oriented 2-gon).

If $e = (u, v)$ is an edge in an edge pair, let e_2 and e_3 be the second and third outgoing edge at v when going counterclockwise around v starting with e. We define the *successor* of e as $\mathrm{succ}(e) = e_2$ if e is a right edge, and $\mathrm{succ}(e) = e_3$ if e is a left edge, see Fig. 6 (b,c). Note that in a corner-edge-labeling $\mathrm{succ}(e)$ is exactly the outgoing edge at v that has the same color as e, see Fig. 4.

Note that $e' = \mathrm{succ}(e)$ may be a loose edge in $\bar{G}$ at the concave vertex for some 5-face in G. For the sake of shorter proofs below, we shall avoid the treatment of this case. To do so, we augment G to a supergraph G' such that starting with any edge in any edge pair and repeatedly taking the successor, we never run into a loose edge pointing to an inner face.

The graph G' is formally obtained from G by stacking a new vertex w into each 5-face f, with an edge to the incoming neighbor v of f in $\bar{G}$ and the vertex u at f that comes second after v in the clockwise order around f in $\bar{G}$. (Indeed, the second vertex in counterclockwise order would be equally good for our purposes.) Let f_1 and f_2 be the resulting 4-face and 5-face incident to w, respectively. We obtain a 4-orientation of the closure $\bar{G}'$ of G' by orienting all edges at f_1 as outgoing, both edges between v and w as right edges (counterclockwise), the remaining three edges at w as outgoing, and the remaining four edges at f_2 as outgoing. See Fig. 7 (left) for an illustration.

Before we augment the 4-orientation of $\bar{G}'$ into a corner-edge-labeling, we need one last observation. Let e and $\mathrm{succ}(e)$ be two edges in edge pairs of $\bar{G}'$ with common vertex v. Consider the wedges at v between e and $\mathrm{succ}(e)$ when going clockwise (left wedge) and counterclockwise (right wedge) around v. Each of e, $\mathrm{succ}(e)$ can be a left edge or right edge, and in a uni-directed pair or a bi-directed pair. This gives us four types of edges and 16 possibilities for the types of e and $\mathrm{succ}(e)$. The graph H in Fig. 6(a) shows for each of these 16 possibilities the number of outgoing edges at v in the left and right wedge at v.

Observation 4. *For every directed closed walk on k edges in the graph H in Fig. 6(a) we have*

$$\#edges \ in \ left \ wedges = \#edges \ in \ right \ wedges = k.$$

Proof. It suffices to check each directed cycle on k edges, $k = 1, 2, 3, 4$.   □

Lemma 3. *Every 4-orientation of $\bar{G}$ induces a corner-edge-labeling of $\bar{G}$.*

A detailed proof of Lemma 3 is given in the full version of this paper [8].

Proof (Sketch). Consider the augmented graph G', its closure $\bar{G}'$ and 4-orientation as defined above. For any edge e in an edge pair in $\bar{G}'$ (and hence every edge of $\bar{G}$ outgoing at some vertex of G) consider the directed walk W_e in $\bar{G}'$ starting with e by repeatedly taking the successor as long as it exists (namely the current edge is in an edge pair).

First we show that W_e is a simple path ending at one of the eight loose edges in the outer face. Indeed, otherwise W_e would contain a simple cycle C where every edge on C, except the first, is the successor of its preceding edge on C. From the graph H of Fig. 6(a) we see that every wedge of C contains at most two outgoing edges. With Observation 4 the number of edges pointing inside C is at least $|V(C)| - 2$ and at most $|V(C)| + 2$, which is a contradiction to Lemma 2.

Now let v_0, v_1, v_2, v_3 be the outer vertices in this counterclockwise order. Define the color of e to be i if W_e ends with the right loose edge at v_i or the left loose edge at v_{i-1}, indices modulo 4. By definition every edge has the same color as its successor in $\bar{G}'$ (if it exists). Thus this coloring is a corner-edge-labeling of $\bar{G}'$ if at every vertex v of G the four outgoing edges are colored 0, 1, 2, and 3, in this counterclockwise order around v.

Claim. Let e_1, e_2 be two outgoing edges at v for which $W_{e_1} \cap W_{e_2}$ consists of more than just v. Then e_1 and e_2 appear consecutively among the outgoing edges around v, say e_1 clockwise after e_2.

Moreover, if $u \neq v$ is a vertex in $W_{e_1} \cap W_{e_2}$ for which the subpaths W_1 of W_{e_1} and W_2 of W_{e_2} between v and u do not share inner vertices, then the last edge e_1' of W_1 is a right edge and the last edge e_2' of W_2 is a left edge, e_1' and e_2' are part of (possibly the same) uni-directed pairs and these pairs sit in the same incoming wedge at u.

To prove this claim, we consider the cycle $C = W_1 \cup W_2$, count the edges pointing inside with the graph H and conclude that neither u nor v may have edges pointing inside C. See Fig. 7 (right) for an illustration.

The claim implies that the two walks W_{e_1} and W_{e_2} can neither cross, nor have an edge in common. Considering the four walks starting in a given vertex, we can argue (with the second part of the claim) that our coloring is a corner-edge-labeling of $\bar{G}'$. Finally, we inherit a corner-edge-labeling of $\bar{G}$ by reverting the stacking of artificial vertices in 5-faces. $\qquad\Box$

4.3 From Corner-Edge-Labelings to Rectangle Contact Arrangements

It remains to compute a rectangle arrangement of G based on a given corner-edge-labeling of $\bar{G}$. That is, we shall prove the following lemma.

Lemma 4. *Every corner-edge-labeling of $\bar{G}$ induces a rectangle contact arrangement of G.*

A detailed proof of Lemma 4 is given in the full version of this paper [8].

Proof (Sketch). Fix a corner-edge-labeling of $\bar{G}$. For every vertex v of G we introduce two pairs of variables $x_1(v), x_2(v)$ and $y_1(v), y_2(v)$ and set up a system of inequalities and equalities such that any solution defines a rectangle contact arrangement $\{R(v) \mid v \in V\}$ of G with $R(v) = [x_1(v), x_2(v)] \times [y_1(v), y_2(v)]$, which is compatible with the given corner-edge-labeling.

For every edge vw of G the way in which $R(v)$ and $R(w)$ are supposed to touch is encoded in the given corner-edge-labeling and this can be described by the inequalities and equalities in Table 1. Here we list the constraint and the conditions (color and orientation) of a single directed edge between v and w or a uni-directed edge pair outgoing at v and incoming at w in $\bar{G}$ under which we have this constraint.

Instead of showing that the system in Table 1 has a solution, we define another set of constraints implying all constraints in Table 1, for which it is easier to prove feasibility.

It suffices to define a system $\mathcal{I}_x$ for x-coordinates and treat the y-coordinates analogously. In $\mathcal{I}_x$ we have $x_1(v) < x_2(v)$ for every vertex v together with all equalities in the left of Table 1, but only those inequalities in the left of Table 1 that arise from edges in bi-directed edge pairs. The inequalities arising from uni-directed edge pairs are implied by the following set of inequalities. For a vertex

Table 1. Constraints encoding the type of contact between $R(v)$ and $R(w)$, defined based on the orientation and color(s) of the edge pair between v and w in $\bar{G}$.

constraint	edge	color	out		constraint	edge	color	out
$x_1(w) < x_1(v) < x_2(w)$	right	2	v		$y_1(w) < y_1(v) < y_2(w)$	right	3	v
	left	1	v			left	2	v
$x_1(w) < x_2(v) < x_2(w)$	right	0	v		$y_1(w) < y_2(v) < y_2(w)$	right	1	v
	left	3	v			left	0	v
$x_1(w) = x_2(v)$	right	1	w		$y_1(w) = y_2(v)$	right	2	w
	left	2	w			left	3	w
	uni	0, 3	v			uni	1, 0	v

v in G let $S_1(v) = a_1, \ldots, a_k$ and $S_2(v) = b_1, \ldots, b_\ell$ be the counterclockwise sequences of neighbors of v in the incoming wedges at v bounded by its outgoing edges of color 0 and 1, and color 2 and 3, respectively. See the left of Fig. 8. Then we have in $\mathcal{I}_x$ the inequalities

$$x_1(a_i) > x_2(a_{i+1}) \text{ for } i = 1, \ldots, k-1 \text{ and } x_2(b_i) < x_1(b_{i+1}) \text{ for } i = 1, \ldots, \ell-1. \quad (1)$$

If $k = 1$ we have no constraint for $S_1(v)$ and if $\ell = 1$ we have no constraint for $S_2(v)$.

We associate the system $\mathcal{I}_x$ with a partially oriented graph I_x whose vertex set is $\{x_1(v), x_2(v) \mid v \in V\}$. For each inequality $a > b$ we have an oriented edge (a, b) in I_x, while for each equality $a = b$ we have an undirected edge ab in I_x, see Fig. 8.

We observe that I_x is planar and prove that I_x has no cycle C in which all directed edges are oriented consistently, which clearly implies that $\mathcal{I}_x$ has a solution. This is done by showing that no inner face is such a cycle, and that for every inner vertex u, vertex $x_1(u)$ has an incident undirected edge or incident outgoing edge and vertex $x_2(u)$ has an incident undirected edge or incident incoming edge.   □

5 MTP Graphs Are Rectangle Contact Graphs – Proofsketch of Theorem 2

Theorem 2 is formally proven in the full version of this paper [8]. The idea is to prove by induction on the number of vertices that for an MTP-graph G we find a 4-orientation of $\bar{G}$. In the inductive step we either have (Case 1) that G has an inner 4-face, or (Case 2) that one can contract an inner edge e, keeping it an MTP-graph. Figures 9 and 10 illustrate how to find a 4-orientation in Cases 1 and 2, respectively.

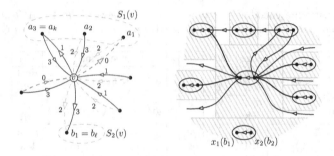

Fig. 8. Illustrating the definition of I_x around a vertex v. On the right a hypothetical rectangle contact arrangement is indicated (Color figure online).

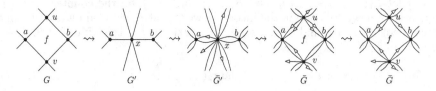

Fig. 9. Collapsing an inner 4-face and inheriting a 4-orientation when uncollapsing.

6   Line-Pierced Rectangle Arrangements and Squarability – Proofsketch of Theorem 3

Recall that a rectangle arrangement $\mathcal{R}$ is line-pierced if there is a horizontal line ℓ that intersects every rectangle in $\mathcal{R}$. Note that by the line-piercing property of $\mathcal{R}$ the intersection graph remains the same if we project each rectangle $R = [a, b] \times [c, d] \in \mathcal{R}$ onto the interval $[a, b] \subseteq \mathbb{R}$. In particular, the intersection graph $G_{\mathcal{R}}$ of a line-pierced rectangle arrangement $\mathcal{R}$ is an *interval graph*, i.e., intersection graph of intervals on the real line.

Line-pierced rectangle arrangements, however, carry more information than one-dimensional interval graphs since the vertical positions of intersection points between rectangles do influence the combinatorial properties of the arrangement. We obtain two squarability results for line-pierced arrangements in Propositions 2 and 3, which yield Theorem 3.

Proposition 2. *Every line-pierced, triangle-free, and cross-free rectangle arrangement $\mathcal{R}$ is squarable.*

There are instances, however, that satisfy the conditions of Proposition 2 and thus have a squaring, but not a line-pierced one. An example is given in Fig. 12.

Proposition 3. *Every line-pierced rectangle arrangement $\mathcal{R}$ restricted to corner intersections is squarable. There even exists a corresponding squaring with unit squares that remains line-pierced.*

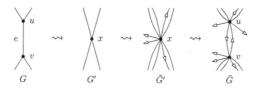

Fig. 10. Contracting an edge and keeping a 4-orientation when uncontracting.

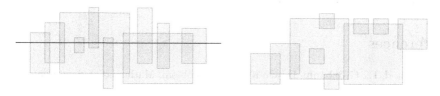

Fig. 11. Constructing a combinatorially equivalent squaring from a line-pierce, triangle-free, and cross-free rectangle arrangement.

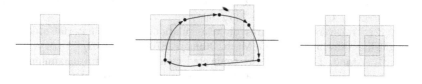

Fig. 12. Left: A line-pierced, triangle-free rectangle arrangement that has no line-pierced squaring. Middle: An unsquarable line-pierced rectangle arrangement due to a forbidden cycle of side-piercing intersections. Right: Squaring the two vertical pairs of rectangles on the right implies that the central square would need to be wider than tall.

Propositions 2 and 3 are proved in the full version of this paper [8]. The crucial observation is that the intersection graph of $\mathcal{R}$ is a caterpillar in the former case (Fig. 11) and a unit-interval graph in the latter case. The results can then be proven by induction on the number of vertices by iteratively removing the "rightmost" rectangle in the representation.

If we drop the restrictions to corner intersections and triangle-free arrangements, we can immediately find unsquarable instances, either by creating cyclic "'smaller than"' relations or by introducing intersection patterns that become geometrically infeasible for squares. Two examples are given in Fig. 12.

7   Conclusions

We have introduced corner-edge-labelings, a new combinatorial structure similar to Schnyder realizers, which captures the combinatorially equivalent maximal rectangle arrangements with no three rectangles sharing a point. Using this, we gave a new proof that every triangle-free planar graph is a rectangle contact graph. We also introduced the squarability problem, which asks for a given rectangle arrangement whether there is a combinatorially equivalent arrangement using only squares. We provide some forbidden configuration for the squarability of an arrangement and show that certain subclasses of line-pierced arrangements are always squarable. It remains open whether the decision problem for general arrangements is NP-complete.

Surprisingly, every unsquarable arrangement that we know has a crossing or a side-piercing. Hence we would like to ask whether every rectangle arrangement with only corner intersections is squarable. Another natural question is whether every triangle-free planar graph is a square contact graph.

References

1. Asplund, E., Grünbaum, B.: On a coloring problem. Mathematica Scandinavica **8**, 181–188 (1960)
2. Felsner, S.: Rectangle and square representations of planar graphs. In: Pach, J. (ed.) Thirty Essays in Geometric Graph Theory, pp. 213–248. Springer, New York (2012)
3. de Fraysseix, H., de Mendez, P.O., Rosenstiehl, P.: On triangle contact graphs. Comb. Probab. Comput. **3**, 233–246 (1994)
4. Fusy, E.: Transversal structures on triangulations: a combinatorial study and straight-line drawings. Discrete Math. **309**(7), 1870–1894 (2009)
5. Imai, H., Asano, T.: Finding the connected components and a maximum clique of an intersection graph of rectangles in the plane. J. Algorithms **4**(4), 310–323 (1983)
6. Kang, R.J., Müller, T.: Arrangements of pseudocircles and circles. Discrete Comput. Geom. **51**, 896–925 (2014)
7. Kant, G., He, X.: Regular edge labeling of 4-connected plane graphs and its applications in graph drawing problems. Theor. Comput. Sci. **172**(1–2), 175–193 (1997)
8. Klawitter, J., Nöllenburg, M., Ueckerdt, T.: Combinatorial properties of triangle-free rectangle arrangements and the squarability problem. CoRR, arXiv:1509.00835, September 2015
9. Koźmiński, K., Kinnen, E.: Rectangular duals of planar graphs. Networks **15**, 145–157 (1985)
10. Schnyder, W.: Planar graphs and poset dimension. Order **5**(4), 323–343 (1989)
11. Schnyder, W.: Embedding planar graphs on the grid. In: 1st ACM-SIAM Symposium on Discrete Algorithms, SODA 1990, pp. 138–148 (1990)
12. Thomassen, C.: Interval representations of planar graphs. J. Comb. Theor. Ser. B **40**(1), 9–20 (1986)
13. Ungar, P.: On diagrams representing graphs. J. London Math. Soc. **28**, 336–342 (1953)
14. Yannakakis, M.: The complexity of the partial order dimension problem. SIAM J. Algebraic Discrete Methods **3**(3), 351–358 (1982)

Applications

Displaying User Behavior in the Collaborative Graph Visualization System OnGraX

Björn Zimmer[(⊠)] and Andreas Kerren

Department of Computer Science, ISOVIS Group, Linnaeus University,
Vejdes Plats 7, 35195 Växjö, Sweden
{bjorn.zimmer,andreas.kerren}@lnu.se

Abstract. The visual analysis of complex networks is a challenging task in many fields, such as systems biology or social sciences. Often, various domain experts work together to improve the analysis time or the quality of the analysis results. Collaborative visualization tools can facilitate the analysis process in such situations. We propose a new web-based visualization environment which supports distributed, synchronous and asynchronous collaboration. In addition to standard collaboration features like event tracking or synchronizing, our client/server-based system provides a rich set of visualization and interaction techniques for better navigation and overview of the input network. Changes made by specific analysts or even just visited network elements are highlighted on demand by heat maps. They enable us to visualize user behavior data without affecting the original graph visualization. We evaluate the usability of the heat map approach against two alternatives in a user experiment.

Keywords: Information visualization · Graph drawing · Network exploration · Interaction · HCI · CSCW · Biological networks · Heat maps

1 Introduction

With the growing size and availability of large and complex data, the cooperative analysis of such data sets is becoming an important new method for many data analysts as cooperation might improve the quality of the analysis process [15] and help to analyze data sets efficiently. One crucial observation is that collaborators—who are often spread across the globe—would like to seamlessly drop in and out of ongoing work [13]. On the one hand, the collaborative analysis process can take place in a joint online session where everybody is working simultaneously on one data set, discussing and changing it together in real-time to create better analysis results. Here, different experts might want to see what the others are doing, and if there are possibilities to coordinate their efforts and find a common ground [3,9]. On the other hand, the experts work on the data set whenever they find the time (i.e., asynchronously) to avoid having to schedule and organize a virtual or physical meeting with a larger group of colleagues. Both situations cause specific problems that should be handled by tools which

© Springer International Publishing Switzerland 2015
E. Di Giacomo and A. Lubiw (Eds.): GD 2015, LNCS 9411, pp. 247–259, 2015.
DOI: 10.1007/978-3-319-27261-0_21

support collaborative work. For instance, while working independently, it would be helpful to see changes of the data performed by other analysts. Another interesting issue is to see which part of the data set has already been explored by others. Here, it is also interesting to know who changed the data: was an established expert working on a specific part of the data, or a new staff member who might not have the same experience as the expert?

To tackle the aforementioned problems in the context of collaborative network analyses, we have developed the visualization tool OnGraX [23–25]. Our system was designed for the *distributed asynchronous* and *synchronous* collaborative exploration of graphs in a modern web browser. Note that we give a detailed explanation about the engineering aspects of OnGraX in paper [25]. In contrast, we here propose interactive visualization techniques that

- help to coordinate work in a collaborative setting for *node-link diagrams* which may change their topology during the analysis process (referred to as dynamic graphs in the following) and
- assist analysts to identify previous activities performed by former users on these networks.

We exemplify our visualization approaches with the help of the collaborative analysis of metabolic networks from the Kyoto Encyclopedia of Genes and Genomes (KEGG) pathway database [1] due to our long lasting research collaborations with biologists/bioinformaticians at several research institutions. Building such biological networks is often based on complex experiments. In consequence, biologists of different domains and experience levels want to explore the resulting networks and check them for wrong entries or missing data and revise the networks wherever it is necessary. Usually, they only check parts of a network that are specific to their own field of expertise or interest. In this case it is important to know, what part of the network has already been checked and what part still needs attention. This can also be used as a kind of quality check: an area which has been investigated by many different experts is likely to have a higher quality than an area only investigated by one scientist. OnGraX supports such analysis tasks by providing methods for data awareness and coordination. Note that we retain this usage scenario in the rest of the paper except in the heat map evaluation (cf. Sect. 5) in order to attract a higher number of test subjects.

The remainder of this paper is organized as follows. In the next section, we discuss related work in collaborative graph visualization. We describe our design decisions in Sect. 3 and explain OnGraX' interaction and visualization techniques for displaying user behavior in Sect. 4. A user experiment to evaluate our heat map approach for identifying previous user activities is discussed in Sect. 5, and we conclude our paper in Sect. 6.

2   Related Work

Isenberg et al. [11] give a good overview of definitions, tasks and sample visualizations in the field of collaborative visualization. The authors define collaborative visualization as "the shared use of computer-supported, (interactive,) visual representations of data by more than one person with the common goal of contribution to joint information processing activities". They also provide an excellent summary of ongoing challenges in this field. All discussed standard systems incl. more recent developments (e.g., ManyEyes [20] or Dashiki [17]) are not suitable for our collaborative analysis problems, since they do not support the interactive visualization of node-link diagrams in a web browser with real-time interactions for collaboration. The benefits of collaborative work were also discussed in an article on social navigation presented by Dieberger et al. [6]. Being able to see the usage history and annotations of former users might help analysts to filter and find relevant information more quickly. In order to be able to work together during a synchronous session, users have to know each other's interactions and views on the data set, usually referred to as "common ground" [3,9]. To find a common ground in node-link visualizations, we apply the techniques from the work of Gutwin and Greenberg [8]. They used secondary viewports and radar views to indicate other users' view areas and mouse cursor positions. We use a similar approach and show the viewports of other users as rectangles in the background of the graph visualization. Another work by Isenberg et al. [12] introduced the concept of collaborative brushing and linking, which "allows users to communicate implicitly, by sharing activities and progress between visualizations". The authors considered sharing activities during synchronous collaborations on a tabletop visualization for document collections. We adapt the concept and utilize it in node-link diagrams with the help of a heat map visualization for the exploration of interaction information in both asynchronous and synchronous, distributed sessions.

Our tool OnGraX utilizes heat maps to analyze and identify highly frequented or edited parts of the graph based on user behavior. Patina [16] uses a similar approach but focuses on visualizing the usage of user interfaces, whereas our tool facilitates heat maps to visualize interactions of users with the data itself. To the best of our knowledge, heat map visualizations for representing data in combination with node-link diagrams are seldomly considered. Usually, they are used to visualize quantitative data in geovisualizations [19], as cluster heat maps [7], or for the visualization of eye tracking data to illustrate the quality of web site designs, user interfaces, or graph layouts [18,21], i.e., for evaluation purposes. One of the few examples where heat maps are used in node-link diagrams is PLATO [22] which employs heat maps to visualize gameplay data.

3   Design Decisions

We carefully designed our system in terms of visual representations, interaction techniques, and analysis processes to support biologists/bioinformaticians in exploring and curating graphs from the KEGG pathway database. We decided to focus our work on node-link diagrams, since this is still the most accepted and preferred graph drawing metaphor, and our users are familiar with this kind of visualization. Our overall goal was to develop a visualization system that allows analysts spatially spread across multiple research labs or even countries to quickly start an analysis session and to work on large and complex networks together. A special problem that arises during the distributed analysis of graphs is that topology and structure of a graph are independent to the layout. Analysts might change the layout drastically during the analysis process, which complicates the task of keeping track of the graph objects and areas that users were most interested in. We also want our tool to support tracking and subsequent visualizing of all actions and graph changes performed by the users. This includes to keep track of the users' camera positions and use this data later to assist users in finding parts of a graph that were interesting to other analysts or have been edited a lot. The reason behind this is that users in a collaborative working environment do not always find the time to work together simultaneously. They would prefer to work on the data set whenever it is convenient for them. And in such a case, they would like to review changes that have been performed on the data set by other analysts in the past. Maybe, they also want to find out which part of the data set another analyst was looking at, since he/she might be an expert in the underlying application field and has another exploration pattern compared to less experienced users. Showing this data—the camera and mouse positions, the logged user views, and changes to specific objects—in the graph without changing the original node-link visualization was an important requirement for our users. Biologists are accustomed to existing layouts and drawing conventions of graphs from the KEGG pathway database. Thus, changing positions, color, or the shape of nodes to show the data which is collected during collaborations is not an option for our analysis tasks.

During their work, analysts would also like to share their thoughts, insights, and questions about specific nodes, edges or regions with other users. This could happen during a *synchronous* session where collaborators want to discuss their findings, or in an *asynchronous* session where users would like to share messages and pointers on specific nodes. Heer and Agrawala discuss these ideas as "Common Ground and Awareness" and "Reference and Deixis" in their work on collaborative visual analytics [9]. In case of graphs that change their topology during the analysis process, single nodes or complete graph regions could be deleted from a graph, rendering old user annotations useless without the possibility to view them in their historical context. Thus, analysts need a way to quickly view the graph in a state when the annotation was originally written. Based on this discussion, we categorize our requirements as described in the following.

Collaboration Requirements (C-R)

1. Users should be aware of the position of other users in the same synchronous session.
2. Users should have possibilities to establish and keep a common ground with other users. Everyone should be aware of performed changes on the graph during a session.
3. They should have an option to discuss ongoing work through persistent chat channels and annotations.

Visualization Requirements (V-R)

1. Annotations should be viewable in their historical context. Thus, it should be possible for users to review old graph states.
2. Provide an easy and intuitive way for analysts to find out which regions of a graph where viewed and/or changed by former users.
3. Additionally, the visualization of this data should not interfere with the original node-link diagram.

4   Interaction and Visualization Techniques

Figure 1 shows an overview of the tool right after joining an ongoing graph analysis session. In this case, the user has joined a session where two other users, Bob and Sue, are already working in. Their viewports are represented as two dashed rectangles: Bob's view is shown in blue (bottom left) and Sue's view is shown in green (bottom right). All users in a session are listed as small icons at the left hand side of the screen. By clicking on one of the user icons, the camera moves to his/her current position in the graph. This feature provides a quick way to join and discuss another user's viewing area. Visualizing the viewports of other users helps us to tackle our first collaboration requirement (cf. C-R 1). An overview of the graph is rendered in the bottom-right corner of the screen. Here, the user's camera position is shown as a blue rectangle. As in many other standard visualizations that use overview+detail [4], this rectangle can be dragged to another position in the overview in order to modify the detail view (the same can be done by simply clicking on the new position in the overview).

We use a standard node-link metaphor to visualize graphs in our system. The visualization uses tapered edges for directed graphs, as suggested by Holten and van Wijk [10], since they provide users with a faster way to find connected nodes as opposed to arrowhead edges. If another user selects one or more nodes, this will be visible to all other participants of the analysis session. An outline in the respective user color is added to a selected node; thereby the system adapts the outline shape to the corresponding node shape. To make graph changes performed by other users during a synchronous session more obvious and to address the second collaboration requirement (cf. C-R 2), we use short animations on the affected objects, similarly to the work of Gutwin and Greenberg [8]. For instance, the outlines for other users' node selections are animated shortly while they are added or removed, nodes are slowly moved to new positions instead of just jumping there after being moved by another user, and deleted nodes slowly vanish instead of just disappearing.

4.1   Annotations and Chat Links

In order to improve the communication among collaborators, our tool has a persistent chat channel for every graph session and offers the possibility to link chat messages to a position or a node in the graph. Users can use those chat links to

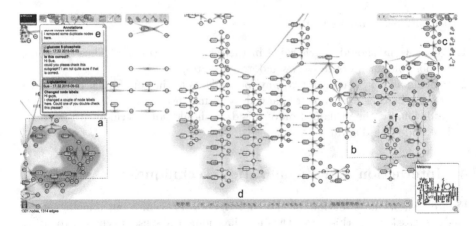

Fig. 1. Overview of our system. The image shows a part of a biochemical network with 1,301 nodes and 1,314 edges. The blue and green dashed rectangles (see (a) and (b)) are the viewing areas (viewports) of two other users who are also exploring this graph simultaneously. In this concrete case, the underlying heat map highlights those nodes that were in the viewing area to all users during the last hour. Symbols in the top-right corner of the screen (c) assist analysts to keep track of recent actions performed by other users. The timeline (d) is used to temporarily revert the graph to a previous state and to replay applied changes. Analysts can pin text annotations to nodes and edges to discuss tasks, insights and questions with each other (e+f) (Color figure online)

move the camera to the linked object or position. A link to an arbitrary position might become obsolete after changes to the graph layout, but a message linked to a node or edge will always be valid as long as the object is not deleted. In addition, users can attach textual annotations directly to nodes or edges (cf. Fig. 1, (e+f)). These annotations work as pointers from the graph visualization to text and vice versa. Clicking on an annotation in the graph visualization opens the annotation dialog and highlights the linked message. A click on an annotation in the dialog moves the camera to the object's position in the graph visualization. With the chat and annotation features, we address our last collaboration requirement (cf. C-R 3).

One problem with textual annotations and chat messages linked to objects is, that the original context in which an annotation or message was initially written could get lost if the respective graph region—where the link is pointing to—is changed during the course of a session or if the object with this link is deleted. We solve this problem by enabling analysts to temporarily revert the complete graph to an old state (similar to the timeline feature, cf. Sect. 4.3) by right clicking on a chat link or an annotation, giving them the possibility to view the graph in a state in which the annotation was originally written. This feature addresses our first visualization requirement (cf. V-R 1).

4.2   Visualizing User Behavior Data with Heat Maps

In order to provide users with a way to quickly find out which nodes or regions of a graph were viewed and/or changed by others (cf. V-R 2), we considered several options. It would be possible to map the corresponding data to the colors or the size of the nodes. Another option would be to use additional glyphs on/around the nodes which represent this data. Using glyphs would also allow us to show both the viewport data and the data for graph changes at the same time, as small bar charts for instance. The third option is a heat map-based visualization in the background of the graph visualization. We decided to omit mapping the data to the size of nodes, as this would interfere too much with the original graph layout and could introduce too many node overlaps. Additional options would have been to use contour lines [2] or bubble sets [5], but for our use case the focus usually lies on finding and marking single nodes instead of bigger regions in a graph. The remaining three options are exemplified in Fig. 2.

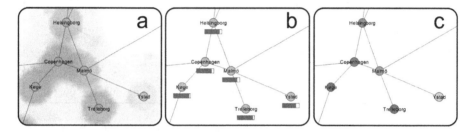

Fig. 2. Heat map visualization (a) and two alternative approaches: glyphs (b) and node color (c). They are used to indicate which parts of the entire graph were viewed or changed by other users (Color figure online)

One disadvantage of glyphs in this context is the increased clutter in the graph visualization. Additionally, depending on the size of the glyphs, it could be hard to see the actual data values in highly zoomed-out views of the graph. Changing the color coding of nodes in a graph as alternative is in conflict with our last visualization requirement (cf. V-R 3), because the color coding can be already mapped to another attribute. Thus, heat maps could provide a good alternative to visualize additional data without directly changing the attributes of objects in a node-link diagram. Users can choose between a colored heat map and a monochrome heat map in case the colored version interferes too much with the actual node colors. We performed a user experiment (cf. Sect. 5) to assess how the heat map approach compares against glyphs and node colors. The actual values, which are mapped to the glyphs, node colors, or heat map can be computed based on two different data sources: viewports and graph changes.

Displaying Viewports. In the first case, values are calculated based on the amount of seconds that nodes have been in the viewing areas of users (visitation rate). For aggregating this data, OnGraX stores each user's viewport together with the time spent on the position whenever the viewport is changed. Additionally, each time a node is moved, the old position is logged. The server correlates all logged user views and node positions to calculate the values, thus making them robust against changes in the layout of the graph. Figure 3 illustrates this approach. In this small example, three stored viewports of one user and two node movements from another user—whose viewports are ignored here—are taken into account. The user arrived at position A at exactly 10:00 AM, stayed there for 10 s, moved his viewport to position B for 5 s and finally stayed 16 s at position C. In viewport A, node 1 was visible for 10 s, but in viewport C, it was only visible for 12 s, as the node was only moved into the viewport 4 s after the user arrived at the position, resulting in a complete viewing time of 22 s for node 1. The viewing time of node 2 is only 2 s, as it was moved into viewport B 13 s after 10:00 AM, and the user arrived there at 10 s after 10:00 AM and left 5 s later.

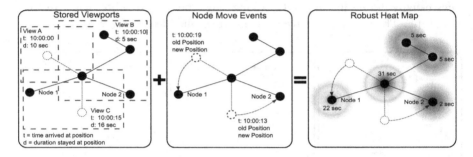

Fig. 3. Illustration for the correlation of all stored viewports with all node move actions to create a heat map that is robust against layout changes of the graph

For zoomed-out views that show a lot of nodes, it is clear that the user does not attend to all nodes in such a view. To solve this issue, users can adjust the settings to filter out these "big views" and only use zoomed-in views to calculate the heat map. Views are also only tracked if the user is actively working on the graph: if a user switches to another window or tab, then the tracking is stopped. It is also stopped if the mouse is not moved for a while (currently 20 s) to avoid tracking views of inactive users. This approach does still include nodes in the views that might not have had attention by an active user, but it gives a better estimate about the viewed graph regions without asking a user to mark every inspected node manually or asking all users to use an eye tracker during the analysis process, for instance.

Displaying Graph Changes. In the second case, OnGraX calculates values based on changes that have been performed on nodes. Seven actions (name changed, shape changed, node moved, node added, node selected, edge added, edge removed) are tracked and can be used to calculate the heat map values in this case. A multiplier is specified in a configuration dialog for each individual action type to give it more or less weight during the calculation. This enables analysts to highlight only nodes that were moved and had their names changed, for instance. The visualization can be configured to only show a specific user or to show the data for all users together (the selection of user groups would also be possible and could easily be added to the system). Furthermore, it is possible to select a time frame, for instance, the last five minutes of the current analysis session, or a specific start and end date. This enables an analyst to review changes done in a collaborative session during a specific time frame or to check the work of a single user.

4.3 Tracking and Replaying User Actions

Actions performed by other users during a *synchronous* session are shown at the right corner of the screen (cf. Fig. 1,(c)) together with the name of the user who initiated the action. A right-click is used to dismiss a recent action and a left-click moves the camera to the location of the action in the graph. Another left-click on the same action moves the camera back to its original position. Thus, users can quickly check what their collaborators are doing and then return to their own work, without having to navigate to every performed action manually. To provide our users with the possibility to keep track of *all* actions that occurred in a session, we use a scrollable timeline at the bottom border of the screen that shows the complete action history of the graph session (cf. Fig. 1,(d)). The mouse tooltip for the symbols in the timeline shows the action time and the name of the user who performed the action. The timeline can also be used to revisit old graph states and replay previous actions. If a user clicks on a symbol, all actions performed since this specific action are replayed in reverse order. The visualization will show the graph in a state before the action was performed. Shortly after the graph has been transformed to its old state, the clicked action is reapplied, animating the graph to the requested point in time. This feature gives users a tool to revisit old graph states and replay old actions allowing them to assess what work has been done by other collaborators. Clicking on the rightmost symbol reverts the graph back to its present state. While viewing an old graph state, it is *not* possible to apply any changes to the graph. We decided against this feature as it would open the possibility to create numerous new branches of different graph states. This is an interesting aspect and actively researched [14], but currently not the focus of our work.

5   Heat Map Evaluation

We performed a user experiment to evaluate the usefulness and acceptance of our heat map approach to visualize user behavior data in comparison to glyphs and node coloring. We recruited 15 participants (7 undergraduate students, 7 graduate students, and 1 post-graduate; average age = 28; 5 female, 10 male). Seven participants had a background in computer science and eight a background in media technology. Eight participants never worked with node-link diagrams before, but everyone was familiar with them.

All 15 sessions were recorded on video and the participants were instructed to employ a think-aloud protocol. Before starting the actual tasks, the tool and the three visualization approaches for user behavior data (glyphs, node color, heat map) and their meaning were introduced by the experimenter and each participant could explore a sample graph to get accustomed to the tool. Each session took about 25–30 min, and we asked the participants to solve each task as quickly as possible, but the time for the tasks was not limited by us. All participants had to solve two tasks for nine different graphs with the help of the three visualization approaches. Both tasks were described as follows:

Task 1 – explore graph changes: Find and count all nodes that were moved by a specific user (9–14 single marked nodes per graph).

Task 2 – explore viewports: Find all regions that a specific user was most interested in (1–3 marked regions per graph).

The experiment was conducted as a within-participants experiment, and users were divided into three different groups. Every group explored all graphs in the same order but with a different sequence of visualization approaches. Six graphs were generated randomly: the first three graphs consisted of 1,000 nodes/edges and the following three of 2,000 nodes/edges. For the last three graphs, we used existing metabolic networks with 1,300 to 1,800 nodes/edges.

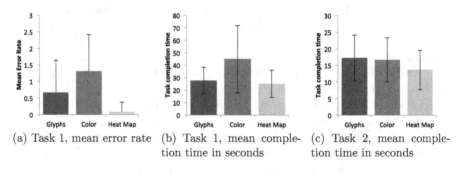

(a) Task 1, mean error rate   (b) Task 1, mean completion time in seconds   (c) Task 2, mean completion time in seconds

Fig. 4. Analysis of the two tasks for the different visualization approaches

Quantitative Results. We started measuring the task time in seconds for each task as soon as the visualization of the user behavioral data was enabled by the participants and stopped the time as soon as they reported a number. For Task 1, we show the number of nodes that were not found by the users (mean error rate). In Task 2, all participants found all marked regions, regardless of the visualization approach. Therefore, we only report the error rate for Task 1. Figure 4 shows the summarized results for all graphs. Initial Friedman tests showed that both tasks had statistically significant differences in task completion time. Task 1: $\chi^2 = 34.881, p < 0.001$. Task 2: $\chi^2 = 16.812, p < 0.001$. We conducted a post hoc analysis with Wilcoxon signed-rank tests for our not normally distributed data. For Task 1, the median interquartile range (IQR) task completion times were 26 (Glyphs), 39 (Node Colors), and 20 (Heat Map). Both, glyphs vs. heat maps ($Z = -3.678, p < 0.001$) and node colors vs. heat maps ($Z = -5.334, p < 0.001$)

had a significant reduction in task completion time. For Task 2, the median (IQR) task completion times were 19 (Glyphs), 15 (Node Colors), and 13 (Heat Map). Here, the heat map approach also performed significantly better in comparison with glyphs ($Z = -3.678, p < 0.001$) and node colors ($Z = -2.406, p = 0.016$).

Qualitative Results. We asked all participants which visualization approach they preferred. Everyone favored the heat map visualization. For them, the heat map was the easiest to perceive, and it also provided the most convenient way to find single nodes with high values, even at lower zoom levels. While performing the second task, four participants mentioned that the glyph approach introduced too much clutter in the view, especially for the metabolic networks. They said that glyphs were hard to distinguish from the actual nodes, because both the nodes and glyphs sometimes had a similar shape.

6   Conclusions

In this paper, we presented a web-based collaborative system for visualizing graphs with several thousands of nodes and edges. Our tool OnGraX provides visualization and interaction techniques for analyzing data sets synchronously *and* asynchronously in a distributed environment. Additionally, all actions performed during a session as well as the users' camera positions are tracked and can be visualized along with the graph data by using heat map representations. We propose using heat maps to efficiently show additional data without affecting the original graph visualization. Based on a user experiment, we show that the heat map-based approach compares better against glyphs or changing the background color of nodes. As future work, we plan to evaluate the other aspects of OnGraX—such as those described in Sects. 4.1 and 4.3—and to use the tool in other contexts. For instance, our collaborators want to use OnGraX for the education of their biology students. The idea is to give students existing metabolic pathways and ask to revise and edit those graphs. Afterwards, the docents could join the online session and discuss those changes with the students. We will use this opportunity to test our tool in another authentic environment and perform a detailed user study during collaborative work in an educational setting. In our specific use case, graph changes are usually limited to a couple of nodes, thus the tracking of *all* actions and visualizing this data is not an issue here. But, it could become problematic if a graph or a subgraph is changed drastically. In this case, additional options to set the granularity for tracked events and alternative visualization techniques would be required incl. a newly designed evaluation.

References

1. KEGG: Kyoto Encyclopedia of Genes and Genomes. http://www.genome.jp/kegg/. Accessed 12 August 2015
2. Alper, B.E., Henry Riche, N., Hollerer, T.: Structuring the space: a study on enriching node-link diagrams with visual references. In: Proceedings of the 32nd Annual ACM Conference on Human Factors in Computing Systems, CHI 2014, pp. 1825–1834. ACM, New York (2014)

3. Chuah, M., Roth, S.: Visualizing common ground. In: Proceedings of the International Conference on Information Visualization (IV 2003), pp. 365–372. IEEE (2003)
4. Cockburn, A., Karlson, A., Bederson, B.B.: A review of overview+detail, zooming, and focus+context interfaces. ACM Comput. Surv. **41**(1), 2:1–2:31 (2009)
5. Collins, C., Penn, G., Carpendale, S.: Bubble sets: revealing set relations with isocontours over existing visualizations. IEEE Trans. Visual Comput. Graphics **15**(6), 1009–1016 (2009)
6. Dieberger, A., Dourish, P., Höök, K.: Social navigation: techniques for building more usable systems. Interactions **7**(6), 36–45 (2000)
7. Wilkinson, L., Friendly, M.: The history of the cluster heat map. Am. Stat. **63**(2), 179–184 (2009). doi:10.1198/tas.2009.0033
8. Gutwin, C., Greenberg, S.: Design for individuals, design for groups: tradeoffs between power and workspace awareness. In: Proceedings of the 1998 ACM Conference on Computer Supported Cooperative Work, CSCW 1998, pp. 207–216. ACM, New York (1998)
9. Heer, J., Agrawala, M.: Design considerations for collaborative visual analytics. Inf. Visual. **7**(1), 49–62 (2008)
10. Holten, D., van Wijk, J.J.: A user study on visualizing directed edges in graphs. In: Proceedings of the 27th International Conference on Human Factors in Computing Systems, CHI 2009, p. 2299 (2009)
11. Isenberg, P., Elmqvist, N., Cernea, D., Scholtz, J., Ma, K.L., Hagen, H.: Collaborative visualization: definition, challenges, and research agenda. Inf. Visual. **10**(4), 310–326 (2011)
12. Isenberg, P., Fisher, D.: Collaborative brushing and linking for co-located visual analytics of document collections. Comput. Graphics Forum **28**(3), 1031–1038 (2009)
13. Isenberg, P., Tang, A., Carpendale, S.: An exploratory study of visual information analysis. In: Proceedings of the SIGCHI Conference on Human Factors in Computing Systems, CHI 2008, pp. 1217–1226. ACM, New York (2008)
14. von Landesberger, T., Fiebig, S., Bremm, S., Kuijper, A., Fellner, D.W.: Interaction taxonomy for tracking of user actions in visual analytics applications. In: Huang, W. (ed.) Handbook of Human Centric Visualization, pp. 653–670. Springer, New York (2014)
15. Mark, G., Kobsa, A.: The effects of collaboration and system transparency on cive usage: an empirical study and model. Presence Teleoper. Virtual Environ. **14**(1), 60–80 (2005)
16. Matejka, J., Grossman, T., Fitzmaurice, G.: Patina: dynamic heatmaps for visualizing application usage. In: Proceedings of the SIGCHI Conference on Human Factors in Computing Systems, CHI 2013, pp. 3227–3236. ACM, New York (2013)
17. McKeon, M.: Harnessing the information ecosystem with wiki-based visualization dashboards. IEEE Trans. Vis. Comput. Graph. **15**(6), 1081–1088 (2009)
18. Pohl, M., Schmitt, M., Diehl, S.: Comparing the readability of graph layouts using eyetracking and task-oriented analysis. In: Computational Aesthetics, pp. 49–56 (2009)
19. Slocum, T.A., Mcmaster, R.B., Kessler, F.C., Howard, H.H.: Thematic Cartography and Geovisualization, 3rd edn. Prentice Hall, Upper Saddle River (2008). (Prentice Hall Series in Geographic Information Science)
20. Viégas, A.B., Wattenberg, M., Ham, F.V., Kriss, J., Mckeon, M.: Many eyes: a site for visualization at internet scale. IEEE Trans. Vis. Comput. Graph. **13**(6), 1121–1128 (2007)

21. Špakov, O., Miniotas, D.: Visualization of eye gaze data using heat maps. Electron. Electr. Eng. **2**(2), 55–58 (2007)
22. Wallner, G., Kriglstein, S.: Plato: a visual analytics system for gameplay data. Comput. Graph. **38**, 341–356 (2013)
23. Zimmer, B., Kerren, A.: Applying heat maps in a web-based collaborative graph visualization. In: Poster Abstract, IEEE Information Visualization (InfoVis 2014) (2014)
24. Zimmer, B., Kerren, A.: Sensemaking and provenance in distributed collaborative node-link visualizations. In: Abstract Papers, IEEE VIS 2014 Workshop: Provenance for Sensemaking (2014)
25. Zimmer, B., Kerren, A.: Harnessing WebGL and websockets for a web-based collaborative graph exploration tool. In: Cimiano, P., Frasincar, F., Houben, G.-J., Schwabe, D. (eds.) ICWE 2015. LNCS, vol. 9114, pp. 583–598. Springer, Heidelberg (2015)

Confluent Orthogonal Drawings
of Syntax Diagrams

Michael J. Bannister[✉], David A. Brown, and David Eppstein

Department of Computer Science, University of California, Irvine, USA
mbannist@uci.edu

Abstract. We provide a pipeline for generating syntax diagrams (also called railroad diagrams) from context free grammars. Syntax diagrams are a graphical representation of a context free language, which we formalize abstractly as a set of mutually recursive nondeterministic finite automata and draw by combining elements from the confluent drawing, layered drawing, and smooth orthogonal drawing styles. Within our pipeline we introduce several heuristics that modify the grammar but preserve the language, improving the aesthetics of the final drawing.

1 Introduction

The languages of computing, such as programming languages and data exchange formats, are typically specified using a finite set of rules called a grammar, and these rules are usually given in Backus–Naur Form or one of its extensions. Backus–Naur Form provides a notation rich enough to express all context-free grammars, and in turn most grammars of practical interest, while being easily machine readable. However, being a purely textual representation, it is perhaps less readable by humans. For this reason, Jensen and Wirth used a graphical representation of context-free grammars, called syntax diagrams, when defining the programming language Pascal [1].[1] We investigate the problem of generating syntax diagrams for context-free grammars and provide several heuristics optimizing the aesthetics of the resulting drawing. Our work provides the first algorithmic study of this problem and the first system that attempts to optimize the resulting diagram for readability rather than directly translating a given grammar into a diagram.

Recall that a *context-free grammar* is defined by four values N, Σ, R, S. In this 4-tuple, N is a set of *nonterminal symbols*, Σ is a set of *terminal symbols*, R is a set of *production rules* of the form $A \to \beta$ where A is a nonterminal symbol and β is a (possibly empty) string of terminal and nonterminal symbols, and S is a nonterminal symbol designated as the *start symbol*. A string σ of terminal symbols belongs to the language defined by the grammar when there exists a

---

Michael Bannister and David Eppstein were supported in part by NSF grant CCF-1228639.

[1] Jensen and Wirth were not the first to use syntax diagrams [2], but they popularized them, and these diagrams have been widely used since.

E. Di Giacomo and A. Lubiw (Eds.): GD 2015, LNCS 9411, pp. 260–271, 2015.
DOI: 10.1007/978-3-319-27261-0_22

Table 1. A context-free grammar for the language of S-expressions in LISP 1.5 [3]

$$\langle\text{S-expression}\rangle \rightarrow \langle\text{atomic symbol}\rangle$$
$$|\ (\langle\text{S-expression}\rangle.\langle\text{S-expression}\rangle)$$
$$|\ (\langle\text{S-expression list}\rangle)$$
$$\langle\text{S-expression list}\rangle \rightarrow \epsilon\ |\ \langle\text{S-expression}\rangle\langle\text{S-expression list}\rangle$$
$$\langle\text{atomic-symbol}\rangle \rightarrow \langle\text{LETTER}\rangle\langle\text{atom part}\rangle$$
$$\langle\text{atom part}\rangle \rightarrow \epsilon\ |\ \langle\text{LETTER}\rangle\langle\text{atom part}\rangle\ |\ \langle\text{number}\rangle\langle\text{atom part}\rangle$$
$$\langle\text{LETTER}\rangle \rightarrow A\ |\ B\ |\ C\ |\ \cdots\ |\ Z$$
$$\langle\text{number}\rangle \rightarrow 0\ |\ 1\ |\ 2\ |\ \cdots\ |\ 9$$

sequence of *rewrite steps* starting from S and ending at σ, each of which replaces a nonterminal symbol A in the current string with a string β such that $A \rightarrow \beta$ is a production rule in the grammar. Table 1 gives an example grammar for the S-expressions in the programming language LISP 1.5.

A *regular grammar* is one in which the production rules all have the form $A \rightarrow b$, $A \rightarrow bC$ or $A \rightarrow \epsilon$, where A and C are nonterminals, b is a terminal, and ϵ is the empty string. An example of a regular grammar is the part of the LISP 1.5 grammar defining $\langle\text{atom part}\rangle$. Languages definable by regular grammars are exactly the regular languages, whose equivalent characterizations include being recognizable by nondeterministic finite automata (NFAs). For these languages, we could use graph drawings of an NFA state graph as a graphical representation, by drawing an st-digraph with edges labeled by terminal symbols. A string σ is in the language if and only if there is a directed path through the graph from s to t such that the concatenation of the edge labels is equal to σ. Unfortunately, such a representation will not work for non-regular languages.

To graphically represent context-free languages we turn to syntax diagrams. Although other authors used syntax diagrams earlier [2], they were popularized by the Pascal User Manual and Report by Jensen and Wirth [1]. The style has been praised for its readability [4] and pedagogical value [5], and has been used by the Smalltalk-80 Blue Book [6], JSON Data Interchange Standard [7], and the W3C technical report on CSS [8]. Several software packages have been created to automate the drawing of syntax diagrams [9–11]. These software packages provide little to no optimization of the drawing, providing only a one-to-one translation of the Extended Backus–Naur grammars into syntax diagrams. Until now, there does not seem to be any algorithmic research involving the generation and optimization of syntax diagrams.

We introduce a new formalization for syntax diagrams consisting of a collection of st-digraphs (see e.g., Fig. 3), each representing the possible expansions of a single nonterminal symbol, with each edge in each graph labeled by either a terminal or a nonterminal symbol. As before a string is in the language if and only if the string can be represented by a directed path from s to t in the start symbol's st-digraph. However, when this path would contain a nonterminal symbol, we recurse into the st-digraph corresponding to that symbol. The concatenation of the terminal symbols in the resulting system of recursively generated paths should match the sequence of terminal symbols in the given string.

Without further optimization this formalization merely gives a new notation for writing production rules, but it has two advantages over extended BNF. Firstly, it gives us additional freedom in our representation: a BNF grammar can only describe syntax diagrams formed by a collection of disjoint paths between the two terminals, and extended BNF can still only describe syntax diagrams in the form of series-parallel graphs, while our diagrams are not restricted in these ways. Secondly, as we describe below, we can use this notation to directly represent the junctions and tracks of a confluent drawing style [12], in which a path through the graph is only valid if it is a smooth path, such as in Fig. 1 (right). It is this drawing style that gives rise to the occasionally used alternative name "railroad diagrams" for syntax diagrams.

Our drawings will combine confluent drawing with Sugiyama-style layered drawing [13,14] using smooth orthogonal edge shapes [15]. The combination of confluent and layered drawing has been studied before [16], but in a different way. Past work considered confluent drawing as a technique for visualizing a specific graph, and involved a search for subgraphs that could be more concisely expressed using confluence. In our application, the graph (NFA) representation that we work with already encodes the confluent features of the drawing: its vertices become confluent junctions in the drawing, and its edges become the boxes and connecting segments of track of the drawing (Fig. 9). Rather than searching for graph features that can become confluent, our focus is on modifying the underlying NFA to produce a simpler and higher-quality drawing while preserving the equivalence of the underlying context-free language described by the drawing.

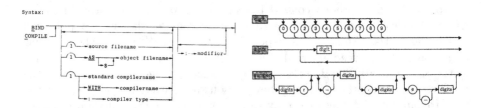

Fig. 1. A syntax diagram from the CANDE Information Manual (left) and a confluent syntax diagram from the Pascal User Manual and Report (right).

1.1   Software Pipeline

We describe our method for producing syntax diagrams with the framework of a generic software pipeline (Fig. 2). In the first step of our pipeline, we convert the grammar to our internal representation, which we will call the *NFA representation*. This representation consists of a family of *st*-digraphs, initially one

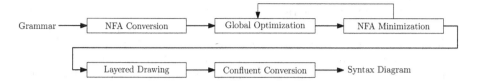

Fig. 2. A flow chart describing our software pipeline.

for each nonterminal symbol, whose edges are labeled by (terminal and non-terminal) symbols in the grammar or ϵ (the empty string). To construct the st-digraph for the nonterminal symbol A we convert each production of the form $A \rightarrow B_0 B_1 \cdots B_{r-1}$ into a directed path of length r labeled by the symbols B_0, B_1, B_{r-1}. Then all of the beginning and ending vertices are respectively merged together. Finally, we add to the graph two extra ϵ-labeled edges, one at the beginning and one at the end. See Fig. 3 for the complete NFA representation of LISP 1.5.

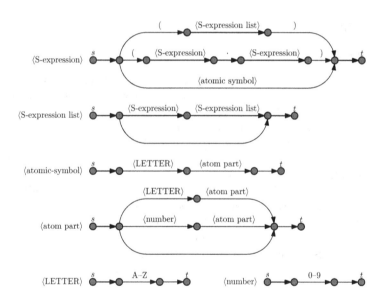

Fig. 3. The initial NFA-representation of S-expressions in the LISP 1.5 grammar.

The second and third steps in the pipeline attempt to reduce the number of total symbols in the NFA representation, through both global optimizations that act on the entire system of graphs and local optimizations that act on a single graph. The local optimization part of the pipeline is a form of the well-studied problem of NFA minimization. In general exact NFA minimization is PSPACE-hard [17,18], and furthermore approximating the minimum NFA efficiently to within an $o(n)$ approximation ratio is also PSPACE-hard [19]. However, since the problem is of practical importance there are many heuristic approaches [20, 21]. In this paper, we use simple heuristics motivated by the structure of real-world grammars and typical simplifications found in hand drawn syntax diagrams, rather than attempting to implement the more complex heuristics devised for minimizing NFAs without regard to their appearance as a diagram.

Once the NFA representation is optimized, we draw each of the st-digraphs in a layered Sugiyama style [13, 14], rotated horizontally to direct edges from left to right. In these graphs, the only directed cycles come from tail recursion elimination, so rather than searching for a small feedback arc set to determine the reversed edges in the drawing, we maintain such a set during the process of NFA minimization and add to it whenever we perform a tail recursion elimination step. In this way, we can ensure that all the tokens in the drawing are traversed from left to right. Standard layered drawing optimizations are applicable in this stage, but were not implemented in our experiments as we were primarily interested in optimizing the NFA representation. Finally, we convert the layered drawing into a confluent syntax diagram.

1.2 Contributions

Our contributions in this paper are summarized below.

- We formalize an abstract representation of syntax diagrams as a collection of mutually recursive NFAs, allowing the application of NFA minimization heuristics beyond what is possible with EBNF.
- We formulate a software pipeline for producing syntax diagrams, based on NFA minimization and confluent layered graph drawing.
- We develop a family of fast and simple NFA minimization heuristics, together with global heuristics that recombine multiple NFAs.
- We describe a geometric layout method based on a horizontal Sugiyama layered drawing, where we reinterpret the vertices and edges in a layered drawing of an NFA as the junctions and vertices of a confluent drawing.
- We provide a proof-of-concept implementation that produces human quality syntax diagrams for real-world context-free languages.
- Finally, we experimentally evaluate the quality of our heuristics.

2   Global Minimization Heuristics

A *global minimization heuristic* seeks to minimize the total number of labeled edges in an NFA representation via the modification of two or more of the st-digraphs in the representation. The only global heuristic that we consider is *nonterminal nesting*, in which a single nonterminal edge in one graph is replaced by the entire graph corresponding to that nonterminal edge. Since the goal is to reduce the total number of symbols in the NFA representation, we enforce the following restrictions when nesting a graph H (corresponding to a nonterminal A) into another graph G:

- A cannot be the start symbol.
- G and H must be two distinct graphs.
- If H has more than one non-ϵ edge, then A must occur only once in the whole system of digraphs, and its occurrence must be in G.
- The number of symbols in the graph produced by nesting H into G must be less than a predefined threshold k.

The final restriction above is intended to keep the size of each individual st-digraph to a human-readable level. The nesting heuristic can be seen to have been used in some hand-drawn syntax diagrams (e.g., the JSON syntax diagrams), but it does not appear to be used by previous syntax diagram software. See Fig. 4 for an example of nesting with the LISP 1.5 grammar.

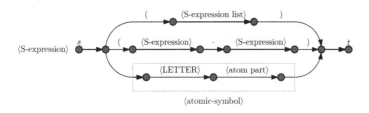

Fig. 4. An example of nesting the ⟨atomic symbol⟩ st-digraph into the ⟨S-expression⟩ st-digraph, within the LISP 1.5 grammar.

3 Local Minimization Heuristics

A *local minimization heuristic* seeks to minimize the total number of labeled edges in a single st-digraph within the NFA representation. Many of these optimizations can be seen in hand-drawn syntax diagrams.

3.1 Tail Recursion Loop Back

The st-digraphs produced from a grammar, before optimization, are acyclic, and nesting preserves acyclicity. However, hand-drawn syntax diagrams typically contain cycles, which we introduce as a replacement for tail-recursive grammars using the *loop back heuristic*. If a nonterminal A appears exactly once in its own st-digraph and the edge on which it appears has t' (the only incoming neighbor of t) as its destination, then we change the destination of the A-labeled edge from t' to s' (the only outgoing neighbor of s) and we change its label from A to ϵ. Although this does not reduce the number of edges in the st-digraph, it does reduce the number of labeled edges and improves the readability of the drawing. In addition, by reducing the number of occurrences of A as a label, it may cause nesting operations to become possible that were previously forbidden. The edges that are modified by this heuristic will be the only ones directed backwards in our eventual drawings. See Fig. 5 for an example of this construction.

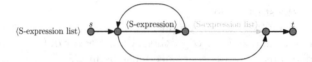

Fig. 5. An example of tail recursion loop back of ⟨S-expression list⟩ in the LISP 1.5 grammar. The removed edge has been colored gray.

3.2 Parallel State Elimination with Squish Heuristic

The *squish forward* heuristic is used to reduce the number of nonempty symbols when there are parallel occurrences of the same symbol. If two edges $e_1 = (u, v_1)$ and $e_2 = (u, v_2)$ are labeled by the same symbol $A \neq \epsilon$, then we replace e_1 and e_2 with $f = (u, t)$ labeled A, $f_1 = (t, v_1)$ labeled ϵ and $f_2 = (t, v_2)$ labeled ϵ. We similarly define the *squish backward* heuristic, to be the squish forward heuristic applied to an st-digraph in which all of the edges have been reversed. See Fig. 6 for an example of this heuristic.

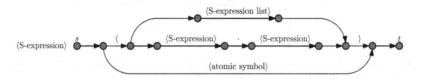

Fig. 6. An example of the squish heuristics applied to ⟨S-expression⟩ in the LISP 1.5 grammar. The squish forward combines the open parenthesis and the squish backward combines the closing parenthesis.

3.3 Epsilon Transition Removal

Our previous optimizations may introduce ϵ-labeled edges. We attempt to remove redundant ϵ-edges using the *epsilon removal* heuristic. If $e = (u, v)$, with $u \neq s$ and $v \neq t$, is an ϵ labeled edge, such that e is not a reversed edge (introduced via the loop back heuristic), and either e is the only outgoing edge of u or the only incoming edge to v, then the edge e is removed by merging u and v. We iteratively find and remove such edges until no such edge exists.

3.4 Confluent Pinch

Our final local optimization would not qualify as an NFA optimization, as it does not attempt to reduce the number of symbols. Instead, the *confluent pinch* heuristic attempts to reduce crossings in the final drawing by removing directed complete bipartite subgraphs (which can be created by the squish heuristic), replacing each one by a single "crossing" vertex. If a digraph contains a set of vertices U and a set of vertices V such that there is an ϵ labeled edge (u, v) for all $u \in U$ and $v \in V$, then we remove all such edges and add ϵ-labeled edges (u, w) for all $u \in U$ and (w, v) for all $v \in V$ (Fig. 7).

Fig. 7. An example of confluent pinch for scientific notation in the JSON grammar.

3.5   Implementing the Heuristics

The application of one heuristic may create new optimization opportunities with respect to a previously applied heuristic. Therefore, we perform multiple rounds of optimization, applying all possible heuristics within each round, until no further optimizations are possible or a maximum number of rounds have been completed. In Fig. 8 we see the optimized NFA representation of S-expressions in LISP 1.5, as produced by our implementation of these heuristics.

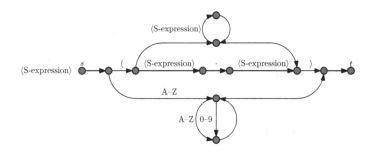

Fig. 8. Optimized NFA representation for S-expressions in LISP 1.5.

4   Sugiyama Layering

Once the NFA representation has been minimized, we give each of the *st*-digraphs a Sugiyama-style layered drawing, using the standard layered-drawing pipeline for layout and crossing minimization. One modification that we make to this pipeline is that it is neither necessary nor desirable to compute a feedback arc set of the *st*-digraphs. Instead, the set of edges introduced during the loop back heuristic already form a feedback arc set with edges which should loop back into the drawing. Since we are using an orthogonal drawing style, we add bends to edges to allow them to shift their vertical positions from one layer to the next, and use an interval-graph coloring algorithm to place the vertical connectors of these bent edges into a small number of columns.

In the final step of our algorithm, we reinterpret the vertices and edges in the resulting orthogonal drawing as the confluent junctions, track segments, and vertices of a confluent drawing. We place a vertex of the confluent drawing at the middle of each edge of the layered drawing whose label is not ϵ, with the confluent vertex being given the same label as the *st*-digraph edge label. We place a confluent junction at each vertex of the layered drawing, connected to a segment of confluent track for each incident edge of the layered drawing. Additionally, confluent junctions are created by the overlapping of edges with a common source. The orientation of the track at each confluent junction is determined by two factors: whether it connects to an earlier or a later layer, and whether it is a forward or reversed edge in the layered drawing. The result of this conversion step is our final syntax diagram. See Fig. 9 for an example of this final conversion step.

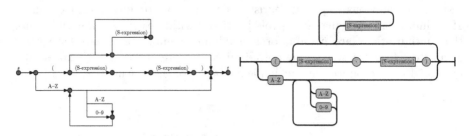

Fig. 9. The final confluent conversion from an orthogonal layered drawing into a syntax diagram for LISP 1.5 S-expressions.

5   Experimental Results

In order to validate the heuristic optimizations performed by our implementation, we tested them on a set of eight real-world context-free grammars collected by Neal Wagner at the web site http://www.cs.utsa.edu/~wagner/CS3723/gram mar/examples2.html together with the Lisp 1.5 and JSON grammars. For each grammar, we measured the area of our drawing (in units of rows and columns), the number of tokens (boxes) in the drawing, and the total number of connected components, both before and after optimization. The results are shown in Table 2.

As these results show, our optimizations were not always effective at reducing the total area of our drawings, and in some cases even increased the area. However, we typically achieved more significant reductions in the numbers of tokens and connected components of the drawings, which we believe to be helpful in reducing their visual clutter. Additionally, it can be seen that our optimizations are typically more effective on grammars with larger numbers of nonterminals, and less effective on grammars that have only a very small number of nonterminals, because in those cases no nesting will be possible.

We did not directly compare the results of other available syntax diagram drawing systems, but the ones we tested all appear to translate the input grammar to a diagram directly, without optimization; therefore, we believe that the results of testing them would be similar to the unoptimized lines of the table.

Table 2. Experimental results

Name	Optimized?	Area	Tokens	Components
Canadian post codes	unoptimized	17	6	1
(simple)	optimized	17	6	1
Canadian post codes	unoptimized	693	69	9
(complex)	optimized	1121	65	5
Ottawa course codes	unoptimized	520	46	15
	optimized	570	36	5
Palindromes	unoptimized	583	105	2
	optimized	583	105	2
Nonempty data files	unoptimized	182	22	8
(repetitive)	optimized	132	11	3
Nonempty data files	unoptimized	143	22	7
(recursive)	optimized	130	7	1
Pascal variable declarations	unoptimized	156	21	7
	optimized	247	12	3
Pascal type declarations	unoptimized	475	52	16
	optimized	486	30	6
LISP 1.5	unoptimized	165	19	6
	optimized	105	9	1
JSON	unoptimized	539	90	15
	optimized	651	42	5

6   Gallery of Examples

We present in Figs. 10 and 11 two complete examples of syntax diagrams of
real-world grammars drawn by our implementation. For the LISP 1.5 grammar,
our optimizations reduce the entire grammar to a single graph. We also present
our results for the JSON grammar, which we believe (despite its obvious flaws)
compares favorably with the official hand-drawn JSON syntax diagrams. Note in
particular that the JSON ⟨number⟩ subgraph is not series-parallel, and therefore
could not be represented by EBNF.

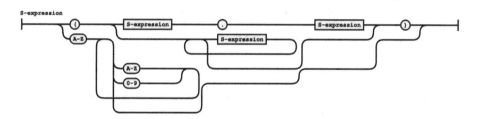

Fig. 10. A syntax diagram for S-expressions in LISP 1.5.

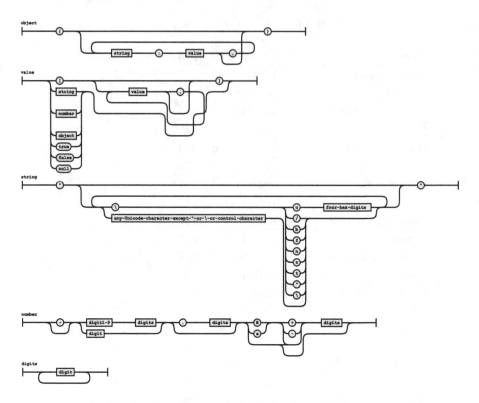

Fig. 11. A syntax diagram for the complete JSON grammar.

References

1. Jensen, K., Wirth, N.: PASCAL User Manual and Report. Springer, New York (1974)
2. Burroughs Corporation: Command and Edit (CANDE) Language Information Manual (1971)
3. McCarthy, J.: LISP 1.5 Programmer's Manual. MIT Press, Cambridge (1965)
4. Braz, L.M.: Visual syntax diagrams for programming language statements. SIGDOC Asterisk J. Comput. Doc. **14**, 23–27 (1990)
5. Bell, S., Gilbert, E.J.: Learning recursion with syntax diagrams. SIGCSE Bull. **6**, 44–45 (1974)
6. Goldberg, A., Robson, D.: Smalltalk-80: The Language and Its Implementation. Addison-Wesley Longman Publishing Co., Inc., Boston (1983)
7. Crockford, D.: Introducing JSON (2015). http://json.org. Accessed: 04 June 2015
8. Atkins, Jr., T., Sapin, S.: CSS Syntax Module Level 3 (2015). http://www.w3.org/TR/css-syntax-3. Accessed: 04 June 2015
9. Dopler, M., Schörgenhumer, S.: EBNF Visualizer (2015). http://dotnet.jku.at/applications/Visualizer. Accessed: 04 June 2015
10. Thiemann, P.: Ebnf2ps: Peter's Syntax Diagram Drawing Tool (2015). http://www2.informatik.uni-freiburg.de/thiemann/haskell/ebnf2ps. Accessed: 04 June 2015

11. Rademacher, G.: Railroad Diagram Generator (2015). http://bottlecaps.de/rr/ui. Accessed: 04 June 2015
12. Dickerson, M.T., Eppstein, D., Goodrich, M.T., Meng, J.Y.: Confluent drawings: visualizing non-planar diagrams in a planar way. In: Liotta, G. (ed.) GD 2003. LNCS, vol. 2912, pp. 1–12. Springer, Heidelberg (2004)
13. Sugiyama, K., Tagawa, S., Toda, M.: Methods for visual understanding of hierarchical system structures. IEEE Trans. Systems Man Cybernet. **11**, 109–125 (1981)
14. Bastert, O., Matuszewski, C.: Layered drawings of digraphs. In: Kaufmann, M., Wagner, D. (eds.) Drawing Graphs. LNCS, vol. 2025, pp. 87–120. Springer, Heidelberg (2001)
15. Bekos, M.A., Kaufmann, M., Kobourov, S.G., Symvonis, A.: Smooth orthogonal layouts. J. Graph Algorithms Appl. **17**, 575–595 (2013)
16. Eppstein, D., Goodrich, M.T., Meng, J.Y.: Confluent layered drawings. In: Pach, J. (ed.) GD 2004. LNCS, vol. 3383, pp. 184–194. Springer, Heidelberg (2005)
17. Hunt III, H.B., Rosenkrantz, D.J., Szymanski, T.G.: On the equivalence, containment, and covering problems for the regular and context-free languages. J. Comput. Syst. Sci. **12**, 222–268 (1976)
18. Stockmeyer, L.J., Meyer, A.R.: Word problems requiring exponential time. In: Proceedings 5th ACM Symposium on Theory of Computing (STOC 1973), pp. 1–9 (1973)
19. Gramlich, G., Schnitger, G.: Minimizing NFA's and regular expressions. J. Comput. Syst. Sci. **73**, 908–923 (2007)
20. Champarnaud, J.M., Coulon, F.: NFA reduction algorithms by means of regular inequalities. Theor. Comput. Sci. **327**, 241–253 (2004)
21. Han, Y.S., Wood, D.: Obtaining shorter regular expressions from finite-state automata. Theor. Comput. Sci. **370**, 110–120 (2007)

KOJAPH: Visual Definition and Exploration of Patterns in Graph Databases

Walter Didimo$^{(\boxtimes)}$, Francesco Giacchè, and Fabrizio Montecchiani

Università Degli Studi di Perugia, Perugia, Italy
{walter.didimo,fabrizio.montecchiani}@unipg.it, fgiacc@gmail.com

Abstract. We present KOJAPH, a new system for the visual definition and exploration of patterns in graph databases. It offers an expressive visual language integrated in a simple user interface, to define complex patterns as a combination of topological properties and node/edge attribute properties. Users can also interact with the query results and visually explore the graph incrementally, starting from such results. From the application perspective, KOJAPH has been designed to run on top of every desired graph database management system (GDBMS). As a proof of concept, we integrated it with NEO4J, the most popular GDBMS.

1 Introduction

Graph databases are of growing interest in the many application domains where data are conveniently modeled as graphs [1,5]. In contrast to relational databases, a graph database allows users to directly execute graph-like queries, such as finding pairs of nodes that are connected by a "short" path, finding the common neighbors of two specific nodes, finding cycles or cliques including a desired subset of nodes, and so on. On graph-structured data, this approach leads to a more efficient extraction process, which does not require the expensive join operations necessary on a relational database. For a survey on query languages for graph databases see, e.g., [10].

Besides the use of new paradigms for storing graph-structured data and retrieving information from them, a complementary line of research focuses on the design of visual languages and systems that allow users to easily define queries for extracting desired information. These tools are particularly valuable when the user wants to look for specific patterns in the data set without learning the native query language of the database management system. Within this research line, GRAPHITE is a system that allows users to visually construct graph patterns and to subsequently apply exact or approximate pattern matching algorithms to extract the results [4]. However, this system has several limitations: (*i*) it is not designed to directly work on top of widely used graph database technologies; (*ii*) its visual language is quite simple and does not allow for the creation of sophisticated patterns; (*iii*) the interaction of the

---

Research supported in part by the MIUR project AMANDA "Algorithmics for MAssive and Networked DAta", prot. 2012C4E3KT_001.

E. Di Giacomo and A. Lubiw (Eds.): GD 2015, LNCS 9411, pp. 272–278, 2015.
DOI: 10.1007/978-3-319-27261-0_23

user with the presented results is rather limited. Conversely, QGRAPH is a more complete visual query language for graphs [3]. It is used in the knowledge discovery system PROXIMITY (https://kdl.cs.umass.edu/display/public/Proximity), which allows users to easily understand and modify large relational data sets. However, PROXIMITY has it own data-structures to efficiently store and retrieve relational data, and it is not conceived to interact with modern and widely used GDBMS, like NEO4J, TITAN, and so on. There are also other technologies in this field, that are however not conceived for graph-structured data. Among them, the system POLARIS offers a visual query language for describing a wide range of table-based graphical presentations of data, extracted from multidimensional relational databases [9]. IMMENS is a system for real-time visual querying of big data [8]. System architectures and algorithms that are mainly designed to efficiently interleave visual query formulation and graph query processing are also described in the literature [2,7].

This paper presents KOJAPH, a new system for the visual definition and exploration of patterns in graph databases. KOJAPH has the following main features: (a) It offers an expressive visual language integrated in an intuitive user interface, to define complex patterns as a combination of topological properties and node/edge attribute properties. (b) Users can interact with the query results, which can be used as seeds for subsequent incremental explorations of the graph. (c) It is designed to work with any graph database management system (GDBMS) and the user can access it with a common Web browser.

Section 2 describes the KOJAPH visual language and user interface. Section 3 presents the system architecture and its integration with NEO4J (http://neo4j. com/). Future work is discussed in Sect. 4. A demo version of KOJAPH is available at: http://mozart.diei.unipg.it:8080/Kojaph/.

2 Visual Language and User Interface

Denote by G the entire graph stored in the graph database. The user can construct a desired pattern to be matched in G, using a graphical interface that integrates all the logical elements of the visual query language. At a high-level view, a pattern P consists of a pair $\langle G_P, R_P \rangle$ of specifications, where $G_P = (V_P, E_P)$ is a graph that defines the topological structure of P, and R_P is a set of rules on the nodes and the edges of G_P. An edge $e \in E_P$ does not necessarily correspond to a single edge of G, but it can also correspond to a path whose length is within a desired range. This correspondence can be established by a specific type of rules of R_P, which we call *path constraints*. The other types of rules in R_P are used to describe desired properties for node/edge attributes of G_P; these properties can then be combined with logical operators AND, OR, NOT to form a binary tree, called the *properties tree*.

Structure of the Interface. The first time the user accesses the graphical interface, the system automatically retrieves from the database all types of node and edge attributes. The interface is shown in Fig. 1. The left-side panel, called

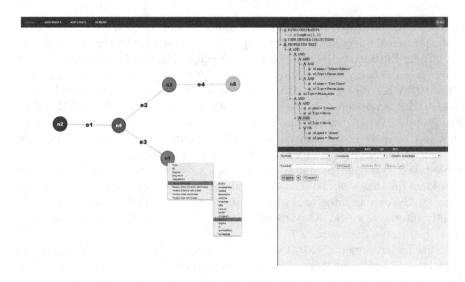

Fig. 1. The interface of KOJAPH for graph pattern definition.

graph-ed panel, is a canvas for editing G_P. Similarly to a common graph editor, it allows users to add, remove, select, or move nodes and edges; multiple edges and self-loops are allowed and are automatically drawn avoiding overlaps. A self-loop on a node v can be useful, for instance, to refer to a cycle that passes through v. Each time a new node is added, it is automatically assigned a unique label (identifier). The right-side of the interface is used to define the rules of R_P and it is further subdivided into a bottom panel and a top panel. The bottom panel, called the *prop-def panel*, is used to define desired properties of node/edge attributes; the top panel is used to group and combine them, so to form the properties tree above mentioned. The top panel also reports the path constraints and user-defined *collections* of attribute values, which can be used to construct properties.

Attributes. Each node or edge of G_P has an associated list of attributes. Generic attributes include the *type* of the element in the database (e.g., in a movie database, the type of a node can be "movie", "actor", "director", etc.), its *identifier* in the database, its *degree* (in/out-degree or total degree) if the element is a node, and its *direction* if the element is an edge. Specific attributes depend on the data modeled by the graph: for instance, in a movie database, a node of type "actor" might have attributes like "name","birthday", or "biography"; an edge connecting an actor to a movie might have an attribute "acts-in", whose value is the character interpreted by the actor in the movie. For any edge e of G_P, there are also additional attributes that are related to path constraints, which can be used when e corresponds to a path Π_e instead of a single edge. In particular, there are attributes whose values define the minimum/maximum length of Π_e and attributes that refer to the nodes and edges in Π_e. For example, the *sub-node properties* of e can be used to define rules on all nodes, any node, a single node, or no nodes of Π_e. Analogously, the attribute *sub-edge properties* of e is used to define rules on all edges, any edge, a single edge, or no edges of Π_e.

Property Definition. To define a property the user must switch to the "select" mode on the graph-ed panel, so to avoid modifications of G_P during the definition process. The property is defined as follows: (*i*) The user can see the list of attributes for a node/edge of G_P by clicking with the mouse right-button on it. (*ii*) A selected attribute is added to the prop-def panel. (*iii*) Attributes can be correlated to specific user-input or constant values, or subset of values like the above mentioned collections; they can also be combined together to form complex expressions, using a variety of operators, accessible from the list of "symbols" in the graphical interface. These symbols consist of *comparison, inclusion,* and *mathematical* operators, including parenthesis to define association rules and arrays. It is also possible to associate an attribute value with a regular expression (using the $\sim=$ operator). Each time the user adds an element (attribute, value, or symbol) to the property, it is visually appended to the right of the previous ones; the user can freely reorder the elements by means of drag-and-drop operations.

Properties Tree. Once a property is defined, the user can add it to the properties tree. The system will just append a new property at the root level. However, in a valid tree, each property must be a leaf node. To this aim, the user can add to the tree a suitable number of internal *operator nodes*, each corresponding to a boolean operator, and then he/she can make a property as a child of an operator node, by means of a drag-and-drop operation. When the user sends the query to the GDBMS, the system first checks the validity of the tree and, in the positive case, it translates the query constructed with the visual language into a query defined with the native language of the GDBMS.

Example. Figure 1 shows a pattern $\langle G_P, R_P \rangle$ defined on a movie database where persons (actors, directors, etc.) are connected to movies. The pattern G_P consists of 5 nodes and 4 edges. The properties tree and the path constraints on the right-hand side describe the set of rules. G_P describes an actor, node n4, such that: (*i*) n4 acted in a movie together with "Monica Bellucci", the node n2; this is expressed by a path constraint that requires that e1 is a path of length 2. (*ii*) n4 acted in at least one "Comedy" (node n1) and at least one "Horror" or "Action" (node n3), together with "Tom Cruise", who is node n5. The rules of R_P are summarized by the following expression:

```
(e1.Length in [2,2]) AND
(n4.Type=Person.Actor AND
 (n2.Type=Person.Actor AND n2.name="Monica Bellucci") AND
 (n5.Type=Person.Actor AND n5.name="Tom Cruise"))
 AND
((n1.Type=Movie AND n1.genre="Comedy") AND
 (n3.Type=Movie AND (n3.genre="Horror" OR n3.genre="Action")))
```

Presentation and Exploration of the Results. The system shows the results of the query in a different window. Multiple results that match the pattern are listed in a pop-up menú and the user can display them one by one. Each result is viewed as a graph isomorphic to G_P, with the same node/edge labels and colors as in G_P, so to preserve the user's mental map. Edges corresponding to paths are depicted as dashed segments. An important option is the possibility of displaying all the results as a unique graph: KOJAPH merges all of them without duplicating elements. Graphs are automatically drawn by the force-directed algorithm of the *D3.js* library (http://d3js.org/); a post-processing procedure is applied to represent multiple edges as non-overlapping curves. Figure 2 shows the results of the query for the pattern of Fig. 1, drawn as a unique graph.

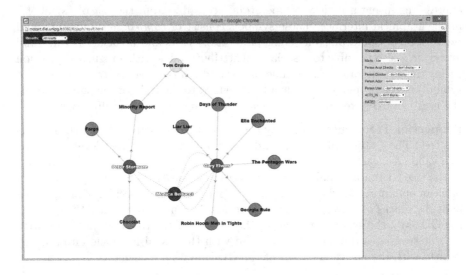

Fig. 2. The results of the query for the pattern of Fig. 1, shown as a unique graph.

The user can interact with the result in different ways. He/she can change at any time the kind of information displayed as node and edge labels. In the figure, actors are labeled with their names and movies with their titles, while edges are not labeled. A mouse-over interaction on an element will show all attribute values of that element. Zooming in/out (using the mouse wheel) and node adjustments are possible. More importantly, the user can explore the displayed result, incrementally enriching it with additional information. Double clicking on a node v, the system visualizes the neighbors of v not already in the drawing. Double clicking on a dashed edge e, the system replaces e with its associated path. To keep clear the original pattern throughout the exploration, new nodes that enter in the drawing have smaller size than the pattern nodes, and the new edges that enter in the drawing for a double click on a node v gets the color of v.

3   System Architecture and Integration with NEO4J

KOJAPH is a client-server Web application. The user interactive graphical environment is implemented using JavaScript, JQuery, and AJAX. From the server-side, client requests are handled by a Java servlet. In order to integrate the KOJAPH server with any GDBMS, we defined a Java abstract class, named DBMSInterface, that describes few methods to interact with the GDBMS, including methods for automatically retrieving node/edge attributes and a method to translate any query constructed with the visual language of KOJAPH into a query defined with the language of the GDBMS. These methods exchange data (input parameters and output values) in the JSON lightweight data-interchange format. A specific implementation of DBMSInterface must be provided for each specific GDBMS to be used.

As a proof of concept, we integrated KOJAPH to the popular NEO4J GDBMS. We implemented the methods of class DBMSInterface using the NEO4J query language *Cypher*, and we exploited the REST API of NEO4J for sending the queries to the GDBMS. The robustness of the integrated system and the correctness of the query answers have been tested on three databases. One is a movie/actor network of $63,042$ nodes and $106,651$ edges (http://neo4j.com/develo per/example-data/); another is a "food network" with approximately the same number of nodes but many more edges, namely $626,641$ (http://blog.bruggen. com/2013/12/fascinating-food-networks-in-neo4j.html). The third database is a portion of the co-authorship network in Computer Science extracted from DBLP (http://dblp.uni-trier.de/); it consists of $198,830$ nodes and $207,793$ edges. The time required to translate a KOJAPH query to the corresponding Cypher query is negligible; hence the response time of the system to a visual query is mainly related to the performances of the graph pattern matching algorithms applied by the GDBMS. Recall that the graph pattern matching problem is NP-hard [6], but the specification of node/edge attribute properties in addition to topological requirements may strongly reduce the search space and consequently greatly improve the GDBMS performances. Furthermore, path constraints in KOJAPH are translated into a pre-processing filtering of the possible results, which often leads to a dramatic reduction of the response time. The query in the example of Fig. 2 took less than 1 s under an Ubuntu Linux OS, within a VMware virtual machine with a 4 vCPU processor and 16 GB RAM.

4   Future Work

In the near future we plan to: (*i*) equip KOJAPH with more functionalities to visualize and explore the results; (*ii*) evaluate the usability of KOJAPH versus similar systems (e.g., PROXIMITY); (*iii*) testing KOJAPH on different GDBMS, other than Neo4J.

References

1. Angles, R., Gutiérrez, C.: Survey of graph database models. ACM Comput. Surv. **40**(1), 1–39 (2008)
2. Bhowmick, S.S., Choi, B., Zhou, S.: VOGUE: towards A visual interaction-aware graph query processing framework. In: CIDR 2013 (2013)

3. Blau, H., Immerman, N., Jensen, D.: A visual language for querying and updating graphs. Technical report UM-CS-2002-037, University of Massachusetts Amherst, Computer Science Department
4. Chau, D.H., Faloutsos, C., Tong, H., Hong, J.I., Gallagher, B., Eliassi-Rad, T.: GRAPHITE: a visual query system for large graphs. In: ICDM 2008, pp. 963–966. IEEE (2008)
5. Dominguez-Sal, D., et al.: Survey of graph database performance on the HPC scalable graph analysis benchmark. In: Shen, H.T., et al. (eds.) WAIM 2010. LNCS, vol. 6185, pp. 37–48. Springer, Heidelberg (2010)
6. Gallagher, B.: Matching structure and semantics: a survey on graph-based pattern matching. Artif. Intell. **6**, 45–53 (2006)
7. Hung, H.H., Bhowmick, S.S., Truong, B.Q., Choi, B., Zhou, S.: QUBLE: towards blending interactive visual subgraph search queries on large networks. VLDB J. **23**(3), 401–426 (2014)
8. Liu, Z., Jiang, B., Heer, J.: imMens: Real-time visual querying of big data. Comput. Graph. Forum **32**(3), 421–430 (2013)
9. Stolte, C., Tang, D., Hanrahan, P.: Polaris: a system for query, analysis, and visualization of multidimensional databases. Commun. ACM **51**(11), 75–84 (2008)
10. Wood, P.T.: Query languages for graph databases. SIGMOD Record **41**(1), 50–60 (2012)

Drawings with Crossings

2-Layer Fan-Planarity: From Caterpillar to Stegosaurus

Carla Binucci[1], Markus Chimani[2], Walter Didimo[1], Martin Gronemann[3],
Karsten Klein[4], Jan Kratochvíl[5], Fabrizio Montecchiani[1(✉)],
and Ioannis G. Tollis[6]

[1] Università Degli Studi di Perugia, Perugia, Italy
{carla.binucci,walter.didimo,fabrizio.montecchiani}@unipg.it
[2] Osnabrück University, Osnabrück, Germany
markus.chimani@uni-osnabrueck.de
[3] University of Cologne, Cologne, Germany
gronemann@informatik.uni-koeln.de
[4] Monash University, Melbourne, Australia
karsten.klein@monash.edu
[5] Charles University, Prague, Czech Republic
honza@kam.mff.cuni.cz
[6] University of Crete and Institute of Computer Science-FORTH,
Crete, Greece
tollis@csd.uoc.gr

Abstract. In a *fan-planar drawing* of a graph there is no edge that crosses two other independent edges. We study *2-layer fan-planar drawings*, i.e., fan-planar drawings such that the vertices are assigned to two distinct horizontal layers and edges are straight-line segments that connect vertices of different layers. We characterize 2-layer fan-planar drawable graphs and describe a linear-time testing and embedding algorithm for biconnected graphs. We also study the relationship between 2-layer fan-planar graphs and 2-layer right-angle crossing graphs.

1 Introduction

In a *2-layer drawing* of a graph, each vertex is drawn as a point of one of two distinct horizontal layers and each edge is drawn as a straight-line segment that connects vertices of different layers. Clearly, a graph admits such a drawing if and only if it is bipartite. The study of 2-layer drawings has a long tradition in Graph Drawing for two main reasons: (*i*) 2-layer drawings are a natural way

Research supported in part by the MIUR project AMANDA "Algorithmics for MAssive and Networked DAta", prot. 2012C4E3KT_001 and by the Australian Research Council through Discovery Project grant DP140100077. Work by Jan Kratochvíl was supported by the grant no. 14-14179S of the Czech Science Foundation GACR. The research in this work initiated at the Bertinoro Workshop on Graph Drawing 2015, supported by the European Science Foundation as part of the EuroGIGA collaborative research program (Graphs in Geometry and Algorithms).

© Springer International Publishing Switzerland 2015
E. Di Giacomo and A. Lubiw (Eds.): GD 2015, LNCS 9411, pp. 281–294, 2015.
DOI: 10.1007/978-3-319-27261-0_24

to visually convey bipartite graphs; (ii) algorithms that compute such drawings represent a building block for the popular *Sugiyama's framework* [17,18], used to draw graphs on multiple horizontal layers.

Since it is commonly accepted that edge crossings negatively affect the readability of a diagram, the study of 2-layer drawings has focused for a long time on the minimization of edge crossings. Eades *et al.* proved that a connected bipartite graph admits a crossing-free 2-layer drawing if and only if it is a *caterpillar* [9], i.e., a tree for which the removal of all vertices of degree one produces a path. Eades and Whitesides proved that the problem of minimizing edge crossings in a 2-layer drawing is NP-hard [11] and, as a consequence, many papers focused on efficient heuristics or exact exponential techniques for computing 2-layer drawings with minimum number of edge crossings; a very limited list of these papers includes [10,12,14,16,19].

More recently, a growing attention has been devoted to the study of graph drawings where edge crossings are allowed under some specific restrictions, which still guarantee a good readability of the layout. In particular, motivated by cognitive experiments of Huang *et al.* [13], several papers investigated *right angle crossing drawings* (*RAC drawings* for short) [7], in which the edges can cross only at right angles (see [8] for a survey on the subject). Di Giacomo *et al.* characterized the class of bipartite graphs that admit a RAC drawing on two layers, and described a linear-time testing and embedding algorithm for 2-layer RAC drawable graphs [5]. Heuristics for computing the maximum 2-layer RAC subgraph of a given graph are also described [6].

In this paper we concentrate on 2-*layer fan-planar drawings*, i.e., 2-layer drawings that are also *fan-planar*. In a fan-planar drawing an edge can only cross edges having a common end-vertex, thus an edge cannot cross two independent edges (see Fig. 1). Fan-planar drawings were introduced by Kaufmann and Ueckerdt [15], who showed that fan-planar graphs with n vertices have at most $5n - 10$ edges, which is a tight bound. Subsequent papers proved that recognizing fan-planar graphs is NP-hard and studied restricted classes of fan-planar graphs in terms of density and recognition algorithms [1,2]. In particular, it is shown that 2-layer fan-planar drawings have at most $2n - 4$ edges (still a tight bound) [2]. From an application perspective, it has been observed that fan-planar drawings may be used to create confluent drawings with few edge crossings per edge [2]. Our contribution is as follows:

(i) We first study biconnected graphs (Sect. 3). We prove that a biconnected graph is 2-layer fan-planar if and only if it is a spanning subgraph of a *snake* graph (Sect. 3.1), which is a chain of complete bipartite graphs $K_{2,h}$ (see Definition 1). We also describe a linear-time algorithm that tests whether a biconnected graph admits a 2-layer fan-planar drawing, and that computes such a drawing if it exists (Sect. 3.2).

(ii) We then give a characterization of the class of graphs that admit a 2-layer fan-planar drawing (Sect. 4). We prove that a connected graph is 2-layer fan-planar if and only if it is a subgraph of a *stegosaurus* graph, a further generalization of a snake (see Definition 2). Since every 2-layer crossing-free drawing is also fan-planar, but not vice versa, caterpillars are a proper subclass of stegosauruses.

(iii) We explore the relationship between 2-layer fan-planar and 2-layer RAC drawable graphs (Sect. 5). We prove that, for biconnected graphs the first class is properly included in the second one, while there is no inclusion relationships for general graphs.

2   Preliminaries

We assume familiarity with basic concepts of graph drawing and planarity [4]. Throughout the paper, a graph with a fixed planar (outerplanar) embedding is also called a *plane* (*outerplane*) graph. Let G be a graph. For each vertex v of G, the set of edges incident to v is called the *fan* of v. Each edge (u, v) of G belongs to the fan of u and to the fan of v at the same time. Two edges that do not share a vertex are called *independent edges*, and always belong to distinct fans. A *fan-planar drawing* Γ of G is a drawing such that: (*a*) no edge is crossed by two independent edges (the forbidden configuration of Fig. 1(a)); (*b*) there are not two adjacent edges (u, v), (u, w) that cross an edge e from different "sides" while moving from u to v and from u to w (the forbidden configuration of Fig. 1(b)). Two allowed configurations of a fan-planar drawing are in Fig. 1(c) and (d). A *fan-planar graph* is a graph that admits a fan-planar drawing. Observe that in a straight-line drawing, the forbidden case (*b*) cannot happen. By definition, a fan-planar drawing does not contain 3 mutually crossing edges.

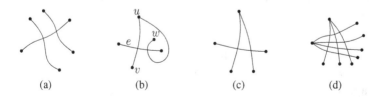

Fig. 1. (a)-(b) Forbidden and (c)-(d) allowed configurations of fan-planar drawings.

In a 2-*layer drawing* of a graph each vertex is drawn as a point on one of two distinct horizontal lines, called *layers*, and each edge is drawn as a straight-line segment that connects vertices of different layers. A 2-*layer fan-planar drawing* is a 2-layer drawing that is also fan-planar. A 2-*layer fan-planar graph* is a graph that admits a 2-layer fan planar drawing. Clearly, every graph that has a 2-layer drawing is bipartite. For a given 2-layer drawing of a bipartite graph $G = (V_1, V_2, E)$, denote by ℓ_i the horizontal line on which the vertices of V_i are drawn ($i = 1, 2$). We always assume that ℓ_1 is above ℓ_2. Two 2-layer drawings of G are equivalent if they have the same left-to-right order π_i of the vertices of V_i along ℓ_i ($i = 1, 2$). A 2-*layer embedding* is an equivalence class of 2-layer drawings and it is described by a pair of linear orderings (i.e., permutations) $\gamma = (\pi_1, \pi_2)$

of the vertices in V_1 and V_2, respectively. Let u and v be two vertices of V_i, we write $u \prec v$ if $\pi_i(u) < \pi_i(v)$ ($i = 1, 2$). Also, the first (last) vertex of π_1 and the first (last) vertex of π_2 are the *leftmost vertices* (*rightmost vertices*) of γ. The edge between the leftmost (rightmost) vertices of γ (if it exists) is called the *leftmost edge* (the *rightmost edge*) of γ. If Γ is a drawing within class γ, we say that γ is the *embedding* of Γ. If Γ is a 2-layer fan-planar drawing, we also say that γ is a *2-layer fan-planar embedding*.

Since any geometric position of the vertices that respects the two linear orderings defined by γ yields a 2-layer fan-planar drawing in linear time, we will concentrate on embeddings in the following. We say that γ is *maximal* if for any two vertices u and v that are not adjacent in G, the embedding obtained from γ by adding the edge (u, v) is no longer 2-layer fan-planar.

3   Biconnected 2-Layer Fan-Planar Graphs

Let G_1 and G_2 be two graphs. The operation of merging G_1 and G_2 by identifying an edge e_1 of G_1 with an edge e_2 of G_2 (in one of the two possible ways) is called an *edge merging*; the resulting graph G is called a *merger* of G_1 and G_2 with respect to e_1, e_2. The end-vertices of the edge obtained by identifying e_1 with e_2 are *merged vertices* of G. In Fig. 2(a), the white vertices in the merger graph are the merged vertices.

Definition 1. *A* snake *is a graph recursively defined as follows: (i) A complete bipartite graph $K_{2,h}$ ($h \geq 2$) is a snake; (ii) A merger of two snakes G_1 and G_2 with respect to edges e_1 of G_1 and e_2 of G_2, with the property that none of the end-vertices of e_i is a merged vertex of G_i ($i = 1, 2$), is a snake.*

Intuitively, a snake is a bipartite planar graph consisting of a chain of complete bipartite graphs $K_{2,h}$ (see Fig. 2(b)). An alternative definition of a snake can be derived from the definition of *ladder*, i.e., a maximal bipartite outerplanar graph consisting of two paths of the same length $\langle u_1, u_2, \ldots, u_{\frac{n}{2}} \rangle$ and $\langle v_1, v_2, \ldots, v_{\frac{n}{2}} \rangle$ plus the edges (u_i, v_i) ($i = 1, 2, \ldots \frac{n}{2}$) (see also [5]); the edges (u_1, v_1) and $\left(u_{\frac{n}{2}}, v_{\frac{n}{2}}\right)$ are called the *extremal edges* of the ladder. A snake is a planar graph obtained from an outerplane ladder, by adding, inside each internal face, an arbitrary number (possibly none) of paths of length two connecting a pair of non-adjacent vertices of the face.

3.1   Characterization

The characterization of the biconnected graphs that admit a 2-layer fan-planar embedding is given by Theorem 3. The proof is based on the next two lemmas.

Lemma 1. *Let G be biconnected graph. If G admits a maximal 2-layer fan-planar embedding γ then G is a snake.*

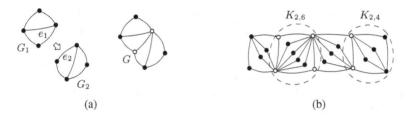

Fig. 2. (a) Edge merging of two graphs. (b) A snake.

Proof Sketch. Due to maximality, the leftmost and the rightmost edges of γ always exist, and do not cross any other edge. Therefore, γ contains at least two uncrossed edges. We prove the statement by induction on the number $l \geq 2$ of uncrossed edges in γ. Recall that, since G is biconnected, it has vertex degree at least two.

Base Case: $l = 2$. In this case, we prove that G is a $K_{2,r}$ for some $r \geq 2$, which implies that G is a snake. Note that G cannot be a $K_{1,r}$, since it has vertex degree at least two. If G contains only four vertices, then G is a $K_{2,2}$, as there are exactly two uncrossed edges. Suppose now that G has more than four vertices.

Claim 1. Let (u,v) and (w,x) be a pair of crossing edges in γ, such that $u \prec w$ on ℓ_1 and $x \prec v$ on ℓ_2. Then the edges (u,x) and (w,v) exist.

Claim 2. If G' is a subgraph of G such that G' is a $K_{2,r'}$ (for some $r' > 2$) and G' contains the leftmost and the rightmost edges of γ, then G is a $K_{2,r}$ (for some $r > r'$).

Using Claims 1 and 2, we now prove that G is a $K_{2,r}$, for some $r > 2$. Consider the rightmost vertex w on ℓ_1 and the rightmost vertex v on ℓ_2 in γ. Due to maximality, edge (w,v) exists and is uncrossed. Also, since w and v have degree at least two, they both have one more incident edge, which we denote by (w,x) and (u,v). Since w and v are the rightmost vertices, (w,x) and (u,v) cross each other, and thus, by Claim 1, edge (u,x) exists. Let H be the $K_{2,2}$ subgraph of G induced by u, v, x, and w. Since we are assuming that G has more than four vertices, there exists a vertex z other than the vertices of H. Without loss of generality, assume that z is on layer ℓ_2.

If (u,x) is the leftmost edge of γ, then $x \prec z \prec v$, and this implies that z can be adjacent to u and w only, as otherwise (w,x), (u,v), and an edge incident to z would form three mutually crossing edges. Also, since z has degree at least two, z is adjacent to both u and w. Thus subgraph G' of G induced by $\{u,v,w,x,z\}$ is a $K_{2,3}$ containing the left- and rightmost edges of γ. By Claim 2, G is a $K_{2,r}$, with $r > 2$.

If (u,x) is not the leftmost edge of γ, then (u,x) is crossed in γ, and, as observed in the proof of Claim 1, it is crossed by an edge having either w or v

as an end-vertex. Without loss of generality, suppose that (u, x) crosses an edge (w, z). By applying Claim 1 to (u, x) and (w, z), edge (u, z) exists. Hence, again, the subgraph G' induced by the vertices of H plus z is a $K_{2,3}$ graph. If (u, z) is the leftmost edge of γ, then by Claim 2, G is a $K_{2,r}$, with $r > 2$. If (u, z) is not the leftmost edge, then again it is crossed by an edge having either w or v as an end-vertex. However, since (u, x) is already crossed by (w, z), (u, z) can only be crossed by edges having w as an end-vertex. Denoted by (w, y) one of the edges that cross (u, z), we have that edge (u, y) exists by Claim 1, and therefore the subgraph induced by the vertices of H plus vertices z and y is a $K_{2,4}$ that contains the rightmost edge of γ. By iterating this argument, we eventually obtain a subgraph $K_{2,r'}$ $(r' > 2)$ of G that contains the rightmost and also the leftmost edge of γ, which by Claim 2 implies that G is a $K_{2,r}$, with $r > 2$.

Inductive Case: $l > 2$. Consider an uncrossed edge (u, v) different from the leftmost and the rightmost edge of γ. Let γ_1 (resp., γ_2) be the embedding induced by the vertices to the left (resp., right) of (u, v) plus u and v. Clearly, γ_1 and γ_2 are 2-layer fan-planar. Let G_i be the subgraph of G consisting of the vertices and edges of γ_i $(i = 1, 2)$. Since (u, v) is uncrossed in γ, G_1 and G_2 are biconnected. Also, each of the two γ_i contains a number $l_i < l$ of uncrossed edges, and thus G_i is a snake by induction. Since G is a merger of G_1 and G_2 with respect to (u, v), G is a snake.   □

Lemma 2. *Every n-vertex snake admits a 2-layer fan-planar embedding, which can be computed in $O(n)$ time.*

Proof Sketch. Let G be a snake. By definition, G is a chain of graphs $G_1, \ldots, G_k$, such that each G_i is a complete bipartite graph K_{2,h_i} that shares a pair of merged vertices with G_{i+1} $(i = 1, \ldots, k-1)$. The idea to construct a 2-layer fan-planar embedding is to put the vertices of each partite set on the corresponding layer, such that those of G_i precede those of G_{i+1}. See Fig. 3 for an illustration.   □

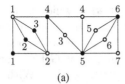

(a)

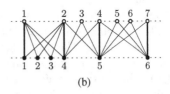

(b)

Fig. 3. Illustration for the proof of Lemma 2. (a) A snake G; the vertices of each partite set are ordered (i.e., numbered) according to the rules given in the proof. (b) A 2-layer fan-planar drawing of G whose embedding reflects the vertex ordering; the uncrossed edges are in bold.

Theorem 3. *A biconnected graph G is 2-layer fan-planar if and only if G is a spanning subgraph of a snake.*

Proof. Suppose first that G has a 2-layer fan-planar embedding γ. If γ is maximal, then G is a snake by Lemma 1. Else, there is a maximal 2-layer fan-planar embedding γ' of a graph G' such that: (i) $G \subset G'$, (ii) G' has the same vertex set of G, and (iii) the restriction of γ' to G coincides with γ. Hence, by Lemma 1, G is a spanning subgraph of a snake. Conversely, let G be a spanning subgraph of a snake. Since any spanning subgraph of a 2-layer fan-planar graph is also 2-layer fan-planar, G is 2-layer fan-planar by Lemma 2. □

3.2 Testing and Embedding Algorithm

We now describe an algorithm to test whether a given biconnected bipartite graph G is 2-layer fan-planar. Since every biconnected 2-layer fan-planar graph is a spanning subgraph of a snake (Theorem 3), the algorithm must check whether G can be augmented to a snake by only adding a suitable set of edges. In what follows we assume that the input graph G is not a simple cycle, as otherwise it is clearly 2-layer fan-planar.

A *chain* $P = \langle u, v_1, v_2, \ldots, v_k, v \rangle$ of G is a maximal path of G such that all its internal vertices v_i have degree 2 in G ($i = 1, \ldots, k$). *Contracting* P is to transform G into a new graph G' obtained from G by replacing P with a single edge $e_P = (u, v)$ of weight $w(e_P) = k$. Reversely, we can say that G is obtained from G' by *expanding* e_P (P is the *expansion* of e_P). Note that G' may have multiple edges that connect u and v. If G is a plane graph, we assume that the contraction of P preserves the embedding of G. The *weighted contraction* of G is the edge-weighted multi-graph $\mathcal{C}(G)$ obtained from G by contracting all inclusion-wise maximal chains of G; all edges of $\mathcal{C}(G)$ that are also in G are assigned weight 0. Figure 4(c) shows the weighted contraction of the graph in Fig. 4(b). Based on weighted contractions, we can reinterpret the characterization of 2-layer fan-planar graphs as follows (cf. Fig. 4):

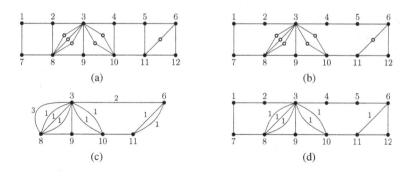

Fig. 4. Illustration for Lemma 4. (a) A plane snake $\overline{G}$ consisting of an outerplane ladder (black vertices) with arbitrary paths of length two inside each internal face. (b) A (plane) biconnected spanning subgraph G of $\overline{G}$. (c) The (plane) weighted contraction $\mathcal{C}(G)$; only edge weights greater than 0 are shown. (d) The plane multi-graph G^* of property (c) in the statement of Lemma 4.

Lemma 4. *Let G be a bipartite biconnected graph that is not a simple cycle. G is a spanning subgraph of a snake if and only if its weighted contraction $\mathcal{C}(G)$ has a planar embedding such that:* **(a)** *All vertices of $\mathcal{C}(G)$ are on the external face;* **(b)** *All edges e_P of $\mathcal{C}(G)$ with $w(e_P) \geq 2$ are on the external face;* **(c)** *Let G^* be the plane multi-graph obtained from $\mathcal{C}(G)$ by expanding all edges e_P of the external face. It is possible to add to G^* internal edges of weight 0, such that the resulting graph H^* is outerplane and the removal of the internal edges of weight 1 from H^* produces a ladder.*

We now give a linear-time algorithm, called `Bic2LFPTest`, that tests whether a bipartite biconnected graph G has a 2-layer fan-planar embedding, and that constructs such an embedding in the positive case. The algorithm checks whether $\mathcal{C}(G)$ admits a planar embedding with the properties (a), (b), and (c) of Lemma 4. If such an embedding exists, a snake for which G is a spanning subgraph is obtained by expanding the edges of weight 1 in the multi-graph H^* of property (c); a 2-layer fan-planar embedding of this snake (and hence of G) is obtained using the construction in the proof of Lemma 2.

Algorithm `Bic2LFPTest` (G)

Step 1. Compute the weighted contraction $\mathcal{C}(G)$ of G, and compute, if any, an outerplanar embedding of $\mathcal{C}(G)$ (i.e. property (a) of Lemma 4). This can be done in linear time: temporarily add to $\mathcal{C}(G)$ a dummy vertex u and a dummy edge (u, v) for every vertex v of $\mathcal{C}(G)$; then run a linear-time planarity testing and embedding algorithm (e.g. [3]) on it. Note that, since $\mathcal{C}(G)$ is still biconnected, the outerplanar embedding of $\mathcal{C}(G)$ is unique (if it exists), except for the permutation of multi-edges. If $\mathcal{C}(G)$ is not outerplanar, the whole test is negative and the algorithm stops, otherwise an outerplanar embedding is found and the algorithm goes to the next step.

Step 2. Check whether the outerplanar embedding can be modified (if needed) so that all edges with weight greater than 1 can be put on the external face (property (b) of Lemma 4), keeping all vertices on the external face. This is possible if and only if: (i) for every pair of consecutive vertices $\{u, v\}$ on the boundary of the external face there is at most one edge $e = (u, v)$ with $w(e) \geq 2$ (which can be then put on the external face), and (ii) there is no chord with weight greater than 1. Both conditions (i) and (ii) can be checked in linear time. If this checking fails, then the whole test is negative, otherwise the new outerplanar embedding with the heaviest edges on the external face is computed and the algorithm goes to the next step.

Step 3. Expand the external edges with weight greater than 0 to get the multi-graph G^* in property (c) of Lemma 4; this can be done in linear time if we suitably store the chain P associated with each edge e_P when $\mathcal{C}(G)$ is computed in Step 1. Then, check whether it is possible to add to G^* a suitable set of internal edges (chords) connecting vertices of the external face such that the resulting multi-graph H^* is still outerplane and becomes a ladder if we subsequently remove the internal edges of weight 1 (property (c)). This can be done with the following procedure. If H^* already contains a chord of weight 0, then: (i) temporarily remove the edges with weight 1; (ii) verify whether the resulting graph can be augmented with extra chords to an outerplane ladder, using the linear-time algorithm described by Di Giacomo et al. [5]. We remark that, if such an augmentation exists it is unique under the assumption that H^* already contains a chord of weight 0; (iii) check whether the removed edges with weight 1 can be reinserted inside the outerplane ladder without violating the planarity (which can be done in linear time). If H^* does not contain a chord with weight 0, then H^* contains at least one chord $e = (u, v)$ with weight 1 (we assumed that G is not a simple cycle, hence H^* contains at least one chord). In this case, consider the two vertices u_1, u_2 that are adjacent to u on the boundary of the external face, and the two vertices v_1, v_2 that are adjacent to v on the boundary of the external face (some of these vertices may coincide). It can be seen that any edge augmentation of H^* that leads to an outerplane ladder with the edges of weight 1 inside its internal faces, must include at least one chord $e' \in C = \{(u, v_1), (u, v_2), (v, u_1), (v, u_2)\}$ (in particular, in the outerplane ladder either two edges of C are chords or one is a chord and one is an extremal edge of the ladder). Hence, for each of these (at most four) chords e', try to add e' to H^* and then repeat the substeps $(i)-(iii)$ described above. If the augmentation fails for all possible choices of e', the whole test is negative, otherwise it is positive and a snake that contains G as a spanning subgraph is obtained. A 2-layer fan-planar embedding of this snake coincides with that of G, and is computed using the construction of Lemma 2.

Theorem 5. *Let G be a bipartite biconnected graph with n vertices. There exists an $O(n)$-time algorithm that tests whether G is 2-layer fan-planar, and that computes a 2-layer fan-planar embedding of G in the positive case.*

4   Simply Connected 2-Layer Fan-Planar Graphs

We saw that a biconnected graph is 2-layer fan-planar if and only if it is a subgraph of a snake. We now show that a (simply) connected graph is 2-layer fan-planar if and only if it is a subgraph of a *stegosaurus*. Clearly, a non-connected graph is 2-layer fan-planar if and only if every connected component is a 2-layer fan-planar graph.

Recall that snakes are obtained by merging edges of a sequence of several $K_{2,h}$ ($h \geq 2$). We may denote the partite set with more than 2 vertices (if any) the *large side* of a $K_{2,h}$. Given a snake G, a vertex in G is *mergeable* if it is an end-vertex of a mergeable edge and belongs to the large side of an original $K_{2,h}$. Note that a snake always has at most two mergeable vertices; by definition, a $K_{2,2}$ on either end of the snake prohibits a mergable vertex. The graph resulting from merging two graphs G_1 and G_2 by identifying a mergeable vertex of G_1 with a mergeable vertex of G_2 is a *vertex merger*.

Definition 2. *A* stegosaurus *is either a fan (a trivial stegosaurus) or a graph recursively defined as follows (Fig. 5(a)): (i) A snake is a stegosaurus, whose* *mergeable vertices are the mergeable vertices of the snake. (ii) The vertex merger of two stegosaurs G_1, G_2 is a stegosaurus. Its mergeable vertices are those (at most one per G_1, G_2) not used in this merging. (iii) Let v be a mergeable vertex of a stegosaurus G_1. Adding a new vertex v' and an edge (v, v') gives a stegosaurus with the same mergeable vertices as G_1.*

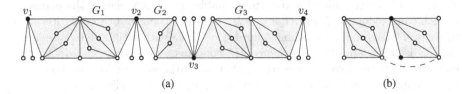

(a)   (b)

Fig. 5. (a) A stegosaurus composed of three snakes G_1, G_2, G_3 that have been merged at v_2, v_3 and several edges have been attached to $v_1, \ldots, v_4$. (b) The result of merging snakes G_1, G_2 using a non-mergeable vertex can be augmented into one snake by adding the dashed edge.

Observation 6. *Consider merging two snakes G_1, G_2 at vertices v_1, v_2. Assume that v_1 is an end-vertex of a mergeable edge but not from a large side; v_2 may be chosen as v_1 or be a mergeable vertex. Then, the merged graph would be a subgraph of a snake (Fig. 5(b)). Thus, only vertices from the large side have to be considered in Definition 2.*

In the following, a block of a graph (i.e., a biconnected component) is called *trivial* if it consists of a single edge. Let an edge e be a trivial block. If e has an end-vertex of degree 1, e is a *stump*, otherwise, it is a *bridge*. A graph is called *maximal 2-layer fan-planar*, if it cannot be augmented by an edge without losing 2-layer fan-planarity. Observe that, in contrast to the biconnected case, we have the situation that an embedding (or drawing) of G is maximal 2-layer fan-planar (i.e., we cannot add an edge within this embedding), but the graph is not maximal 2-layer fan-planar; it "simply" requires a different 2-layer fan-planar embedding into which we can add another edge. Figure 6(a),(b) show an example. By definition and Theorem 3, a biconnected graph is 2-layer fan-planar if and only if it is the subgraph of a snake, and thus, of a stegosaurus. Also, a simply connected graph that is a subgraph of a snake is 2-layer fan-planar. We will first show that stegosaurs are 2-layer fan-planar. Then, we will show that every 2-layer fan-planar graph is a subgraph of a stegosaurus.

Lemma 7. *Every stegosaurus has a 2-layer fan-planar embedding.*

Proof. Figure 6(c) outlines the idea. We already know that snakes are 2-layer fan-planar and how to draw them, and, by definition, that the non-trivial blocks of a stegosaurus are snakes. Drawing a stegosaurus hence means drawing the individual snakes and realizing that we can draw additional trivial blocks (arising from (iii) in the definition) at the left and right "ends" of the stegosaurus, as well as at its cut vertices.   □

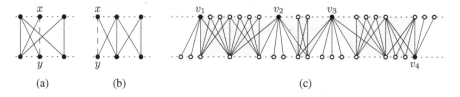

(a)   (b)   (c)

Fig. 6. (a) A maximal 2-layer fan-planar drawing and (b) a different embedding to which one may add the edge (x, y). (c) A 2-layer fan-planar drawing of the stegosaurus from Fig. 5(a).

It should be understood that a trivial stegosaurus is a maximal 2-layer fan-planar graph. In the following, we only have to consider non-trivial stegosaurs. We start with proving a property that holds for all 2-layer fan-planar drawings, not only for maximal ones:

Lemma 8. *Let B be a non-trivial block of a 2-layer fan-planar graph G, and e an independent edge, i.e., none of its end-vertices belongs to B. No edge of B can be crossed by e in any 2-layer fan-planar embedding of G.*

Proof. Assume there is an embedding where some edge $b \in E(B)$ is crossed by e. Since B is a non-trivial block, b is part of a cycle $C \subseteq E(B)$ with $|C| \geq 4$. Hence, by the properties of 2-layer embeddings, e needs to cross another edge $c \in C$ as well. The edges b, c need to be adjacent, as otherwise we would get pairwise crossings between three independent edges. Embedding a cycle, in our case C, on two layers, requires a crossing of every edge except for two non-adjacent edges. Hence either b or c will have a crossing with another edge of C and the independent edge e, a contradiction. □

From the above lemma, we obtain a simple but useful observation:

Corollary 9. *In a 2-layer fan-planar embedding, two non-trivial blocks cannot cross.*

Hence we know that in a 2-layer fan-planar drawing, non-trivial blocks are "nicely" placed next to each other from left to right without crossings between them. We now show several properties of maximal 2-layer fan-planar graphs. Clearly, a maximal 2-layer fan-planar graph will be connected.

Lemma 10. *Let G be a maximal 2-layer fan-planar graph. There exists an embedding γ of G in which no stump is crossed.*

Proof Sketch. One may assume that a vertex is incident to at most one stump. Choosing an embedding γ with the least crossing count between stumps, yields the result. □

Lemma 11. *A maximal 2-layer fan-planar graph G does not contain bridges.*

(a)   (b)

Fig. 7. (a) The tree T_3; it is 2-layer RAC but not 2-layer fan-planar. (b) A 2-layer RAC embedding (not drawing) of T_3 In both figures the path connecting u to v has bold edges.

Proof Sketch. Using the embedding from Lemma 10, one can show that if a bridge exists, it is not crossed, and one may insert an edge, contradicting maximality of G.   □

Corollary 12. *Let G be a maximal 2-layer fan-planar graph. There exists an embedding in which no two blocks cross. Any cut vertex is either contained in two non-trivial blocks, or is a left- or rightmost vertex in this embedding.*

Hence we have that a maximal 2-layer fan-planar graph allows a drawing where non-trivial blocks are neither crossed by other non-trivial nor by trivial blocks. Furthermore, in contrast to the non-biconnected case, if an embedding of a biconnected graph G is maximal 2-layer fan-planar, then G is maximal 2-layer fan-planar. We can deduce:

Corollary 13. *Let G be a maximal 2-layer fan-planar graph. Its non-trivial blocks are maximal 2-layer fan-planar biconnected graphs, i.e., snakes.*

Lemma 7 and Corollary 13 imply the following.

Theorem 14. *A graph is 2-layer fan-planar if and only if it is a subgraph of a stegosaurus.*

5   Relationship with 2-Layer RAC Drawings

It is natural to ask for the relationship between 2-layer fan-planarity and 2-layer RAC. Di Giacomo *et al.* proved that a 2-layer embedding γ is RAC (i.e., there exists a 2-layer RAC drawing w.r.t. γ) if and only if γ has neither 3 mutually crossing edges nor two adjacent edges crossed by a third one [5]. For example, the embedding in Fig. 7(b) is 2-layer RAC. They also showed that a biconnected graph has a 2-layer RAC embedding if and only if it is a subgraph of a ladder. Since a ladder is a special snake (but not vice versa), we deduce from Theorem 3:

Corollary 15. *The biconnected 2-layer RAC graphs are a proper subclass of the biconnected 2-layer fan-planar graphs.*

For general graphs, however, there is no inclusion relationship between those two concepts. In particular, we exhibit infinitely many trees T_k ($k \geq 3$) that are 2-layer RAC but not 2-layer fan-planar. T_k consists of two vertices u and v connected by a path of length $k \geq 3$, and such that each u and v have further (disjoint) three paths of length $k + 1$ attached to them. Figure 7(a) depicts T_3. Using the characterization of 2-layer RAC trees [5], one can verify that T_k has a 2-layer RAC embedding, see Fig. 7(b).

By Theorem 14, we can show that T_k is not 2-layer fan-planar by observing that it cannot be a subgraph of a stegosaurus. Indeed, suppose that G is some stegosaurus that contains T_k, and suppose that Γ is a planar drawing of G as in Fig. 5(a), where all vertices of degree greater than two lie on the external face and are suitably placed on two distinct horizontal lines. Since u and v have degree 4 in T_k, they are external vertices of Γ. Denote by P_{uv} the path from u to v in Γ that corresponds to the path from u to v in T_k. Consider the three paths of length $k+1$ attached to u in T_k. Since they only share vertex u, and also share only vertex u with P_{uv}, one of them, call it P_u, is necessarily "routed towards" v in Γ, while the other two can be routed away from v. Analogously, one of the three paths of length $k + 1$ attached to v, call it P_v, must be routed towards u in Γ, while the other two can be routed away from u. Since P_{uv} has length k, it is not difficult to verify that either P_u and P_v must share a vertex or at least one of them share a vertex with P_{uv}; a contradiction. Thus, G cannot exist.

6   Open Problems

The main open problem of our study is to provide, if any, an efficient 2-layer fan-planarity testing algorithm for general (i.e., not necessarily biconnected) graphs, which exploits Theorem 14. Another interesting research line is designing algorithms that compute 2-layer drawings that are "as fan-planar as possible", i.e., whose number of forbidden configurations (two independent edges crossed by a third one) is minimized.

References

1. Bekos, M.A., Cornelsen, S., Grilli, L., Hong, S.-H., Kaufmann, M.: On the recognition of fan-planar and maximal outer-fan-planar graphs. In: Duncan, C., Symvonis, A. (eds.) GD 2014. LNCS, vol. 8871, pp. 198–209. Springer, Heidelberg (2014)
2. Binucci, C., Di Giacomo, E., Didimo, W., Montecchiani, F., Patrignani, M., Symvonis, A., Tollis, I.G.: Fan-planarity: properties and complexity. Theor. Comput. Sci. **589**, 76–86 (2015)
3. Booth, K., Lueker, G.: Testing for the consecutive ones property, interval graphs and graph planarity using PQ-trees. J. Comput. Syst. Sci. **13**, 335–379 (1976)
4. Di Battista, G., Eades, P., Tamassia, R., Tollis, I.G.: Graph Drawing. Prentice Hall, Upper Saddle River (1999)
5. Di Giacomo, E., Didimo, W., Eades, P., Liotta, G.: 2-layer right angle crossing drawings. Algorithmica **68**(4), 954–997 (2014)
6. Di Giacomo, E., Didimo, W., Grilli, L., Liotta, G., Romeo, S.A.: Heuristics for the maximum 2-layer RAC subgraph problem. Comput. J. **58**(5), 1085–1098 (2015)
7. Didimo, W., Eades, P., Liotta, G.: Drawing graphs with right angle crossings. Theor. Comput. Sci. **412**(39), 5156–5166 (2011)

8. Didimo, W., Liotta, G.: The crossing angle resolution in graph drawing. In: Pach, J. (ed.) Thirty Essays on Geometric Graph Theory, pp. 167–184. Springer, New York (2012)

9. Eades, P., McKay, B., Wormald, N.: On an edge crossing problem. In: ACSC 1986, pp. 327–334 (1986)

10. Eades, P., Kelly, D.: Heuristics for drawing 2-layered networks. Ars Comb. **21**, 89–98 (1986)

11. Eades, P., Whitesides, S.: Drawing graphs in two layers. Theor. Comput. Sci. **131**(2), 361–374 (1994)

12. Eades, P., Wormald, N.C.: Edge crossings in drawings of bipartite graphs. Algorithmica **11**(4), 379–403 (1994)

13. Huang, W., Eades, P., Hong, S.: Larger crossing angles make graphs easier to read. J. Vis. Lang. Comput. **25**(4), 452–465 (2014)

14. Jünger, M., Mutzel, P.: 2-layer straightline crossing minimization: performance of exact and heuristic algorithms. J. Graph Algorithms Appl. **1**, 1–25 (1997)

15. Kaufmann, M., Ueckerdt, T.: The density of fan-planar graphs. CoRR abs/1403.6184 (2014). http://arxiv.org/abs/1403.6184

16. Mutzel, P.: An alternative method to crossing minimization on hierarchical graphs. SIAM J. Optim. **11**(4), 1065–1080 (2001)

17. Sugiyama, K.: Graph Drawing and Applications for Software and Knowledge Engineers. World Scientific, Singapore (2002)

18. Sugiyama, K., Tagawa, S., Toda, M.: Methods for visual understanding of hierarchical system structures. IEEE Trans. Syst. Man Cybern. **11**(2), 109–125 (1981)

19. Valls, V., Martí, R., Lino, P.: A branch and bound algorithm for minimizing the number of crossing arcs in bipartite graphs. Eur. J. Oper. Res. **90**(2), 303–319 (1996)

Recognizing and Drawing IC-Planar Graphs

Franz J. Brandenburg[1], Walter Didimo[2], William S. Evans[3],
Philipp Kindermann[4(✉)], Giuseppe Liotta[2], and Fabrizio Montecchiani[2]

[1] Universität Passau, Passau, Germany
brandenb@fim.uni-passau.de
[2] Università Degli Studi di Perugia, Perugia, Italy
{walter.didimo,giuseppe.liotta,fabrizio.montecchiani}@unipg.it
[3] University of British Columbia, Vancouver, Canada
will@cs.ubc.ca
[4] Universität Würzburg, Würzburg, Germany
philipp.kindermann@uni-wuerzburg.de

Abstract. IC-planar graphs are those graphs that admit a drawing where no two crossed edges share an end-vertex and each edge is crossed at most once. They are a proper subfamily of the 1-planar graphs. Given an embedded IC-planar graph G with n vertices, we present an $O(n)$-time algorithm that computes a straight-line drawing of G in quadratic area, and an $O(n^3)$-time algorithm that computes a straight-line drawing of G with right-angle crossings in exponential area. Both these area requirements are worst-case optimal. We also show that it is NP-complete to test IC-planarity both in the general case and in the case in which a rotation system is fixed for the input graph. Furthermore, we describe a polynomial-time algorithm to test whether a set of matching edges can be added to a triangulated planar graph such that the resulting graph is IC-planar.

1 Introduction

The study of graphs that are, in some sense, "nearly-planar", is an emerging topic in graph theory, graph algorithms, and network visualization. The general framework is to relax the planarity constraint by allowing edge crossings but still forbidding those configurations that would affect the readability of the drawing too much. Different types of forbidden edge-crossing configurations give rise to different families of nearly-planar graphs. For example, if the number of crossings per edge is bounded by a constant k, we have the family of *k-planar graphs* (see, e.g., [2,23]). The *k-quasi-planar graphs* admit drawings with no k pairwise crossing edges (see, e.g., [15,20]). RAC *(Right Angle Crossing) graphs* can be drawn such that edges cross only at right angles (see, e.g., [17,19]). Generalizations

The research is supported in part by the Deutsche Forschungsgemeinschaft (DFG), grant Br835/18-1; the MIUR project AMANDA "Algorithmics for MAssive and Networked DAta", prot. 2012C4E3KT_001; the ESF EuroGIGA project GraDR (DFG grant Wo 758/5-1).

© Springer International Publishing Switzerland 2015
E. Di Giacomo and A. Lubiw (Eds.): GD 2015, LNCS 9411, pp. 295–308, 2015.
DOI: 10.1007/978-3-319-27261-0_25

of RAC drawings are ACE_α and ACL_α drawings, where the edges can cross only at an angle that is *exactly* α or *at least* α, respectively, where $\alpha \in (0, \pi/2]$; see [18] for a survey. Further families of nearly-planar graphs are *fan-crossing free graphs* [11] and *fan-planar graphs* [6,7,24]. Most of the existing literature on nearly-planar graphs can be classified according to the study of the following problems (see also [27] for additional references).

Coloring Problem: While the chromatic number of planar graphs is four, it is rather natural to ask what restrictions on the crossing configurations force the chromatic number of a graph to be relatively small. For example, Borodin [8] proves that the chromatic number of 1-planar graphs is six.

Turán-type Problem: The question here is to determine how many edges a nearly-planar graph can have. In particular, it is known that all the families of nearly-planar graphs mentioned above are rather sparse (see, e.g., [1,9,17,20,24, 28]).

Recognition Problem: In contrast to planarity testing, recognizing a nearly-planar graph has often been proved to be NP-hard. This is for example the case for 1-planar graphs [25], RAC graphs [3], and fan-planar graphs [6,7]. For some constrained classes of nearly-planar graphs, polynomial-time tests exist (e.g., [4,22]).

Drawing Algorithms: Some recent papers describe drawing algorithms for different families of nearly-planar graphs; the majority of them focuses on drawings with straight-line edges and often considers the interplay with other readability constraints, such as compact area. A limited list of examples includes [2,14,16].

Inclusion/intersection Relationships: Relationships between different classes of nearly-planar graphs are also proved, as a fundamental step towards developing a comprehensive theory of graph drawing beyond planarity (see, e.g., [7,19]).

This paper studies *IC-planar graphs*, which stands for Independent Crossings graphs, i.e., graphs that admit a drawing where no two crossed edges share an end-vertex and each edge is crossed at most once. They are 1-planar graphs with the additional property that all crossing edges form an independent set. Král and Stacho [26] exploited this property to show that they have chromatic number at most five. Zhang and Liu [31] study the Turán-type problem and prove that they have at most $13n/4 - 6$ edges, which is a tight bound. Zhang [30] studies so-called *plane graphs with near independent crossings (NIC-planar graphs)*, that is, each pair of crossing edges shares at most one endpoint, and states the computational complexity of recognizing IC-planar graphs as an open problem.

We extend the theory on IC-planarity beyond the already studied coloring and Turán-type problems. We investigate drawing algorithms, the complexity of the recognition problem, and the interplay between IC-planar graphs and other families of nearly-planar graphs. Our results are as follows.

(i) We present an $O(n)$-time algorithm that computes a straight-line drawing of an embedded IC-planar graph with n vertices in $O(n^2)$ area, which is worst-case optimal (Theorem 1). It may be worth recalling that not all 1-planar graphs

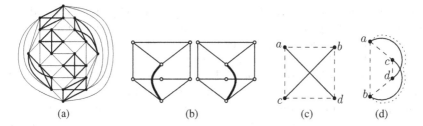

Fig. 1. (a) An IC-planar drawing. (b) Two different IC-planar embeddings of the same graph with the same rotation system. (c) An X-configuration. (d) A B-configuration

admit a straight-line drawing [29] and that there are embedded 1-planar graphs that require $\Omega(2^n)$ area [23].

(ii) We prove that IC-planarity testing is NP-complete both in the variable embedding setting (Theorem 2) and when the rotation system of the graph is fixed (Theorem 3). Note that 1-planarity testing is already known to be NP-complete in general [25], even if the rotation system is fixed [5]. In addition to the hardness result, we present a polynomial-time algorithm that tests whether a set of matching edges can be added to a triangulated plane graph such that the resulting graph is IC-planar (Theorem 4). We remark that in any IC-planar drawing the set of crossing edges form a matching.

(iii) We study the interplay between IC-planar graphs and RAC graphs. Namely, we show that every IC-planar graph is a RAC graph (Theorem 5), which sheds new light on an open problem about the relationship between 1-planar graphs and RAC graphs [19]. We also prove that a straight-line RAC drawing of an IC-planar graph may require $\Omega(q^n)$ area, for a suitable constant $q > 1$ (Theorem 6).

2 Preliminaries

We consider simple undirected graphs G. A *drawing* Γ of G maps the vertices of G to distinct points in the plane and the edges of G to simple Jordan curves between their end-points. If the vertices are drawn at integer coordinates, Γ is a *grid drawing*. Γ is *planar* if no edges cross, and *1-planar* if each edge is crossed at most once. Γ is *IC-planar* if it is 1-planar and there are no crossing edges that share a vertex (see Fig. 1(a)).

A planar drawing Γ of a graph G induces an *embedding*, which is the class of topologically equivalent drawings. In particular, an embedding specifies the regions of the plane, called *faces*, whose boundary consists of a cyclic sequence of edges. The unbounded face is called the *outer face*. For a 1-planar drawing, we can still derive an embedding considering that the boundary of a face may consist also of edge segments from a vertex to a crossing point. A graph with a given planar (1-planar, IC-planar) embedding is called a *plane* (*1-plane, IC-plane*) graph. A *rotation system* $\mathcal{R}(G)$ of a graph G describes a possible cyclic ordering of the edges around the vertices. $\mathcal{R}(G)$ is planar (1-planar, IC-planar) if G admits a planar (1-planar, IC-planar) embedding that preserves $\mathcal{R}(G)$. Observe that $\mathcal{R}(G)$ can directly be retrieved from a drawing or an embedding. The converse does not necessarily hold, as shown in Fig. 1(b).

A *kite K* is a graph isomorphic to K_4 with an embedding such that all the vertices are on the boundary of the outer face, the four edges on the boundary are planar, and the remaining two edges cross each other; see Fig. 1(c). Thomassen [29] characterized the possible crossing configurations that occur in a 1-planar drawing. Applying this characterization to IC-planar drawings gives rise to the following property, where an X-crossing is of the type described in Fig. 1(c) (the crossing is "inside" the cycle), and a B-crossing is of the type described in Fig. 1(d) (without the dotted edge; the crossing is "outside" the cycle).

Property 1. Every crossing of an IC-planar drawing is either an X- or a B-crossing.

Let G be a plane (1-plane, IC-plane) graph. G is *maximal* if no edge can be added without violating planarity (1-planarity, IC-planarity). A planar (1-planar, IC-planar) graph G is maximal if every planar (1-planar, IC-planar) embedding is maximal. If we restrict to 1-plane (IC-plane) graphs, we say that G is *planar-maximal* if no edge can be added without creating at least an edge crossing on the newly added edge (or making the graph not simple). We call the operation of adding edges to G until it becomes planar-maximal a *planar-maximal augmentation*.

3   Straight-Line Drawings of IC-Planar Graphs

We show that every IC-planar graph admits an IC-planar straight-line grid drawing in quadratic area, and this area is worst-case optimal (Theorem 1). The result is based on first using a new technique that possibly augments the input graph to a maximal IC-plane graph (the resulting embedding might be different from the original one) with specific properties (Lemma 1), and then suitably applying a drawing algorithm by Alam *et al.* for triconnected 1-plane graphs [2] on the augmented graph. We say that a kite (a, b, c, d) with crossing edges (a, d) and (b, c) is *empty* if it contains no other vertices, that is, the edges (a, c), (a, d), and (a, b) are consecutive in the counterclockwise order around a; see Fig. 2(b). The condition for the edges around b, c, and d is analogous.

Lemma 1. *Let $G = (V, E)$ be an IC-plane graph with n vertices. There exists an $O(n)$-time algorithm that computes a planar-maximal IC-plane graph $G^+ = (V, E^+)$ with $E \subseteq E^+$ such that the following conditions hold:*

(c1) *The four endvertices of each pair of crossing edges induce a kite.*
(c2) *Each kite is empty.*

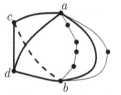
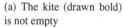
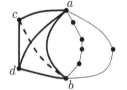
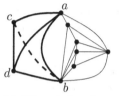

(a) The kite (drawn bold) is not empty

(b) Rerouting edge (a, b) to make the kite empty

(c) Triangulating the remaining faces

Fig. 2. Illustration for the proof of Lemma 1

(**c3**) *Let C be the set of crossing edges in G^+. Let $C^* \subset C$ be a subset containing exactly one edge for each pair of crossing edges. Then $G^+ \setminus C^*$ is plane and triangulated.*

(**c4**) *The outer face of G^+ is a 3-cycle of non-crossed edges.*

Proof. Let G be an IC-plane graph; we augment G by adding edges such that for each pair of crossing edges (a, d) and (b, c) the subgraph induced by vertices $\{a, b, c, d\}$ is isomorphic to K_4; see the dashed edges in Figs. 1(c) and (d). Next, we want to make sure that this subgraph forms an X-configuration and the resulting kite is empty. Since G is IC-planar, it has no two B-configurations sharing an edge. Thus, we remove a B-configuration with vertices $\{a, b, c, d\}$ by rerouting the edge (a, b) to follow the edge (a, d) from vertex a until the crossing point, then edge (b, c) until vertex b, as shown by the dotted edge in Fig. 1(d). This is always possible, because edges (a, c) and (b, d) only cross each other; hence, following their curves, we do not introduce any new crossing. The resulting IC-plane graph satisfies (**c1**) (recall that, by Property 1, only X- and B-configurations are possible). Now, assume that a kite $\{a, b, c, d\}$ is not empty; see Fig. 2(a). Following the same argument as above, we can reroute the edges (a, b), (b, d), (c, d) and (a, d) to follow the crossing edges (a, d) and (b, c); see Fig. 2(b). The resulting IC-plane graph is denoted by G' and satisfies (**c2**).

We now augment G' to G^+, such that (**c3**) is satisfied. Let C be the set of all pairs of crossing edges in G'. Let C^* be a subset constructed from C by keeping only one (arbitrary) edge for each pair of crossing edges. The graph $G' \setminus C^*$ is clearly plane. To ensure (**c3**), graph $G^+ \setminus C^*$ must be plane and triangulated. Because G' satisfies (**c2**), each removed edge spans two triangular faces in $G' \setminus C^*$. Thus, no face incident to a crossing edge has to be triangulated. We internally triangulate the other faces by picking any vertex on its boundary and connecting it to all other vertices (avoiding multiple edges) of the boundary; see e.g. Fig. 2(c). Graph G^+ is then obtained by reinserting the edges in C^* and satisfies (**c3**). To satisfy (**c4**), notice that G^+ is IC-plane, hence, it has a face f whose boundary contains only non-crossed edges. Also, f is a 3-cycle by construction. Thus, we can re-embed G^+ such that f is the outer face. Since IC-planar graphs are sparse [31], each step can clearly be done in $O(n)$ time. □

Theorem 1. *There is an $O(n)$-time algorithm that takes an IC-plane graph G with n vertices as input and constructs an IC-planar straight-line grid drawing of G in $O(n) \times O(n)$ area. This area is worst-case optimal.*

Sketch of Proof. Augment G into a planar-maximal IC-plane graph G^+ in $O(n)$ time using Lemma 1. Graph G^+ is triconnected, as it contains a triangulated plane subgraph. Draw G^+ with the algorithm by Alam *et al.* [2] which takes as input a 1-plane triconnected graph with n vertices and computes a 1-planar drawing on the $(2n-2) \times (2n-3)$ grid in $O(n)$ time; this drawing is straight-line, but for the outer face, which may contain a bent edge if it has two crossing edges. By Lemma 1 the outer face of G^+ has no crossed edges, so Γ is straight-line and IC-planar. Dummy edges are then removed from Γ. The proof that the area is worst-case optimal is given in the full version [10].   □

4   Recognizing IC-Planar Graphs

The *IC-planarity testing* problem asks if a graph G admits an IC-planar embedding.

Hardness of the Problem. The next theorem shows that IC-planarity testing is NP-complete. The full proof is given in the full version of the paper [10].

Theorem 2. *IC-planarity testing is NP-complete.*

Sketch of Proof. IC-planarity is in NP, as one can guess an embedding and check whether it is IC-planar [21]. For the hardness proof, the reduction is from the *1-planarity testing* problem, which asks whether a given graph is 1-planar or not. The reduction uses a 3-cycle gadget and exploits the fact that at most one edge of a 3-cycle is crossed in an IC-planar drawing. We transform an instance G of 1-planarity testing into an instance G^* of IC-planarity testing, by replacing each edge (u, v) of G with a graph G_{uv} consisting of two 3-cycles, T_{uv} and T_{vu}, with vertices $\{u, c_{uv}, a_{uv}\}$ and $\{v, c_{vu}, a_{vu}\}$, respectively, plus edge (a_{uv}, a_{vu}), called the *attaching edge* of u and v; see Fig. 3.

Let Γ be a 1-planar drawing of G. An IC-planar drawing Γ^* of G^* can be easily constructed by replacing each curve representing an edge (u, v) in Γ with a drawing of G_{uv} where T_{uv} and T_{vu} are drawn planar and sufficiently small, such that the possible crossing that occurs on the edge (u, v) in Γ occurs on the attaching edge (a_{uv}, a_{vu}) in Γ^*. Hence, since all the attaching edges are independent, Γ^* is IC-planar.

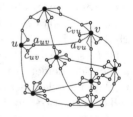

Fig. 3. Reduction from 1-planarity to IC-planarity. Dummy vertices are unfilled

Let Γ^* be an IC-planar drawing of G^*. We show that it is possible to transform the drawing in such a way that all crossings occur only between attaching edges. Once this condition is satisfied, in order to construct a 1-planar drawing Γ of G, it suffices to remove, for each edge (u, v), the vertices c_{uv} and c_{vu}, and to

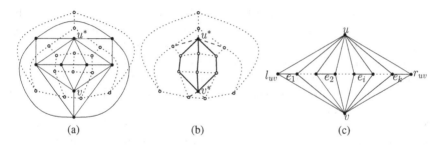

Fig. 4. (a) A triconnected graph T (solid) and its dual T^* (dotted), (b) The extended graph $T^* \cup \{u^*, v^*\}$ and the three length-3 paths between u^* and v^* (bold). (c) The ordered routing edges $e_1, \ldots, e_k$ lie inside the quadrangle (u, l_{uv}, v, r_{uv})

replace a_{uv} and a_{vu} with a bend point. Namely, as already observed, no more than one edge can be crossed for every gadget T_{uv} of G^*. If an edge incident to a dummy vertex is crossed, then we can reroute this edge to remove the crossing. $\square$

Note that this construction does not work for IC-planarity testing with a given rotation system since the rerouting step changes the rotation system. However, we show that IC-planarity testing is NP-hard even if the rotation system of the input graph is fixed. The reduction is from planar-3SAT. We exploit the *membrane technique* introduced in [5] for the hardness proof of 1-planarity testing with fixed rotation system. The main issue is to design suitable gadgets for IC-planar graphs. The proof is given in the full version of the paper [10].

Theorem 3. *IC-planarity testing with given rotation system is* NP-*complete.*

Polynomial-time Test for a Triangulated Plane Graph Plus a Matching.
On the positive side, we now describe an $O(n^3)$-time algorithm to test whether a graph $G = (V, E_T \cup E_M)$ that consists of a triangulated plane graph $T = (V, E_T)$ and a matching $M = (V_M, E_M)$ with $V_M \subseteq V, E_M \cap E_T = \emptyset$ admits an IC-planar drawing that preserves the embedding of T. In the positive case the algorithm also computes an IC-planar drawing. An outline of the algorithm is as follows. (1) Check for every matching edge if there is a way to draw it such that it crosses only one edge of T. (2) Split T into subgraphs that form a hierarchical tree structure. (3) Traverse the 4-block tree bottom-up and solve a 2SAT formula for each tree node.

In order to check whether there is a valid placement for each matching edge $(u, v) \in M$, we have to find two adjacent faces, one of which is incident to u, while the other one is incident to v. To this end, we consider the dual T^* of T that contains a vertex for each face in T that is not incident to a vertex $w \in V_M \setminus \{u, v\}$, and an edge for each edge in T that separates two faces. Further, we add two additional vertices u^* and v^* to T^* that are connected to all faces that are incident to u and v, respectively. In the resulting graph $T^* \cup \{u^*, v^*\}$, we look for all paths of length 3 from u^* to v^*. These paths are equivalent to routing (u, v) through two faces that are separated by a single edge. Note that no path of length 1 or 2 can exist, since (*i*) by construction u^*

and v^* are not connected by an edge and (ii) if there was a path of length 2 between u^* and v^*, then u and v would lie on a common face in the triangulated graph T; thus, the edge (u,v) would exist both in E_T and in E_M, which is not possible since $E_T \cap E_M = \emptyset$. See Fig. 4 for an illustration. If there is an edge that has no valid placement, then G is not IC-planar and the algorithm stops. Otherwise, we save each path that we found as a possible routing for the corresponding edge in M.

Now, we make some observations on the structure of the possible routings of an edge $(u,v) \in M$ that we can use to get a hierarchical tree structure of the graph T. Every routing is uniquely represented by an edge that separates a face incident to u and a face incident to v and that might be crossed by (u,v). We call these edges *routing edges*. Let there be k routing edges for the pair (u,v). Each of these edges forms a triangular face with u. From the embedding, we can enumerate the edges by the counterclockwise order of their corresponding faces at u. This gives an ordering $e_1, \ldots, e_k$ of the routing edges. Let $e_1 = (l_{uv}, l'_{uv})$ and $e_k = (r'_{uv}, r_{uv})$ such that the edge (u, l_{uv}) comes before the edge (u, l'_{uv}), and the edge (u, r'_{uv}) comes before (u, r_{uv}) in the counterclockwise order at u. Then, all edges $e_1, \ldots, e_k$ lie within the *routing quadrangle* (u, l_{uv}, v, r_{uv}); see Fig. 4(c). Note that there may be more complicated structures between the edges, but they do not interfere with the ordering. Denote by $Q_{uv} = (u, l_{uv}, v, r_{uv})$ the routing quadrilateral of the matching edge $(u,v) \in M$. We define the *interior* $\mathcal{I}_{uv} = (\mathcal{V}_{uv}, \mathcal{E}_{uv})$ as the maximal subgraph of T such that, for every vertex $w \in \mathcal{V}_{uv}$, each path from w to a vertex on the outer face of T contains u, l_{uv}, v, or r_{uv}. Consequently, $Q_{uv} \in \mathcal{V}_{uv}$. The following lemma states that two interiors cannot overlap. The proof is given in the full version of the paper [10].

Lemma 2. *For each pair of interiors $\mathcal{I}_{uv}, \mathcal{I}_{ab}$, exactly one of the following conditions holds: (a) $\mathcal{I}_{uv} \cap \mathcal{I}_{ab} = \emptyset$ (b) $\mathcal{I}_{uv} \subset \mathcal{I}_{ab}$ (c) $\mathcal{I}_{ab} \subset \mathcal{I}_{uv}$ (d) $\mathcal{I}_{uv} \cap \mathcal{I}_{ab} = Q_{uv} \cap Q_{ab}$.*

By using Lemma 2, we can find a hierarchical structure on the routing quadrilaterals. We construct a directed graph $H = (V_H, E_H)$ with $V_H = \{\mathcal{I}_{uv} \mid (u,v) \in M\} \cup \{G\}$. For each pair $\mathcal{I}_{uv}, \mathcal{I}_{xy}$, E_H contains a directed edge $(\mathcal{I}_{uv}, \mathcal{I}_{xy})$ if and only if $\mathcal{V}_{uv} \subset \mathcal{V}_{xy}$ and there is no matching edge (a,b) with $\mathcal{V}_{uv} \subset \mathcal{V}_{ab} \subset \mathcal{V}_{xy}$. Finally, we add an edge from each subgraph that has no outgoing edges to G. Each vertex but G only has one outgoing edge. Obviously, this graph contains no (undirected) cycles. Thus, H is a tree.

We will now show how to construct a drawing of G based on H in a bottom-up fashion. We will first look at the leaves of the graph. Let $\mathcal{I}_{uv}$ be a vertex of H whose children are all leaves. Let $\mathcal{I}_{u_1v_1}, \ldots, \mathcal{I}_{u_kv_k}$ be these leaves. Since these interiors are all leaves in H, we can pick any of their routing edges. However, the interiors may touch on their boundary, so not every combination of routing edges can be used. Assume that a matching edge $(u_i, v_i), 1 \leq i \leq k$ has more than two valid routing edges. Then, we can always pick a *middle* one, that is, a routing edge that is not incident to $l_{u_iv_i}$ and $r_{u_iv_i}$, since this edge will not interfere with a routing edge of another matching edge.

Now, we can create a 2SAT formula to check whether there is a valid combination of routing edges as follows. For the sake of clarity, we will create several redundant variables and formulas. These can easily be removed or substituted by shorter structures to improve the running time. For each matching edge $(u_i, v_i), 1 \leq i \leq k$, we create two binary variables l_i and r_i, such that l_i is **true** if and only if the routing edge incident to $l_{u_i v_i}$ is picked, and r_i is **true** if and only if the routing edge incident to $r_{u_i v_i}$ is picked. If (u_i, v_i) has only one routing edge, then it is obviously incident to $l_{u_i v_i}$ and $r_{u_i v_i}$, so we set $l_{u_i v_i} = r_{u_i v_i} = $ **true** by adding the clauses $l_{u_i v_i} \vee$ **false** and $r_{u_i v_i} \vee$ **false**. If (u_i, v_i) has exactly two routing edges, the picked routing edge has to be incident to either $l_{u_i v_i}$ or $r_{u_i v_i}$, so we add the clauses $l_{u_i v_i} \vee r_{u_i v_i}$ and $\neg l_{u_i v_i} \vee \neg r_{u_i v_i}$. If (u_i, v_i) has more than two routing edges, we can pick a middle one, so we set $l_{u_i v_i} = r_{u_i v_i} = $ **false** by adding the clauses $\neg l_{u_i v_i} \vee$ **false** and $\neg r_{u_i v_i} \vee$ **false**. Next, we have to add clauses to forbid pairs of routing edges that can not be picked simultaneously, i.e., they share a common vertex. Consider a pair of matching edges $(u_i, v_i), (u_j, v_j), 1 \leq i, j \leq k$. If $r_{u_i v_i} = l_{u_j v_j}$, we add the clause $\neg r_i \vee \neg l_j$. For the other three cases, we add an analogue clause.

Now, we use this 2SAT to decide whether the subgraph I_{uv} is IC-planar, and which routing edges can be used. For each routing edge (a, b) of I_{uv}, we solve the 2SAT formula given above with additional constraints that forbid the use of routing edges incident to a and b. To that end, add the following additional clauses: If $l_{u_i v_i} = a$, add the clause $\neg l_i \vee$ **false**. For the other three cases, we add an analogue clause. If this 2SAT formula has no solution, then the subgraph $\mathcal{I}_{uv}$ is not IC-planar. Otherwise, there is a solution where you pick the routing edges corresponding to the binary variables. To decide whether a subgraph I_{uv} whose children are not all leaves is IC-planar, we first compute which of their routing edges can be picked by recursively using the 2SAT formula above. Then, we use the 2SAT formula for I_{uv} to determine the valid routing edges of I_{uv}. Finally, we can decide whether G is IC-planar and, if yes, get a drawing by solving the 2SAT formula of all children of G. Hence, we give the following (see the full version of the paper [10] for the proof of the time complexity).

Theorem 4. *Let $T = (V, E_T)$ be a triangulated plane graph with n vertices and let $M = (V, E_M)$ be a matching. There exists an $O(n^3)$-time algorithm to test if $G = (V, E_T \cup E_M)$ admits an IC-planar drawing that preserves the embedding of T. If the test is positive, the algorithm computes a feasible drawing.*

5 IC-Planarity and RAC Graphs

It is known that every n-vertex maximally dense RAC graph (i.e., RAC graph with $4n - 10$ edges) is 1-planar, and that there exist both 1-planar graphs that are not RAC and RAC graphs that are not 1-planar [19]. Here, we further investigate the intersection between the classes of 1-planar and RAC graphs, showing that all IC-planar graphs are RAC. To this aim, we describe a polynomial-time constructive algorithm. The computed drawings may require exponential area, which is however worst-case optimal; indeed, we exhibit IC-planar graphs that require exponential area in any possible IC-planar straight-line RAC drawing. Our construction extends the linear-time algorithm by de Fraysseix *et al.* that computes a planar straight-line grid drawing of a maximal (i.e., triangulated) plane graph in quadratic area [13]; we call it the dFPP algorithm. We need to recall the idea behind dFPP before describing our extension.

Algorithm dFPP. Let G be a maximal plane graph with $n \geq 3$ vertices. The dFPP algorithm first computes a suitable linear ordering of the vertices of G, called a *canonical ordering* of G, and then incrementally constructs a drawing of G using a technique called *shift method*. This method adds one vertex per time, following the computed canonical ordering and shifting vertices already in the drawing when needed. Namely, let $\sigma = (v_1, v_2, \ldots, v_n)$ be a linear ordering of the vertices of G. For each integer $k \in [3, n]$, denote by G_k the plane subgraph of G induced by the k vertices $v_1, v_2, \ldots, v_k$ $(G_n = G)$ and by C_k the boundary of the outer face of G_k, called the *contour* of G_k. Ordering σ is a canonical ordering of G if the following conditions hold for each integer $k \in [3, n]$: (i) G_k is biconnected and internally triangulated; (ii) (v_1, v_2) is an outer edge of G_k; and (iii) if $k + 1 \leq n$, vertex v_{k+1} is located in the outer face of G_k, and all neighbors of v_{k+1} in G_k appear on C_k consecutively.

We call *lower neighbors* of v_k all neighbors v_j of v_k for which $j < k$. Following the canonical ordering σ, the shift method constructs a drawing of G one vertex per time. The drawing Γ_k computed at step k is a drawing of G_k. Throughout the computation, the following invariants are maintained for each Γ_k, with $3 \leq k \leq n$: **(I1)** $p_{v_1} = (0, 0)$ and $p_{v_2} = (2k - 4, 0)$; **(I2)** $x(w_1) < x(w_2) < \cdots < x(w_t)$, where $w_1 = v_1, w_2, \ldots, w_t = v_2$ are the vertices that appear along C_k, going from v_1 to v_2. **(I3)** Each edge (w_i, w_{i+1}) (for $i = 1, 2, \ldots, t - 1$) is drawn with slope either $+1$ or -1.

More precisely, Γ_3 is constructed placing v_1 at $(0, 0)$, v_2 at $(2, 0)$, and v_3 at $(1, 1)$. The addition of v_{k+1} to Γ_k is executed as follows. Let $w_p, w_{p+1}, \ldots, w_q$ be the lower neighbors of v_{k+1} ordered from left to right. Denote by $\mu(w_p, w_q)$ the intersection point between the line with slope $+1$ passing through w_p and the line with slope -1 passing through w_q. Point $\mu(w_p, w_q)$ has integer coordinates and thus it is a valid placement for v_{k+1}. With this placement, however, (v_{k+1}, w_p) and (v_{k+1}, w_q) may overlap with (w_p, w_{p+1}) and (w_{q-1}, w_q), respectively. To avoid this, a *shift* operation is applied: $w_{p+1}, w_{p+2}, \ldots, w_{q-1}$ are shifted to the right by 1 unit, and $w_q, w_{q+1}, \ldots, w_t$ are shifted to the right by 2 units. Then v_{k+1} is placed at point $\mu(w_p, w_q)$ with no overlap. We recall that, to keep planarity, when the algorithm shifts a vertex w_i $(p + 1 \leq i \leq t)$ of C_k, it also shifts some of the inner vertices together with it; for more details on this point refer to [12, 13]. By Invariants **(I1)** and **(I2)**, the area of the final drawing is $(2n - 4) \times (n - 2)$.

Our Extension. Let G be an IC-plane graph, and assume that G^+ is the planar-maximal IC-plane graph obtained from G by applying the technique of Lemma 1. Our drawing algorithm computes an IC-planar drawing of G^+ with right-angle crossings, by extending algorithm dFPP. It adds to the classical shift operation *move* and *lift* operations to guarantee that one of the crossing edges of a kite is vertical and the other is horizontal. We now give an idea of our technique, which we call `RAC-drawer`. Details are given in the proof of Theorem 5. Let σ be a canonical ordering constructed from the underlying maximal plane graph of G^+. Vertices are incrementally added to the drawing, according to σ, following the same approach as for `dFPP`. However, suppose that $K = (a, b, c, d)$ is a kite of G^+, and that a and d are the first and the last vertex of σ among the vertices in K, respectively. Once d has been added to the drawing, the algorithm applies a suitable combination of move and lift operations to the vertices of the kite to rearrange their positions so to guarantee a right-angle crossing. Note that, following the `dFPP` technique, a was placed at a y-coordinate smaller than the y-coordinate of d. A move operation is then used to shift d horizontally to the same x-coordinate as a (i.e., (a, d) becomes a vertical segment in the drawing); a lift operation is used to vertically shift the lower between b and c, such that these two vertices get the same y-coordinates. Both operations are applied so to preserve planarity and to maintain Invariant (**I3**) of `dFPP`; however, they do not maintain Invariant (**I1**), thus the area can increase more than in the `dFPP` algorithm and may be exponential. The application of move/lift operations on the vertices of two distinct kites do not interfere each other, as the kites do not share vertices in an IC-plane graph. The main operations of the algorithm are depicted in Fig. 5.

Theorem 5. *There is a $O(n^3)$-time algorithm that takes an IC-plane graph G with n vertices as input and constructs a straight-line IC-planar RAC grid drawing of G.*

Sketch of Proof. Let G^+ be the augmented graph constructed from G by using Lemma 1. Call G' the subgraph obtained from G^+ by removing one edge from each pair of crossing edges; G' is a maximal plane graph (see condition (**c3**) of Lemma 1). We apply on G' the shelling procedure used by de Fraysseix *et al.* to compute a canonical ordering σ of G' in $O(n)$ time [13]; it goes backwards, starting from a vertex on the outer face of G' and successively removing a vertex per time from the current contour. However, during this procedure, some edges of G' can be replaced with some other edges of G^+ that were previously excluded, although G' remains maximal planar. Namely, whenever the shelling procedure encounters the first vertex d of a kite $K = (a, b, c, d)$, it marks d as top(K), and considers the edge e of K that is missing in G'. If e is incident to d in K, the procedure reinserts it and removes from G' the other edge of K that crosses e in G^+. If e is not incident to d, the procedure continues without varying G'. We say that $u \prec v$ if $\sigma(u) < \sigma(v)$.

We then compute a drawing of G^+ by using the `RAC-drawer` algorithm. Let vertex $v = v_{k+1}$ be the next vertex to be placed according to σ. Let $\mathcal{U}(v)$ be the set of lower neighbors of v, and let $\lambda(v)$ and $\rho(v)$ be the leftmost and the rightmost vertex in $\mathcal{U}(v)$, respectively. Also, denote by $\mathcal{A}_l(v)$ the vertices to the top-left of v, and by $\mathcal{A}_r(v)$ the vertices to the top-right of v. If v is not top(K) for some kite K, then v is placed by following the rules of `dFPP`, that is, at the intersection of the ± 1 diagonals through $\lambda(v)$ and $\rho(v)$ after applying a suitable shift operation. If $v = top(K)$ for some kite K, the algorithm proceeds as follows. Let $K = (a, b, c, d)$ with $v = d = top(K)$. The remaining three vertices of K are in G_k and are consecutive along the contour C_k, as they all belong to $\mathcal{U}(d)$ (by construction, G' contains edge (a, d)). W.l.o.g., assume that they are encountered in the order $\{b, a, c\}$ from left to right. Two cases are now possible:

Case 1: $a \prec b$ and $a \prec c$. This implies that $a = \rho(b)$ and $a = \lambda(c)$. The edges (a, b) and (a, c) have slope -1 and $+1$, respectively, as they belong to C_k. We now aim at having b and c at the same y-coordinate, by applying a lift operation. Suppose first that $r = y(c) - y(b) > 0$; see Fig. 5(a). We apply the following steps: (i) Temporarily undo the placement of b and of all vertices in $\mathcal{A}_l(b)$. (ii) Apply the shift operation to $\rho(b) = a$ by $2r$ units to the right, which implies that the intersection of the diagonals through $\lambda(b)$ and $\rho(b)$ is moved by r units to the right and by r units above their former intersection point. Hence, b and c are placed at the same y-coordinate; see also Fig. 5(b). (iii) Reinsert the vertices of $\mathcal{A}_l(b)$ and modify σ accordingly. Namely, by definition, each vertex in $\mathcal{A}_l(b)$ does not belong to $\mathcal{U}(b)$ and it is not an inner vertex below b; therefore, vertices in $\mathcal{A}_l(b)$ can be safely removed. Hence, σ can be modified such that $b \prec w$ for each $w \in \mathcal{A}_l(b)$. If $r = y(c) - y(b) < 0$, we apply a symmetric operation: (i) Undo the placement of c and of all vertices in $\mathcal{A}_r(c)$. (ii) Apply the shift operation to $\rho(c)$ by $|2r|$ additional units to the right. (iii) Reinsert the vertices of $\mathcal{A}_r(c)$.

Finally, we place d vertically above a. To this aim, we first apply the shift operation according to the insertion step of dFPP. After that, we may need to apply a move operation; see Fig. 5(c). If $s = x(d) - x(a) > 0$, then we apply the shift operation to vertex $\rho(d) = c$ by $2s$ units to the right and then place d (see Fig. 5(d)). If $s = x(d) - x(a) < 0$, then we apply the shift operation to vertex $\lambda(d) = b$ by $2s$ units to the left and then place d (clearly, the shift operation can be used to operate in the left direction with a procedure that is symmetric to the one that operates in the right direction). Edges (a, d) and (b, c) are now vertical and horizontal, respectively. In the next steps, their slopes do not change, as their endvertices are shifted only horizontally (they do not belong to other kites); also, a is shifted along with d, as it belongs to $\mathcal{U}(d)$.

Case 2: $b \prec a \prec c$. The proof for this case is given in the full version of the paper [10]. Case $c \prec a \prec b$ is symmetric to **Case 2**; case $b \prec c \prec a$ is impossible as K is a kite. The time complexity is also analyzed in the full version.   □

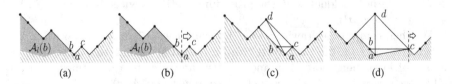

(a)   (b)   (c)   (d)

Fig. 5. (a-b) The lift operation: (a) Vertex b is r units below c. (b) Lifting b. (c-d) The move operation: (c) Vertex d is s units to the left of b. (d) Moving d

Theorem 5 and the fact that there exist n-vertex RAC graphs with $4n - 10$ edges [17] while an n-vertex IC-planar graph has at most $13n/4 - 6$ edges [31] imply that IC-planar graphs are a proper subfamily of RAC graphs. The next result is proved in the full version of the paper [10].

Theorem 6. *There exists an infinite family $\mathcal{G}$ of graphs such that every IC-planar straight-line RAC drawing of an n-vertex graph $G \in \mathcal{G}$ requires area $\Omega(q^n)$, for some constant $q > 1$.*

6 Conclusion

We have shown that every IC-planar graph can be drawn straight-line in quadratic area, although the angle formed by any two crossing edges can be small. Conversely, straight-line RAC drawings of IC-planar graphs may require exponential area. It would be interesting to design algorithms that draw IC-planar graphs in polynomial area and good crossing resolution. Also, although IC-planar graphs are both 1-planar and RAC, a full characterization of the intersection between these two classes is still an open problem. For example, are NIC-planar graphs [30] also RAC graphs?.

References

1. Ackerman, E.: On the maximum number of edges in topological graphs with no four pairwise crossing edges. Discrete Comput. Geom. **41**(3), 365–375 (2009)
2. Alam, M.J., Brandenburg, F.J., Kobourov, S.G.: Straight-line grid drawings of 3-connected 1-planar graphs. In: Wismath, S., Wolff, A. (eds.) GD 2013. LNCS, vol. 8242, pp. 83–94. Springer, Heidelberg (2013)
3. Argyriou, E.N., Bekos, M.A., Symvonis, A.: The straight-line RAC drawing problem is NP-hard. J. Graph Algorithms Appl. **16**(2), 569–597 (2012)
4. Auer, C., Bachmaier, C., Brandenburg, F.J., Gleißner, A., Hanauer, K., Neuwirth, D., Reislhuber, J.: Recognizing outer 1-planar graphs in linear time. In: Wismath, S., Wolff, A. (eds.) GD 2013. LNCS, vol. 8242, pp. 107–118. Springer, Heidelberg (2013)
5. Auer, C., Brandenburg, F.J., Gleiner, A., Reislhuber, J.: 1-planarity of graphs with a rotation system. J. Graph Algorithms Appl. **19**(1), 67–86 (2015)
6. Bekos, M.A., Cornelsen, S., Grilli, L., Hong, S.-H., Kaufmann, M.: On the recognition of fan-planar and maximal outer-fan-planar graphs. In: Duncan, C., Symvonis, A. (eds.) GD 2014. LNCS, vol. 8871, pp. 198–209. Springer, Heidelberg (2014)
7. Binucci, C., Di Giacomo, E., Didimo, W., Montecchiani, F., Patrignani, M., Symvonis, A., Tollis, I.G.: Fan-planarity: properties and complexity. Theor. Comput. Sci. **589**, 76–85 (2015)
8. Borodin, O.V.: Solution of the Ringel problem on vertex-face coloring of planar graphs and coloring of 1-planar graphs. Metody Diskret Analiz. **41**, 12–26 (1984)
9. Brandenburg, F.J., Eppstein, D., Gleißner, A., Goodrich, M.T., Hanauer, K., Reislhuber, J.: On the density of maximal 1-planar graphs. In: Didimo, W., Patrignani, M. (eds.) GD 2012. LNCS, vol. 7704, pp. 327–338. Springer, Heidelberg (2013)

10. Brandenburg, F.J., Didimo, W., Evans, W.S., Kindermann, P., Liotta, G., Montecchiani, F.: Recognizing and drawing IC-planar graphs. Arxiv report (2015). Available at http://arxiv.org/abs/1509.00388

11. Cheong, O., Har-Peled, S., Kim, H., Kim, H.-S.: On the number of edges of fan-crossing free graphs. In: Cai, L., Cheng, S.-W., Lam, T.-W. (eds.) Algorithms and Computation. LNCS, vol. 8283, pp. 163–173. Springer, Heidelberg (2013)

12. Chrobak, M., Payne, T.H.: A linear-time algorithm for drawing a planar graph on a grid. Inf. Process. Lett. **54**(4), 241–246 (1995)

13. de Frayseix, H., Pach, J., Pollack, R.: How to draw a planar graph on a grid. Combinatorica **10**(1), 41–51 (1990)

14. Di Giacomo, E., Didimo, W., Eades, P., Liotta, G.: 2-layer right angle crossing drawings. Algorithmica **68**(4), 954–997 (2014)

15. Di Giacomo, E., Didimo, W., Liotta, G., Montecchiani, F.: h-Quasi planar drawings of bounded treewidth graphs in linear area. In: Golumbic, M.C., Stern, M., Levy, A., Morgenstern, G. (eds.) WG 2012. LNCS, vol. 7551, pp. 91–102. Springer, Heidelberg (2012)

16. Di Giacomo, E., Didimo, W., Liotta, G., Montecchiani, F.: Area requirement of graph drawings with few crossings per edge. Comput. Geom. **46**(8), 909–916 (2013)

17. Didimo, W., Eades, P., Liotta, G.: Drawing graphs with right angle crossings. Theoretical Comput. Sci. **412**(39), 5156–5166 (2011)

18. Didimo, W., Liotta, G.: The crossing angle resolution in graph drawing. In: Pach, J. (ed.) Thirty Essays on Geometric Graph Theory, pp. 167–184. Springer, New York (2012)

19. Eades, P., Liotta, G.: Right angle crossing graphs and 1-planarity. Discrete Appl. Math. **161**(7–8), 961–969 (2013)

20. Fox, J., Pach, J., Suk, A.: The number of edges in k-quasi-planar graphs. SIAM J. Discrete Math. **27**(1), 550–561 (2013)

21. Garey, M.R., Johnson, D.S.: Crossing number is NP-complete. SIAM J. Algebraic Discrete Methods **4**(3), 312–316 (1983)

22. Hong, S., Eades, P., Katoh, N., Liotta, G., Schweitzer, P., Suzuki, Y.: A linear-time algorithm for testing outer-1-planarity. Algorithmica **46**, 1–22 (2014)

23. Hong, S.-H., Eades, P., Liotta, G., Poon, S.-H.: Fáry's theorem for 1-planar graphs. In: Gudmundsson, J., Mestre, J., Viglas, T. (eds.) COCOON 2012. LNCS, vol. 7434, pp. 335–346. Springer, Heidelberg (2012)

24. Kaufmann, M., Ueckerdt, T.: The density of fan-planar graphs. Arxiv report, (2014). Available at http://arxiv.org/abs/1403.6184

25. Korzhik, V.P., Mohar, B.: Minimal obstructions for 1-immersions and hardness of 1-planarity testing. J. Graph Theor. **72**(1), 30–71 (2013)

26. Král, D., Stacho, L.: Coloring plane graphs with independent crossings. J. Graph Theor. **64**(3), 184–205 (2010)

27. Liotta, G.: Graph drawing beyond planarity: some results and open problems. In: Theoretical Computer Science (ICTCS 2014), pp. 3–8 (2014)

28. Pach, J., Tóth, G.: Graphs drawn with few crossings per edge. Combinatorica **17**(3), 427–439 (1997)

29. Thomassen, C.: Rectilinear drawings of graphs. J. Graph Theor. **12**(3), 335–341 (1988)

30. Zhang, X.: Drawing complete multipartite graphs on the plane with restrictions on crossings. Acta Math. Sinica **30**(12), 2045–2053 (2014)

31. Zhang, X., Liu, G.: The structure of plane graphs with independent crossings and its applications to coloring problems. Central Eur. J. Math. **11**(2), 308–321 (2013)

Simple Realizability of Complete Abstract Topological Graphs Simplified

Jan Kynčl[1,2(✉)]

[1] Department of Applied Mathematics and Institute for Theoretical Computer Science, Faculty of Mathematics and Physics, Charles University, Malostranské nám. 25, 118 00 Praha 1, Czech Republic
kyncl@kam.mff.cuni.cz
[2] Chair of Combinatorial Geometry, EPFL-SB-MATHGEOM-DCG, École Polytechnique Fédérale de Lausanne, Station 8,1015 Lausanne, Switzerland

Abstract. An *abstract topological graph* (briefly an *AT-graph*) is a pair $A = (G, \mathcal{X})$ where $G = (V, E)$ is a graph and $\mathcal{X} \subseteq \binom{E}{2}$ is a set of pairs of its edges. The AT-graph A is *simply realizable* if G can be drawn in the plane so that each pair of edges from $\mathcal{X}$ crosses exactly once and no other pair crosses. We characterize simply realizable complete AT-graphs by a finite set of forbidden AT-subgraphs, each with at most six vertices. This implies a straightforward polynomial algorithm for testing simple realizability of complete AT-graphs, which simplifies a previous algorithm by the author.

1 Introduction

A *topological graph* $T = (V(T), E(T))$ is a drawing of a graph G in the plane such that the vertices of G are represented by a set $V(T)$ of distinct points and the edges of G are represented by a set $E(T)$ of simple curves connecting the corresponding pairs of points. We call the elements of $V(T)$ and $E(T)$ the *vertices* and the *edges* of T, respectively. The drawing has to satisfy the following general position conditions: (1) the edges pass through no vertices except their endpoints, (2) every pair of edges has only a finite number of intersection points, (3) every intersection point of two edges is either a common endpoint or a proper crossing ("touching" of the edges is not allowed), and (4) no three edges pass through the same crossing. A topological graph or a drawing is *simple* if every pair of edges has at most one common point, which is either a common endpoint or a crossing. Simple topological graphs appear naturally as crossing-minimal drawings; it is well known that if two edges in a topological graph have more than one common point, then a local redrawing decreases the total number of crossings. A topological graph is *complete* if it is a drawing of a complete graph.

J. Kynčl—Supported by Swiss National Science Foundation Grants 200021-137574 and 200020-14453, by the ESF Eurogiga project GraDR as GAČR GIG/11/E023, by the grant no. 14-14179S of the Czech Science Foundation (GAČR) and by the grant GAUK 1262213 of the Grant Agency of Charles University.

E. Di Giacomo and A. Lubiw (Eds.): GD 2015, LNCS 9411, pp. 309–320, 2015.
DOI: 10.1007/978-3-319-27261-0_26

An *abstract topological graph* (briefly an *AT-graph*), a notion introduced by Kratochvíl, Lubiw and Nešetřil [10], is a pair $(G, \mathcal{X})$ where G is a graph and $\mathcal{X} \subseteq \binom{E(G)}{2}$ is a set of pairs of its edges. Here we assume that $\mathcal{X}$ consists only of independent (that is, nonadjacent) pairs of edges. For a simple topological graph T that is a drawing of G, let $\mathcal{X}_T$ be the set of pairs of edges having a common crossing. A simple topological graph T is a *simple realization* of $(G, \mathcal{X})$ if $\mathcal{X}_T = \mathcal{X}$. We say that $(G, \mathcal{X})$ is *simply realizable* if $(G, \mathcal{X})$ has a simple realization.

An *AT-subgraph* of an AT-graph $(G, \mathcal{X})$ is an AT-graph $(H, \mathcal{Y})$ such that H is a subgraph of G and $\mathcal{Y} = \mathcal{X} \cap \binom{E(H)}{2}$. Clearly, a simple realization of $(G, \mathcal{X})$ restricted to the vertices and edges of H is a simple realization of $(H, \mathcal{Y})$.

We are ready to state our main result.

Theorem 1. *Every complete AT-graph that is not simply realizable has an AT-subgraph on at most six vertices that is not simply realizable.*

We also show that AT-subgraphs with five vertices are not sufficient to characterize simple realizability.

Theorem 2. *There is a complete AT-graph A with six vertices such that all its induced AT-subgraphs with five vertices are simply realizable, but A itself is not.*

Theorem 1 implies a straightforward polynomial algorithm for simple realizability of complete AT-graphs, running in time $O(n^6)$ for graphs with n vertices. It is likely that this running time can be improved relatively easily. However, compared to the first polynomial algorithm for simple realizability of complete AT-graphs [13], the new algorithm may be more suitable for implementation and for practical applications, such as generating all simply realizable complete *AT*-graphs of given size or computing the crossing number of the complete graph [5,15]. On the other hand, the new algorithm does not directly provide the drawing itself, unlike the original algorithm [13]. The explicit list of realizable AT-graphs on six vertices can be generated using the database of small simple complete topological graphs created by Ábrego et al. [1].

For general noncomplete graphs, no such finite characterization by forbidden *AT*-subgraphs is possible. Indeed, in the special case when $\mathcal{X}$ is empty, the problem of simple realizability is equivalent to planarity, and there are nonplanar graphs of arbitrarily large girth, such as subdivisions of K_5. Moreover, simple realizability for general *AT*-graphs is NP-complete [11]. See [13] for an overview of other similar realizability problems.

The proof of Theorem 1 is based on the polynomial algorithm for simple realizability of complete AT-graphs from [13]. The main idea is very simple: every time the algorithm rejects the input, it is due to an obstruction of constant size.

Theorem 1 is an analogue of a similar characterization of simple monotone drawings of K_n by forbidden 5-tuples, and pseudo linear drawings of K_n by forbidden 4-tuples [4].

Ábrego et al. [1,2] independently verified that simple complete topological graphs with up to nine vertices can be characterized by forbidden rotation systems of five-vertex subgraphs; see Sect. 2 for the definition. They conjectured that the same characterization is true for all simple complete topological graphs [2]. This conjecture now follows by combining their result for six-vertex graphs with Theorem 1. This gives a finite characterization of *realizable abstract rotation systems* defined in [14, Sect. 3.5], where it was also stated that such a characterization was not likely [14, p. 739]. The fact that only 5-tuples are sufficient for the characterization by rotation systems should perhaps not be too surprising, as rotation systems characterize simple drawings of K_n more economically, using only $O(n^2 \log n)$ bits, whereas AT-graphs need $\Theta(n^4)$ bits.

2   Preliminaries

Topological graphs G and H are *weakly isomorphic* if they are realizations of the same abstract topological graph.

The *rotation* of a vertex v in a topological graph is the clockwise cyclic order in which the edges incident with v leave the vertex v. The *rotation system* of a topological graph is the set of rotations of all its vertices. Similarly we define the *rotation* of a crossing x of edges uv and yz as the clockwise order in which the four parts xu, xv, xy and xz of the edges uv and yz leave the point x. Note that each crossing has exactly two possible rotations. We will represent the rotation of a vertex v as an ordered sequence of the endpoints of the edges incident with v. The *extended rotation system* of a topological graph is the set of rotations of all its vertices and crossings.

Assuming that T and T' are drawings of the same abstract graph, we say that their rotation systems are *inverse* if for each vertex $v \in V(T)$, the rotation of v and the rotation of the corresponding vertex $v' \in V(T')$ are inverse cyclic permutations. If T and T' are weakly isomorphic simple topological graphs, we say that their extended rotation systems are *inverse* if their rotation systems are inverse and, in addition, for every crossing x in T, the rotation of x and the rotation of the corresponding crossing x' in T' are inverse cyclic permutations. For example, if T' is a mirror image of T, then T and T' have inverse extended rotation systems.

We say that two cyclic permutations of sets A, B are *compatible* if they are restrictions of a common cyclic permutation of $A \cup B$.

Simple complete topological graphs have the following key property.

Proposition 3 [7,13].

(1) *If two simple complete topological graphs are weakly isomorphic, then their extended rotation systems are either the same or inverse.*
(2) *For every edge e of a simple complete topological graph T and for every pair of edges $f, f' \in E(T)$ that have a common endpoint and cross e, the AT-graph of T determines the order of crossings of e with the edges f, f'.*

By inspecting simple drawings of K_4, it can be shown that the converse of Proposition 3 also holds: the rotation system of a simple complete topological graph determines which pairs of edges cross [12,16].

3 Proof of Theorem 2

We use the shortcut ij to denote the edge $\{i,j\}$. Let $A = ((V,E), \mathcal{X})$ be the complete AT-graph with vertex set $V = \{0,1,2,3,4,5\}$ and with

$$\mathcal{X} = \{\{02,13\},\{02,14\},\{02,15\},\{02,35\},\{03,14\},\{03,15\},\{03,24\},$$
$$\{04,15\},\{04,25\},\{04,35\},\{13,24\},\{24,35\},\{35,14\},\{14,25\},\{25,13\}\}.$$

Every complete AT-subgraph of A with five vertices is simply realizable; see Fig. 1.

On the other hand, we show that A is not simply realizable. Suppose that T is a simple realization of A. Without loss of generality, assume that the rotation of 5 in $T[\{1,2,3,5\}]$ is $(1,2,3)$. By Proposition 3 and by the first drawing in Fig. 1, the rotation of 5 in $T[\{1,2,3,4,5\}]$ is $(1,2,3,4)$, since the inverse would not be compatible with $(1,2,3)$. Similarly, by the second drawing in Fig. 1 the rotation of 5 in $T[\{0,2,3,4,5\}]$ is $(2,3,0,4)$, since the inverse would not be compatible with $(1,2,3,4)$. By the third drawing in Fig. 1, the rotation of 5 in $T[\{0,1,3,4,5\}]$ is $(0,1,3,4)$ or $(0,4,3,1)$, but neither of them is compatible with both $(1,2,3,4)$ and $(2,3,0,4)$; a contradiction.

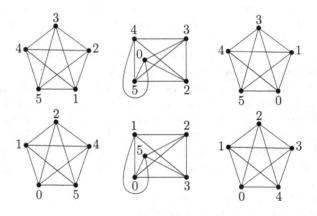

Fig. 1. Simple realizations of all six complete subgraphs of A with five vertices.

4 Proof of Theorem 1

Let $A = (K_n, \mathcal{X})$ be a given complete abstract topological graph with vertex set $[n] = \{1,2,\ldots,n\}$. The algorithm from [13] for deciding simple realizability of A has the following three main steps: computing the rotation system, determining the homotopy class of every edge with respect to the edges incident with one chosen vertex v, and computing the number of crossings of every pair of edges in a crossing-optimal drawing with the rotation system and homotopy class fixed from the previous steps. We follow the algorithm and analyze each step in detail.

Step 1: Computing the Extended Rotation System
This step is based on the proof of Proposition 3; see [13, Proposition 3].

1(a) Realizability of 5-tuples. For every 5-tuple Q of vertices of A, the algorithm tests whether $A[Q]$ is simply realizable. If not, then the 5-tuple certifies that A is not simply realizable. If $A[Q]$ is simply realizable, then by Proposition 3, the algorithm computes a rotation system $\mathcal{R}(Q)$ such that the rotation system of every simple realization of $A[Q]$ is either $\mathcal{R}(Q)$ or the inverse of $\mathcal{R}(Q)$.

1(b) Orienting 5-tuples. For every 5-tuple $Q \subseteq [n]$, the algorithm selects an orientation $\Phi(\mathcal{R}(Q))$ of $\mathcal{R}(Q)$ so that for every pair of 5-tuples Q, Q' having four common vertices and for each $x \in Q \cap Q'$, the rotations of x in $\Phi(\mathcal{R}(Q))$ and $\Phi(\mathcal{R}(Q'))$ are compatible. If there is no such orientation map Φ, the AT-graph A is not simply realizable. We show that in this case there is a set S of six vertices of A that certifies this.

Let Q_1, Q_2 be two 5-tuples with four common elements, let $\mathcal{R}_1$ be a rotation system on Q_1 and let $\mathcal{R}_2$ be a rotation system on Q_2. We say that $\mathcal{R}_1$ and $\mathcal{R}_2$ are *compatible* if for every $x \in Q_1 \cap Q_2$, the rotations of x in $\mathcal{R}_1$ and $\mathcal{R}_2$ are compatible.

Let $\mathcal{G}$ be the graph with vertex set $\binom{[n]}{5}$ and edge set consisting of those pairs $\{Q, Q'\}$ whose intersection has size 4. For every edge $\{Q, Q'\}$ of $\mathcal{G}$, at most one orientation of $\mathcal{R}(Q')$ is compatible with $\mathcal{R}(Q)$. If no orientation of $\mathcal{R}(Q')$ is compatible with $\mathcal{R}(Q)$, then the 6-tuple $S = Q \cup Q'$ certifies that A is not simply realizable. We may thus assume that for every edge $\{Q, Q'\}$ of $\mathcal{G}$, exactly one orientation of $\mathcal{R}(Q')$ is compatible with $\mathcal{R}(Q)$. Let $\mathcal{E}$ be the set of those edges $\{Q, Q'\}$ of $\mathcal{G}$ such that $\mathcal{R}(Q)$ and $\mathcal{R}(Q')$ are not compatible.

Call a set $W \subseteq \binom{[n]}{5}$ *orientable* if there is an orientation map Φ assigning to every rotation system $\mathcal{R}(Q)$ with $Q \in W$ either $\mathcal{R}(Q)$ itself or its inverse $(\mathcal{R}(Q))^{-1}$, such that for every pair of 5-tuples $Q, Q' \in W$ with $|Q \cap Q'| = 4$, the rotation systems $\Phi(\mathcal{R}(Q))$ and $\Phi(\mathcal{R}(Q'))$ are compatible.

Lemma 4. *If $\binom{[n]}{5}$ is not orientable, then there is a 6-tuple $S \subseteq [n]$ such that $S5$ is not orientable.*

Proof. Clearly, $\binom{[n]}{5}$ is not orientable if and only if $\mathcal{G}$ has a cycle with an odd number of edges from $\mathcal{E}$. Call such a cycle a *nonorientable* cycle. We claim that if $\mathcal{G}$ has a nonorientable cycle, then $\mathcal{G}$ has a nonorientable triangle. Let $\mathcal{C}(\mathcal{G})$ be the cycle space of $\mathcal{G}$. The parity of the number of edges of $\mathcal{E}$ in $\mathcal{K} \in \mathcal{C}(\mathcal{G})$ is a linear form on $\mathcal{C}(\mathcal{G})$. Hence, to prove our claim, it is sufficient to show that $\mathcal{C}(\mathcal{G})$ is generated by triangles.

Suppose that $\mathcal{K} = F_1 F_2 \ldots F_k$, with $k \geq 4$, is a shortest cycle in $\mathcal{G}$ that is not a sum of triangles in $\mathcal{C}(\mathcal{G})$. Then $\mathcal{K}$ is an induced cycle in $\mathcal{G}$, that is, $|F_i \cap F_j| \leq 3$ if $2 \leq |i - j| \leq k - 2$. Let $z \in F_1 \setminus F_2$. Then $z \in F_k$, otherwise $|F_k \cap F_1 \cap F_2| = 4$. Let i be the smallest index such that $i \geq 3$ and $z \in F_i$. We have $i \geq 4$, otherwise $|F_1 \cap F_3| = |(F_1 \cap F_2 \cap F_3) \cup \{z\}| = 4$. For every

$j \in \{2, \ldots, i-2\}$, let $F'_j = (F_j \cap F_{j+1}) \cup \{z\}$. Then $\mathcal{K}$ is the sum of the closed walk $\mathcal{K}' = F_1 F'_2 \ldots F'_{i-2} F_i \ldots F_k$ and the triangles $F_1 F_2 F'_2, F_{i-1} F_i F'_{i-2}, F_j F_{j+1} F'_j$ for $j = 2, \ldots, i-2$ and $F_{j+1} F'_j F'_{j+1}$ for $j = 2, \ldots, i-3$; see Fig. 2. Since the length of $\mathcal{K}'$ is $k-1$, we have a contradiction with the choice of $\mathcal{K}'$.

Let $Q_1 Q_2 Q_3$ be a nonorientable triangle in $\mathcal{G}$. The 5-tuples Q_1, Q_2, Q_3 have either three or four common elements. Suppose that $|Q_1 \cap Q_2 \cap Q_3| = 4$ and let $\{u, v, w, z\} = Q_1 \cap Q_2 \cap Q_3$. Then we may orient the rotation systems $\mathcal{R}(Q_1), \mathcal{R}(Q_2)$ and $\mathcal{R}(Q_3)$ so that the rotation of u in each of the orientations is compatible with (v, w, z). This implies that the rotations of u in the resulting rotation systems are pairwise compatible. Thus, the resulting rotation systems are pairwise compatible, a contradiction. Hence, we have $|Q_1 \cap Q_2 \cap Q_3| = 3$, which implies that $|Q_1 \cup Q_2 \cup Q_3| = 6$. Setting $S = Q_1 \cup Q_2 \cup Q_3$, the set $\binom{S}{5}$ is not orientable. $\qquad\square$

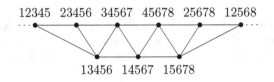

Fig. 2. Triangulating a cycle in $\mathcal{G}$. The vertices in the first row represent the vertices $F_1, \ldots, F_i$ of the original cycle $\mathcal{K}$, the vertices in the second row represent the vertices $F'_2, \ldots, F'_{i-2}$.

If $\binom{[n]}{5}$ is orientable, there are exactly two possible solutions for the orientation map. We will assume that the rotation of 1 in $\Phi(\mathcal{R}(\{1, 2, 3, 4, 5\}))$ is compatible with $(2, 3, 4)$, so that there is at most one solution Φ.

1(c) Computing the Rotations of Vertices. Having oriented the rotation system of every 5-tuple, the algorithm now computes the rotation of every $x \in [n]$, as the cyclic permutation compatible with the rotation of x in every $\Phi(\mathcal{R}(Q))$ such that $x \in Q \in \binom{[n]}{5}$. We show that this is always possible. The following lemma forms the core of the argument.

Lemma 5. Let $k \geq 4$. For every $F \in \binom{[k+1]}{k}$, let π_F be a cyclic permutation of F such that for every pair $F, F' \in \binom{[k+1]}{k}$, the cyclic permutations π_F and π'_F are compatible. Then there is a cyclic permutation $\pi_{[k+1]}$ of $[k+1]$ compatible with all the cyclic permutations π_F with $F \in \binom{[k+1]}{k}$.

1(d) Computing the Rotations of Crossings. For every pair of edges $\{\{u, v\}, \{x, y\}\} \in \mathcal{X}$, the algorithm determines the rotation of their crossing from the rotations of the vertices u, v, x, y. This finishes the computation of the extended rotation system.

Step 2: Determining the Homotopy Classes of the Edges

Let v be a fixed vertex of A and let $S(v)$ be a topological star consisting of v and all the edges incident with v, drawn in the plane so that the rotation of v agrees with the rotation computed in the previous step. For every edge $e = xy$ of A not incident with v, the algorithm computes the order of crossings of e with the subset $E_{v,e}$ of edges of $S(v)$ that e has to cross. By Proposition 3 (2), the five-vertex AT-subgraphs of A determine the relative order of crossings of e with every pair of edges of $E_{v,e}$. Define a binary relation $\prec_{x,y}$ on $E_{v,e}$ so that $vu \prec_{x,y} vw$ if the crossing of e with vu is closer to x than the crossing of e with vw. If $\prec_{x,y}$ is acyclic, it defines a total order of crossings of e with the edges of $E_{v,e}$. If $\prec_{x,y}$ has a cycle, then it also has an oriented triangle vu_1, vu_2, vu_3. This means that the AT-subgraph of A induced by the six vertices v, u_1, u_2, u_3, x, y is not simply realizable.

We recall that the *homotopy class* of a curve φ in a surface Σ relative to the boundary of Σ is the set of all curves that can be obtained from φ by a continuous deformation within Σ, keeping the boundary of Σ fixed.

The *homotopy class of e* is determined by the following combinatorial data: the set $E_{v,e}$, the total order $\prec_{x,y}$ in which the edges of $E_{v,e}$ cross e, the rotations of these crossings, and the rotations of the vertices x and y. Consider the star $S(v)$ drawn on the sphere. Cut circular holes around the points representing all the vertices except v, and let Σ be the resulting surface with boundary. Let x_e and y_e be fixed points on the boundaries of the two holes around x and y, respectively, so that the orders of these points corresponding to all the edges of A on the boundaries of the holes agree with the computed rotation system. Draw a curve φ_e with endpoints x_e and y_e satisfying all the combinatorial data of e. Now the homotopy class of e is defined as the homotopy class of φ_e in Σ relative to the boundary of Σ.

Step 3: Computing the Minimum Crossing Numbers

For every pair of edges e, f, let $\mathrm{cr}(e, f)$ be the minimum possible number of crossings of two curves from the homotopy classes of e and f. Similarly, let $\mathrm{cr}(e)$ be the minimum possible number of self-crossings of a curve from the homotopy class of e. The numbers $\mathrm{cr}(e, f)$ and $\mathrm{cr}(e)$ can be computed in polynomial time in any 2-dimensional surface with boundary [3,17]. In our special case, the algorithm is relatively straightforward [13].

We use the key fact that from the homotopy class of every edge, it is possible to choose a representative such that the crossing numbers $\mathrm{cr}(e, f)$ and $\mathrm{cr}(e)$ are all realized simultaneously [13]. This is a consequence of the following facts.

Lemma 6 [9]. *Let γ be a curve on an orientable surface S with endpoints on the boundary of S that has more self-intersections than required by its homotopy class. Then there is a singular 1-gon or a singular 2-gon bounded by parts of γ.*

Here a *singular 1-gon* of a curve $\gamma : [0, 1] \to S$ is an image $\gamma[\alpha]$ of an interval $\alpha \subset [0, 1]$ such that γ identifies the endpoints of α and the resulting loop is contractible in S. A *singular 2-gon* of γ is an image of two disjoint intervals

$\alpha, \beta \subset [0,1]$ such that γ identifies the endpoints of α with the endpoints of β and the resulting loop is contractible in S.

Lemma 7 [6,9]. *Let C_1 and C_2 be two simple curves on a surface S such that the endpoints of C_1 and C_2 lie on the boundary of S. If C_1 and C_2 have more intersections than required by their homotopy classes, then there is an innermost embedded 2-gon between C_1 and C_2, that is, two subarcs of C_1 and C_2 bounding a disc in S whose interior is disjoint with C_1 and C_2.*

Whenever there is a singular 1-gon, a singular 2-gon, or an embedded innermost 2-gon in a system of curves on S, it is possible to eliminate the 1-gon or 2-gon by a homotopy of the corresponding curves, which decreases the total number of crossings.

For the rest of the proof, we fix a drawing D of A such that its rotation system is the same as the rotation system computed in Step 1, the edges of $S(v)$ do not cross each other, every other edge is drawn as a curve in its homotopy class computed in Step 2, and under these conditions, the total number of crossings is the minimum possible. Then every edge f of $S(v)$ crosses every other edge e at most once, and this happens exactly if $\{e, f\} \in \mathcal{X}$. Moreover, for every pair of edges e_1, e_2 not incident with v, the corresponding curves in D cross exactly $cr(e_1, e_2)$ times, and the curve representing e_1 has $cr(e_1)$ self-crossings. Hence, A is simply realizable if and only if all the edges e_1, e_2 not incident with v satisfy $cr(e_1) = 0$, $cr(e_1, e_2) \leq 1$, and $cr(e_1, e_2) = 1 \Leftrightarrow \{e_1, e_2\} \in \mathcal{X}$. Moreover, if A is simply realizable, then D is a simple realization of A.

We further proceed in four substeps. Due to space limitations, we only include a short sketch of the substeps 3(b)–3(d).

3(a) Characterization of the Homotopy Classes. Let $w_1, w_2, \ldots, w_{n-1}$ be the vertices of A adjacent to v so that the rotation of v is $(w_1, w_2, \ldots, w_{n-1})$. Let $w_a w_b$ be an edge such that $1 \leq a < b \leq n-1$. Since every AT-subgraph of A with 4 or 5 vertices is simply realizable, we have the following conditions on the homotopy class of $w_a w_b$. Refer to Fig. 3.

Observation 8. *Suppose that $\{w_a w_b, v w_c\} \in \mathcal{X}$; that is, $v w_c \in E_{v, w_a w_b}$. If $a < c < b$, then the rotation of the crossing of $w_a w_b$ with $v w_c$ is (w_a, w_c, w_b, v). If $c < a$ or $b < c$, then the rotation of the crossing is (w_b, w_c, w_a, v).* □

Observation 8 implies that the homotopy class of the edge $w_a w_b$ is determined by a permutation of $E_{v, w_a w_b}$ that determines the order in which $w_a w_b$ crosses the edges in $E_{v, w_a w_b}$. The next observation further restricts this permutation.

Observation 9. *Suppose that $v w_c, v w_d \in E_{v, w_a w_b}$. If $a < c < d < b$, then $v w_c \prec_{w_a, w_b} v w_d$. If (c, d, a, b) is compatible with $(1, 2, \ldots, n-1)$ as cyclic permutations, then $v w_d \prec_{w_a, w_b} v w_c$.* □

On the other hand, it is easy to see that every homotopy class satisfying Observations 8 and 9 has a representative that is a simple curve. Therefore, $cr(w_a w_b) = 0$.

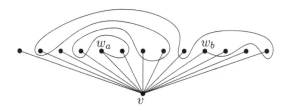

Fig. 3. A drawing of a "typical" edge $w_a w_b$ and the star $S(v)$.

3(b) The Parity of the Crossing Numbers. It can be shown that if e_1 and e_2 are independent edges not incident with v, then $cr(e_1, e_2)$ is odd if and only if $\{e_1, e_2\} \in \mathcal{X}$. It can also be shown that if e_1 and e_2 are adjacent edges not incident with v, then $cr(e_1, e_2)$ is even. It follows that A is realizable if and only if every pair of edges in D crosses at most once.

3(c) Multiple Crossings of Adjacent Edges. Next we show that adjacent edges do not cross in D, otherwise some AT-subgraph of A with five vertices is not simply realizable. This part is rather straightforward, although the full proof is not short. Let $w_a w_b$ and $w_a w_c$ be two adjacent edges. By symmetry, we may assume that $a < b < c$. We will consider cyclic intervals (a, b), (b, c) and $(c, a) = (c, n-1] \cup [1, a)$. We define the following subsets of $E_{v, w_a w_b}$ and $E_{v, w_a w_c}$. For each of the three cyclic intervals (i, j), let $F_b(i, j) = \{vw_k \in E_{v, w_a w_b}; k \in (i, j)\}$ and $F_c(i, j) = \{vw_k \in E_{v, w_a w_c}; k \in (i, j)\}$. We will also write $\prec_b$ as a shortcut for $\prec_{w_a w_b}$ and $\prec_c$ as a shortcut for $\prec_{w_a w_c}$. By symmetry, we have two general cases: (I) $w_a w_b$ does not cross vw_c and $w_a w_c$ does not cross vw_b, and (II) $w_a w_b$ does not cross vw_c and $w_a w_c$ crosses vw_b.

For case (I), one can observe the following conditions; we omit the proofs.

Observation 10.

(1) We have $F_b(c, a) \subseteq F_c(c, a)$ and $F_c(a, b) \subseteq F_b(a, b)$.
(2) The sets $F_b(b, c)$ and $F_c(b, c)$ are disjoint.
(3) If $vw_d \in F_b(b, c)$ and $vw_e \in F_c(b, c)$, then $d < e$.
(4) Let $vw_d \in F_b(a, b) \cap F_c(a, b)$ and $vw_e \in F_b(c, a) \cap F_c(c, a)$. Then $vw_d \prec_b vw_e \Leftrightarrow vw_d \prec_c vw_e$.
(5) Let $vw_d \in F_c(a, b)$ and $vw_e \in F_b(b, c)$. Then $vw_d \prec_b vw_e$. Similarly, if $vw_d \in F_b(c, a)$ and $vw_e \in F_c(b, c)$, then $vw_d \prec_c vw_e$.

We show that Observation 10 implies that $cr(w_a w_b, w_a w_c) = 0$. Refer to Fig. 4. Start with drawing the edges $w_a w_b$ and $w_a w_c$ without crossing. Conditions (2) and (4) imply that there is a total order $\prec$ on $E_{v, w_a w_b} \cup E_{v, w_a w_c}$ that is a common extension of $\prec_b$ and $\prec_c$. Let vw_i be the $\prec$-largest element of $F_b(c, a) \cup F_c(a, b)$. Condition (5) implies that all edges vw_j from $F_b(b, c) \cup F_c(b, c)$ satisfy $vw_i \prec vw_j$. Condition (1) implies that we can draw the edges vw_j with $vw_j \preceq vw_i$ like in the figure. Conditions (2), (3) and (5) imply that we can draw the edges vw_j with $vw_i \prec vw_j$ like in the figure. The remaining edges of $S(v)$ can be drawn easily. In this way we obtain a simple drawing with noncrossing representatives of the homotopy classes of $w_a w_b$ and $w_a w_c$.

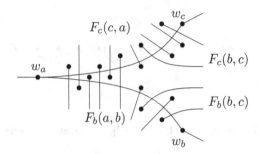

Fig. 4. A drawing of the edges $w_a w_b$, $w_a w_c$ and parts of edges of $S(v)$ in case (I), where $w_a w_b$ and $w_a w_c$ do not cross. The rotation of v is compatible with the counterclockwise cyclic order of the parts of the edges drawn.

The analysis of case (II) is similar.

3(d) Detecting Multiple Crossings of Independent Edges. Finally, we show by induction that if two independent edges cross more than once, then there is a five-vertex AT-subgraph that forces this, possibly for a different pair of edges. In this part, we strongly rely on the established fact that adjacent edges do not cross in D. We avoid a tedious case analysis by not continuing in the approach chosen for adjacent pairs of edges.

Let $e = w_a w_b$ and $f = w_c w_d$ be two independent edges that cross more than once in D. In the subgraph of D formed by the two edges e and f, the vertices w_a, w_b, w_c, w_d are incident to a common face, since adjacent edges do not cross in D and every pair of the four vertices w_a, w_b, w_c, w_d is connected by an edge. We assume without loss of generality that w_a, w_b, w_c, w_d are incident to the outer face. That is, we may draw a simple closed curve γ containing the vertices w_a, w_b, w_c, w_d but no interior points of e or f, such that the relative interiors of e and f are inside γ.

Suppose that e and f cross an even number of times. The edge e splits the region inside γ into two regions, $R_0(e)$ and $R_1(e)$, where $R_0(e)$ is the region that does not contain the endpoints of f on its boundary. Similarly, f splits the region inside γ into regions $R_0(f)$ and $R_1(f)$ where $R_0(f)$ is the region that does not contain the endpoints of e on its boundary.

By Lemma 7, there is an innermost embedded 2-gon between e and f. For brevity, we call an innermost embedded 2-gon shortly a *bigon*. For a bigon B, by B^o we denote the open region inside B and we call it the *inside of B*. There are four possible types of bigons between e and f, according to the regions $R_i(e)$ and $R_j(f)$ in which their insides are contained. For $i, j \in \{0, 1\}$, we call a bigon B an *ij-bigon* if $B^o \subseteq R_i(e) \cap R_j(f)$; see Fig. 5.

Since D is a drawing realizing the crossing number $\mathrm{cr}(e, f)$, there is at least one vertex of D inside every bigon. The graph induced by v, the endpoints of e and f, and a set of vertices intersecting all bigons, certifies that e and f have at least $\mathrm{cr}(e, f)$ crossings forced by their homotopy classes.

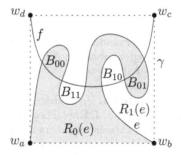

Fig. 5. Four types of bigons between e and f. An ij-bigon is denoted by B_{ij}.

The following lemma quickly solves the case when there is at least one 00-bigon between e and f.

Lemma 11. *If e and f cross evenly and there is a 00-bigon B between e and f in D, then there is a vertex w_i inside B, and the AT-subgraph of A induced by the 5-tuple $Q = \{w_a, w_b, w_c, w_d, w_i\}$ is not simply realizable.*

We are left with the case that there is no 00-bigon between e and f.

Observation 12. *If e and f cross evenly and at least twice in D, and there is no 00-bigon between e and f, then there is a 01-bigon and a 10-bigon between e and f.*

For a subset W of vertices of A containing v and the endpoints of two edges g and f, let $\mathrm{cr}_W(g, f)$ be the minimum possible number of crossings of two curves from the homotopy classes of g and f determined by $A[W]$, by a procedure analogous to the one in Step 2.

The following lemma proves the induction step in the case when e and f cross an even number of times.

Lemma 13. *If e and f cross evenly and at least twice in D, and there is no 00-bigon between e and f, then there is a proper subset W of vertices of A and an edge g independent from f such that $\mathrm{cr}_W(g, f) \geq 2$.*

If e and f cross an odd number of times, one can easily find another pair of independent edges crossing evenly and more than once. We omit the details.

Acknowledgements. I thank Martin Balko for his comments on an earlier version of the manuscript. I also thank all the reviewers for their suggestions for improving the presentation.

References

1. Ábrego, B.M., Aichholzer, O., Fernández-Merchant, S., Hackl, T., Pammer, J., Pilz, A., Ramos, P., Salazar, G., Vogtenhuber, B.: All good drawings of small complete graphs, EuroCG 2015, Book of abstracts, pp. 57–60 (2015)
2. Aichholzer, O.: Personal communication. 2014
3. Armas-Sanabria, L., González-Acuña, F., Rodríguez-Viorato, J.: Self-intersection numbers of paths in compact surfaces. J. Knot Theor. Ramif. **20**(3), 403–410 (2011)
4. Balko, M., Fulek, R., Kynčl, J.: Crossing numbers and combinatorial characterization of monotone drawings of K_n. Discrete Comput. Geom. **53**(1), 107–143 (2015)
5. Chimani, M.: Facets in the crossing number polytope. SIAM J. Discrete Math. **25**(1), 95–111 (2011)
6. Farb, B., Thurston, B.: Homeomorphisms and simple closed curves, unpublished manuscript
7. Gioan, E.: Complete graph drawings up to triangle mutations. In: Kratsch, D. (ed.) WG 2005. LNCS, vol. 3787, pp. 139–150. Springer, Heidelberg (2005)
8. Harborth, H., Mengersen, I.: Drawings of the complete graph with maximum number of crossings, In: Proceedings of the Twenty-third Southeastern International Conference on Combinatorics, Graph Theory, and Computing, (Boca Raton, FL, 1992), Congressus Numerantium, 88 pp. 225–228 (1992)
9. Hass, J., Scott, P.: Intersections of curves on surfaces. Isr. J. Math. **51**(1–2), 90–120 (1985)
10. Kratochvíl, J., Lubiw, A., Nešetřil, J.: Noncrossing subgraphs in topological layouts. SIAM J. Discrete Math. **4**(2), 223–244 (1991)
11. Kratochvíl, J., Matoušek, J.: NP-hardness results for intersection graphs. Commentationes Math. Univ. Carol. **30**, 761–773 (1989)
12. Kynčl, J.: Enumeration of simple complete topological graphs. Eur. J. Comb. **30**(7), 1676–1685 (2009)
13. Kynčl, J.: Simple realizability of complete abstract topological graphs in P. Discrete Comput. Geom. **45**(3), 383–399 (2011)
14. Kynčl, J.: Improved enumeration of simple topological graphs. Discrete Comput. Geom. **50**(3), 727–770 (2013)
15. Mutzel, P.: Recent advances in exact crossing minimization (extended abstract). Electron. Notes Discrete Math. **31**, 33–36 (2008)
16. Pach, J., Tóth, G.: How many ways can one draw a graph? Combinatorica **26**(5), 559–576 (2006)
17. Schaefer, M., Sedgwick, E., Štefankovič, D.: Computing Dehn twists and geometric intersection numbers in polynomial time, In: Proceedings of the 20th Canadian Conference on Computational Geometry, CCCG 2008, pp. 111–114 2008

The Utility of Untangling

Vida Dujmović[(⊠)]

School of Computer Science and Electrical Engineering, University of Ottawa,
Ottawa, Canada
vida.dujmovic@uottawa.ca

Abstract. In this paper we show how techniques developed for untangling planar graphs by Bose *et al.* [Discrete & Computational Geometry 42(4): 570–585 (2009)] and Goaoc *et al.* [Discrete & Computational Geometry 42(4): 542–569 (2009)] imply new results about some recent graph drawing models. These include column planarity, universal point subsets, and partial simultaneous geometric embeddings (with or without mappings). Some of these results answer open problems posed in previous papers.

1 Introduction

A *geometric graph* is a graph whose vertex set is a set of distinct points in the plane and each pair of adjacent vertices $\{v, w\}$ is connected by a line segment $\overline{vw}$ that intersects only the two vertices. A geometric graph is *planar* if its underlying combinatorial graph is planar. It is *plane* if no two edges cross other than in a common endpoint. A *straight-line crossing-free drawing* of a planar graph is a representation of that graph by a plane geometric graph.

Given a geometric planar graph, possibly with many crossings, to *untangle* it, means to move some of its vertices to new locations (that is, change their coordinates) such that the resulting geometric graph is plane. The goal is to do so by moving as few vertices as possible, or in other words, by keeping the locations of as many vertices as possible unchanged (that is, *fixed*). A series of papers have studied untangling of planar graphs or subclasses of planar graphs [9,11,14,24,26,28,30]. The best known (lower) bound for general planar graphs is due to Bose *et al.* [9] who proved that every n-vertex geometric planar graph can be untangled while keeping the locations of at least $\Omega(n^{1/4})$ vertices fixed. On the other hand, Cano *et al.* [11] showed that for all large enough n, there exists an n-vertex geometric planar graph that cannot be untangled while keeping the locations of more than $\omega(n^{0.4948})$ vertices fixed.

The purpose of this paper is to highlight how the techniques developed by Bose *et al.* [9] and Goaoc *et al.* [24] can be used to establish new results on several recently studied graph drawing problems. Before presenting the new results we state the two key lemmas that are at the basis of all the results. The statements of these two lemmas are new, but their proofs are contained in and directly inferred by the work described in [9,24].

© Springer International Publishing Switzerland 2015
E. Di Giacomo and A. Lubiw (Eds.): GD 2015, LNCS 9411, pp. 321–332, 2015.
DOI: 10.1007/978-3-319-27261-0_27

Let G be a plane triangulation (that is, an embedded simple planar graph each of whose faces is bounded by a 3-cycle). Canonical orderings of plane triangulations were introduced by de Fraysseix *et al.* [22]. They proved that G has a vertex ordering $\sigma = (v_1 := x, v_2 := y, v_3, \ldots, v_n := z)$, called a *canonical ordering*, with the following properties. Define G_i to be the embedded subgraph of G induced by $\{v_1, v_2, \ldots, v_i\}$. Let C_i be the subgraph of G induced by the edges on the boundary of the outer face of G_i. Then

- x, y and z are the vertices on the outer face of G.
- For each $i \in \{3, 4, \ldots, n\}$, C_i is a cycle containing xy.
- For each $i \in \{3, 4, \ldots, n\}$, G_i is biconnected and *internally 3-connected*; that is, removing any two interior vertices of G_i does not disconnect it.
- For each $i \in \{3, 4, \ldots, n\}$, v_i is a vertex of C_i with at least two neighbours in C_{i-1}, and these neighbours are consecutive on C_{i-1}.

The following structure was defined first in Bose *et al.* [9]. Using the above notation, a *frame* $\mathcal{F}$ of G is the oriented subgraph of G with vertex set $V(\mathcal{F}) := V(G)$, where:

- Edges xy, xv_1 and v_1y are in $E(\mathcal{F})$ where xy is oriented from x to y, xv_1 is oriented from x to v_1 and v_1y is oriented from v_1 to y.
- For each $i \in \{4, 5 \ldots, n\}$ in the canonical ordering σ of G, edges pv_i and v_ip' are in $E(\mathcal{F})$, where p and p' are the first and the last neighbour, respectively, of v_i along the path in C_{i-1} from x to y not containing edge xy. Edge pv_i is oriented from p to v_i, and edge v_ip' is oriented from v_i to p'.

By definition, $\mathcal{F}$ is a directed acyclic graph with one source x and one sink y. $\mathcal{F}$ defines a partial order $<_{\mathcal{F}}$ on $V(\mathcal{F})$, where $v <_{\mathcal{F}} w$ whenever there is a directed path from v to w in $\mathcal{F}$.

Subsequently, it has been observed that a frame of G can also be obtained by taking the union of any two trees in Schnyder 3-tree-decompositions where the orientation of the edges in one of the two trees is reversed. See, for example, page 13 in Di Giacomo *et al.* [16] for this alternative formulation.

Recall that a *chain* (*antichain*) in a partial order is a subset of its elements that are pairwise comparable (incomparable). Given a partial order $(V, \leq)$ on a set of vertices V of some graph, we will often refer to a *chain* $V' \subseteq V$ (or antichain) and by that mean a subset of vertices of V that form a chain (antichain) in the given partial order $(V, \leq)$. We also say that a chain V' contains a chain V'' if V' and V'' are both chains in $(V, \leq)$ and $V'' \subseteq V'$.

Consider an n-vertex planar graph G and a set P of $k \leq n$ points in the plane together with a bijective mapping from a set V_k of k vertices in G to P. Let D be a straight-line crossing-free drawing of G. We say that D *respects* the given mapping if each vertex of V_k is represented in D by its image point as determined by the given mapping.

The following two lemmas are implicit in the work of Bose *et al.* [9] and Goaoc *et al.* [24]. Parts (b), (c) and consequently (d), in Lemma 1, are due to Goaoc *et al.* [24]. Note that, unlike here, the results of Goaoc *et al.* [24] are not expressed in terms of a chain in the frame of G but an equivalent structure: a simple path L in a plane triangulation, connecting two vertices x and y on the outer face x, y, z with the property that all chords of L lie on one side of L and z lies on the other.

Consider a graph G, a set $S \subseteq V(G)$ and a set P of $|S|$ points in the plane together with a bijective mapping from S to P. For a vertex $v \in S$ mapped to a point $p \in P$, let $\mathsf{x}(v)$ denote the x-coordinate of p.

Lemma 1. *[9,24] Let G be an n-vertex plane triangulation with a partial order $<_\mathcal{F}$ associated with a frame $\mathcal{F}$ of G. Let $C \subseteq V$ be a chain in $<_\mathcal{F}$. Let H be the graph induced in G by a maximal chain that contains C in $<_\mathcal{F}$. The embedding of H is implied by the embedding of G. Then:*

(a) H is a 2-connected outerplane graph, i.e. a 2-connected embedded outerplanar graph all of whose vertices lie on the cycle bounding the infinite face.
(b) Let $I \subseteq V(H)$ such that if $v, w \in I$ and $vw \in E(H)$ then vw lies on the outer face of H. Let P be any set of $|I|$ points in the plane where no two points of P have the same x-coordinate. Given a bijective mapping from I to P such that, for every two vertices $v, w \in I$, $v <_\mathcal{F} w$ if and only if $\mathsf{x}(v) < \mathsf{x}(w)$, there exists a straight-line crossing-free drawing of G that respects the given mapping.
(c) There exists such a set I with at least $(V(H) + 1)/2$ vertices.
(d) There exists such a set I with at least $|C|/3$ vertices of C.

While the lower bound in part (c) is stronger than the lower bound in part (d), part (d) ensures that a fraction of vertices of C are used. That will be critical for some applications (see Theorem 2 in Sect. 2 and Theorem 6 in Sect. 4.2). Part (d) follows from (b) as follows. Consider the graph H' induced in H by the vertices of C. By part (a), H' is outerplanar. Thus its vertices can be coloured with three colours such that adjacent vertices in H' receive distinct colours. Thus there exists an independent set I in H' that contains at least $|C|/3$ vertices of C. The condition imposed on the vertex set I in part (b) are immediate since I is an independent set in H' and H.

Note that, in an interesting recent development, Di Giacomo *et al.* [17] proved that every n-vertex plane triangulation has a frame where some chain has size at least $n^{1/3}$. Thus by part (a), $|V(H)| \geq n^{1/3}$ in that frame and consequently, every n-vertex plane triangulation has a 2-connected outerplane graph of size at least $n^{1/3}$ as an embedded induced subgraph.

The following is the second key lemma.

Lemma 2. *[9] Let G be an n-vertex plane triangulation with a partial order $<_\mathcal{F}$ associated with a frame $\mathcal{F}$ of G and the total order $<_\sigma$ associated with the corresponding canonical ordering. Let $A \subseteq V$ be an antichain in $<_\mathcal{F}$. Let P be any set of $|A|$ points in the plane where no two points of P have the same x-coordinate. Given a bijective mapping from A to P such that, for every two vertices $v, w \in A$, $v <_\sigma w$ if and only if $\mathsf{x}(v) < \mathsf{x}(w)$, then there exists a straight-line crossing-free drawing of G that respects the given mapping.*

2   Column Planarity

Given a planar graph G, a set $R \subset V(G)$ is *column planar* in G if the vertices of R can be assigned x-coordinates such that given any arbitrary assignment of y-coordinates to R, there exists a straight-line crossing free drawing of G that respects the implied mapping of vertices of R to the plane.

The column planar sets were first defined by Evans *et al.* [21]. A slightly stronger notion[1] was used earlier (although not named) in [9] (see Lemma 1 and Lemma 6 in [9]) where such sets were studied and used to prove Lemma 2 in the previous section. In particular, define a set $R \subset V(G)$ as *strongly column planar* if the following holds: there exists a total order μ on R such that

(a) given any set P of $|R|$ points in the plane where no two points have the same x-coordinate; and,
(b) given a bijective mapping from R to P such that, for every two vertices $v, w \in R$, $v <_\mu w$ if and only if $\mathsf{x}(v) < \mathsf{x}(w)$,

then there exists a straight-line crossing-free drawing of G that respects the given mapping. Being strongly column planar implies being column planar but not the converse. We use this slightly stronger notion as it is needed in the later sections.

Notions similar to column planarity were studied by Estrella-Balderrama *et al.* [20] and Di Giacomo *et al.* [15].

It is implicit in the work of Bose *et al.* [9] (see the proof of Lemma 2 in [9]) that every tree has a strongly column planar set of size at least $n/2$. For column planar sets, this result is improved to $14n/17$ by Evans *et al.* [21]. Having a bound greater than $n/2$ is critical for an application of column planarity to partial simultaneous geometric embedding with mapping [21]. Barba *et al.* [6] prove that every n-vertex outerplanar graph has a column planar set of size at least $n/2$.[2]

Evans *et al.* [21] pose as an open problem the question of developing any bound for column planar sets in general planar graphs. We provide here the first non-trivial (that is, better than constant) bound for this problem.

Theorem 1. *For every n, every n-vertex planar graph G has a (strongly) column planar set of size at least $\sqrt{n/2}$.*

Proof. If $|V(G)| \leq 2$, the result is trivially true. Thus we may assume that G is a triangulated plane graph. Let $\mathcal{F}$ be a frame of G, let $<_{\mathcal{F}}$ be its associated partial order, and let σ be the associated canonical ordering. Consider a chain in $<_{\mathcal{F}}$ of maximum size. (Hence, the chain starts with x and ends with y). Let H be the subgraph of G induced by that chain, as defined in Lemma 1. Let $I \subseteq V(H)$ be as defined in Lemma 1(b). Consider any set P of $|I|$ points in the plane where no two points have the same x-coordinate and consider a bijective mapping from

---

[1] With the roles of x and y coordinates reversed.
[2] We suspect that the results and proofs in both [6] and [21] also hold for strongly column planar sets but we have not verified that.

I to P such that, for every two vertices $v, w \in I$, it holds that $v <_{\mathcal{F}} w$ if and only if $\mathsf{x}(v) < \mathsf{x}(w)$. By Lemma 1(b), there exists a straight-line crossing-free drawing of G that respects the given mapping and thus I, as ordered by $<_{\mathcal{F}}$, is a strongly column planar set. By Lemma 1(c), $|I| \geq |V(H)|/2$. Thus if the size of the maximum chain in $<_{\mathcal{F}}$ is at least $\sqrt{2n}$, and thus $|V(H)| \geq \sqrt{2n}$, we are done. Otherwise, by Dilworth's theorem [18], $<_{\mathcal{F}}$ has a partition into at most $\sqrt{2n}$ antichains. By the pigeon-hole principle, there is an antichain in that partition with at least $n/\sqrt{2n} = \sqrt{n/2}$ vertices. Let $A \subseteq V(G)$ be the maximum antichain in $<_{\mathcal{F}}$. Consider any set P of $|A|$ points in the plane where no two points have the same x-coordinate and consider a bijective mapping from A to P such that, for every two vertices $v, w \in A$, it holds that $v <_{\sigma} w$ if and only if $\mathsf{x}(v) < \mathsf{x}(w)$. By Lemma 2, there exists a straight-line crossing-free drawing of G that respects the given mapping and thus A, as ordered by $<_{\sigma}$, is a strongly column planar set. This completes the proof since $|A| \geq \sqrt{n/2}$. $\qquad\square$

We conclude this section by proving a slightly stronger statement (with a slightly weaker bound when $S = V$) than Theorem 1. This stronger statement relies on part (d) of Lemma 1, and is a critical strengthening for some applications, such as partial simultaneous geometric embeddings with mappings (see Theorem 6 in Sect. 4.2).

Theorem 2. *Given any planar graph G and any subset $S \subseteq V$, there exists $R \subseteq S$ such that R is a strongly column planar set of G and $|R| \geq \sqrt{|S|/3}$.*

Proof. If $|V(G)| \leq 2$, the result is trivially true. Thus we may assume that G is a triangulated plane graph. Let $\mathcal{F}$ be a frame of G, let $<_{\mathcal{F}}$ be its associated partial order, and let σ be the associated canonical ordering. Assume first that $<_{\mathcal{F}}$ has a chain C such that $C \subseteq S$ and $|C| \geq \sqrt{3|S|}$. Let H be the subgraph of G induced by a maximal chain that contains C in $<_{\mathcal{F}}$, as defined in Lemma 1. Let $I \subseteq S$ be as defined in Lemma 1(b) and (d). Consider any set P of $|I|$ points in the plane where no two points have the same x-coordinate and consider a bijective mapping from I to P such that, for every two vertices $v, w \in I$, it holds that $v <_{\mathcal{F}} w$ if and only if $\mathsf{x}(v) < \mathsf{x}(w)$. By Lemma 1(b), there exists a straight-line crossing-free drawing of G that respects the given mapping and thus I, as ordered by $<_{\mathcal{F}}$, is a strongly column planar set. By Lemma 1(d), $I \subseteq C$ and $|I| \geq |C|/3$. Thus if $<_{\mathcal{F}}$ has a chain C such that $C \subseteq S$ and $|C| \geq \sqrt{3|S|}$, we are done. Otherwise, by Dilworth's theorem [18], $<_{\mathcal{F}}$, when restricted to S, has a partition into at most $\sqrt{3|S|}$ antichains. By the pigeon-hole principle, there is an antichain $A \subseteq S$ in that partition that has at least $|S|/\sqrt{3|S|} = \sqrt{|S|/3}$ elements. Consider any set P of $|A|$ points in the plane where no two points have the same x-coordinate and consider a bijective mapping from A to P, such that for every two vertices $v, w \in A$, $v <_{\sigma} w$ if and only if $\mathsf{x}(v) < \mathsf{x}(w)$. By Lemma 2, there exists a straight-line crossing-free drawing of G that respects the given mapping and thus A, as ordered by $<_{\sigma}$, is a strongly column planar set. This completes the proof since $|A| \geq \sqrt{|S|/3}$ and $A \subseteq S$. $\qquad\square$

3   Universal Point Subsets

A set of points P is *universal* for a set of planar graphs if every graph from the set has a straight-line crossing-free drawing where each of its vertices maps to a distinct point in P. It is known that, for all large enough n, universal pointsets of size n do not exist for all n-vertex planar graphs – as first proved by de Fraysseix *et al.* [22]. The authors also proved that the $O(n) \times O(n)$ integer grid is universal for all n-vertex planar graphs and thus a universal pointsets of size $O(n^2)$ exists. Currently the best known lower bound on the size of a smallest universal pointset for n-vertex planar graphs is $1.235n - o(n)$ [27] and the best known upper bound is $n^2/4 - O(n)$ [5]. Closing the gap between $\Omega(n)$ and $O(n^2)$ is a major, and likely difficult, graph drawing problem, open since 1988 [22,23].

This motivated the following notion introduced by Angelini *et al.* [2]. A set P of $k \leq n$ points in the plane is a *universal point subset* for all n-vertex planar graphs if the following holds: every n-vertex planar graph G has a subset $S \subseteq V(G)$ of k vertices and a bijective mapping from S to P such that there exists a straight-line crossing-free drawing of G that respects that mapping.

Angelini *et al.* [2] proved that for every n there exists a set of points of size at least $\sqrt{n}$ that is a universal point subset for all n-vertex planar graphs. Di Giacomo *et al.* [16] continued this study and showed that for every n, *every* set P of at most $(\sqrt{\log_2 n} - 1)/4$ points in the plane is a universal point subset for *all* n-vertex planar graphs. They also showed that *every* one-sided convex point set P of at most $n^{1/3}$ points in the plane is a universal point subset for *all* n-vertex planar graphs. The following theorem improves all these results.

Theorem 3. *Every set P of at most $\sqrt{n/2}$ points in the plane is a universal point subset for all n-vertex planar graphs.*

The proof of this lemma can be derived directly from Lemmas 1 and 2, similarly to the proof of Theorem 1, but we will instead prove it using Theorem 1.

Proof. Rotate P to obtain a new pointset P' where no two points of P' have the same x-coordinate. By Theorem 1, every n-vertex planar graph has a strongly column planar set R of size $|P|$. Thus, by the definition of strongly column planar sets, there exists a total order μ on R such that given a bijective mapping from R to P' where for every two vertices $v, w \in R$, $v <_\mu w$ if and only if $x(v) < x(w)$, there exists a straight-line crossing-free drawing of G that respects the given mapping. Such a mapping clearly exists since no two points of P' have the same x-coordinate. Rotating P' back to the original pointset completes the proof. $\square$

It is not known if, for all n, there exist a universal point subset of size $n^{1/2+\epsilon}$ for some $\epsilon > 0$. Better bounds are only known for outerplanar graphs. Namely, every pointset of size n in general position is universal for all n-vertex outerplanar graphs [8,13,25]. Should the results of Barba *et al.* [6] apply to strongly column planar sets, then arguments equivalent to those above would show that every pointset of size $n/2$ is a universal point subset for all n-vertex outerplanar graphs.

4   (Partial) Simultaneous Geometric Embeddings

Simultaneous Geometric Embeddings were introduced by Braß *et al.* [10]. Initially there were two main variants of this problem, one in which the mapping between the vertices of the two graphs is given and another in which the mapping is not given. Since then there has been a plethora of work on the subject for various variants of the problem – see, for example a survey by Bläsius *et al.* [7].

4.1   Without Mapping

Whether the following statement, on simultaneous geometric embeddings, is true is an open question asked by Braß *et al.* [10] in 2003: For all n and for any two n-vertex planar graphs there exists a pointset P of size n such that each of the two graphs has a straight-line crossing-free drawing with its vertices mapped to distinct points of P. The statement is known *not* to be true when "two" is replaced by 7393 and $n = 35$ [12].

This motivates a study of (partial) geometric simultaneous embeddings – various versions of which have been proposed and studied in the literature [7]. We start with the following version.

Two graphs G_1 and G_2, where $|V(G_1)| \geq |V(G_2)|$ are said to have a *geometric simultaneous embedding with no mapping* if there exists a pointset P of size $|V(G_1)|$ such that each of the two graphs has a straight-line crossing-free drawing where all of its vertices are mapped to distinct points in P. Angelini *et al.* [3] write: "What is the largest $k \leq n$ such that every n-vertex planar graph and every k-vertex planar graph admit a geometric simultaneous embedding with no mapping? Surprisingly, we are not aware of any super-constant lower bound for the value of k."

The following theorem answers their questions.

Theorem 4. *For every n and every $k \leq \sqrt{n/2}$, every n-vertex planar graph and every k-vertex planar graph admit a geometric simultaneous embedding with no mapping.*

Proof. Let G_1 and G_2 be the two given planar graphs with $|V(G_1)| = n$ and $|V(G_2)| = k$. By Fáry's theorem, G_2 has a straight-line crossing-free drawing on some set, P_2, of k points. By Theorem 3, G_1 has a straight-line crossing-free drawing where $|P_2|$ vertices of G_1 are mapped to distinct points in P_2. Consider now the set of points, P, defined by the vertices in the drawing of G_1. This set is our desired pointset as it is a set of n points such that each of G_1 and G_2 has a straight-line crossing-free drawing where all of its vertices are mapped to the points in P.   □

Here is another variant of the (partial) geometric simultaneous embedding problem. For $k \leq n$, two n-vertex planar graphs G_1 and G_2 are said to have a *k-partial simultaneous geometric embedding with no mapping* (*k-PSGENM*) if there exists a set P of at least k points in the plane such that each of the two graphs has a straight-line crossing-free drawing where $|P|$ of its vertices are mapped to distinct points of P. Recall that Angelini *et al.* [2] proved that for every n there exists a set of points of size at least $\sqrt{n}$ that is a universal point subset for all n-vertex planar graphs. This implies that, for all n, any two n-vertex planar graphs have an $\sqrt{n}$-partial simultaneous geometric embedding with no mapping. Note however that this does not imply Theorem 4. Namely, if one starts with a straight-line crossing-free drawing of the smaller graph G_2 (say on $\sqrt{n}$ vertices), there is no guarantee with this result that the bigger, n-vertex graph, G_1 can be drawn while using all the points generated by the drawing of G_2.

4.2   With Mapping

The notion of k-*partial simultaneous geometric embedding with mapping* (k-PSGE) is the same as k-PSGENM except that a bijective mapping between $V(G_1)$ and $V(G_2)$ is given and the two drawings have a further restriction that if $v \in V(G_1)$ is mapped to a point in P then the vertex w in $V(G_2)$ that v maps to, has to be mapped to the same point in P. In other words, two n-vertex planar graphs G_1 and G_2 on *the same vertex set, V,* are said to have a k-*partial simultaneous geometric embedding with mapping* (k-PSGE) if there exists a straight-line crossing free drawing D_1 of G_1 and D_2 of G_2 such that there exists a subset $V' \subseteq V$ with $|V'| \geq k$ and each vertex $v \in V'$ is represented by the same point in D_1 and D_2.

It is known that, for every large enough n, there are pairs of n-vertex planar graphs that do not have an n-partial simultaneous geometric embedding with mapping, that is, an n-PSGE [10]. In fact the same is true for simpler families of planar graphs, for example for a tree and a path [4], for a planar graph and a matching [4] and for three paths [10].

k-PSGE was introduced by Evans *et al.* [21] who proved (using their column planarity result) that any two n-vertex trees have an $11n/17$-PSGE. Barba *et al.* [6] proved that any two n-vertex outerplanar graphs have an $n/4$-PGSE. Evans *et al.* [21] also observed that the main untangling result by Bose *et al.* [9] implies that every pair of n-vertex planar graphs has an $\Omega(n^{1/4})$-PSGE. Namely, start with a straight-line crossing-free drawing of G_1. Since the vertex sets of G_1 and G_2 are the same, the drawing of G_1 (or rather the drawing of its vertex set) defines a straight-line drawing of G_2. Untangling G_2 such that $\Omega(n^{1/4})$ of its vertices remain fixed (which is possible by [9]) gives the result.

Theorem 5. *[6] Every pair of n-vertex planar graphs has an $\Omega(n^{1/4})$-partial simultaneous geometric embedding with mapping, that is, it has an $\Omega(n^{1/4})$-PGSE.*

However, the above untangling argument fails if we try to apply it one more time. Namely, consider the following generalization of the k-PGSE problem. Given any set $\{G_1, \ldots, G_p\}$ of $p \geq 2$ n-vertex planar graphs on *the same vertex set, V,* we say that $G_1, \ldots, G_p$ have a k-*partial simultaneous geometric embedding with mapping* (k-PSGE) if there exists a straight-line crossing-free drawing D_i

of each G_i, $i \in \{1, \ldots, p\}$ such that there exists a subset $V' \subseteq V$ with $|V'| \geq k$ and each vertex $v \in V'$ is represented by the same point in all drawings D_i, $i \in \{1, \ldots, p\}$.

If we try to mimic the earlier untangling argument that proves Theorem 5, it fails for $p = 3$ already since we cannot guarantee that when G_3 is untangled the set of its vertices that stays fixed has a non-empty intersection with the set that remained fixed when untangling G_2. It is here that part (d) of Lemma 1 is needed, or rather the stronger result on column planarity from Theorem 2.

Theorem 6. *Any set of $p \geq 2$ n-vertex planar graphs has an $\Omega(n^{1/4^{(p-1)}})$-partial simultaneous geometric embedding with mapping, that is, it has an $\Omega(n^{1/4^{(p-1)}})$-PGSE.*

Proof. Let $\{G_1, \ldots, G_p\}$ be the given set of p n-vertex planar graphs. The proof is by induction on p. The base case, $p = 2$, is true by Theorem 5. Let $p \geq 3$ and assume by induction that the set $\{G_1, \ldots, G_{p-1}\}$ has an $\Omega(n^{1/4^{(p-2)}})$-PGSE. Let $V' \subseteq V$ be the set from the definition of k-PSGE and let P' be the set of $|V'|$ points that V' is mapped to in the drawings $D_1, \ldots, D_{p-1}$. Thus $|V'| \in \Omega(n^{1/4^{(p-2)}})$ by induction. We may assume that no pair of points in P' has the same x-coordinate as otherwise we can just rotate the union of $D_1, \ldots, D_{p-1}$. By Theorem 2, there exists $R \subseteq V'$ that is strongly column planar in G_p and $|R| \geq \sqrt{|V'|/3}$. Since the vertices of V' are bijectively mapped to P', that mapping defines a bijective mapping from R to a subset P_R of P'. Consider the total order μ of R (the total order from the definition of strongly column planar sets) and the total order ϕ of R as defined by the x-coordinates of P_R. By the Erdős–Szekeres theorem [19, 31], there exists a subset R' of R of at least $\sqrt{|R|} \geq (|V'|/3)^{1/4}$ vertices such that the order of R' in μ is the same or reverse as the order of R' in ϕ. In the second case the union of all the drawings of $D_1, \ldots, D_p$ can be mirrored such that the order of R' in μ is the same as the order of R' in ϕ. Thus in both cases, we can apply Theorem 2. Since the vertices of R are bijectively mapped to P_R, this defines a bijective mapping from R' to a subset P'_R of P_R. Since R' is strongly column planar in G_p, we can apply Theorem 2 to conclude that G_p has a straight-line crossing-free drawing D_p that respects the mapping from R' to P'_R and thus each vertex $v \in R'$ is represented by the same point in all drawings D_i, $i \in \{1, \ldots, p\}$. Since $|V'| \in \Omega(n^{1/4^{(p-2)}})$, and $|R'| \geq (|V'|/3)^{1/4}$, the lower bound holds. $\square$

Note that the definition of k-PSGE, as introduced in Evans *et al.* [21], has one additional requirement, as compared with the definition used here. Namely, the additional requirement states that if $v, w \in V$ are mapped to a same point in D_i and D_j, then $v = w$. However this additional requirement can always be met by the fact that it is possible to perturb any subset of vertices of a geometric plane graph without introducing crossings. More precisely, for any geometric plane graph there exists a value $\epsilon > 0$ such that each vertex can be moved any distance of at most ϵ, and the resulting geometric graph is also crossing-free.[3]

5   Conclusion

The main purpose of this paper is to draw attention to Lemmas 1 and 2 in the current form as they seem to have applications to numerous, some seemingly unrelated, graph drawing problems as evidenced by the results highlighted in the previous sections. The two lemmas appear in the current form for the first time here. Their original formulation was tailored towards specific application (untangling) and not directly applicable to any of the above mentioned problems.

Acknowledgements. Many thanks to Pat Morin and David R. Wood for very helpful comments on the preliminary version of this article. Similarly, many thanks to the anonymous referees, especially the one who painstakingly corrected my ever random selection from {the, a, {}}.

The author is supported by NSERC and Ontario Ministry of Research and Innovation.

References

1. Abellanas, M., Hurtado, F., Ramos, P.: Tolerancia de arreglos de segmentos. In: Proceedings of VI Encuentros de Geometría Computacional, pp. 77–84 (1995)
2. Angelini, P., Binucci, C., Evans, W., Hurtado, F., Liotta, G., Mchedlidze, T., Meijer, H., Okamoto, Y.: Universal point subsets for planar graphs. In: Chao, K.-M., Hsu, T., Lee, D.-T. (eds.) ISAAC 2012. LNCS, vol. 7676, pp. 423–432. Springer, Heidelberg (2012)
3. Angelini, P., Evans, W., Frati, F., Gudmundsson, J.: SEFE with no mapping via large induced outerplane graphs in plane graphs. In: Cai, L., Cheng, S.-W., Lam, T.-W. (eds.) Algorithms and Computation. LNCS, vol. 8283, pp. 185–195. Springer, Heidelberg (2013)
4. Angelini, P., Geyer, M., Kaufmann, M., Neuwirth, D.: On a tree and a path with no geometric simultaneous embedding. J. Graph Algorithms Appl. **16**(1), 37–83 (2012)
5. Bannister, M.J., Cheng, Z., Devanny, W.E., Eppstein, D.: Superpatterns and universal point sets. J. Graph Algorithms Appl. **18**(2), 177–209 (2014)
6. Barba, L., Hoffmann, M., Kusters, V.: Column planarity and partial simultaneous geometric embedding for outerplanar graphs. In: Abstracts of the 31st European Workshop on Computational Geometry (EuroCG), pp. 53–56 (2015)
7. Bläsius, T., Kobourov, S.G., Rutter, I.: Simultaneous embedding of planar graphs. In: Tamassia, R. (ed.) Handbook of Graph Drawing and Visualizationpp, pp. 349–381. CRC Press, Boca Raton (2013)

---

[3] The maximum value ϵ for which this property holds is called the tolerance of the arrangement of segments. This concept, both for the geometric realization and the combinatorial meaning of the graphs was systematically studied in [1,29].

8. Bose, P.: On embedding an outer-planar graph in a point set. Comput. Geom. **23**(3), 303–312 (2002)
9. Bose, P., Dujmović, V., Hurtado, F., Langerman, S., Morin, P., Wood, D.R.: A polynomial bound for untangling geometric planar graphs. Discrete Comput. Geom. **42**(4), 570–585 (2009)
10. Brass, P., Cenek, E., Duncan, C.A., Efrat, A., Erten, C., Ismailescu, D., Kobourov, S.G., Lubiw, A., Mitchell, J.S.B.: On simultaneous planar graph embeddings. In: Dehne, F., Sack, J.-R., Smid, M. (eds.) WADS 2003. LNCS, vol. 2748, pp. 243–255. Springer, Heidelberg (2003)
11. Cano, J., Tóth, C.D., Urrutia, J.: Upper bound constructions for untangling planar geometric graphs. SIAM J. Discrete Math. **28**(4), 1935–1943 (2014)
12. Cardinal, J., Hoffmann, M., Kusters, V.: On universal point sets for planar graphs. In: Akiyama, J., Kano, M., Sakai, T. (eds.) TJJCCGG 2012. LNCS, vol. 8296, pp. 30–41. Springer, Heidelberg (2013)
13. Castañeda, N., Urrutia, J.: Straight line embeddings of planar graphs on point sets. In: Proceedings of the 8th Canadian Conference on Computational Geometry, (CCCG), pp. 312–318 (1996)
14. Cibulka, J.: Untangling polygons and graphs. Discrete Comput. Geom. **43**(2), 402–411 (2010)
15. Di Giacomo, E., Didimo, W., van Kreveld, M.J., Liotta, G., Speckmann, B.: Matched drawings of planar graphs. J. Graph Algorithms Appl. **13**(3), 423–445 (2009)
16. Di Giacomo, E., Liotta, G., Mchedlidze, T.: How many vertex locations can be arbitrarily chosen when drawing planar graphs? CoRR abs/1212.0804 (2012)
17. Di Giacomo, E., Liotta, G., Mchedlidze, T.: Lower and upper bounds for long induced paths in 3-connected planar graphs. In: Brandstädt, A., Jansen, K., Reischuk, R. (eds.) WG 2013. LNCS, vol. 8165, pp. 213–224. Springer, Heidelberg (2013)
18. Dilworth, R.P.: A decomposition theorem for partially ordered sets. Ann. Math. **2**(51), 161–166 (1950)
19. Erdős, P., Szekeres, G.: A combinatorial problem in geometry. Compositio Mathematica **2**, 463–470 (1935)
20. Estrella-Balderrama, A., Fowler, J.J., Kobourov, S.G.: Characterization of unlabeled level planar trees. Comput. Geom. **42**(6–7), 704–721 (2009)
21. Evans, W., Kusters, V., Saumell, M., Speckmann, B.: Column planarity and partial simultaneous geometric embedding. In: Duncan, C., Symvonis, A. (eds.) GD 2014. LNCS, vol. 8871, pp. 259–271. Springer, Heidelberg (2014)
22. de Fraysseix, H., Pach, J., Pollack, R.: How to draw a planar graph on a grid. Combinatorica **10**(1), 41–51 (1990)
23. de Fraysseix, H., Pach, J., Pollack, R.: Small sets supporting fary embeddings of planar graphs. In: Proceedings of the Twentieth Annual ACM Symposium on Theory of Computing, STOC '88, pp. 426–433 (1988)
24. Goaoc, X., Kratochvíl, J., Okamoto, Y., Shin, C., Spillner, A., Wolff, A.: Untangling a planar graph. Discrete Comput. Geom. **42**(4), 542–569 (2009)
25. Gritzmann, P., Mohar, B., Pach, J., Pollack, R.: Embedding a planar triangulation with vertices at specified points (solution to problem e3341). Am. Math. Monthly **98**, 165–166 (1991)
26. Kang, M., Pikhurko, O., Ravsky, A., Schacht, M., Verbitsky, O.: Untangling planar graphs from a specified vertex position - hard cases. Discrete Appl. Math. **159**(8), 789–799 (2011)

27. Kurowski, M.: A 1.235 lower bound on the number of points needed to draw all n-vertex planar graphs. Inf. Process. Lett. **92**(2), 95–98 (2004)
28. Pach, J., Tardos, G.: Untangling a polygon. Discrete Comput. Geom. **28**(4), 585–592 (2002)
29. Ramos, P.: Tolerancia de estructuras geométricas y combinatorias. Ph.D. thesis, Universidad Politécnica de Madrid, Madrid, Spain (1995)
30. Ravsky, A., Verbitsky, O.: On collinear sets in straight-line drawings. In: Kolman, P., Kratochvíl, J. (eds.) WG 2011. LNCS, vol. 6986, pp. 295–306. Springer, Heidelberg (2011)
31. Steele, J.M.: Variations on the monotone subsequence theme of Erdös and Szekeres. In: Aldous, D., Diaconis, P., Spencer, J., Steele, J.M. (eds.) Discrete probability and algorithms, pp. 111–131. Springer, Heidelberg (1995)

Polygons and Convexity

Representing Directed Trees
as Straight Skeletons

Oswin Aichholzer[1], Therese Biedl[2], Thomas Hackl[1], Martin Held[3],
Stefan Huber[4], Peter Palfrader[3(✉)], and Birgit Vogtenhuber[1]

[1] Institut für Softwaretechnologie, Technische Universität Graz, 8010 Graz, Austria
{oaich,thackl,bvogt}@ist.tugraz.at
[2] Cheriton School of Computer Science, University of Waterloo,
Waterloo, ON N2L 1A2, Canada
biedl@uwaterloo.ca
[3] FB Computerwissenschaften, Universität Salzburg, 5020 Salzburg, Austria
{held,palfrader}@cosy.sbg.ac.at
[4] Institute of Science and Technology Austria, 3400 Klosterneuburg, Austria
stefan.huber@ist.ac.at

Abstract. The straight skeleton of a polygon is the geometric graph
obtained by tracing the vertices during a mitered offsetting process. It
is known that the straight skeleton of a simple polygon is a tree, and
one can naturally derive directions on the edges of the tree from the
propagation of the shrinking process.

In this paper, we ask the reverse question: Given a tree with directed
edges, can it be the straight skeleton of a polygon? And if so, can we
find a suitable simple polygon? We answer these questions for all directed
trees where the order of edges around each node is fixed.

1 Introduction

Many geometric structures on sets of points, line segments, or polygons, e.g.
Delaunay triangulations, Voronoi diagrams, straight skeletons, and rectangle-of-
influence graphs can be represented as graphs. The *graph representation* problem
(for each of these geometric structures) asks which graphs can be represented
in this way. That is, given a graph G, can we find a suitable input set S of
points, segments, or polygons such that the geometric structure induced by S is
equivalent to G?

Graph representation has been studied for numerous geometric structures in
the past. To name just a few examples: Every planar graph is the intersection
graph of line segments [6], every wheel is a rectangle-of-influence graph [11], and
all 4-connected planar graphs are Delaunay triangulations [8]. See also [7] for
many results on proximity drawability of graphs.

Of particular interest to our paper are two results. First, Liotta and
Meijer [12] studied when a tree can be represented as the Voronoi diagram of

---

OA and BV supported by Austrian Science Fund (FWF) I 648-N18; TB by NSERC;
TH by FWF P23629-N18; MH and PP by FWF P25816-N15.

E. Di Giacomo and A. Lubiw (Eds.): GD 2015, LNCS 9411, pp. 335–347, 2015.
DOI: 10.1007/978-3-319-27261-0_28

a set of points, and showed that this is always possible (and the points are in convex position). Secondly, Aichholzer et al. [4] studied when a tree can be represented as the straight skeleton of a polygon, and showed that this is always possible (and the polygon is convex).

1.1 Background

The *straight skeleton* $\mathcal{S}(P)$ of a simple polygon P is defined via a wavefront-propagation process: Each edge of P emits a wavefront edge moving in a self-parallel manner at unit speed towards the interior of the polygon.

Initially, at time $t = 0$, this wavefront is a single polygon that is identical to P. As the propagation process continues, however, the wavefront will change due to self-interaction: (i) In *edge events*, an edge of the wavefront shrinks to zero length and is removed from the wavefront. (ii) In *split events*, a vertex of the wavefront meets the interior of a previously non-incident wavefront edge. This split partitions the edge and the polygon into two parts that now propagate independently. (iii) If the input is not in general position, more complex interactions are possible. For example, entire portions of the wavefront collapse at once when parallel wavefronts that were emanated by parallel polygon edges meet, or new reflex wavefront vertices are created when multiple reflex vertices interact in a *vertex-event*. The propagation process ends once all components of the wavefront have collapsed. Therefore, the set of wavefront edges at any time t form one or more polygons, which we call the wavefront and denote by $\mathcal{W}(t)$.

The straight skeleton $\mathcal{S}(P)$, introduced by Aichholzer et al. [2], is then defined as the geometric graph whose edges are the traces of the vertices of $\mathcal{W}(t)$ over time; see Fig. 1. For simple polygons, $\mathcal{S}(P)$ always is a tree [2], with the leaves corresponding to vertices of P and interior vertices having degree 3 or more. Several algorithms are known to construct the straight skeleton [1,9,10].

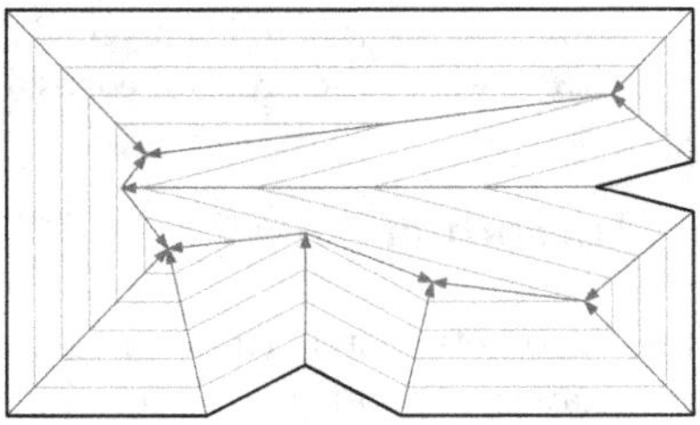

Fig. 1. The straight skeleton $\mathcal{S}(P)$ of an input polygon P (bold) is the union of the traces of wavefront vertices. Wavefront polygons at different times are shown in gray.

We can distinguish between convex and reflex vertices of P or $\mathcal{W}(t)$. A vertex v is *reflex* (*convex*) if the interior angle at v is greater (less) than π. We call an arc of $\mathcal{S}(P)$ *reflex* (*convex*) if it was traced out by a reflex (convex) vertex of the wavefront. When discussing the wavefront propagation process, we will often interchangeably refer to wavefront vertices and straight skeleton arcs.

The roof model [2] represents a convenient means to study the wavefront over the entire propagation period. It embeds the wavefront in three-space, where the z-axis represents time: $\mathcal{T}(P) := \bigcup_{t \geq 0}(\mathcal{W}(t) \times \{t\})$. The inner edges and vertices of this polytope correspond to arcs and nodes of the straight skeleton $\mathcal{S}(P)$, and the z-coordinate of each element corresponds to the time it was traced out by the wavefront-propagation process. Reflex arcs correspond to valleys and convex arcs to ridges.

If we exclude polygons where parallel polygon edges cause entire wavefront segments to collapse at one time, resulting in horizontal roof edges, then arcs of $\mathcal{S}(P)$ will have been traced out by the wavefront during its propagation process, and we can assign a natural direction to these arcs: make them point into their trace direction. This assignment gives rise to the *directed straight skeleton*, $\mathcal{S}_d(P)$.

A *directed tree* T is a directed graph whose underlying undirected graph is a tree, i.e., connected and acyclic. A *labeled tree* T_ℓ is a tree with assignments of labels to its arcs. For most of this paper, trees are *ordered*, i.e., for every node there is a fixed circular order in which the arcs appear around this node.

It is customary to refer to the edges and vertices of the straight skeleton as *arcs* and *nodes*, and to reserve *edges* and *vertices* for elements of input or wavefront polygons. We also use *arcs* and *nodes* to refer to elements of trees.

1.2 Our Results

Our paper was inspired by the work in [4], which studies undirected trees. However, the structure of the straight skeleton imposes directions on the arcs, except in degenerate cases. Hence, the natural question to ask is:

Probelm 1 (Directed straight-skeleton realizability). Given a directed tree T, (i) is there a polygon P such that $\mathcal{S}_d(P)$ shares the structure of T (we denote this by $\mathcal{S}_d(P) \sim T$), and (ii) if yes, can we reconstruct such a polygon P from T?

Having directions assigned to the arcs makes the straight-skeleton realizability problem significantly harder: For example, one easily sees that in a convex polygon the straight skeleton is a rooted tree (with exactly one sink), and so not all directed trees can be represented via convex polygons. Hence, the results from [4] do not transfer to directed trees.

The directed-straight-skeleton-realizability question can be asked for multiple meanings of "directed tree": It could be a *geometric tree* (nodes are given with coordinates), an *ordered tree* (the clockwise order of arcs around each node is specified) or an *unordered tree* (we have the nodes and arcs but nothing else). For a geometric tree, the problem is trivial, since the locations of the leaves specify the vertices of the only polygon for which this could be the straight skeleton. (If leaves are not specified as points but only as "being on a ray", then the geometric setting is non-trivial, but can be solved in polynomial time [5].)

In this paper, we consider the variant of the problem for ordered trees. In the case of polygons in general position, we give three obviously necessary conditions and show that these are also always sufficient. It turns out that the order of arcs around nodes is not important, so the algorithm also works for unordered directed trees. We then turn to polygons without restrictions on vertex-positions. In this case the directed straight skeletons can be significantly more complicated, and in particular, have arbitrary degrees. Testing whether a directed tree can be represented as straight skeleton requires deeper insight into the structure of straight skeletons, and we can exploit these to develop such a testing algorithm and, in case of a positive answer, find a suitable polygon.

2 Trees from Polygons in General Position

In a first step we restrict the problem to polygons in general position. By general position we mean that no four edges have supporting lines which are tangent to a common circle. In particular this means that during the wavefront propagation process only standard edge and split events are observed, resulting in straight skeletons where all interior nodes are of degree exactly three.

Investigating the structure of such directed straight skeletons enables us to establish a number of necessary conditions for a directed tree to be a directed straight skeleton of a polygon in general position.

Necessary Conditions. Let P be a polygon in general position and let T be the directed tree such that $\mathcal{S}_d(P) \sim T$.

The leaves of T correspond to the vertices of P. In the roof model, these vertices have zero z-coordinate, while all other nodes have positive z-coordinates since they will have been created by an event at some time $t > 0$. Thus, any arc incident to a vertex v of P increases in elevation as it moves away from v. As such, all leaves of T must have in-degree 0 and out-degree 1.

The interior nodes all have degree 3 and are classified by their in- and out-degrees as follows:

in-degree 3: *(peak nodes)* A collapse of a wavefront component (of triangular shape) at the end of its propagation process is witnessed by a local maximum in the roof. These local maxima correspond to nodes with in-degree three.

in-degree 2: *(collapse nodes)* Edge events in the propagation process, i.e., collapsed wavefront edges, result in a node with two incoming arcs and one outgoing arc.

in-degree 1: *(split nodes)* Split events will cause a node that has only one incoming arc and two outgoing arcs.

in-degree 0: Since the roof model will have no local minima except at the edges of P [2], nodes with in-degree zero and out-degree three cannot exist.

Of these, the case of a split event requires some more attention since it imposes additional restrictions on the incoming arc. Recall that we can distinguish between reflex and convex vertices of P, and note that any vertex is either reflex or convex by the general position assumption. For a split event to occur, a reflex vertex of the wavefront must crash into a previously non-incident part of the wavefront.

In the absence of vertex events, which create skeleton-nodes of degree at least four and therefore do not happen when the polygon is in general position, no reflex vertex can ever be created by an event. Thus, any reflex vertex that is part of an event must have been emanating from a reflex vertex of the input polygon itself. Accordingly, the incoming arc in a split event node must have a leaf at its other end.

We summarize these conditions in the following lemma.

Lemma 1. *Let P be a simple polygon in general position and let T be the directed tree such that $S_d(P) \sim T$. Then in T the following hold:*

(G1) *the incident arc of each leaf is outgoing,*
(G2) *every interior node has degree 3 and at most two outgoing arcs,*
(G3) *if an interior node has out-degree two, then the incoming arc connects directly from a leaf.*

Observations. Let T be a directed tree that satisfies conditions (G1–G3). Classify the interior nodes as split nodes, collapse nodes and peaks as above. The goal is to show that any such tree can indeed be realized as a straight skeleton. For this, we split the tree into multiple subtrees in a particular way (also illustrated in the example in Fig. 3). We have the following observations. Full proofs of the next four lemmas can be found in Appendix A in the arXiv-version of this work [3].

Lemma 2. *Let T be a directed tree that satisfies conditions (G1–G3). If T has no split nodes, then T has exactly one peak.*

Lemma 3. *Let T be a tree that satisfies conditions (G1–G3). Create a forest F as follows: At any split node s of T, remove s, remove the leaf incident to the incoming arc of s, and replace the two outgoing arcs of s by two new leaves that are connected to the other ends of these arcs. Then each component of F satisfies conditions (G1–G3) and has exactly one peak.*

Sufficient Conditions. It remains to be shown that the necessary conditions (G1–G3) from Lemma 1 are also sufficient. We show this by constructing a simple polygon P such that $S_d(P) \sim T$, given a directed tree T that satisfies (G1–G3). We start by showing this for trees that have no split nodes.

Lemma 4. *For any directed tree T that satisfies (G1–G3) and has no split nodes, there is a convex polygon P such that $S_d(P) \sim T$.*

Proof. We show this by constructive induction. Any triangle is a polygon such that its straight skeleton shares the structure of the peak node of T.

To construct a polygon P for a tree T with k interior nodes, we first construct a polygon P' for a tree T' with $k-1$ interior nodes. We obtain T' by replacing a node of T and its two adjacent leaves with a single leaf ℓ. (There always is such a node.) To obtain P, we compute an exterior offset of P' and replace the vertex that corresponds to ℓ with a sufficiently small edge such that it collapses before the wavefront of P reaches P'.

This polygon P will then satisfy $S_d(P) \sim T$. See Fig. 2 for an illustration. $\square$

Furthermore, it is possible to add a constraint on one interior angle of the polygon:

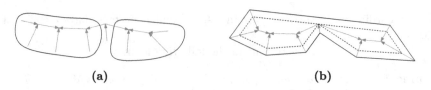

(a)   **(b)**

Fig. 3. (a) Given a directed tree we split it into a forest F of subtrees without split nodes. (b) Recursing on the structure of F, we can create convex polygons for each element (dotted) and then merge them into ever larger polygons.

Lemma 5. *Let T be a directed tree without split nodes and let ℓ be a leaf of T. Further, let α be an arbitrary angle with $0 < \alpha < \pi$. Then there exists a convex polygon P such that $S_d(P) \sim T$ and such that the interior angle at the vertex that corresponds to ℓ is α.*

Now we are ready to consider trees with split nodes.

Lemma 6. *Let T be a directed tree that satisfies conditions (G1–G3). Then there exists a polygon P such that $S_d(P) \sim T$.*

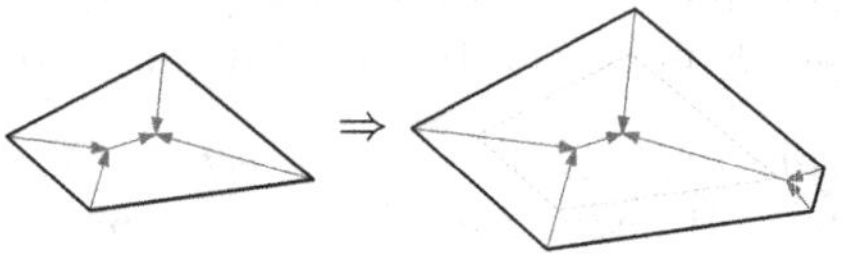

Fig. 2. Creating the polygon for a tree with $k + 1$ interior nodes from a polygon for a tree with one less.

Proof. As in Lemma 3, we split T at split nodes, also dropping the incoming reflex arc and its incident leaf. We obtain a forest $F = \{T_1, T_2, \ldots, T_n\}$ where each T_i is a tree without split nodes.

This forest can in turn be considered an undirected tree, where each T_i gives rise to a node and nodes are connected if and only if the corresponding trees originally had a split node in common. We pick an arbitrary root for F, say T_1, and construct a convex polygon P_1 such that $S_d(P_1) \sim T_1$.

This root T_1 is connected to one or more children in F via split nodes. Let T_2 be such a child, and let $\ell_1 \in T_1$ and $\ell_2 \in T_2$ be the two leaves obtained when splitting at the split node common to T_1 and T_2. Let v_1 be the vertex in P_1 corresponding to ℓ_1, and assume it has angle $\alpha_1 < \pi$.

We construct a convex polygon P_2 such that $S_d(P_2) \sim T_2$ and the vertex v_2 corresponding to ℓ_2 has angle $\alpha < \pi - \alpha_1$. This enables us to merge P_1 and P_2 in the following way: We place P_2 in the plane such that v_1 of P_1 and v_2 of P_2 occupy the same locus. We rotate P_2 such that the angle between a pair of edges of P_1 and P_2 is exactly π. Which pair of edges is chosen depends on where in the cyclic order the incident reflex leaf at the split node in T lies. The layout of P_1 and P_2 then corresponds to the layout of the wavefront at the split-event time. If we now compute a small outer offset and designate this to be P, then the directed straight skeleton of P has the same structure as the subtree of T that is made up by T_1, T_2, the split node, and its incident leaf; see Fig. 3.

We then repeat this process for another child of T_1 or T_2. Note that it may be necessary to scale the polygon that we add to a sufficiently small size so that it does not conflict with other parts of the polygon already constructed. This is always possible since for each vertex of a polygon there exists a disk that intersects the polygon only in the wedge defined by the vertex.

Once all elements of the forest have been processed, we obtain a polygon P whose straight skeleton has the same structure as T, i.e., $\mathcal{S}_d(P) \sim T$.   □

Notice that conditions (G1–G3) do not depend on the order of arcs around nodes; we can construct a polygon for any such order. So in particular if T is an unordered tree that satisfies (G1–G3), then we can pick an arbitrary order and the lemma holds. Hence, we get the following theorem:

Theorem 1. *An (ordered or unordered) directed tree T is the directed straight skeleton of a simple polygon P in general position if and only if T satisfies conditions (G1–G3).*

3   Realizing Trees with Labeled Arcs

Recall that an arc of the straight skeleton is called *reflex* (*convex*) if it was traced out by a reflex (convex) vertex of the wavefront. One can easily see that the construction in Lemma 6 creates a polygon where all arcs of the straight skeleton are convex, with the exception of arcs from leaves to split nodes.

For later constructions (for trees with higher degrees), it will be important that we test not only whether an ordered tree can be realized, but additionally we want to impose onto each arc whether it is reflex or convex in the straight skeleton. We study this question here first for trees with maximum degree 3.

So assume we have a directed tree T that satisfies (G1–G3). Additionally we now label each arc of T with either "reflex" or "convex", and we ask whether there exists a polygon P that realizes this labeled directed tree in the sense that $\mathcal{S}_d(P) \sim T$ and the type of skeleton-arc in $\mathcal{S}_d(P)$ matches the label of the arc in T. We denote this by $\mathcal{S}_d(P) \sim_\ell T$.

We observe that a peak node is created when a wavefront of three edges, a triangle, collapses. Therefore, all incident arcs at a peak node are convex.

For collapse nodes, we know that the outgoing arc is convex. (Recall that reflex arcs in a straight skeleton are only created in vertex events, which cannot exist when all interior nodes have degree three.) At least one of the incoming arcs needs to be convex, as two reflex wavefront vertices meeting in an event will result in a node of degree at least four.

We have already established that the incoming arc at a split node needs to be reflex. Furthermore, it is easy to see that the two outgoing arcs of a split node need to be convex. We summarize the necessary conditions for a labeled directed tree to correspond to a straight skeleton in the following lemma:

Lemma 7. *Let P be a simple polygon in general position and let T_ℓ be the labeled directed tree such that $S_d(P) \sim_\ell T_\ell$. Then*

(L1) *for peak nodes, all incoming arcs are convex;*
(L2) *for collapse nodes, at least one incoming arc and the outgoing arc are convex;*
(L3) *for split nodes, the incoming arc is reflex and both outgoing arcs are convex.*

We will now show that (L1–L3) are also sufficient:

Lemma 8. *Any labeled directed tree T_ℓ that meets conditions (G1–G3) from Lemma 1 and (L1–L3) from Lemma 7 is realizable by a simple polygon P.*

Proof. Since reflex arcs in T_ℓ only originate at leaves and terminate at collapse or split nodes (interior nodes never have outgoing reflex arcs, and peak nodes have no incoming reflex arcs), we can create a tree T_ℓ' by replacing each collapse node that has an incoming reflex arc with a leaf and dropping the two incident incoming arcs and their leaves. This resulting tree T_ℓ' will have no reflex arcs except for those leading from a leaf to split nodes by (L1), (L3), and (G3). Thus, we can create a polygon P' such that $S_d(P') \sim_\ell T_\ell'$ by the process described in the proof of Lemma 6, respecting all labels.

We now obtain P by offsetting P' slightly to the outside. Then, we modify P at each vertex v that corresponds to a leaf in T_ℓ' that was the result of replacing a collapse node of T_ℓ. Note that each such vertex v is convex since the outgoing arc of a collapse node is convex by (L2). We insert a small edge in place of v, replacing it with v_1, the edge, and v_2. We choose the angles at v_1 and v_2 such that one of them is reflex and the other is convex, in order to match the labeling of T_ℓ. Figure 4 illustrates this operation.

By making the new edge sufficiently small, we can ensure that these events happen before the wavefront becomes identical to P', and thus before all remaining events of the wavefront propagation.   □

4   Arbitrary Node Degrees

Once we allow for straight skeletons where interior nodes can have degrees larger than three, a number of previous constraints no longer hold. Most importantly, during the wavefront propagation vertex events can happen, resulting in new reflex vertices in the wavefront after the event. Consequently, for instance, split nodes no longer need to be adjacent to leaves. Larger node degrees also result in more complex variants of split, collapse, and peak nodes. Note that we continue

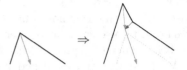

Fig. 4. Extending a convex vertex of P such that a leaf in its tree is replaced by a collapse node where one incoming arc is reflex and one is convex.

to restrict polygons from having parallel edges as those might cause skeleton arcs which have no direction (when they get created as a result of two wavefront edges crashing into each other) or straight skeleton arcs that are neither reflex nor convex (when two parallel wavefronts moving in the same direction become incident at an event).

In order to understand what combinations of reflex and convex incoming and outgoing arcs may exist at a node in a directed straight skeleton, we study the different shapes that a wavefront may have at an event at locus p and time t. At a time $t - \delta$ immediately prior to the event, the wavefront will consist of a combination of reflex and convex vertices, tracing out reflex and convex arcs, all of which will meet at p at time t. We choose δ sufficiently small such that no event will happen in the interval $[t - \delta, t)$.

Consider the wavefront around a locus p at an event, and consider the *wedges* that have been already swept by the wavefront. With wedge we mean the area near p swept over by a continuous portion of the wavefront polygon until time t.

If a single wedge covers the entire area around p, we call it a *full* wedge. The interior angles of other wedges might be less than π, greater than π, or exactly π as illustrated in Fig. 5. We call the first type of wedge *reflex* and the second type of wedge *convex*, after their corresponding wavefront vertices in the simple case. The third type of wedge we simply call π-wedge.

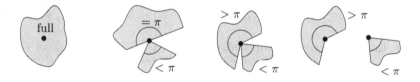

Fig. 5. Wavefront wedges at event times are either full wedges or can be classified by their interior angle. The wedges (already-swept areas) are gray, the remaining white sectors are covered by the wavefront polygon(s).

A single wedge may have been traced out by just one wavefront edge if a wedge at an event has an interior angle of exactly π, but in all other cases it is the area covered by two or more wavefront edges and their incident wavefront vertices, which have traced out one or more incoming arcs at p. Note that all but the two outermost edges of this part of the wavefront collapse at time t.

We will establish arc-*patterns* to describe combinations of arcs at a node. Such a pattern is a string consisting of the types of arcs: r for an incoming reflex arc, c for an incoming convex arc, and $\hat{r}$ and $\hat{c}$ for their outgoing counterparts. We will use operators known from language theory, such as parentheses to group blocks, the asterisk (*) to indicate the preceding character or group may occur zero or more times, the plus sign ($^+$) to indicate the preceding block may occur one or more times, and the question mark ($^?$) to indicate it may exist zero times or once. When defining patterns we give them variable names in capitals.

We now investigate which combination of arcs may trace out which types of wedges. For this purpose we first characterize single wedges and provide their describing arc-patterns. Then we examine all possible single wedge combinations and provide allowed arc-patterns for interior nodes.

Due to lack of space we give here only an overview on the arc-patterns of single wedges and the arc-patterns for possible combinations of these wedges at interior nodes. For a detailed analysis please see Appendix B (see footnote 1).

As mentioned, single wedges can be reflex wedges, convex wedges, π-wedges, and so called *full wedges*, the latter being traced out by a wavefront that collapses completely around a locus p. The arc-pattern for a reflex wedge is $R := r\,(cr)^*$, i.e., one reflex vertex, followed by zero or more (convex, reflex) pairs of vertices and thus arcs. If a reflex wedge is created by a single reflex wavefront vertex (and its incident edges) then we call it *trivial*. Otherwise, we call it *non-trivial*, with $R_+ = r\,(cr)^+$ specifying the arc-pattern of a non-trivial reflex wedge.

The arc-pattern for a convex wedge is $C := r^?\,c\,(r^?c)^*\,r^?$. Like for reflex wedges we distinguish trivial (traced out by a single convex vertex) and non-trivial convex wedges, the latter (any pattern matching C and having length at least two) being denoted by C_+.

The case of π-wedges is related to the case of reflex wedges. A π-wedge is either traced out by exactly one wavefront edge (no incoming arcs), or it has the same pattern as for a non-trivial reflex wedge. Hence, we have as arc-pattern $P := \emptyset \mid R_+$. Note that since we have explicitly excluded parallel polygon edges for this problem setting, only trivial π-wedges can exist. Therefore, we will set $P := \emptyset$ here.

Last, the arc-pattern of a full wedge is $F := c\,(r^?c)^*\,c\,(r^?c)^*\,c\,(r^?c)^*$.

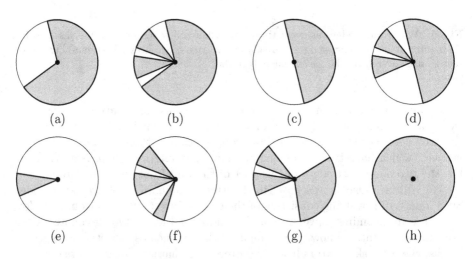

(a)   (b)   (c)   (d)

(e)   (f)   (g)   (h)

Fig. 6. All wedge combinations possible at a node.

Analyzing possible combinations of single wedges at interior nodes we get seven different allowed arc-patterns (see Fig. 6(a–h)): $N_a := C_+\,\hat{c}$, $N_b := C\,\hat{c}\,(R\,\hat{c})^+$, $N_d := P\,\hat{c}\,(R\,\hat{c})^+$, $N_e := R_+\,\hat{r}$, $N_f := (R\,\hat{c})^+ R\,\hat{r}$, $N_g := (R\,\hat{c})^+ R\,\hat{c}$, and

$N_h := F$. (Note that the pattern from Fig. 6(c) will not result in an event without parallel polygon edges, which we have excluded.) A simple split node is matched via N_d, a simple collapse node is handled by N_a (with C_+ being either cc, rc, or cr), and a simple peak node is matched by N_h.

Since these are all possible wavefront/wedge combinations, any interior node of the straight skeleton will have to match $N := N_a|N_b|N_d|N_e|N_f|N_g|N_h$. We can state the following lemma:

Lemma 9. *Let P be a simple polygon and let T_ℓ be the labeled directed tree such that $\mathcal{S}_d(P) \sim_\ell T_\ell$. Then the cyclic order of arcs of any interior node of T_ℓ needs to match the pattern specification N defined above.*

After the necessity of the discussed conditions we now prove their sufficiency.

Lemma 10. *Let T_ℓ be any labeled directed tree for which (i) the cyclic order of arcs of each interior node matches the pattern specification N defined above and (ii) each leaf has out-degree one. Then T_ℓ is realizable by some simple polygon P, that is, $\mathcal{S}_d(P) \sim_\ell T_\ell$.*

Proof. We will construct P in a manner similar to the one described in Lemma 6. We start by identifying maximally connected components $C_1, C_2, \ldots, C_n$ of T_ℓ containing nodes with out-degree either zero or one. These components take the place of the subtrees of our forest from Lemma 6, and they are connected in T_ℓ via split-nodes, i.e., nodes with out-degree ≥ 2.

We pick an arbitrary component C_1 and create a polygon P_1 such that $\mathcal{S}_d(P_1) \sim_\ell C_1$ as follows. If there are no outgoing arcs from C_1, we start at its unique peak node. Otherwise, there is exactly one outgoing arc, and we begin at the node it is incident to. Constructing a polygon for this node is straightforward by applying the concepts learned from considering convex, reflex, and full wavefront wedges. We proceed by extending this initial polygon step by step like in Lemma 4, treating each reflex or convex vertex of the polygon as a wavefront wedge to be constructed, until we have a polygon for the entire component.

Next, we pick one of the split nodes connected to C_1. That split node, we call it n, will have one of the forms from Fig. 6 that have at least two white sectors. The polygon we just created will cover one of these white sectors. (Depending on the type of arc connecting C_1 to n, the polygon will either have a reflex or convex vertex for this arc.) We continue by constructing polygons for all remaining white sectors in the same fashion we used for constructing P_1. Note that this process allows us to force at least one angle, and therefore we can construct polygons that fit into the white sectors for n.

Now we have the wavefront polygon as it should be when the event at n happens. We compute a small exterior offset, joining all polygons into one larger polygon. The new reflex or convex vertices at this point are then further subdivided as required by the incoming arcs for n.

We repeat this process until we have covered all split nodes and thus all components and have thereby created a polygon whose structure matches T_ℓ. □

Please see Appendix C in the arXiv-version of this work [3].

Combining Lemmas 9 and 10, we obtain the following theorem:

Theorem 2. *An ordered labeled directed tree T_ℓ is the directed straight skeleton of a simple polygon P without parallel edges if and only if (i) the cyclic order of arcs of each interior node of T_ℓ matches the pattern specification N defined above and (ii) each leaf has out-degree one.*

5   Conclusion

In this work we developed a complete characterization of necessary and also sufficient conditions such that a given directed and labeled ordered tree can be represented as the straight skeleton of a simple polygon. This extends previous work on representing trees via related geometric structures [4,12].

We leave the algorithmic question – how efficient suitability of a given input tree can be tested and, in case of an affirmative answer, a corresponding simple polygon can be computed – for future research. We conjecture that both is possible in time linear in the size of the given tree.

References

1. Aichholzer, O., Aurenhammer, F.: Straight skeletons for general polygonal figures in the plane. In: Samoilenko, A. (ed.) Voronoi's Impact on Modern Sciences II, vol. 21, pp. 7–21. Institute of Mathematics of the National Academy of Sciences of Ukraine, Kiev, Ukraine (1998)
2. Aichholzer, O., Aurenhammer, F., Alberts, D., Gärtner, B.: A novel type of skeleton for polygons. J. Univ. Comput. Sci. **1**(12), 752–761 (1995)
3. Aichholzer, O., Biedl, T., Hackl, T., Held, M., Huber, S., Palfrader, P., Vogtenhuber, B.: Representing directed trees as straight skeletons [cs.CG] (2015). http://arxiv.org/abs/1508.01076
4. Aichholzer, O., Cheng, H., Devadoss, S.L., Hackl, T., Huber, S., Li, B., Risteski, A.: What makes a tree a straight skeleton? In: Proceedings of the 24th Canadian Conference on Computational Geometry, (CCCG 2012), pp. 253–258. Charlottetown, PE, Canada (2012)
5. Biedl, T., Held, M., Huber, S.: Recognizing straight skeletons and Voronoi diagrams and reconstructing their input. In: Gavrilova, M., Vyatkina, K. (eds.) Proceedings of the 10th International Symposium on Voronoi Diagrams in Science & Engineering (ISVD 2013), pp. 37–46. IEEE Computer Society, Saint Petersburg, Russia (2013)
6. Chalopin, J., Gonçalves, D.: Every planar graph is the intersection graph of segments in the plane (Extended Abstract). In: Proceedings of 41st Annual ACM Symposium Theory Computing (STOC 2009), pp. 631–638. ACM, Bethesda, MD, USA (2009)
7. Di Battista, G., Lenhart, W., Liotta, G.: Proximity drawability: a survey. In: Tamassia, R., Tollis, I.G. (eds.) GD '94. LNCS, vol. 894, pp. 328–339. Springer, Princeton, NJ, USA (1995)
8. Dillencourt, M.B., Smith, W.D.: Graph-theoretical conditions for inscribability and delaunay realizability. Discrete Math. **161**(1–3), 63–77 (1996)

9. Eppstein, D., Erickson, J.: Raising roofs, crashing cycles, and playing pool: applications of a data structure for finding pairwise interactions. Discrete Comput. Geom. **22**(4), 569–592 (1999)
10. Huber, S., Held, M.: A fast straight-skeleton algorithm based on generalized motorcycle graphs. Int. J. Comput. Geom. Appl. **22**(5), 471–498 (2012)
11. Liotta, G., Lubiw, A., Meijer, H., Whitesides, S.: The rectangle of influence drawability problem. Comput. Geom. **10**(1), 1–22 (1998)
12. Liotta, G., Meijer, H.: Voronoi drawings of trees. Comput. Geom. **24**(3), 147–178 (2003)

Drawing Graphs with Vertices and Edges in Convex Position

Ignacio García-Marco[1]($\boxtimes$) and Kolja Knauer[2]

[1] LIP, ENS Lyon - CNRS - UCBL - INRIA, Université de Lyon UMR 5668,
Lyon, France
ignacio.garcia-marco@ens-lyon.fr
[2] Aix-Marseille Université, CNRS, LIF UMR 7279, Marseille, France
kolja.knauer@lif.univ-mrs.fr

Abstract. A graph has strong convex dimension 2, if it admits a straight-line drawing in the plane such that its vertices are in convex position and the midpoints of its edges are also in convex position. Halman, Onn, and Rothblum conjectured that graphs of strong convex dimension 2 are planar and therefore have at most $3n - 6$ edges. We prove that all such graphs have at most $2n - 3$ edges while on the other hand we present a class of non-planar graphs of strong convex dimension 2. We also give lower bounds on the maximum number of edges a graph of strong convex dimension 2 can have and discuss variants of this graph class. We apply our results to questions about large convexly independent sets in Minkowski sums of planar point sets, that have been of interest in recent years.

1 Introduction

A point set $X \subseteq \mathbb{R}^2$ is in *(strictly) convex position* if all its points are vertices of their convex hull. A point set X is said to be in *weakly convex position* if X lies on the boundary of its convex hull. A *drawing* of a graph G is an injective mapping $f : V(G) \to \mathbb{R}^2$ such that edges are straight line segments connecting vertices and neither midpoints of edges, nor vertices, nor midpoints and vertices coincide. Through most of the paper we will not distinguish between (the elements of) a graph and their drawings.

For $i, j \in \{s, w, a\}$ we define $\mathcal{G}_i^j$ as the class of graphs admitting a drawing such that the vertices are in $\begin{cases} \text{strictly convex} & \text{if } i = s \\ \text{weakly convex} & \text{if } i = w \\ \text{arbitrary} & \text{if } i = a \end{cases}$ position and the

I. García-Marco—Supported by ANR project CompA (project number: ANR-13-BS02-0001-01)

K. Knauer—Supported by ANR EGOS grant ANR-12-JS02-002-01 and PEPS grant EROS.

E. Di Giacomo and A. Lubiw (Eds.): GD 2015, LNCS 9411, pp. 348–359, 2015.
DOI: 10.1007/978-3-319-27261-0_29

$$\text{midpoints of edges are in} \begin{cases} \text{strictly convex} & \text{if } j = s \\ \text{weakly convex} & \text{if } j = w \\ \text{arbitrary} & \text{if } j = a \end{cases} \text{position. Further, we define}$$

$g_i^j(n)$ to be the maximum number of edges an n-vertex graph in $\mathcal{G}_i^j$ can have.

Clearly, all $\mathcal{G}_i^j$ are closed under taking subgraphs and $\mathcal{G}_s^a = \mathcal{G}_w^a = \mathcal{G}_a^a$ is the class of all graphs.

Previous Results and Related Problems: Motivated by a special class of convex optimization problems [4], Halman, Onn, and Rothblum [3] studied drawings of graphs in $\mathbb{R}^d$ with similar constraints as described above. In particular, in their language a graph has convex dimension 2 if and only if it is in $\mathcal{G}_a^s$ and strong convex dimension 2 if and only if it is in $\mathcal{G}_s^s$. They show that all trees and cycles are in $\mathcal{G}_s^s$, while $K_4 \in \mathcal{G}_a^s \setminus \mathcal{G}_s^s$ and $K_{2,3} \notin \mathcal{G}_a^s$. Moreover, they show that $n \leq g_s^s(n) \leq 5n - 8$. Finally, they conjecture that all graphs in $\mathcal{G}_s^s$ are planar and thus $g_s^s(n) \leq 3n - 6$.

The problem of computing $g_a^s(n)$ and $g_s^s(n)$ was rephrased and generalized in the setting of convexly independent subsets of Minkowski sums of planar point sets by Eisenbrand et al. [2] and then regarded as a problem of computational geometry in its own right. We introduce this setting and give an overview of known results before explaining its relation to the original graph drawing problem.

Given two point sets $A, B \subseteq \mathbb{R}^d$ their *Minkowski sum* $A + B$ is defined as $\{a + b \mid a \in A, b \in B\}$. Define $M(m, n)$ as the largest cardinality of a convexly independent set $X \subseteq A + B$, for A and B planar point sets with $|A| = m$ and $|B| = n$. In [2] it was shown that $M(m, n) \in O(m^{2/3}n^{2/3} + m + n)$, which was complemented with an asymptotically matching lower bound by Bílka et al. [1] even under the assumption that A itself is in convex position, i.e., $M(m, n) \in \Theta(m^{2/3}n^{2/3} + m + n)$. Notably, the lower bound works also for the case $A = B$, as shown by Swanepoel and Valtr [5]. In [6] Tiwary gives an upper bound of $O((m + n) \log(m + n))$ for the largest cardinality of a convexly independent set $X \subseteq A + B$, for A and B planar convex point sets with $|A| = m$ and $|B| = n$. Determining the asymptotics in this case remains open.

The graph drawing problem of Halman et al. is related to the largest cardinality of a convexly independent set $X \subset A + A$, for A some planar point set. In fact, from X and A one can deduce a graph $G \in \mathcal{G}_a^s$ on vertex set A, with an edge aa' for all $a \neq a'$ with $a + a' \in X$. The midpoint of the edge aa' then just is $\frac{1}{2}(a + a')$. Conversely, from any $G \in \mathcal{G}_a^s$ one can construct X and A as desired. The only trade-off in this translation are the pairs of the form aa, which are not taken into account by the graph-model, because they correspond to vertices. Hence, they do not play a role from the purely asymptotic point of view. Thus, the results of [1,2,5] yield $g_a^s(n) = \Theta(n^{4/3})$. Conversely, the bounds for $g_s^s(n)$ obtained in [3] give that the largest cardinality of a convexly independent set $X \subseteq A + A$, for A a planar convex point set with $|A| = n$ is in $\Theta(n)$.

Our Results: In this paper we study the set of graph classes defined in the introduction. We endow the list of properties of point sets considered in earlier works with *weak* convexity. We completely determine the inclusion relations on the resulting classes. We prove that $\mathcal{G}_s^s$ contains non-planar graphs, which disproves a conjecture of Halman et al. [3], and that $\mathcal{G}_s^w$ contains cubic graphs, while we believe this to be false for $\mathcal{G}_s^s$. We give new bounds for the parameters $g_i^j(n)$: we show that $g_s^w(n) = 2n - 3$, which is an upper bound for $g_s^s(n)$ and therefore improves the upper bound of $3n - 6$ conjectured by Halman et al. [3]. Furthermore we show that $\lfloor \frac{3}{2}(n-1) \rfloor \leq g_s^s(n)$.

For the relation with Minkowski sums we show that the largest cardinality of a weakly convexly independent set $X \subseteq A + A$, for A some convex planar point set of $|A| = n$ is $2n$ and of a strictly convex set is between $\frac{3}{2}n$ and $2n - 2$.

2   Graph Drawings

Given a graph G drawn in the plane with straight line segments as edges, we denote by P_V the convex hull of its vertices and by P_E the convex hull of the midpoints of its edges. Clearly, P_E is strictly contained in P_V.

2.1   Inclusions of Classes

We show that most of the classes defined in the introduction coincide and determine the exact set of inclusions among the remaining classes.

Theorem 1. *We have $\mathcal{G}_s^s = \mathcal{G}_w^s \subsetneq \mathcal{G}_s^w \subsetneq \mathcal{G}_w^w = \mathcal{G}_a^w = \mathcal{G}_s^a = \mathcal{G}_w^a = \mathcal{G}_a^a$ and $\mathcal{G}_s^s \subsetneq \mathcal{G}_a^s \subsetneq \mathcal{G}_w^w$. Moreover, there is no inclusion relationship between $\mathcal{G}_a^s$ and $\mathcal{G}_s^w$. See Fig. 1 for an illustration.*

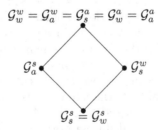

$$\mathcal{G}_w^w = \mathcal{G}_a^w = \mathcal{G}_s^a = \mathcal{G}_w^a = \mathcal{G}_a^a$$

$$\mathcal{G}_a^s \qquad\qquad\qquad \mathcal{G}_s^w$$

$$\mathcal{G}_s^s = \mathcal{G}_w^s$$

Fig. 1. Inclusions and identities among the classes $\mathcal{G}_i^j$.

Proof. Let us begin by proving that $\mathcal{G}_s^s = \mathcal{G}_w^s$, the inclusion $\mathcal{G}_s^s \subset \mathcal{G}_w^s$ is obvious. Take $G \in \mathcal{G}_w^s$ drawn in the required way. We observe that there exists $\delta > 0$ such that if we move every vertex a distance $< \delta$, then the midpoints of the edges are still in convex position. Thus, whenever there are vertices $z_1, \ldots, z_k$ in the interior of the segment connecting two vertices x, y, we do the following construction. We assume without loss of generality that x is in the point $(0,0)$, y is in $(1,0)$ and that P_V is entirely contained in the closed halfplane $\{(a, b) \mid b \leq 0\}$.

We take $s_1, s_2 \in \mathbb{R} \cup \{\pm\infty\}$ the slopes of the previous and following edge of the boundary of P_V. Now we consider $\epsilon : 0 < \epsilon < \min\{\delta, |s_1|, |s_2|\}$, we observe that the set $P' := P_V \cup \{(a, b) \mid 0 \leq a \leq 1 \text{ and } 0 \leq b \leq \epsilon a(1-a)\}$ is convex. Then, for all $i \in \{1, \ldots, k\}$, if z_i is in $(\lambda_i, 0)$ with $0 < \lambda_i < 1$, we translate z_i to the point $(\lambda_i, \epsilon \lambda_i (1 - \lambda_i))$. We observe that the point z_i has been moved a distance $< \epsilon/4 < \delta$ and, then, the set of midpoints of the edges is still in convex position. Moreover, now $z_1, \ldots, z_k$ are in the boundary of P'. Repeating this argument when necessary we get that $G \in \mathcal{G}_s^s$.

To prove the strict inclusion $\mathcal{G}_s^s \subsetneq \mathcal{G}_s^w$ we show that the graph $K_4 - e$, i.e., the graph obtained after removing an edge e from the complete graph K_4 belongs to $\mathcal{G}_s^w$ but not to $\mathcal{G}_s^s$. Indeed, if we take x_0, x_1, x_2, x_3 the 4 vertices of $K_4 - e$ and assume that $e = x_2 x_3$, it suffices to draw $x_0 = (0, 1)$, $x_1 = (0, 0)$, $x_2 = (1, 0)$ and $x_3 = (1, 2)$ to get that $K_4 - e \in \mathcal{G}_s^w$. Let us prove that $K_4 - e \notin \mathcal{G}_s^s$. Take x_0, x_1, x_2, x_3 in convex position, by means of an affine transformation we may assume that $x_0 = (0, 1)$, $x_1 = (0, 0)$, $x_2 = (1, 0)$ and $x_3 = (a, b)$, with $a, b > 0$ and $a + b > 1$. If $x_i x_{i+1 \bmod 4}$ is an edge for all $i \in \{0, 1, 2, 3\}$, then clearly P_E is not convex because the midpoint of $x_0 x_3$ is in the convex hull of the midpoints of the other 4 edges. So, assume that $x_2 x_3$ is not an edge, so the midpoints of the edges are in positions $m_0 = (0, 1/2)$, $m_1 = (1/2, 0)$, $m_2 = (1/2, 1/2)$, $m_3 = (a/2, b/2)$, $m_4 = (a/2, (b+1)/2)$. If m_0, m_1, m_2, m_3 are in convex position, then we deduce that $a < 1$ or $b < 1$ but not both. However, if $a < 1$, then m_3 belongs to the convex hull of $\{m_0, m_1, m_2, m_4\}$, and if $b < 1$, then m_2 belongs to the convex hull of $\{m_0, m_1, m_3, m_4\}$. Hence, we again have that P_E is not convex and we conclude that $K_4 - e \notin \mathcal{G}_s^s$.

The strict inclusion $\mathcal{G}_s^w \subsetneq \mathcal{G}_a^a$ comes as a direct consequence of Theorem 2.

Let us see that every graph belongs to $\mathcal{G}_w^w$, for this purpose it suffices to show that $K_n \in \mathcal{G}_w^w$. Indeed, drawing the vertices in the points $(0, 0)$, and $(1, 2^i)$ for $i \in \{1, \ldots, n-1\}$ gives the result. Then, we clearly have that $\mathcal{G}_w^w = \mathcal{G}_a^w = \mathcal{G}_s^a = \mathcal{G}_w^a = \mathcal{G}_a^a$.

The strict inclusions $\mathcal{G}_s^s \subsetneq \mathcal{G}_a^s \subsetneq \mathcal{G}_w^w$ come from the facts that $g_s^s = \Theta(n^{4/3})$ and that, $g_s^s(n) \leq g_s^w(n) \leq 2n - 3$ by Theorem 2. This also proves that $\mathcal{G}_s^a \not\subset \mathcal{G}_s^w$.

To prove that $\mathcal{G}_s^w \not\subset \mathcal{G}_a^s$ it suffices to consider the complete bipartite graph $K_{2,3}$. Indeed, if $\{x_1, x_2, x_3\}$, $\{y_1, y_2\}$ is the vertex partition, it suffices to draw x_1, x_2, x_3 in $(0, 0)$, $(4, 0)$, $(3, 2)$, respectively, and y_1, y_2 in $(1, 1)$, $(4, 1)$, respectively, to get that $K_{2,3} \in \mathcal{G}_s^w$. Finally, $K_{2,3} \notin \mathcal{G}_a^s$ was already shown in [3]. $\square$

2.2   Bounds on Numbers of Edges

We show that $\lfloor \frac{3}{2}(n-1) \rfloor \leq g_s^s(n) \leq g_s^w(n) = 2n - 3$.

Whenever P_V is weakly convex, for every vertex x, one can order the neighbors of x according to their clockwise appearance around the border of P_V starting at x. If in this order the neighbors of x are $y_1, \ldots, y_k$, then we say that $xy_2, \ldots, xy_{k-1}$ are the *interior edges of* x. Non-interior edges of x are called *exterior edges of* x. Clearly, any vertex has at most two exterior edges. A *vertex* v *sees an edge* e if the straight-line segment connecting v and the midpoint m_e of e does not intersect the interior of P_E.

Lemma 1. *If $G \in \mathcal{G}_s^w$, then no vertex sees its interior edges. In particular, any vertex sees at most 2 incident edges.*

Proof. Assume that there exists a vertex x seeing an interior edge xu_i. Take u_1, u_k such that xu_1, xu_k are the exterior edges of x. We consider G' the induced graph with vertex set $V' = \{v, u_1, u_i, u_k\}$ and denote by E' its corresponding edge set. Clearly $P_{V'} \subset P_V$ and $P_{E'} \subset P_E$, so x sees xu_i in $P_{E'}$. Moreover, xu_i is still an interior edge of x in G'. Denote by m_j the midpoint of the edge vu_j, for $j \in \{1, i, k\}$. Since x sees xu_i, the closed halfplane supported by the line passing through m_1, m_k containing x also contains m_i.

However, since $P_{V'}$ is strictly convex u_i and x are separated by the line passing through u_1, u_k. This is a contradiction because $m_j = (u_j + x)/2$. See Fig. 2.   □

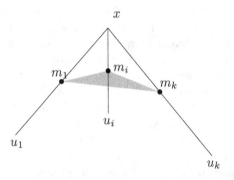

Fig. 2. The construction in Lemma 1

Theorem 2. *If a graph $G \in \mathcal{G}_s^w$ has n vertices, then it has at most $2n - 3$ edges, i.e., $g_s^w(n) \leq 2n - 3$.*

Proof. Take $G \in \mathcal{G}_s^w$. Since the midpoints of the edges are in weakly convex position, every edge has to be seen by one of its vertices. Lemma 1 guarantees that interior edges cannot be seen. Hence, no edge can be interior to both endpoints. This proves that G has at most $2n$ edges.

We improve this bound by showing that at least three edges are exterior by both their endpoints, i.e., are counted twice in the above estimate. During the proof let us call such edges *doubly exterior*.

Since deleting leaves only decreases the ratio of vertices and edges, we can assume that G has no leaves. Clearly, we can also assume that G has at least three edges. For an edge e, we denote by H_e^+ and H_e^- the open halfplanes supported by the line containing e. We claim that whenever an edge $e = xy$ is an interior edge of x, then $H_e^+ \cup \{x\}$ and $H_e^- \cup \{x\}$ contain a doubly exterior edge. This follows by induction on the number of vertices in $H_e^+ \cap P_V$. Since e is interior to x, there is an edge $f = xz$ contained in $H_e^+ \cup \{x\}$ and exterior of x. If f is doubly exterior we are done. Otherwise, we set H_f^+ the halfplane supported by the line containing f

and not containing y. We claim that $(H_f^+ \cup \{z\}) \cap P_V \subset (H_e^+ \cup \{x\}) \cap P_V$. Indeed, if there is a point $v \in (H_f^+ \cup \{z\}) \cap P_V$ but not in $H_e^+ \cup \{x\}$, then x is in the interior of the triangle with vertices $v, y, z \in P_V$, a contradiction. Thus, $(H_f^+ \cup \{z\}) \cap P_V$ is contained in $(H_e^+ \cup \{x\}) \cap P_V$ and has less vertices of P_V, in particular, it does not contain x. By induction, we can guarantee that $(H_e^+ \cup \{x\}) \cap P_V$ contains a doubly exterior edge. The same works for $H_e^- \cup \{x\}$.

Applying this argument to any edge e which is not doubly exterior gives already two doubly exterior edges f, g contained in $H_e^+ \cup \{x\}$ and $H_e^- \cup \{x\}$, respectively. Choose an endpoint z of f, which is not an endpoint of g. Let $h = zw$ be the other exterior edge of z. If h is doubly exterior we are done. Otherwise, none of $H_h^+ \cup \{w\}$ and $H_h^- \cup \{w\}$ contains f because $z \notin H_h^+$ and $z \notin H_h^-$; moreover one of $H_h^+ \cup \{w\}$ and $H_h^- \cup \{w\}$ does not contain g. Thus, there must be a third doubly exterior edge. ☐

Definition 1. *For every $n \geq 2$, we denote by L_n the graph consisting of two paths $P = (u_1, \ldots, u_{\lfloor \frac{n}{2} \rfloor})$ and $Q = (v_1, \ldots, v_{\lceil \frac{n}{2} \rceil})$ and the edges $u_1 v_1$ and $u_i v_{i-1}$ and $v_j u_{j-1}$ for $1 < i \leq \lfloor \frac{n}{2} \rfloor$ and $1 < j \leq \lceil \frac{n}{2} \rceil$. We observe that L_n has $2n - 3$ edges.*

Theorem 3. *For all $n \geq 2$ we have $L_n \in \mathcal{G}_s^w$, i.e., $g_s^w(n) \geq 2n - 3$.*

Proof. For every $k \geq 1$ we are constructing $L_{4k+2} \in \mathcal{G}_s^w$ (the result for other values of n follows by suppressing degree 2 vertices). We take $0 < \epsilon_0 < \epsilon_1 < \cdots < \epsilon_{2k}$ and set $\delta_j := \sum_{i=j}^{2k} \epsilon_i$ for all $j \in \{1, \ldots, 2k\}$. We consider the graph G with vertices $r_i = (i, \delta_{2i}), r_i' = (i, -\delta_{2i})$ for $i \in \{0, \ldots, k\}$ and $\ell_i = (-i, \delta_{2i-1}), \ell_i' = (-i, -\delta_{2i-1})$ for $i \in \{1, \ldots, k\}$; and edge set

$$\{r_0 r_0'\} \cup \{r_i \ell_i, r_i \ell_i', r_i' \ell_i, r_i' \ell_i' \mid 1 \leq i \leq k\} \cup \{r_{i-1} \ell_i, r_{i-1} \ell_i', r_{i-1}' \ell_i, r_{i-1}' \ell_i' \mid 1 \leq i \leq k\}.$$

See Fig. 3 for an illustration of the final drawing. By construction, the midpoints of the edges never coincide and they lie on the vertical lines $x = 0$ and $x = -1/2$; thus they are in weakly convex position. It is straight-forward to verify that the constructed graph is L_{4k+2}. ☐

Definition 2. *For every odd $n \geq 3$, we denote by B_n the graph consisting of an isolated C_3 and $\frac{n-3}{2}$ copies of C_4 altogether identified along a single edge uv. We observe that B_n has $\frac{3}{2}(n-1)$ edges and deleting a degree 2 vertex from B_{n+1} one obtains an n-vertex graph with $\frac{3}{2}(n-1) - \frac{1}{2}$ edges.*

Theorem 4. *For all odd $n \geq 3$ we have $B_n \in \mathcal{G}_s^s$, i.e., $g_s^s(n) \geq \lfloor \frac{3}{2}(n-1) \rfloor$.*

Proof. Let $n \geq 3$ be such that $n - 3$ is divisible by 4 (if $n - 3$ is not divisible by 4, then B_n is an induced subgraph of B_{n+1}). We will first draw B_n in an unfeasible way and then transform it into another one proving $B_n \in \mathcal{G}_s^s$.

See Fig. 4 for an illustration of the final drawing.

We draw the $C_3 = (uvw)$ as an isosceles triangle with horizontal base uv. Let $u = (-1, 0)$, $v = (1, 0)$, and $w = (0, \frac{n-1}{2})$. There are $n - 3$ remaining points.

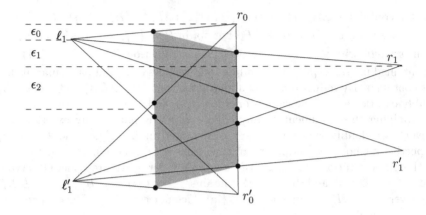

Fig. 3. The graph L_6 is in $\mathcal{G}_s^w$.

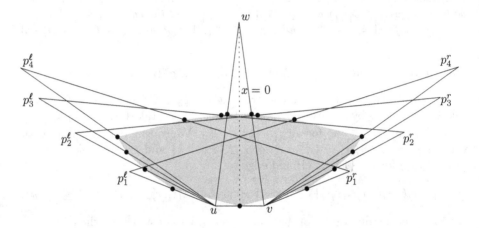

Fig. 4. The graph B_{11} is in $\mathcal{G}_s^s$.

Draw one half of them on coordinates $p_i^\ell = (-1 - i, i)$ for $1 \leq i \leq \frac{n-3}{2}$ and the other half mirrored along the y-axis, i.e., $p_i^r = (1 + i, i)$ for $1 \leq i \leq \frac{n-3}{2}$.

Now we add all edges $p_i^\ell u$ (left edges), $p_i^r v$ (right edges), for $1 \leq i \leq \frac{n-3}{2}$ and edges of the form $p_i^\ell p_{\frac{n-3}{2}+1-i}^r$ (diagonal edges) for all $1 \leq i \leq \frac{n-3}{2}$.

We observe that the points p_i^ℓ and u lie on the line $x + y = -1$, the points p_i^r and v lie on the line $x - y = 1$ and all midpoints of diagonal edges have y-coordinate $\frac{n-1}{4}$. In order to bring P_V and P_E into strict convex position, we simultaneously decrease the y-coordinates of points $p_{\frac{n-3}{2}+1-i}^\ell, p_{\frac{n-3}{2}+1-i}^r$ by $2^i \epsilon$ for $i \in \{1, \ldots, \frac{n-3}{2}\}$ for a sufficiently small value $\epsilon > 0$. It suffices to conveniently decrease the y-coordinate of w to get a drawing witnessing that $B_n \in \mathcal{G}_s^s$. $\quad\square$

2.3   Further Members of $\mathcal{G}_s^s$ and $\mathcal{G}_s^w$

We show that there are non-planar graphs in $\mathcal{G}_s^s$ and cubic graphs in $\mathcal{G}_s^w$.

Definition 3. *For all $k \geq 2$, we denote by H_k the graph consisting of a $2k$-gon with vertices $v_1, \ldots, v_{2k}$ and a singly subdivided edge from v_i to $v_{i+3 \bmod 2k}$ for all i even, i.e., there are k degree 2 vertices $u_1, \ldots, u_k$ and edges $u_i v_{2i}$ for all $i \in \{1, \ldots, k\}$, $u_i v_{2i+3}$ for all $i \in \{1, \ldots, k-2\}$, $u_{k-1} v_1$ and $u_k v_3$. We observe that H_k is planar if and only if k is even.*

Theorem 5. *For every $k \geq 2$, $H_k \in \mathcal{G}_s^s$. In particular, for every $n \geq 9$ there is a non-planar n-vertex graph in $\mathcal{G}_s^s$.*

Proof. We start by drawing C_{2k} as a regular $2k$-gon. Take an edge $e = xy$ and denote by x', y' the neighbors of x and y, respectively. For convenience consider e to be of horizontal slope with the $2k$-gon below it. Our goal is to place v_e a new vertex and edges $v_e x', v_e y'$ preserving the convexity of vertices and midpoints of edges. We consider the upward ray r based at the midpoint m_e of e and the upward ray s of points whose x-coordinate is the average between the x-coordinates of m_e and x'. We denote by Δ the triangle with vertices the midpoint $m_{x'x}$ of the edge $x'x$, the point x and m_e. Since $s \cap \Delta$ is nonempty, we place v_e such that the midpoint of $v_e x'$ is in $s \cap \Delta$. Clearly v_e is in r. Hence, the middle point of $v_e y'$ is in the corresponding triangle Δ' and the convexity of vertices and midpoints of edges is preserved. See Fig. 5 for an illustration. Since we only have to add a vertex on alternating edges of C_{2k}, these choices are independent of each other. It is easy to verify that the constructed graph is H_k.   □

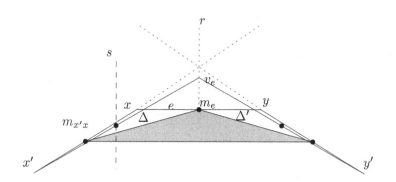

Fig. 5. The construction in Theorem 5

Definition 4. *For all $k \geq 3$, we denote by P_k the graph consisting of a prism over a k-cycle. We observe that P_k is a 3-regular graph.*

Theorem 6. *For every $k \geq 3$, $P_k \in \mathcal{G}_s^w$. In particular, for every even $n \geq 6$ there is a 3-regular n-vertex graph in $\mathcal{G}_s^w$.*

Proof. Let $k \geq 3$. In order to draw P_k, place $2k$ vertices $v_0, \ldots, v_{2k-1}$ as the vertices of a $2k$-gon in the plane, in which all inner angles are the same and at most two different side lengths occur in alternating fashion around it. (Apart from this, these lengths do not matter for the construction.) Add all *inner* edges of the form $v_i v_{i+2 \bmod 2k}$ for all i and *outer* edges $v_i v_{i+1 \bmod 2k}$ for i even. Clearly, the midpoints of outer edges are in strictly convex position and their convex hull is a regular k-gon. Now, consider four vertices say $v_0, \ldots, v_3$. They induce two outer edges, $v_0 v_1$ and $v_2 v_3$ and two inner edges $v_0 v_2$ and $v_1 v_3$. Now, the triangles $v_0 v_1 v_2$ and $v_1 v_2 v_3$ share the base segment $v_1 v_2$. Hence, the segments $m_{v_2 v_3} m_{v_1 v_3}$ and $m_{v_2 v_0} m_{v_1 v_0}$ share the slope of $v_1 v_2$. Now, since the angle between $v_1 v_2$ and $v_2 v_3$ equals the angle between $v_1 v_2$ and $v_0 v_1$ and $v_0 v_1$ and $v_2 v_3$ are of equal length, the segment $m_{v_2 v_3} m_{v_1 v_0}$ also has the same slope. Thus, all the midpoint lie on a line and all midpoints lie on the boundary of the midpoints of outer edges. See Fig. 6 for an illustration. □

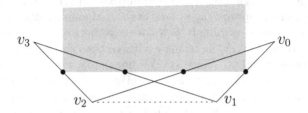

Fig. 6. The construction in Theorem 6

One can show that P_k is not in $\mathcal{G}_s^s$. More generally we believe that:

Conjecture 1. If $G \in \mathcal{G}_s^s$ then G is 2-degenerate.

2.4 Structural Questions

One can show that adding a leaf at the vertex r_1 of L_8 (see Definition 1) produces a graph not in $\mathcal{G}_s^w$. Under some conditions it is possible to add leafs to graphs in $\mathcal{G}_s^s$. We say that an edge is *V-crossing* if it intersects the interior of P_V.

Proposition 1. *Let $G \in \mathcal{G}_s^s$ be drawn in the required way. If uv is not V-crossing, then attaching a new vertex w to v yields a graph in $\mathcal{G}_s^s$.*

Proof. Let $G \in \mathcal{G}_s^s$ with at least 3 vertices and let $e = uv$ be the edge of G from the statement. For convenience consider that uv come in clockwise order on the boundary of P_V. Consider the supporting hyperplane of P_E through the midpoint m_e of e, whose side containing P_E contains v. A new midpoint can go inside the triangle Δ defined by the two supporting hyperplanes containing m_e and the additional supporting hyperplane containing the clockwisely consecutive midpoint m'. Since P_E is contained in P_V a part of Δ lies outside P_V. Choosing the midpoint of a new edge attached to v inside this region very close to e preserves strict convexity of vertices and midpoints. See Fig. 7 for an illustration. □

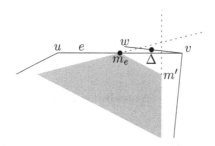

Fig. 7. The construction in Proposition 1

We wonder whether the class $\mathcal{G}_s^s$ is closed under adding leafs.

Despite the fact that $K_{2,n} \notin \mathcal{G}_s^s$, we have found in Theorem 4 a subdivision of $K_{2,n}$ which belongs to G_s^s. Similarly, Theorem 5 gives that a subdivision of $K_{3,3}$ is in G_s^s while $K_{3,3}$ is not. We have the impression that subdividing edges facilitates drawings in $\mathcal{G}_s^s$. Even more, we believe that:

Conjecture 2. The edges of every graph can be (multiply) subdivided such that the resulting graph is in $\mathcal{G}_s^s$.

3   Minkowski Sums

We show that the largest cardinality of a weakly convexly independent set X, which is a subset of the Minkowski sum of a convex planar n-point set A with itself is $2n$. If X is required to be in strict convex position then its size lies between $\frac{3}{2}n$ and $2n - 2$.

As mentioned in the introduction there is a slight trade-off when translating the graph drawing problem to the Minkowski sum problem. Since earlier works have been considering only asymptotic bounds this was neglected. Here we are fighting for constants, so we want to deal with it. Recall that a point in $x \in X \subseteq A + A$ is not captured by the graph model if $x = a + a$ for some $a \in A$. Thus, the point x corresponds to a vertex in the drawing of the graph. It is now clear, that in order to capture the trade-off we define $\tilde{g}_i^j(n)$ as the maximum of $n' + m$, where m is the number of edges of an n-vertex graph in $\mathcal{G}_i^j$ such that n' of its vertices can be added to the set of midpoints, such that the resulting set is in

$$\begin{cases} \text{strictly convex} & \text{if } j = s \\ \text{weakly convex} & \text{if } j = w \text{ position.} \\ \text{arbitrary} & \text{if } j = a \end{cases}$$

Lemma 2. *Let $G \in \mathcal{G}_s^w$ be drawn in the required way and $v \in G$. If v can be added to the drawing of G such that v together with the midpoints of G is in weakly convex position, then every edge $vw \in G$ is seen by w.*

Proof. Otherwise the midpoint of vw will be in the convex hull of v together with parts of P_E to the left and to the right of vw, see Fig. 8.   □

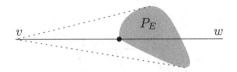

Fig. 8. The contradiction in Lemma 2

We say that an edge is *good* if it can be seen by both of its endpoints.

Theorem 7. *For every $n \geq 3$ we have $\widetilde{g}_s^w(n) = 2n$. This is, the largest cardinality of a weakly convexly independent set $X \subseteq A + A$, for A a convex planar n-point set, is $2n$.*

Proof. The lower bound comes from drawing C_n as the vertices and edges of a convex polygon. The set of vertices and midpoints is in weakly convex position.

For the upper bound let $G \in \mathcal{G}_s^w$ with n vertices and m edges, we denote by n_i the number of vertices of G that see i of its incident edges for $i \in \{0, 1, 2\}$. Since every edge is seen by at least one of its endpoints and every vertex sees at most 2 of its incident edges (Lemma 1), we know that $m = n_1 + 2n_2 - m_g$, where m_g is the number of good edges.

Let n' be the number of vertices of G that can be added to the drawing such that together with the midpoints they are in weakly convex position. Denote by n'_i the number of these vertices that see i of its incident edges for $i \in \{0, 1, 2\}$. By Lemma 2 the edges seen by an added vertex have to be good. Thus, $m_g \geq \frac{1}{2}(n'_1 + 2n'_2)$. This yields

$$m + n' \leq n_1 + 2n_2 - \frac{1}{2}(n'_1 + 2n'_2) + n'_0 + n'_1 + n'_2 \leq n_0 + \frac{3}{2}n_1 + 2n_2 \leq 2n. \quad \square$$

Theorem 8. *For every $n \geq 3$ we have $\lfloor \frac{3}{2}n \rfloor \leq \widetilde{g}_s^s(n) \leq 2n - 2$. This is, the largest cardinality of a convexly independent set $X \subseteq A + A$, for A a convex planar n-point, lies within the above bounds.*

Proof. The lower bound comes from drawing C_n as the vertices and edges of a convex polygon. The set formed by an independent set of vertices and all midpoints is in convex position.

Take $G \in \mathcal{G}_s^s$ with n vertices and m edges. The upper bound is very similar to Theorem 7. Indeed, following the same notations we also get that $m = n_1 + 2n_2 - m_g$. Again, the edges seen by an added vertex have to be good. Since now moreover the set of addable vertices has to be independent, we have $m_g \geq n'_1 + 2n'_2$. This yields

$$m + n' \leq n_1 + 2n_2 - n'_1 - 2n'_2 + n'_0 + n'_1 + n'_2 \leq n + n_2 - n'_2.$$

If $n + n_2 - n'_2 > 2n - 2$ then either $n_2 = n$ and $n'_2 < 2$, or $n_2 = n - 1$ and $n'_2 = 0$. In both cases we get that $n' \leq 1$. By Theorem 2 we have $m \leq 2n - 3$, then it follows that $m + n' \leq 2n - 2$. $\square$

4   Conclusions

We have improved the known bounds on $g_s^s(n)$, the number of edges an n-vertex graph of strong convex dimension can have. Still describing this function exactly is open. Confirming our conjecture that graphs in $\mathcal{G}_s^s$ have degeneracy 2 would not improve our bounds. Similarly, the exact largest cardinality $\widetilde{g}_s^s(n)$ of a convexly independent set $X \subseteq A + A$ for A a convex planar n-point set, remains to be determined. Curiously, in both cases we have shown that the correct answer lies between $\frac{3}{2}n$ and $2n$. The more general family $\mathcal{G}_s^w$ seems to be easier to handle, in particular we have provided the exact value for both g_s^w and $\widetilde{g}_s^w$.

From a more structural point of view we wonder what graph theoretical measures can ensure that a graph is in $\mathcal{G}_s^s$ or $\mathcal{G}_s^w$. The class $\mathcal{G}_s^w$ is not closed under adding leafs. We do not know if the same holds for $\mathcal{G}_s^s$. Finally, we believe that subdividing a graph often enough ensures that it can be drawn in $\mathcal{G}_s^s$.

References

1. Bílka, O., Buchin, K., Fulek, R., Kiyomi, M., Okamoto, Y., Tanigawa, S., Tóth, C.D.: A tight lower bound for convexly independent subsets of the Minkowski sums of planar point sets. Electron. J. Combin. **17**(1), Note 35, 4 (2010)
2. Eisenbrand, F., Pach, J., Rothvoß, T., Sopher, N.B.: Convexly independent subsets of the Minkowski sum of planar point sets, Electron. J. Combin. **15**(1), Note 8, 4 (2008)
3. Halman, N., Onn, S., Rothblum, U.G.: The convex dimension of a graph. Discrete Appl. Math. **155**(11), 1373–1383 (2007)
4. Onn, S., Rothblum, U.G.: Convex combinatorial optimization. Discrete Comput. Geom. **32**(4), 549–566 (2004)
5. Swanepoel, K.J., Valtr, P.: Large convexly independent subsets of Minkowski sums. Electron. J. Combin. **17**(1) (2010). Research Paper 146, 7
6. Hans Raj Tiwary: On the largest convex subsets in Minkowski sums. Inf. Process. Lett. **114**(8), 405–407 (2014)

Drawing Graphs Using a Small Number of Obstacles

Martin Balko[✉], Josef Cibulka, and Pavel Valtr

Department of Applied Mathematics, Faculty of Mathematics and Physics,
Charles University, Malostranské nám. 25, 118 00 Praha 1, Czech Republic
{balko,cibulka}@kam.mff.cuni.cz

Abstract. An *obstacle representation* of a graph G is a set of points in the plane representing the vertices of G, together with a set of polygonal obstacles such that two vertices of G are connected by an edge in G if and only if the line segment between the corresponding points avoids all the obstacles. The *obstacle number* obs(G) *of* G is the minimum number of obstacles in an obstacle representation of G.

We provide the first non-trivial general upper bound on the obstacle number of graphs by showing that every n-vertex graph G satisfies obs(G) $\leq 2n \log n$. This refutes a conjecture of Mukkamala, Pach, and Pálvölgyi. For bipartite n-vertex graphs, we improve this bound to $n-1$. Both bounds apply even when the obstacles are required to be convex. We also prove a lower bound $2^{\Omega(hn)}$ on the number of n-vertex graphs with obstacle number at most h for $h < n$ and an asymptotically matching lower bound $\Omega(n^{4/3}M^{2/3})$ for the complexity of a collection of $M \geq \Omega(n)$ faces in an arrangement of n^2 line segments with $2n$ endpoints.

Keywords: Obstacle number · Geometric drawing · Obstacle representation · Arrangement of line segments

1 Introduction

In a *geometric drawing of a graph* G, the vertices of G are represented by distinct points in the plane and each edge e of G is represented by the line segment between the pair of points that represent the vertices of e. As usual, we identify the vertices and their images, as well as the edges and the line segments representing them.

Let P be a finite set of points in the plane in *general position*, that is, there are no three collinear points in P. The *complete geometric graph* K_P is the

---

The first and the third author acknowledge the support of the project CE-ITI (GAČR P202/12/G061) of the Czech Science Foundation and the grant GAUK 1262213 of the Grant Agency of Charles University. The first author was also supported by the grant SVV–2015–260223. Part of the research was conducted during the workshop Homonolo 2014 supported by the European Science Foundation as a part of the EuroGIGA collaborative research program (Graphs in Geometry and Algorithms).

© Springer International Publishing Switzerland 2015
E. Di Giacomo and A. Lubiw (Eds.): GD 2015, LNCS 9411, pp. 360–372, 2015.
DOI: 10.1007/978-3-319-27261-0_30

geometric drawing of the complete graph $K_{|P|}$ with vertices represented by the points of P.

An *obstacle* is a polygon in the plane. An *obstacle representation* of a graph G is a geometric drawing D of G together with a set $\mathcal{O}$ of obstacles such that two vertices of G are connected by an edge e if and only if the line segment representing e in D is disjoint from all obstacles in $\mathcal{O}$. The *obstacle number* $\mathrm{obs}(G)$ of G is the minimum number of obstacles in an obstacle representation of G. The *convex obstacle number* $\mathrm{obs}_c(G)$ of a graph G is the minimum number of obstacles in an obstacle representation of G in which all the obstacles are required to be convex. Clearly, we have $\mathrm{obs}(G) \leq \mathrm{obs}_c(G)$ for every graph G.

In this paper, we provide the first nontrivial general upper bound on the obstacle number of graphs (Theorem 2). We also show a lower bound for the number of graphs with small obstacle number (Theorem 3) and a matching lower bound for the complexity of a collection of faces in an arrangement of line segments that share endpoints (Theorem 4). All proofs of our results are based on so-called ε-*dilated bipartite drawings* of $K_{m,n}$, which we introduce in Sect. 2.

In the following, we make no serious effort to optimize the constants. All logarithms in this paper are base 2.

1.1   Bounding the Obstacle Number

The obstacle number of a graph was introduced by Alpert, Koch, and Laison [1] who showed, among several other results, that for every positive integer h there is a graph G with $\mathrm{obs}(G) \geq h$. Using extremal graph theoretic tools, Pach and Sariöz [11] proved that the number of labeled n-vertex graphs with obstacle number at most h is at most $2^{o(n^2)}$ for every fixed integer h. This implies that there are bipartite graphs with arbitrarily large obstacle number.

Mukkamala, Pach, and Sariöz [10] established more precise bounds by showing that the number of labeled n-vertex graphs with obstacle number at most h is at most $2^{O(hn \log^2 n)}$ for every fixed positive integer h. It follows that, for every n, there is a graph G on n vertices with $\mathrm{obs}(G) \geq \Omega(n/\log^2 n)$. Later, Mukkamala, Pach, and Pálvölgyi [9] improved the lower bound to $\mathrm{obs}(G) \geq \Omega(n/\log n)$. Currently, the strongest lower bound on the obstacle number is due to Dujmović and Morin [4] who showed that there is a graph G with n vertices and $\mathrm{obs}(G) \geq \Omega(n/(\log \log n)^2)$ for every n.

Surprisingly, not much has been done for the general upper bound on the obstacle number. We are only aware of the trivial bound $\mathrm{obs}(G) \leq \binom{n}{2}$ for every graph G on n vertices. This follows easily, as we can consider the complete geometric graph K_P for some point set P of size n and place a small obstacle O_e on every *non-edge* e of G such that O_e intersects only e in K_P. A non-edge of a graph $G = (V, E)$ is an element of $\binom{V}{2} \setminus E$.

Concerning special graph classes, Fulek, Saeedi, and Sariöz [6] showed that the convex obstacle number is at most five for every outerplanar graph, and at most four for every bipartite permutation graph.

Alpert, Koch, and Laison [1] asked whether the obstacle number of every graph on n vertices can be bounded from above by a linear function of n. We show that this is true for bipartite graphs, even for the convex obstacle number.

Theorem 1. *For every pair of positive integers* m, n *and every bipartite graph* $G \subseteq K_{m,n}$ *and its complement* $\overline{G}$, *we have*

$$\mathrm{obs}_c(G), \mathrm{obs}_c(\overline{G}) \leq m + n - 1.$$

In contrast, Mukkamala, Pach, and Pálvölgyi [9] conjectured that the maximum obstacle number of n-vertex graphs is around n^2. We refute this conjecture by showing the first non-trivial general upper bound on the obstacle number of graphs. In fact, we prove a stronger result that provides a general upper bound for the convex obstacle number.

Theorem 2. *For every positive integer* n *and every graph* G *on* n *vertices, the convex obstacle number of* G *satisfies*

$$\mathrm{obs}_c(G) \leq 2n \log n.$$

By a more careful approach, which we omit in this paper, the bound in Theorem 2 can be improved to $n\lceil \log n \rceil - n + 1$. The question whether the upper bound on $\mathrm{obs}(G)$ can be improved to $O(n)$ for every n-vertex graph G remains open.

1.2 Number of Graphs with Small Obstacle Number

For positive integers h and n, let $g(h, n)$ be the number of labeled n-vertex graphs with obstacle number at most h. The lower bounds on the obstacle number by Mukkamala, Pach, and Pálvölgyi [9] and by Dujmović and Morin [4] are both based on the upper bound $g(h, n) \leq 2^{O(hn \log^2 n)}$. In fact, any improvement on the upper bound for $g(h, n)$ will translate into an improved lower bound on the obstacle number [4]. Dujmović and Morin [4] conjectured $g(h, n) \leq 2^{f(n) \cdot o(h)}$ where $f(n) \leq O(n \log^2 n)$. We show the following lower bound on $g(h, n)$.

Theorem 3. *For every pair of integers* n *and* h *satisfying* $0 < h < n$, *we have*

$$g(h, n) \geq 2^{\Omega(hn)}.$$

1.3 Complexity of Faces in Arrangements of Line Segments

An *arrangement* $\mathcal{A}$ *of line segments* is a finite collection of line segments in the plane. The line segments of $\mathcal{A}$ partition the plane into *vertices*, *edges*, and *cells*. A vertex is a common point of two or more line segments. Removing the vertices from the line segments creates a collection of subsegments which are called edges. The cells are the connected components of the complement of the line segments. A *face* of $\mathcal{A}$ is a closure of a cell.

Note that every geometric drawing of a graph is an arrangement of line segments and vice versa. The edges of the graph correspond to the line segments of the arrangement and the vertices of the graph correspond to the endpoints of the line segments.

A line segment s of $\mathcal{A}$ is *incident* to a face F of $\mathcal{A}$ if s and F share an edge of $\mathcal{A}$. The *complexity of a face* F is the number of the line segments of $\mathcal{A}$ that are incident to F. If $\mathcal{F}$ is a set of faces of $\mathcal{A}$, then the *complexity of* $\mathcal{F}$ is the sum of the complexities of F taken over all $F \in \mathcal{F}$.

An *arrangement of lines* is a finite collection of lines in the plane with faces and their complexity defined analogously.

Edelsbrunner and Welzl [5] constructed an arrangement of m lines having a set of M faces with complexity $\Omega(m^{2/3}M^{2/3}+m)$ for every m and $M \le \binom{m}{2}+1$. Wiernik and Sharir [13] constructed an arrangement of m line segments with a single face of complexity $\Omega(m\alpha(m))$. These two constructions can be combined to provide the lower bound $\Omega(m^{2/3}M^{2/3} + m\alpha(m))$ for the complexity of M faces in an arrangement of m line segments, where $M \le \binom{m}{2}+1$. The best upper bound for the complexity of M faces in an arrangement of m line segments is $O(m^{2/3}M^{2/3} + m\alpha(m) + m\log M)$ by Aronov et al. [3].

Arkin et al. [2] studied arrangements whose line segments share endpoints. That is, they considered the maximum complexity of a face when we bound the number of endpoints of the line segments instead of the number of the line segments. They showed that the complexity of a single face in an arrangement of line segments with n endpoints is at most $O(n \log n)$. An $\Omega(n \log n)$ lower bound was then proved by Matoušek and Valtr [8].

Arkin et al. [2] posed as an open problem to determine the maximum complexity of a set of M faces in an arrangement of line segments with n endpoints. Since every arrangement of line segments with n endpoints contains at most $\binom{n}{2}$ line segments, the upper bound $O(n^{4/3}M^{2/3} + n^2\alpha(n) + n^2 \log M)$ can be deduced from the upper bound of Aronov et al. [3]. We give a lower bound that, whenever $M \ge n \log^{3/2} n$, matches this upper bound up to a multiplicative factor.

Theorem 4. *There is constant C such that for every sufficiently large integer n, there is an arrangement $\mathcal{A}$ of n^2 line segments with $2n$ endpoints such that for every M satisfying $Cn \le M \le n^4/C$ there is a set of at most M faces of $\mathcal{A}$ with complexity $\Omega(n^{4/3}M^{2/3})$.*

Taking only the faces with the highest complexity from the construction from the proof of Theorem 4 gives the following lower bound for smaller values of M.

Corollary 1. *For every sufficiently large integer n, there is an arrangement $\mathcal{A}$ of n^2 line segments with $2n$ endpoints such that for every M satisfying $M \le O(n)$ there is a set of at most M faces of $\mathcal{A}$ with complexity $\Omega(nM)$.*

Consequently, for every value of M, the lower bounds differ from the best known upper bounds by at most an $O(\log n)$ multiplicative factor.

2   Dilated Bipartite Drawings

For a point $p \in \mathbb{R}^2$, let $x(p)$ and $y(p)$ denote the x- and the y-coordinate of p, respectively. An *intersection point* in a geometric drawing D of a graph G is a common point of two edges of G that share no vertex.

Let m and n be positive integers. We say that a geometric drawing of $K_{m,n}$ is *bipartite* if the vertices of the same color class of $K_{m,n}$ lie on a common vertical line and not all vertices of $K_{m,n}$ lie on the same vertical line. For the rest of this section, we let D be a bipartite drawing of $K_{m,n}$ and use $P := \{p_1, \ldots, p_m\}$ and $Q := \{q_1, \ldots, q_n\}$ with $y(p_1) < \cdots < y(p_m)$ and $y(q_1) < \cdots < y(q_n)$ to denote the point sets representing the color classes of $K_{m,n}$ in D. We let ℓ_P and ℓ_Q be the vertical lines that contain the points of P and Q, respectively. The *width* w of D is $|x(q_1) - x(p_1)|$. In the following, we assume that ℓ_P is to the left of ℓ_Q and that $p_1 = (0,0)$, $q_1 = (w,0)$. We set $d_i := y(p_{i+1}) - y(p_i)$ for $i = 1, \ldots, m-1$ and $h_j := y(q_{j+1}) - y(q_j)$ for $j = 1, \ldots, n-1$. We call d_1 the *left step* of D and h_1 the *right step* of D.

We say that D is *regular* if we have $d_1 = \cdots = d_{m-1}$ and $h_1 = \cdots = h_{n-1}$. Note that every regular drawing of $K_{m,n}$ is uniquely determined by its width, left step, and right step. A *regularization* of a (possibly non-regular) bipartite drawing D is the regular bipartite drawing of $K_{m,n}$ with the vertices $\pi(p_i) := (0, (i-1)d_1)$ and $\pi(q_j) := (w, (j-1)h_1)$ for $i = 1, \ldots, m$ and $j = 1, \ldots, n$.

For $1 \le k \le m + n - 1$, the *kth level* of D is the set of edges $p_i q_j$ with $i + j = k + 1$. Note that the levels of D partition the edge set of $K_{m,n}$ and that the kth level of D contains $\min\{k, m, n, m + n - k\}$ edges. If D is regular, then, for every $1 < k < m + n - 1$, the edges of the kth level of D share a unique intersection point that lies on the vertical line $\{\frac{d_1}{d_1 + h_1} w\} \times \mathbb{R}$.

For an integer $l \ge 2$, an ordered l-tuple $(p_{i_1} q_{j_1}, \ldots, p_{i_l} q_{j_l})$ of edges of D is *uniformly crossing* if we have $0 < i_2 - i_1 = \cdots = i_l - i_{l-1}$ and $j_2 - j_1 = \cdots = j_l - j_{l-1} < 0$. In particular, a set of edges forming a level of D, ordered by their decreasing slopes, is uniformly crossing. Note that if $(p_{i_1} q_{j_1}, \ldots, p_{i_l} q_{j_l})$ is uniformly crossing, then the edges $\pi(p_{i_1})\pi(q_{j_1}), \ldots, \pi(p_{i_l})\pi(q_{j_l})$ of the regularization of D share a common intersection point, which we call the *meeting point* of $(p_{i_1} q_{j_1}, \ldots, p_{i_l} q_{j_l})$. In the other direction, if D is regular and $(e_1, \ldots, e_l)$ is a maximal set of edges of D that share a common intersection point and are ordered by their decreasing slopes, then $(e_1, \ldots, e_l)$ is uniformly crossing.

Let $\varepsilon > 0$ be a real number. We say that D is *ε-dilated* if we have $d_1 < \cdots < d_{m-1} < (1 + \varepsilon)d_1$ and $h_1 < \cdots < h_{n-1} < (1 + \varepsilon)h_1$.

In a geometric drawing D' of a (not necessarily bipartite) graph, let $(e_1, \ldots, e_l)$ be an ordered l-tuple of edges of D' such that e_i and e_{i+1} share an intersection point r_i for $i = 1, \ldots, l-1$. We say that $(e_1, \ldots, e_l)$ *forms a cap*, if $x(r_1) < \cdots < x(r_{l-1})$ and the slopes of $e_1, \ldots, e_l$ are strictly decreasing. A *cap* C is then the component of the lower envelope of $e_1 \cup \cdots \cup e_l$ that contains $r_1, \ldots, r_{l-1}$. The points r_i are *vertices* of C and $e_1 \cap C, \ldots, e_l \cap C$ are *edges* of C. See part a) of Fig. 1. A cap C is *good in D'*, if the edges of C are incident to the same bounded face of D' or if C has only one edge. If D' is bipartite and the edges of one of its levels form a cap C, then we call C a *level-cap* of D'.

The following lemma is crucial in the proofs of all our main results.

Lemma 1. *(i) If D satisfies $d_1 < \cdots < d_{m-1}$ and $h_1 < \cdots < h_{n-1}$, then, for every $l \ge 2$, every uniformly crossing l-tuple of edges of D forms a cap.*

(ii) For all $w, d_1, h_1 \in \mathbb{R}^+$ and $m, n \in \mathbb{N}$, there is an $\varepsilon = \varepsilon_{m,n}(w, d_1, h_1) > 0$ such that if D is an ε-dilated bipartite drawing of $K_{m,n}$ with width w, left step d_1, and right step h_1, then for every $l \geq 2$ every uniformly crossing l-tuple of edges of D forms a good cap in D.

Proof. For part (i), let $(e_1, \ldots, e_l)$ be a uniformly crossing l-tuple of edges of D with $e_k := p_{i_k} q_{j_k}$ for every $k = 1, \ldots, l$. Consider edges e_k, e_{k+1}, e_{k+2} and let r_k and r_{k+1} be the points $e_k \cap e_{k+1}$ and $e_{k+1} \cap e_{k+2}$, respectively. The points r_k and r_{k+1} exist, as $y(p_{i_k}) < y(p_{i_{k+1}}) < y(p_{i_{k+2}})$ and $y(q_{j_{k+2}}) < y(q_{j_{k+1}}) < y(q_{j_k})$.

Consider the midpoint p of $p_{i_k} p_{i_{k+2}}$ and the midpoint q of $q_{j_k} q_{j_{k+2}}$. Since $(e_1, \ldots, e_l)$ is uniformly crossing and $d_1 < \cdots < d_{m-1}$ and $h_1 < \cdots < h_{n-1}$, we have $y(p_{i_{k+1}}) < y(p)$ and $y(q_{j_{k+1}}) < y(q)$. See part b) of Fig. 1. The edges pq, e_k, and e_{k+2} share a common point that lies above e_{k+1}. Since r_k and r_{k+1} lie on e_{k+1}, we obtain $x(r_k) < x(r_{k+1})$. The slopes of e_k, e_{k+1}, e_{k+2} are strictly decreasing, thus $(e_1, \ldots, e_l)$ forms a cap.

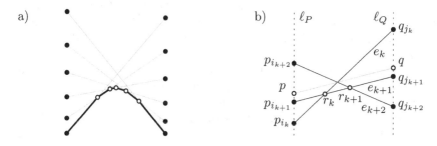

Fig. 1. (a) An example of a cap with vertices denoted by empty circles and with edges denoted black. (b) A situation in the proof of Lemma 1.

To show (ii), we use the following claim. For all $w, d_1, h_1, \delta \in \mathbb{R}^+$ and $m, n \in \mathbb{N}$, there is an $\varepsilon = \varepsilon_{m,n}(w, d_1, h_1, \delta) > 0$ such that if D is ε-dilated, then the intersection point between any two edges $p_i q_j$ and $p_{i'} q_{j'}$ of D lies in distance less than δ from the intersection point $\pi(p_i)\pi(q_j) \cap \pi(p_{i'})\pi(q_{j'})$.

This follows from the fact that for fixed w, d_1, h_1, all ε'-dilated drawings of $K_{m,n}$ with width w, left step d_1, and right step h_1 converge to their common regularization as $\varepsilon' > 0$ tends to zero.

Let $\delta_{m,n}(w, d_1, h_1) = \delta > 0$ be the half of the minimum distance between two intersection points of the regular drawing of $K_{m,n}$ with width w, left step d_1, and right step h_1. For $\varepsilon = \varepsilon_{m,n}(w, d_1, h_1, \delta)$, let D be an ε-dilated drawing of $K_{m,n}$ with width w, left step d_1, and right step h_1. According to (i), every uniformly crossing l-tuple $(e_1, \ldots, e_l)$ of edges of D forms a cap. It follows from the claim that the vertices of a cap C formed by $(e_1, \ldots, e_l)$ are contained in an open disc B with the center in the meeting point s of $(e_1, \ldots, e_l)$ and radius δ. In particular, there is a connected component K of $B \setminus (e_1 \cup \cdots \cup e_l)$ such that every edge of C is incident to the closure $\overline{K}$ of K.

Suppose for a contradiction that C is not good in D. Then there is an edge pq of D that divides $\overline{K}$ into two parts, each incident to some edge of C and each having an empty intersection with some edge of C. Otherwise all edges of C are incident to a single face of D, implying that C is good. The edge pq intersects some edge $p_{i_k} q_{j_k} \cap C$ of C in a point $r \in B$. By (i), the intersection point $r' := \pi(p)\pi(q) \cap \pi(p_{i_k})\pi(q_{j_k})$ is different from s, since edges of C and the edge pq do not form a cap. The distance of r and s is less than δ, as $r \in B$. By the claim, the distance of r and r' is also less than δ. On the other hand, the distance of r' and s is at least 2δ from the choice of δ. This gives us a contradiction with the triangle inequality.   □

3   Proof of Theorem 1

Let $G \subseteq K_{m,n}$ be a bipartite graph and $\overline{G}$ be its complement. Using Lemma 1, we can easily show $\mathrm{obs}_c(\overline{G}) \leq m + n - 1$. Let $\varepsilon > 0$ be chosen as in Lemma 1 for $K_{m,n}$ and $w = d_1 = h_1 = 1$. Consider an ε-dilated drawing D of $K_{m,n}$ with $w = d_1 = h_1 = 1$, $p_1 = (0,0)$, and $q_1 = (1,0)$. Since edges of every level of D are uniformly crossing, part (ii) of Lemma 1 implies that the edges of the kth level of D form a good level-cap C_k in D for every $1 \leq k \leq m + n - 1$. That is, there is a bounded face F_k of D such that each edge of C_k is incident to F_k or C_k contains only one edge.

For every integer k satisfying $1 \leq k \leq m+n-1$, we construct a single convex obstacle O_k. If C_k contains only one edge e, the obstacle O_k is an arbitrary point of e or an empty set. Otherwise every edge $p_i q_{k+1-i}$ of the kth level of D shares a line segment s_k^i of positive length with F_k. The obstacle O_k is defined as the convex hull of the midpoints of the line segments s_k^i where $p_i q_{k+1-i}$ is not an edge of $\overline{G}$. See part a) of Fig. 2. The levels partition the edge set of $K_{m,n}$, therefore we block every non-edge of $\overline{G}$. Since every bounded face of D is convex, we have $O_k \subseteq F_k$. Therefore no edge of $\overline{G}$ is blocked and we obtain an obstacle representation of $\overline{G}$. In total, we produce at most $m + n - 1$ obstacles.

To show $\mathrm{obs}_c(G) \leq m + n - 1$, we proceed analogously as above, except the vertices of D are suitably perturbed before obstacles O_k are defined, which allows to add two (long and skinny) convex obstacles O_P and O_Q blocking all the edges $p_i p_{i'}$ and $q_j q_{j'}$, respectively. The addition of the obstacles O_P and O_Q may be compensated by using a single convex obstacle to block non-edges in the first and the second level and in the $(m + n - 2)$th and the $(m + n - 1)$th level.

4   Proof of Theorem 2

We show that the convex obstacle number of every graph G on n vertices is at most $2n \log n$. The high-level overview of the proof is as follows. We partition the edges of G to edge sets of $O(n)$ induced bipartite subgraphs of G by iteratively partitioning the vertex set of G into two (almost) equal parts and considering the corresponding induced bipartite subgraphs of G. For every $j = 0, \ldots \lfloor \log n \rfloor$, the number of such bipartite subgraphs of size about $n/2^j$ is 2^j. Then we construct an obstacle representation of G whose restriction to every such bipartite subgraph resembles the obstacle representation from the proof of Theorem 1.

This is achieved by choosing a variant of the well-known *Horton sets* [12] as the underlying vertex set. Since the obstacle representation of every bipartite subgraph of size about $n/2^j$ uses about $n/2^j$ obstacles, we have $O(n \log n)$ obstacles in total.

Let S be a finite set of points on a vertical line. We say that a point p of S is an *odd point of* S if p has an odd-numbered position in the ordering of S by increasing y-coordinates. Otherwise p is said to be an *even point of* S.

Let $N \geq 2$ be the least power of two such that $N \geq n$. If $N > n$, then we add $N - n$ isolated vertices to G. Clearly, this does not decrease the obstacle number. Let $\varepsilon > 0$ be chosen as in Lemma 1 for $K_{N,N}$ and $w = d_1 = h_1 = 1$. Let D be an ε-dilated bipartite drawing of $K_{N,N}$ with width, left step, and right step equal to 1 and with $d_i = h_i$ for every $i = 1, \ldots, N - 1$. We let $P := \{p_1, \ldots, p_N\}$ and $Q := \{q_1, \ldots, q_N\}$ be the color classes of D ordered by increasing y-coordinates such that $p_1 = (0,0)$ and $q_1 = (1,0)$. By part (ii) of Lemma 1, edges of each level of D form a good cap in D. For the rest of the proof, the y-coordinates of all points remain fixed. Let $\alpha = \alpha(\varepsilon) > 0$ be a real number to be determined later.

First, we let D_1 be the drawing obtained from D by removing the even points from P and the odd points from Q. We use P_1^1 and P_1^2 to denote the left and the right color class of D_1, respectively. We map the vertices of G to the vertices of D_1 arbitrarily. Let $\mathcal{C}_1$ be the set of the level-caps of D_1. Since every level-cap in D is good in D, every cap in $\mathcal{C}_1$ is good in D_1.

The drawing D_1 is a first step towards making an obstacle representation of G. In fact, we can now block a large portion of non-edges of G by placing obstacles in D_1 as in the proof of Theorem 1. Then we take care of the edges between vertices in the left color class P_1^1 of $K_{N/2,N/2}$ (edges between vertices in the right color class P_1^2 of $K_{N/2,N/2}$ are dealt with analogously). We slightly shift the even points in P_1^1 horizontally to the right. Only some of the edges of a copy of $K_{N/4,N/4}$ between the even and the odd points of P_1^1 belong to G. Hence we can place convex obstacles along the level-caps of this $K_{N/4,N/4}$, again, same as in the bipartite case. To take care of the edges between vertices in the same color class of $K_{N/4,N/4}$, and for each of the color classes we proceed similarly as above.

We now describe this iterative process formally. Having chosen point sets $P_{j-1}^1, \ldots, P_{j-1}^{2^{j-1}}$ for some $2 \leq j \leq \log N$, we define $P_j^1, \ldots, P_j^{2^j}$ as follows. For $1 \leq k \leq 2^{j-1}$, let P_j^{2k-1} be the set of odd points of P_{j-1}^k and let P_j^{2k} be the set of even points of P_{j-1}^k. Let $\varepsilon_j > 0$ be a small real number. If k is odd, we move the points from P_j^{2k} to the right by ε_j. If k is even, we move the points from P_j^{2k-1} to the left by ε_j. We slightly abuse the notation by using D_{j-1} and $\mathcal{C}_{j-1}$ to denote the modified drawing D_{j-1} and the set of modified caps from the original set $\mathcal{C}_{j-1}$, respectively.

For $1 \leq k \leq 2^{j-1}$, we add all edges between points from P_j^{2k-1} and P_j^{2k} to create a bipartite drawing D_j^k of $K_{N/2^j,N/2^j}$. We let $\mathcal{C}_j$ be the union of $\mathcal{C}_{j-1}$ with a set of level-caps of the drawings D_j^k for $1 \leq k \leq 2^{j-1}$. We also set $D_j := D_j^1 \cup \cdots \cup D_j^{2^{j-1}} \cup D_{j-1}$.

We choose ε_j small enough so that each cap $C \in \mathcal{C}_{j-1}$, which is good in D_{j-1}, is good in D_j after the translations by ε_j. Such ε_j exists, as every geometric drawing of a graph is compact and the distance of two points is a continuous function. We choose ε_j small enough such that for every edge e of the modified drawing D_{j-1}, the portion of e between P_j^{2k-1} and P_j^{2k} is contained in the horizontal strip $\mathbb{R} \times (y(p) - \alpha, y(p) + \alpha)$ for some endpoint p of e. This can be done, as the vertical strips between $P_{j-1}^{2k'-1}$ and $P_{j-1}^{2k'}$ for $1 \le k' \le 2^{j-2}$ do not change during the translations by ε_j.

After $\log N$ steps, the drawings $D_{\log N}^k$ contain two vertices and the construction stops. We show that we can add at most $2n \log n$ convex obstacles to the drawing $D_{\log N}$ to obtain an obstacle representation of G.

For $2 \le j \le \log N$ and $1 \le k \le 2^{j-1}$, let $f_{j,k} \colon \mathbb{R}^2 \to \mathbb{R}^2$ be the affine mapping $f_{j,k}(x,y) := (x/\varepsilon_j - c_{j,k}, y)$ where $c_{j,k} \in \mathbb{R}$ is chosen such that the left color class of $f_{j,k}(D_j^k)$ lies on $\{0\} \times \mathbb{R}$. Note that the drawing $f_{j,k}(D_j^k)$ is contained in the drawing D and thus edges of the levels of $f_{j,k}(D_j^k)$ form good caps in $f_{j,k}(D_j^k)$. Since $f_{j,k}$ does not change the edge-face incidences in D_j^k, edges of the levels of D_j^k form good caps in D_j^k.

Let C be a level-cap formed by edges of a level L of $f_{j,k}(D_j^k)$ and let F_C be the bounded face of $f_{j,k}(D_j^k)$ such that all edges of C are incident to F_C. Edges of L are also edges of a level L' of D. Since the indices of edges of L have the same parity, L' contains an edge $p_i q_i$ for some $1 \le i \le N$. Let ℓ_C be the horizontal line containing $p_i q_i$. No vertex of the level-cap formed by edges of L' lies strictly above ℓ_C and no edge of C is contained in ℓ_C. Thus there is $\alpha_C > 0$ such that every edge of C is incident to $F_C \cap (\mathbb{R} \times (-\infty, y(p_i) - \alpha_C))$. See part b) of Fig. 2. We choose $\alpha = \alpha(\varepsilon)$ to be the minimum of α_C over all level-caps C of $f_{j,k}(D_j^k)$ with $2 \le j \le \log N$ and $1 \le k \le 2^{j-1}$. Since $f_{j,k}(D_j^k)$ is a drawing contained in D and determined by j and k, we see that α depends only on ε.

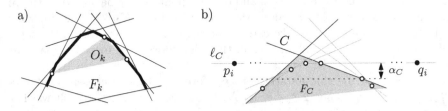

Fig. 2. (a) Placing a convex obstacle O_k that blocks three edges of $K_{m,n}$. (b) All edges of a cap C are incident to a part of a face F_C strictly below $p_i q_i$.

Since $f_{j,k}$ does not change the y-coordinates, for every level-cap C of D_j^k, there is a bounded face F_C of D_j^k such that all edges of C are incident to the part of F_C that lies below $\ell_{f_{j,k}(C)}$ in the vertical distance larger than α.

By induction on j, $1 \le j \le \log N$, we show that every cap from $\mathcal{C}_j$ is good in D_j in the jth step of the construction. We already observed that this is true for $j = 1$. Suppose for a contradiction that there is a cap $C \in \mathcal{C}_j$ that is not good in D_j for $j > 1$. Using the inductive hypothesis and the choice of ε_j, C is not in C_{j-1}. Therefore there is a drawing D_j^k for $1 \leq k \leq 2^{j-1}$ such that C is a level-cap of D_j^k. Since C is good in D_j^k, all edges forming C are incident to a single bounded face F_C of D_j^k. However, C is not good in D_j, thus some edge e of $D_j \setminus D_j^k$ divides F_C into two parts, each incident to an edge of C and each having an empty intersection with some edge of C. The drawings $D_j^1, \ldots, D_j^{2^{j-1}}$ are contained in pairwise disjoint vertical strips, thus e is an edge of D_{j-1}. It follows from the proof of Lemma 1 that all edges of C are incident to F_C in a 2δ-neighborhood of $\ell_{f_{j,k}(C)}$ for some $\delta = \delta(\varepsilon) > 0$. Therefore e intersects this 2δ-neighborhood. By the choice of ε_j, the portion of e between P_j^{2k-1} and P_j^{2k} is contained in the horizontal strip $\mathbb{R} \times (y(p) - \alpha, y(p) + \alpha)$ for an endpoint p of e. Assuming α and δ are sufficiently small with respect to the left and the right step of D, say $\alpha, \delta < 1/8$, we see that p lies on $\ell_{f_{j,k}(C)}$. Thus the portion of e between P_j^{2k-1} and P_j^{2k} lies in the α-neighborhood of $\ell_{f_{j,k}(C)} = \mathbb{R} \times \{y(p)\}$. On the other hand, all edges of C are incident to the part of F_C that is strictly below $\mathbb{R} \times \{y(p) - \alpha\}$. Thus e cannot divide F_C, a contradiction.

For every (modified) drawing D_j^k, we place the obstacles as in the first part of the proof of Theorem 1 with respect to the whole drawing $D_{\log N}$. Using the fact that bounded faces of every geometric drawing of K_N are convex, it follows from the construction of $D_{\log N}$ that we obtain an obstacle representation of G. For every $1 \leq j \leq \log N$ and $1 \leq k \leq 2^{j-1}$, we place at most $N/2^{j-1} - 1$ convex obstacles in the drawing D_j^k of $K_{N/2^j, N/2^j}$. For every j, we thus use at most $2^{j-1}(N/2^{j-1} - 1) = N - 2^{j-1}$ obstacles. Summing over j, we obtain an obstacle representation of G with at most $\sum_{j=1}^{\log N}(N - 2^{j-1}) = N(\log N - 1) + 1$ convex obstacles. Since $N < 2n$, we have less than $2n \log n + 1$ convex obstacles.

5   Proof of Theorem 3

Let h and n be given positive integers with $h < n$. We show that the number $g(h, n)$ of labeled n-vertex graphs of obstacle number at most h is at least $2^{\Omega(hn)}$.

For a point set $P \subseteq \mathbb{R}^2$ in general position, let $e(h, P)$ be the maximum integer for which there is a set $\mathcal{F}$ of at most h bounded faces of K_P and a set of $e(h, P)$ edges of K_P that are incident to at least one face from $\mathcal{F}$. Let $e(h, n)$ be the maximum of $e(h, P)$ over all sets P of n points in the plane in general position.

Claim. We have $g(h, n) \geq 2^{e(h,n)}$.

To prove the claim, let P be a set of n points in the plane in general position for which $e(h, P) = e(h, n)$. Let $\mathcal{F}$ be the set of at most h bounded faces of K_P such that $e(h, n)$ edges of K_P are incident to at least one face from $\mathcal{F}$. For a face $F \in \mathcal{F}$, let E_F denote the set of edges of K_P that are incident to F. We use G to denote the graph with the vertex set P and with two vertices connected by an edge if and only if the corresponding edge of K_P is incident to no face F of $\mathcal{F}$.

We show that every subgraph G' of K_P containing G satisfies $\mathrm{obs}(G') \leq h$. The claim then follows, as the number of such subgraphs G' is $2^{e(h,n)}$.

Let G' be a subgraph of K_P such that $G \subseteq G'$. For every face $F \in \mathcal{F}$, we define a convex obstacle O_F as the convex hull of midpoints of line segments $e \cap F$ for every $e \in E_F$ that represents a non-edge of G'. Note that, since all bounded faces of K_P are convex, the obstacle O_F is contained in F and thus O_F blocks only non-edges of G'. Since every non-edge of G' is contained in E_F for some $F \in \mathcal{F}$, we obtain an obstacle representation of G' with at most h convex obstacles. This finishes the proof of the claim.   □

Since $h < n$, the following and the previous claim give Theorem 3.

Claim. For $n \geq 3$, we have $e(h, n) \geq \frac{2hn - h^2 - 1}{4}$.

Let $\varepsilon > 0$ be chosen as in Lemma 1 for $K_{\lceil n/2 \rceil, \lfloor n/2 \rfloor}$ and $w = d_1 = h_1 = 1$. Let D be an ε-dilated drawing of $K_{\lceil n/2 \rceil, \lfloor n/2 \rfloor}$ with $w = d_1 = h_1 = 1$, $p_1 = (0,0)$, and $q_1 = (1,0)$. By part (ii) of Lemma 1, the edges of the kth level of D form a good cap C_k in D for every $k = 1, \ldots, n - 1$.

We perturb the vertices of D such that the vertex set of the resulting geometric drawing D' of $K_{\lceil n/2 \rceil, \lfloor n/2 \rfloor}$ is in general position. We let K_P be the geometric drawing of K_n obtained from D' by adding the missing edges. Note that if the perturbation is sufficiently small, then every good cap C_k in D corresponds to a good cap C'_k in K_P.

Let $\mathcal{F} := \{F_1, \ldots, F_h\}$ be the set of (not necessarily distinct) bounded faces of K_P such that, for $i = 1, \ldots, h$, all edges of the cap $C'_{\lfloor n/2 \rfloor - \lceil h/2 \rceil + i}$ are incident to F_i. That is, $F_1, \ldots, F_h$ are faces incident to edges of h middle caps C'_k. Since caps $C'_{\lfloor n/2 \rfloor - \lceil h/2 \rceil + i}$ are good in K_P and $n \geq 3$, the faces F_i exist.

Every cap C'_k is formed by $\min\{k, n - k\}$ edges for every $k = 1, \ldots, n - 1$. Therefore, for every $i = 1, \ldots, h$, the face F_i is incident to at least $\min\{\lfloor n/2 \rfloor - \lceil h/2 \rceil + i, \lceil n/2 \rceil + \lceil h/2 \rceil - i\}$ edges of K_P. Summing over $i = 1, \ldots, h$, we obtain at least $(2hn - h^2 - 1)/4$ edges of K_P incident to at least one face of $\mathcal{F}$. This implies $e(h, n) \geq (2hn - h^2 - 1)/4$ and proves the claim.   □

6   Proof of Theorem 4

For a sufficiently large constant C and every sufficiently large integer n, we find a bipartite drawing D of $K_{n,n}$ such that for every integer M satisfying $Cn \leq M \leq n^4/C$ there is a set of at most M faces of D with complexity at least $\Omega(n^{4/3} M^{2/3})$. Theorem 4 then follows, as D can be treated as an arrangement of n^2 line segments with $2n$ endpoints.

Let D' be the regular bipartite drawing of $K_{n,n}$ with width, left step, and right step equal to 1, $p_1 = (0,0)$, and $q_1 = (1,0)$. For integers i and k satisfying $1 \leq i < k \leq n/2$ and $\gcd(i, k) = 1$, every intersection point of a uniformly crossing l-tuple of edges $(p_{i_1} q_{j_1}, \ldots, p_{i_l} q_{j_l})$ of D' with $i_2 - i_1 = i$ and $j_2 - j_1 = i - k$ is called a *uniform (i, k)-crossing*. A point that is a uniform (i, k)-crossing for some integers i and k is called a *uniform crossing*.

Note that all uniform (i, k)-crossings lie on the vertical line $\{\frac{i}{k}\} \times \mathbb{R}$ and that no uniform (i, k)-crossing is a uniform (i', k')-crossing for any pair $(i', k') \neq (i, k)$, as $\gcd(i,k) = 1$. Since the y-coordinate of every uniform (i,k)-crossing equals j/k for some $0 \le j \le kn - k$, the number of uniform (i,k)-crossings is at most kn. There is also at least $n^2 - 2in > n^2 - 2kn$ edges of D' that contain a uniform (i,k)-crossing. This follows easily, as for every edge $p_{i'}q_{j'}$ of D' with $i < i' \le n-i$ and $1 \le j' \le n$ either $p_{i'-i}q_{j'+k-i}$ or $p_{i'+i}q_{j'-k+i}$ is an edge of D' and forms a uniform (i,k)-crossing with $p_{i'}q_{j'}$. Here we use the fact $k \le n/2$.

We choose $\varepsilon > 0$ as in Lemma 1 for $K_{n,n}$ and $w = d_1 = h_1 = 1$. Let D be an ε-dilated drawing of $K_{n,n}$ with width, left step, and right step equal to 1, with the left lowest point $(0,0)$, and with the right lowest point $(1,0)$. By part (ii) of Lemma 1, every uniformly crossing l-tuple of edges of D forms a good cap in D. In particular, every uniform crossing c in D' is the meeting point of edges of D that form a good cap C_c. Let F_c be a bounded face of D such that all edges of C_c are incident to F_c. Note that the faces F_c and $F_{c'}$ of D are distinct for distinct uniform crossings c and c' in D'.

Let $K \le n/2$ be a positive integer whose value we specify later. For integers i and k satisfying $1 \le i < k \le K$ and $\gcd(i,k) = 1$, let $\mathcal{F}_{i,k}$ be the set of faces F_c where c is a uniform (i,k)-crossing in D'. It follows from our observations that $\mathcal{F}_{i,k}$ contains at most kn faces and that the complexity of $\mathcal{F}_{i,k}$ is at least $n^2 - 2kn$. We let $\mathcal{F} := \bigcup_{i,k} \mathcal{F}_{i,k}$ where the union is taken over all integers i and k satisfying $1 \le i < k \le K$ and $\gcd(i,k) = 1$. Then $\mathcal{F}$ contains at most

$$\sum_{k=2}^{K} \sum_{\substack{i=1 \\ \gcd(i,k)=1}}^{k-1} kn = n \sum_{k=2}^{K} k\varphi(k-1) = n \sum_{j=1}^{K-1} (j+1)\varphi(j) < \frac{nK^3}{2}$$

faces where $\varphi(j)$ denotes the Euler's totient function. The last inequality follows from $\varphi(j) < K$ for every positive integer $j < K$.

Since the sets $\mathcal{F}_{i,k}$ are pairwise disjoint, the complexity of $\mathcal{F}$ is at least

$$\sum_{k=2}^{K} \sum_{\substack{i=1 \\ \gcd(i,k)=1}}^{k-1} (n^2 - 2kn) = \sum_{k=2}^{K} \varphi(k-1)(n^2 - 2kn) > n^2 \sum_{j=1}^{K-1} \varphi(j) - nK^3.$$

The totient summary function satisfies $\sum_{j=1}^{m} \varphi(j) \ge \frac{3m^2}{\pi^2} - O(m \log m)$ [7, pp. 268–269]. Thus the complexity of $\mathcal{F}$ is at least $\frac{3n^2 K^2}{\pi^2} - nK^3 - O(n^2 K \log K)$.

Let M be a given integer that satisfies $8n \le M \le n^4/8$. We set $K := (M/n)^{1/3}$. We may assume that K is an integer, as it does not affect the asymptotics. For $8n \le M \le n^4/8$, we have $2 \le K \le n/2$. The set $\mathcal{F}$ then contains at most M faces and its complexity is at least $\frac{3}{\pi^2} n^{4/3} M^{2/3} - M - O(M^{1/3} n^{5/3} \log (M/n))$, which is $\Omega(n^{4/3} M^{2/3})$ for a sufficiently large absolute constant C and $Cn \le M \le n^4/C$.

References

1. Alpert, H., Koch, C., Laison, J.D.: Obstacle numbers of graphs. Discrete Comput. Geom. **44**(1), 223–244 (2010)
2. Arkin, E.M., Halperin, D., Kedem, K., Mitchell, J.S.B., Naor, N.: Arrangements of segments that share endpoints: single face results. Discrete Comput. Geom. **13**(1), 257–270 (1995)
3. Aronov, B., Edelsbrunner, H., Guibas, L.J., Sharir, M.: The number of edges of many faces in a line segment arrangement. Combinatorica **12**(3), 261–274 (1992)
4. Dujmović, V., Morin, P.: On obstacle numbers. Electron. J. Combin. **22**(3), P3.1 (2015)
5. Edelsbrunner, H., Welzl, E.: On the maximal number of edges of many faces in an arrangement. J. Combin. Theory Ser. A **41**(2), 159–166 (1986)
6. Fulek, R., Saeedi, N., Sarıöz, D.: Convex obstacle numbers of outerplanar graphs and bipartite permutation graphs. In: Pach, J. (ed.) Thirty Essays on Geometric Graph Theory, pp. 249–261. Springer, New York (2013)
7. Hardy, G.H., Wright, E.M.: An introduction to the theory of numbers, 5th edn. Clarendon Press, Oxford (1979)
8. Matoušek, J., Valtr, P.: The complexity of lower envelope of segments with h endpoints. Intuitive Geom. Bolyai Soc. Math. Stud. **6**, 407–411 (1997)
9. Mukkamala, P., Pach, J., Pálvölgyi, D.: Lower bounds on the obstacle number of graphs. Electron. J. Combin. **19**(2), P32 (2012)
10. Mukkamala, P., Pach, J., Sarıöz, D.: Graphs with large obstacle numbers. In: Thilikos, D.M. (ed.) WG 2010. LNCS, vol. 6410, pp. 292–303. Springer, Heidelberg (2010)
11. Pach, J., Sarıöz, D.: On the structure of graphs with low obstacle number. Graphs Combin. **27**(3), 465–473 (2011)
12. Valtr, P.: Convex independent sets and 7-holes in restricted planar point sets. Discrete Comput. Geom. **7**(1), 135–152 (1992)
13. Wiernik, A., Sharir, M.: Planar realizations of nonlinear Davenport-Schinzel sequences by segments. Discrete Comput. Geom. **3**(1), 15–47 (1988)

Vertical Visibility Among Parallel Polygons in Three Dimensions

Radoslav Fulek[1]([✉]) and Rados Radoicic[2]

[1] IST Austria, Am Campus 1, Klosterneuburg 3400, Austria
{radoslav.fulek,radosrr}@gmail.com
[2] Baruch College, CUNY, New York City, NY, USA

Abstract. Let $\mathcal{C} = \{C_1, \ldots, C_n\}$ denote a collection of translates of a regular convex k-gon in the plane with the stacking order. The collection $\mathcal{C}$ forms a *visibility clique* if for every $i < j$ the intersection C_i and C_j is not covered by the elements that are stacked between them, i.e., $(C_i \cap C_j) \setminus \bigcup_{i<l<j} C_l \neq \emptyset$.

We show that if $\mathcal{C}$ forms a visibility clique its size is bounded from above by $O(k^4)$ thereby improving the upper bound of 2^{2^k} from the aforementioned paper. We also obtain an upper bound of $2^{2\binom{k}{2}+2}$ on the size of a visibility clique for homothetes of a convex (not necessarily regular) k-gon.

1 Introduction

In a visibility representation of a graph $G = (V, E)$ we identify the vertices of V with sets in the Euclidean space, and the edge set E is defined according to some visibility rule. Investigation of visibility graphs, driven mainly by applications to VLSI wire routing and computer graphics, goes back to the 1980s [12,14]. This also includes a significant interest in three-dimensional visualizations of graphs [3,4,8,10].

Babilon et al. [1] studied the following three-dimensional visibility representations of complete graphs. The vertices are represented by translates of a regular convex polygon lying in distinct planes parallel to the xy-plane and two translates are joined by an edge if they can *see* each other, which happens if it is possible to connect them by a line segment orthogonal to the xy-plane avoiding all the other translates. They showed that the maximal size $f(k)$ of a clique represented by regular k-gons satisfies $\lfloor \frac{k+1}{2} \rfloor + 2 \leq f(k) \leq 2^{2^k}$ and that $f(3) \geq 14$. Hence, $\lim_{k\to\infty} f(k) = \infty$. Fekete et al. [8] proved that $f(4) = 7$ thereby showing that $f(k)$ is not monotone in k. Nevertheless, it is plausible that $f(k+2) \geq f(k)$ for every k, and surprisingly enough this is stated as an open problem in [1]. Another interesting open problem from the same paper is to decide if the limit

The research leading to these results has received funding from the People Programme (Marie Curie Actions) of the European Union's Seventh Framework Programme (FP7/2007-2013) under REA grant agreement no [291734].

E. Di Giacomo and A. Lubiw (Eds.): GD 2015, LNCS 9411, pp. 373–379, 2015.
DOI: 10.1007/978-3-319-27261-0_31

$\lim_{k\to\infty} \frac{f(k)}{k}$ exists. In the present note we improve the above upper bound on $f(k)$ to $O(k^4)$[1] and we extend our investigation to families of homothetes of general convex polygons. The main tool to obtain the result is Dilworth Theorem [6], which was also used by Babilon et al. to obtain the doubly exponential bound in [1]. Roughly speaking, our improvement is achieved by applying Dilworth Theorem only once whereas Babilon et al. used its k successive applications.

Fekete et al. [8] observed that a clique of arbitrary size can be represented by translates of a disc. Their construction can be adapted to translates of any convex set whose boundary is partially smooth, or to translates of possibly rotated copies of a convex polygon. The same is true for non-convex shapes, see Fig. 1.

Fig. 1. A visibility clique formed by translates of a non-convex 4-gon.

An analogous question was extensively studied for arbitrary, i.e. not necessarily translates or homothetes of, axis parallel rectangles [3,8], see also [11]. Bose et al. [3] showed that in this case a clique on 22 vertices can be represented. On the other hand, they showed that a clique of size 57 cannot be represented by rectangles.

For convenience, we restate the problem of Babilon et al. as follows. Let $\mathcal{C} = \{C_1, \ldots, C_n\}$ denote a collection of sets in the plane with the *stacking order* given by the indices of the elements in the collection. By a standard perturbation argument, we assume that the boundaries of no three sets in $\mathcal{C}$ pass through a common point. The collection $\mathcal{C}$ forms a *visibility clique* if for every i and j, $i < j$, the intersection C_i and C_j is not covered by the elements that are stacked between them, i.e., $(C_i \cap C_j) \setminus \bigcup_{i<k<j} C_k \neq \emptyset$. Note that reversing the stacking order of $\mathcal{C}$ does not change the property of $\mathcal{C}$ forming a visibility clique. We are interested in the maximum size of $\mathcal{C}$, if $\mathcal{C}$ is a collection of translates and homothetes, resp., of a convex k-gon. We prove the following.

Theorem 1. *If $\mathcal{C}$ is a collection of translates of a regular convex k-gon forming a visibility clique, the size of $\mathcal{C}$ is bounded from above by $O(k^4)$.*

Theorem 2. *If $\mathcal{C}$ is a collection of homothetes of a convex k-gon forming a visibility clique, the size of $\mathcal{C}$ is bounded from above by $2^{2\binom{k}{2}+2}$.*

The paper is organized as follows. In Sect. 2 we give a proof of Theorem 1. In Sect. 3 we give a proof of Theorem 2. We conclude with open problems in Sect. 4.

---

[1] After acceptance of the paper the authors became aware of the fact that the upper bound of $O(k^4)$ was previously proven by Štola [13].

2   Proof of Theorem 1

We let $\mathcal{C} = \{C_1, \ldots, C_n\}$ denote a collection of translates of a regular convex k-gon C in the plane with the stacking order given by the indices of the elements in the collection. Let $\mathbf{c_i}$ denote the center of gravity of C_i. We assume that $\mathcal{C}$ forms a visibility clique. We label the vertices of C by natural numbers starting in the clockwise fashion from the topmost vertex, which gets label 1. We label in the same way the vertices in the copies of C. The proof is carried out by successively selecting a large and in some sense regular subset of $\mathcal{C}$. Let W_i be the convex wedge with the apex $\mathbf{c_1}$ bounded by the rays orthogonal to the sides of C_1 incident to the vertex with label i. The set $\mathcal{C}$ is *homogenous* if for every $1 \leq i \leq k$ all the vertices of C_j's with label i are contained in W_i. We remark that already in the proof of the following lemma our proof falls apart if C can be arbitrary or only centrally symmetric convex k-gon.

Lemma 1. *If C is a regular k-gon then $\mathcal{C}$ contains a homogenous subset of size at least $\Omega\left(\frac{n}{k^2}\right)$.*

Let $(C_{i_1}, \ldots, C_{i_n})$ be the order in which the ray bounding W_i orthogonal to the segment $i[(i-1) \bmod k]$ of C_1 intersects the boundaries of C_j's. The set $\mathcal{C}$ forms an *i-staircase* if the order $(C_{i_1}, \ldots, C_{i_n})$ is the stacking order. As a direct consequence of Dilworth Theorem or Erdős–Szekeres Lemma [6,7] we obtain that if $\mathcal{C}$ is homogenous, it contains a subset of size at least $\sqrt{|\mathcal{C}|}$ forming an i-staircase.

A graph $G = (\{1, \ldots, n\}, E)$ is a *permutation graph* if there exists a permutation π such that $ij \in E$, where $i < j$, iff $\pi(i) > \pi(j)$. Let $G_i = (\mathcal{C}', E)$ denote a graph such that $\mathcal{C}'$ is a homogenous subset of $\mathcal{C}$, and two vertices C_j' and C_k' of G_i are joined by an edge if and only if the orders in which the rays bounding W_i intersect the boundaries of C_j' and C_k' are reverse of each other. In other words, the boundaries of C_j' and C_k' intersect inside W_i, see Fig. 2(a). Thus, G_i's form a family of permutation graphs sharing the vertex set. Note that every pair of boundaries of elements in $\mathcal{C}'$ cross exactly twice.

Since for an even k a regular k-gon is centrally symmetric the graphs G_i and $G_{i+k/2 \bmod k}$ are identical. For an odd k, we only have $G_i \subseteq G_{i+\lceil k/2 \rceil \bmod k} \cup G_{i+\lfloor k/2 \rfloor \bmod k}$. The notion of the i-staircase and homogenous set is motivated by the following simple observation illustrated by Fig. 2(b).

Observation 1. *If $\mathcal{C}'$ forms an i-staircase then there do not exist two indices i and j, $i \neq j$, such that both G_i and G_j contain the same clique of size three.*

The following lemma lies at the heart of the proof of Theorem 1.

Lemma 2. *Suppose that $\mathcal{C}'$ forms an i-staircase, and that there exists a pair of identical induced subgraphs $G_i' \subseteq G_i$ and $G_j' \subseteq G_j$, where $i \neq j$, containing a matching of size two. Then $\mathcal{C}'$ does not form a visibility clique.*

Proof. The lemma can be proved by a simple case analysis as follows. There are basically two cases to consider depending on the stacking order of the elements

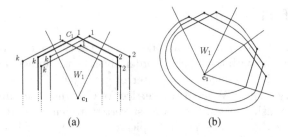

Fig. 2. (a) The wedge W_1 containing all the copies of vertex 1. (b) The 1-staircase giving rise to a clique of size three in G_1 and G_j for some j that cannot appear in a visibility clique.

of $\mathcal{C}'$ supporting the matching M of size two in G_i'. Let u_1, v_1 and u_2, v_2, respectively, denote the vertices (or elements of $\mathcal{C}'$) of the first and the second edge in M, such that u_1 is the first one in the stacking order. By symmetry and without loss of generality we assume that the ray R bounding W_i orthogonal to the segment $i[(i-1) \mod k]$ of C_1 intersects the boundary of u_1 before intersecting the boundaries of u_2, v_1 and v_2, and the boundary of u_2 before v_2.

First, we assume that R intersects the boundary of u_2 before the boundary of v_1. In the light of Observation 1, u_1, v_1 and u_2 look combinatorially like in the Fig. 3(a). Then all the possibilities for the position of v_2 cause that the first and last element in the stacking order do not see each other. Otherwise, R intersects the boundary of v_1 before the boundary of u_2. In the light of Observation 1, u_1, v_1 and u_2 look combinatorially like in the Fig. 3(b), but then v_2 cannot see u_1. ■

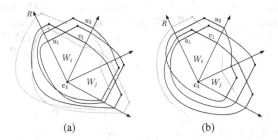

Fig. 3. The case analysis of possible combinatorial configurations of the boundaries of u_1, v_1, u_2 and v_2, after the first three boundaries were fixed. (a) If R intersects the boundary of u_2 before v_1 the first and the last element in the stacking order cannot see each other. (b) If R intersects the boundary of v_1 before u_2 then u_1 cannot see v_2.

Finally, we are in a position to prove Theorem 1. We consider two cases depending on whether k is even or odd. First, we treat the case when k is even which is easier.

Thus, let C be a regular convex k-gon for an even k. By Lemma 1 and Dilworth Theorem we obtain a homogenous subset $\mathcal{C}'$ of $\mathcal{C}$ of size at least $\Omega(\sqrt{\frac{n}{k^2}})$

forming a 1-staircase. Note that for $\mathcal{C}'$ the hypothesis of Lemma 2 is satisfied with $i = 1$ and $j = 1 + k/2$. Since $\mathcal{C}'$ forms a visibility clique, the graph G_1 does not contain a matching of size two. Hence, $G_1 = (\mathcal{C}' = \mathcal{C}_1, E)$ contains a dominating set of vertices $\mathcal{C}_1'$ of size at most two. Let $\mathcal{C}_2 = \mathcal{C}_1 \setminus \mathcal{C}_1'$. Note that $\mathcal{C}_2$ forms a 2-staircase and that the hypothesis of Lemma 2 is satisfied with $\mathcal{C}' = \mathcal{C}_2, i = 2$ and $j = 2 + k/2 \mod k$. Thus, $G_2 = (\mathcal{C}_2, E)$ contains a dominating set of vertices $\mathcal{C}_2'$ of size at most two. Hence, $\mathcal{C}_3 = \mathcal{C}_2 \setminus \mathcal{C}_2'$ forms a 3-staircase. In general, $\mathcal{C}_i = \mathcal{C}_{i-1} \setminus \mathcal{C}_{i-1}'$ forms an i-staircase and the hypothesis of Lemma 2 is satisfied with $\mathcal{C}' = \mathcal{C}_i, i = i$ and $j = i + k/2 \mod k$. Note that $|\mathcal{C}_{k/2+1}| \leq 1$. Thus, $|\mathcal{C}'| \leq k + 1$. Consequently, $n = O(k^4)$.

In the case when k is odd we proceed analogously as in the case when k was even except that for $\mathcal{C}'$ as defined above the hypothesis of Lemma 2 might not be satisfied, since we cannot guarantee that G_i and G_j are identical for some $i \neq j$. Nevertheless, since the two tangents between a pair of intersecting translates of a convex k-gon in the plane are parallel we still have $G_i \subseteq G_{i+\lceil \frac{k}{2} \rceil \mod k} \cup G_{i+\lfloor \frac{k}{2} \rfloor \mod k}$. The previous property will help us to find a pair of identical induced subgraphs in G_i, and $G_{i+\lceil \frac{k}{2} \rceil \mod k}$ or $G_{i+\lfloor \frac{k}{2} \rfloor \mod k}$ to which Lemma 2 can be applied, if G_i contains a matching M of size c, where c is a sufficiently big constant determined later. It will follow that G_i does not contain a matching of size c, and thus, the inductive argument as in the case when k was even applies. (Details will appear in the full version.)

3   Homothetes

The aim of this section is to prove Theorem 2. Let C denote a convex polygon in the plane. Let $\mathcal{C} = \{C_1, C_2, \ldots, C_n\}$ denote a finite set of homothetes of C with the stacking order. Unlike as in previous sections, this time we assume that the indices correspond to the order of the centers of gravity of C_i's from left to right. Let $\mathbf{c_i}$ denote the center of gravity of C_i. Let $x(\mathbf{p})$ and $y(\mathbf{p})$, resp., denote x and y-coordinate of $\mathbf{p}$. Thus, we assume that $x(\mathbf{c_1}) < x(\mathbf{c_2}) < \ldots < x(\mathbf{c_n})$

Suppose that $\mathcal{C}$ forms a visibility clique. Similarly as in the previous sections we label the vertices of C by natural numbers starting in the clockwise fashion from the topmost vertex, which gets label 1. We label in the same way the vertices in the copies of C. Consider the poset $(\mathcal{C}, \subset)$ and note that it contains no chain of size five. By Dilworth theorem it contains an anti-chain of size at least $\frac{1}{4}|\mathcal{C}|$. Since we are interested only in the order of magnitude of the size of the biggest visibility clique, from now on we assume that no pair of elements in $\mathcal{C}$ is contained one in another.

Every pair of elements in $\mathcal{C}$ has exactly two common tangents, since every pair intersect and no two elements are contained one in another. We color the edges of the clique $G = (\mathcal{C}, \binom{\mathcal{C}}{2})$ as follows. Each edge $C_i C_j$, $i < j$, is colored by an ordered pair, in which the first component is an unordered pair of vertices of G supporting the common tangents of C_i and C_j, and the second pair is an indicator equal to one if C_i is below C_j in the stacking order, and zero otherwise.

Lemma 3. *The visibility clique G does not contain a monochromatic path of length two of the form $C_i C_j C_k$, $i < j < k$.*

We say that a path $P = C_1 C_2 \ldots C_k$ in G is monotone if $x(\mathbf{c_1}) < x(\mathbf{c_2}) < \ldots < x(\mathbf{c_k})$. It was recently shown [9, Theorem 2.1] that if we color the edges of an ordered complete graph on $2^c + 1$ vertices with c colors we obtain a monochromatic monotone path of length two. We remark that this result is tight and generalizes Erdős–Szekeres Lemma [7]. Thus, if G contains more than $2^{2\binom{k}{2}+2}$ vertices it contains a monochromatic path of length two which is a contradiction by Lemma 3.

4 Open Problems

Since we could not improve the lower bound from [1] even in the case of homothetes, we conjecture that the polynomial upper bound in k on the size of the visibility clique holds also for any family of homothetes of an arbitrary convex k-gon. To prove Theorem 2 we used a Ramsey-type theorem [9, Theorem 2.1] for ordered graphs. We wonder if the recent developments in the Ramsey theory for ordered graphs [2,5] could shed more light on our problem.

Acknowledgement. We would like to thank Martin Balko for telling us about [9].

References

1. Babilon, R., Nyklová, H., Pangrác, O., Vondrák, J.: Visibility representations of complete graphs. In: Kratochvíl, J. (ed.) GD 1999. LNCS, vol. 1731, pp. 333–341. Springer, Heidelberg (1999)
2. Balko, M., Cibulka, J., Král, K., Kynčl, J.: Ramsey numbers of ordered graphs. arXiv:1310.7208v3
3. Bose, P., Everett, H., Fekete, S.P., Houle, M.E., Lubiw, A., Meijer, H., Romanik, K., Rote, G., Shermer, T.C., Whitesides, S., Zelle, C.: A visibility representation for graphs in three dimensions. J. Graph Algorithm Appl. **2**(3), 1–16 (1998)
4. Cohen, R.F., Eades, P., Lin, T., Ruskey, F.: Three-dimensional graph drawing. Algorithmica **17**(2), 199–208 (1997)
5. Conlon, D., Fox, J., Lee, C., Sudakov, B.: Ordered Ramsey numbers. arXiv:1410.5292v1
6. Dilworth, R.P.: A decomposition theorem for partially ordered sets. Ann. Math. **51**, 161–166 (1950)
7. Erdős, P., Szekeres, G.: A combinatorial problem in geometry. In: Gessel, I., Rota, G.-C. (eds.) Classic Papers in Combinatorics. Modern Birkhäuser Classics, pp. 49–56. Birkhäuser Boston, Boston (1987)
8. Fekete, S.P., Houle, M.E., Whitesides, S.: New results on a visibility representation of graphs in 3D. In: Brandenburg, F.J. (ed.) GD 1995. LNCS, vol. 1027, pp. 234–241. Springer, Heidelberg (1996)
9. Milans, K.G., Stolee, D., West, D.B.: Ordered ramsey theory and track representations of graphs. http://www.math.illinois.edu/stolee/Papers/MSW12-OrderedRamsey.pdf

10. Robertson, G.G., Mackinlay, J.D., Card, S.K.: Cone trees: animated 3D visualizations of hierarchical information. In: Proceedings of the SIGCHI Conference on Human Factors in Computing Systems, CHI 1991, pp. 189–194. ACM, New York, NY, USA (1991)

11. Romanik, K.: Directed VR-representable graphs have unbounded dimension. In: Tamassia, R., Tollis, I.G. (eds.) Graph Drawing. LNCS, vol. 894, pp. 177–181. Springer, Berlin Heidelberg (1995)

12. Tamassia, R., Tollis, I.G.: A unified approach to visibility representations of planar graphs. Discrete Comput. Geom. 1(1), 321–341 (1986)

13. Štola, J.: 3D visibility representations by regular polygons. In: Eppstein, D., Gansner, E.R. (eds.) GD 2009. LNCS, vol. 5849, pp. 323–333. Springer, Heidelberg (2010)

14. Wismath, S.K.: Characterizing bar line-of-sight graphs. In: Proceedings of the First Annual Symposium on Computational Geometry, SCG 1885, pp. 147–152. ACM, New York, NY, USA (1985)

Drawing Graphs on Point Sets

Alternating Paths and Cycles of Minimum Length

William S. Evans[1], Giuseppe Liotta[2], Henk Meijer[3], and Stephen Wismath[4](✉)

[1] University of British Columbia, Vancouver, Canada
[2] Universitá degli Studi di Perugia, Perugia, Italy
[3] U. C. Roosevelt, Middelburg, The Netherlands
[4] University of Lethbridge, Lethbridge, Canada
wismath@uleth.ca

Abstract. Let R be a set of n red points and B be a set of n blue points in the Euclidean plane. We study the problem of computing a planar drawing of a cycle of minimum length that contains vertices at points $R \cup B$ and alternates colors. When these points are collinear, we describe a $\Theta(n \log n)$-time algorithm to find such a shortest alternating cycle where every edge has at most two bends. We extend our approach to compute shortest alternating paths in $O(n^2)$ time with two bends per edge and to compute shortest alternating cycles on 3-colored point-sets in $O(n^2)$ time with $O(n)$ bends per edge. We also prove that for arbitrary k-colored point-sets, the problem of computing an alternating shortest cycle is NP-hard, where k is any positive integer constant.

1 Introduction

A recent paper by Chan et al. [5] studies the problem of computing a planar drawing of an n-vertex planar graph such that the vertex locations are given as part of the input and the drawing has minimum total edge length. The problem is known to be NP-hard [4] in general and Chan et al. describe different polynomial time approximation algorithms for paths, matchings, and general planar graphs. They also give a polynomial time exact algorithm for paths on fixed positions that lie on a line, which computes a planar drawing where all edges are monotone in a common direction and each edge can be represented by a poly-line having $O(n)$ bends.

In this paper we consider a variant of the problem by Chan et al. where the position for each vertex is not fixed, but it can be chosen by the algorithm as one in a given subset of a point set. To be precise, we are given a k-colored graph (i.e., a graph where each vertex is one of k different colors) and we want to compute a planar drawing of the graph on a given k-colored point-set so that vertices are mapped to distinct points of the same color and the total edge length is minimized.

The research reported in this paper started at the 2015 Bertinoro workshop, sponsored by the EuroGIGA Project. Research also supported by NSERC, and by MIUR of Italy under project AlgoDEEP prot. 2008TFBWL4.

E. Di Giacomo and A. Lubiw (Eds.): GD 2015, LNCS 9411, pp. 383–394, 2015.
DOI: 10.1007/978-3-319-27261-0_32

We mainly focus on drawing shortest alternating 2-colored (bicolored) paths and cycles and collinear point-sets. But we also consider the case of more than two colors and the case that the points are non-collinear. Our main results are:

- Let R be a set of n red points and B be a set of n blue points such that $R \cup B$ is a set of distinct and collinear points. We describe a $\Theta(n \log n)$-time sweep-line algorithm to compute a planar drawing of an alternating cycle of minimum length on $R \cup B$ such that every edge is a poly-line with at most two bends.
- We adapt the approach for cycles to the problem of computing a shortest alternating path on a bicolored set of collinear points. We describe an $O(n^2)$-time algorithm that solves the problem by computing drawings with at most two bends per edge.
- We extend the study to 3-colored collinear point-sets and describe an $O(n^2)$-time algorithm to compute shortest alternating cycles (visiting the colors in cyclic order) such that every edge has $O(n)$ bends.
- We consider non-collinear point-sets and prove that computing a shortest alternating cycle is NP-hard in the general case of k-colored point-sets, where $k \geqslant 1$ is a given constant.

From a technical point of view, our drawing algorithms are based on the idea of computing an alternating topological book embedding of a path or cycle such that the number of edges that are intersected by any cut is minimum. This approach seems to be specific for two and three colors, since we also present an example with four colors where an alternating cycle of minimum length cannot match the cut lower bound that we use for fewer than four colors.

1.1 Related Work and Paper Organization

The problem of computing a planar alternating path or a planar alternating cycle on $R \cup B$ has a long tradition in graph drawing and computational geometry. While the interested reader may refer to the survey by Kaneko and Kano [11] for a list of early references, we briefly recall here some of the milestone results. Akiyama and Urrutia [2] study straight-line alternating paths when $R \cup B$ is in convex position; they exhibit a set of sixteen points for which a straight-line alternating path does not exist and present an $O(n^2)$-time algorithm to test when a straight-line alternating path on points in convex position exists. Abellanas et al. [1] show that if either the convex hull of $R \cup B$ consists of all red points and no blue points or there exists a line that separates all blue points from red ones, then a straight-line alternating path always exists. Kaneko, Kano, and Suzuki [12] characterize those point sets in general position for which a straight-line alternating path always exists: If $R \cup B$ consists of at most twelve points or if it consists of exactly fourteen points, then a straight-line alternating path always exists; for all other cases, there exist configurations of red and blue points for which a straight-line alternating path does not exist. These early results about straight-line alternating paths have motivated further research on computing alternating paths and cycles when the edges can bend. Di Giacomo et al. [8] proved that every point set admits an alternating path and an alternating cycle with at most one bend per edge; the result is based on projecting the points on a horizontal line and then computing a book embedding on this line before mapping the edges back to the original points.

The results above have motivated further research where either bicolored graph families other than paths or cycles have been studied or more than two colors have been considered, or both. For graph families other than paths or cycles, the input is a bicolored planar graph G together with a bicolored set of points in the plane and the goal is to compute a planar drawing of G such that every red vertex is mapped to a red point and every blue vertex is mapped to a blue point and either each edge is a straight-line segment or it has a constant number of bends. See, for example, [6,9,10,13,14]. For more than two colors (see, e.g., [7,8]), the input is k point sets each with the same cardinality and the goal is to compute an alternating path/cycle on the entire set of points; that is, a planar drawing of a path/cycle containing the given points and such that the ith point on the cycle comes from the $(i \bmod k)$th set. In the extreme case, one is given n colors modeled as n numbers from 1 to n and the goal is to compute a planar drawing of a path or cycle that touches the vertices in increasing order. In other words, the n-colored version of the problem is the same as asking for a planar drawing of a graph where the location of the vertices is specified as part of the input. A seminal result in this context is due to Pach and Wenger [15] who prove that linearly many bends per edge are always sufficient and sometimes necessary for n-colored paths and n-colored point-sets in convex position. Their drawing technique applies to general n-colored planar graphs and the number of bends per edge was improved by Badent et al. [3].

The remainder of the paper is organized as follows. An overview of our algorithmic approach is presented in Sect. 2. Section 3 describes the algorithm for shortest alternating cycles on collinear red-blue points. Shortest alternating paths on collinear red-blue points are studied in Sect. 4. Shortest alternating cycles on more than two colors and the proof of hardness for general k-colored point-sets is in Sect. 5. Finally, open problems are listed in Sect. 6. For reasons of space, some proofs have been sketched and will be available in the full version of the paper.

2   Overview of the Algorithmic Approach

Let R be a set of n distinct red points and B be a set of n distinct blue points in the Euclidean plane. An *alternating cycle* (*alternating path*) is a drawing of a cycle (path) such that the vertex set of the drawing is the set $R \cup B$ and such that no two vertices having the same color are adjacent. The drawing is *planar* if no two edges cross. The length of the cycle (path) is the sum of the lengths of its edges. A *shortest* alternating cycle (path) is one of minimum length. In this paper we are interested in computing shortest planar alternating cycles (paths). Since the problem is NP-hard for general point sets (Sect. 5), we focus on collinear point-sets and assume that the line through the point set, called the *spine*, is horizontal.

A set of n blue points and n red points on a line define $2n + 1$ intervals, two of which are infinite. Assume we have a (not necessarily optimal) planar drawing of an alternating cycle (path). Consider a vertical line in any interval and count how many edges of the cycle (path) are intersected by the line. If we multiply the length of each finite interval by the number of edges that are intersected by a vertical line through the interval and then sum up all the obtained numbers, we obtain a lower bound on the length of the cycle. Therefore, we aim at computing an alternating cycle (path) C such that for any vertical line ℓ, the number of edges of C cut by ℓ is the minimum over all alternating cycles (paths). In addition, no two edges of C cross and every edge is a poly-line consisting of at most three segments (i.e., it has at most two bends). For brevity, in what follows we will often say alternating cycle (path) to mean planar alternating cycle (path).

Based on the observation above, the problem turns into the computation of a special type of topological book embedding, such that every edge can cross the spine at most once and such that the number of edges that span any interval between two consecutive points along the spine is minimum. Every edge of such a topological book embedding can be represented as a poly-line with at most two bends. We recall that a topological book embedding is a planar drawing of a graph such that all vertices are points of a line called the *spine* and the edges are simple Jordan arcs.

It is worth remarking that we are interested in solving the combinatorial problem of finding an order in which a shortest alternating cycle (path) visits the colored points, and the embedding of its edges. Once this is found, a planar alternating cycle (path) of minimum length can be computed by making the edges "as flat as possible" around the spine, that is by making the distance between each edge and the spine tend to zero. Hence when we say that we "compute the shortest cycle (path)", we mean that we compute an ordering and embedding for which such a cycle (path) exists.

3   Shortest Alternating Cycle on Collinear Red-Blue Points

Following the approach of Sect. 2, we start by giving a lower bound on the number of edges of any alternating cycle intersected by a vertical line. Next, we present a sweep-line algorithm to compute a topological book embedding such that every interval is spanned by the minimum number of edges and such that every edge crosses the spine at most once.

A Lower Bound Lemma. The following lemma establishes the lower bound that will be used to prove the optimality of the alternating cycles.

Lemma 1. *Let R be a set of n red points and let B be a set of n blue points such that all points are distinct and lie on the x-axis. Let ℓ be a vertical line that intersects the x-axis between two points of $R \cup B$. If there are r red points and b blue points to the left of ℓ, then any alternating cycle on $R \cup B$ crosses ℓ at least $2 \max\{1, |r - b|\}$ times.*

Proof. Let C be an alternating cycle on $R \cup B$ and ℓ^- be the halfplane to the left of ℓ. In each component of $C \cap \ell^-$, the number of points of one color can be at most one more than the number of points of another color. Thus, the minimum number of components of C to the left of ℓ is $|r - b|$. If the line ℓ lies between two vertices of C (i.e. it is not to the left of the leftmost vertex of C and it is not to the right of the rightmost vertex of C), then the number of components to the left of ℓ is also at least one, and the number of edges of C that intersect ℓ is twice the number of components in $C \cap \ell^-$. □

A Sweep-Line Algorithm. We now describe a sweep-line algorithm that computes the shortest alternating cycle of a set of n red points and n blue points lying on the horizontal line $y = 0$, called the *spine*. We call our algorithm Spine-Sweep.

Spine-Sweep first orders the points by increasing x-coordinate and then it sweeps a vertical cut line ℓ across the points. The algorithm maintains a set of disjoint curves to the left of ℓ, each of which has both endpoints, called *terminals*, on ℓ. These curves are the connected components of the intersection of some shortest alternating cycle with the halfplane, ℓ^-, to the left of ℓ. The terminals are colored red or blue depending on the color of the closest colored point on the curve. Terminals above the spine are *positive* and those below are *negative*; this is called the *sign of the terminal*. If both terminals of a component have the same sign, then this is the *sign of the component*, otherwise the component *straddles* the spine. The *distance* of a component to the spine is the minimum number of terminals between one of its terminals and the spine. Two terminals are *adjacent* if the segment connecting them contains no other terminals. Note that these definitions are with respect to the current sweep-line ℓ; the distance of a component to the spine, for example, may change as the line ℓ moves.

During the sweep, components are created or merged when ℓ encounters a colored point. By carefully selecting which components to create and merge and how to merge them, the algorithm maintains the following invariants:

P1. If there is exactly one component and its terminals have different colors, then its terminals have different signs.

P2. If there are more than two terminals, then they all have the same color.

P3. The two closest components to the spine do not have the same sign.

When the algorithm encounters a colored point, p, it either *forks* a new component, if p's color matches the color of all terminals (or there are no terminals), otherwise it *merges* p with one or two existing components creating a single new component. We describe these two cases under the assumption that the encountered point p is blue. Symmetric operations hold if p is red. In the next two figures, the terminals are drawn as squares; also, light/dark vertices are red/blue.

Fork. If there are no red terminals, we create a new component containing (blue) p that straddles the spine and has adjacent (blue) terminals. See Fig. 1(a).

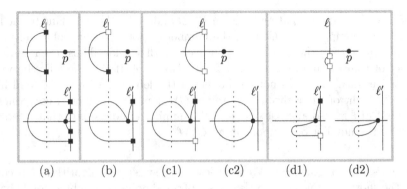

Fig. 1. The top row shows the initial configurations (the symmetric versions of (b) and (d) are not shown). The bottom row shows the possible configurations after merging. (a) Fork. One component is shown in the initial configuration but there may be many or none as long as all terminals are blue. The two new terminals are closest to the spine. (b-d) Merge with one component. All components are shown. Cases (c2) and (d2) only occur when p is the last colored point (Color figure online)

Merge. If there are red terminals, we create a new component that contains (blue) p. If there is only one component, we add p to that component by extending the edge from the closest red terminal to p. If p is not the last colored point, we add a new edge from p to a new (blue) terminal so that the new component straddles the spine. If p is the last colored point, we extend the edge from the other (red) terminal to p. See Fig. 1.

If there are at least two components, then by property P2 all their terminals are red. Let K and J be the two closest components to the spine. We extend the edges from a terminal from K and a terminal from J to p. We choose the terminals and route the edges from all terminals to ensure that our invariant properties remain true.

By property P3 and the fact that components do not intersect, the configuration of the two closest components to the spine is one of the four shown schematically in the top row of Fig. 2. For two of these configurations, (c) and (d) in Fig. 2, we extend the edges from the closest terminals to the spine from K and J to p, and extend all other terminals horizontally. In the other two configurations, we choose how to merge based on the sign of the closest component to the spine after the merge that is not the newly merged component. If this component is negative, we merge to form configuration (a1) or (b1) in Fig. 2. Otherwise, we merge to form configuration (a2) or (b2) in Fig. 2. This is used to preserve property P3.

Notice that forming configuration (b1) in Fig. 2 causes an edge of the alternating cycle (shown in bold) to cross the spine. We ensure that this edge is not forced to cross the spine again which implies that each edge of the alternating cycle produced by the algorithm can be drawn with at most two bends.

Main Theorem. We prove that Spine-Sweep computes an alternating cycle C such that each edge crosses the spine at most once and no two edges cross each

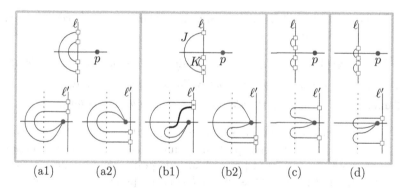

Fig. 2. Merge with two components. The top row shows the four basic configurations (the symmetric versions of (b) and (d) are not shown). The bottom row shows the possible configurations after merging (Color figure online).

other. Also, any vertical line ℓ that intersects C, does so exactly $2 \max\{1, |r - b|\}$ times, where r and b denote the number of red and blue points to the left of ℓ. Therefore, by using Lemma 1 and the observations that: (i) each fork/merge operation can be executed in constant time (for example by using a stack to maintain the components sorted according to their distance from the spine); and (ii) the red and blue points must be sorted in increasing x-order, we obtain the following.

Theorem 1. *Let R be a set of n red points and let B be a set of n blue points such that all points are distinct and collinear. There exists an optimal $\Theta(n \log n)$-time algorithm that computes a planar alternating cycle of minimum length with at most two bends per edge.*

Note that the time complexity of Theorem 1 is worst-case optimal. Namely, if the red and blue points alternate along the spine, computing a shortest alternating cycle is equivalent to computing a circular sorting of the point set.

4   Shortest Alternating Paths on Collinear Red-Blue Points

We can also obtain a shortest alternating path on a set of n red and n blue points that are all distinct and collinear, provided we are given the endpoints. The approach is the same as in the cycle case: we prove a lower bound on the number of times any alternating path with these endpoints intersects a vertical line ℓ and use essentially the same algorithm to find an alternating path that matches the bound. The lower bound is complicated slightly by the path endpoints.

Lemma 2. *Let R be a set of n red points and let B be a set of n blue points such that all points are distinct and lie on the x-axis. Let ℓ be a vertical line that intersects the x-axis between two points of $R \cup B$. If there are r red points and*

b *blue points to the left of* ℓ, *then the number of times any alternating path on* $R \cup B$ *crosses* ℓ *is at least:*

$$2 \max\{1, |r - b|\} \text{ if both path endpoints are on the same side of } \ell,$$
$$1 + 2 \max\{b - r, r - b - 1\} \text{ if only the red path endpoint is left of } \ell,$$
$$1 + 2 \max\{r - b, b - r - 1\} \text{ if only the blue path endpoint is left of } \ell.$$

Proof. If both path endpoints are on the same side of ℓ, the proof is the same as in the cycle case. If only the red path endpoint is left of ℓ, then its component can have at most one more red point than blue points, and at most zero more blue points than red points. Thus, if $r > b$ this component can account for one of the excess $r - b$ red points, while if $b \geqslant r$, it cannot account for any of the excess $b - r$ blue points. This component crosses ℓ once; all others cross twice. If only the blue path endpoint is left of ℓ, a symmetric argument applies.   □

To find a shortest alternating path between two given endpoints, we can use a modification of **Spine-Sweep**. More precisely, a fork operation on a vertex of degree one gives rise to a component with only one terminal and a merge operation on a vertex of degree one joins the closest terminal of the closest component. The main difference with the approach described in the previous section is that we may have an odd number of terminals during some steps of the sweep-line procedure, which however does not change the reasoning behind either the proof of correctness or the time complexity.

Lemma 3. *Given a set R of n red points and a set B of n blue points such that all points are distinct and collinear, and given $u \in R$ and $v \in B$, there exists a $\Theta(n \log n)$-time algorithm that computes a planar alternating path of minimum length with at most two bends per edge that starts at u and ends at v.*

Suppose we want to find the shortest alternating path but it may start and end at any pair of points. While one might think that a shortest alternating path always starts at the leftmost red (blue) point and always ends at the rightmost blue (red) point, this is not always the case. For example, the point set of Fig. 3 has an alternating path in Fig. 3(a) whose length is minimal if it is required that both endpoints are extremal but it is not as short as the one in Fig. 3(b).

To find the best endpoints, we may use the fact that our algorithm matches the lower bound described in Lemma 2. Let r_i and b_i be the number of red and blue points, respectively, among the first (leftmost) i colored points in $R \cup B$. (Note: $r_i + b_i = i$.) Let $c_i = 2 \max\{1, |r_i - b_i|\}$, $s_i = 1 + 2 \max\{b_i - r_i, r_i - b_i - 1\}$, and $t_i = 1 + 2 \max\{r_i - b_i, b_i - r_i - 1\}$. Let d_i be the distance between the ith and $(i+1)$st colored points. If the path starts at the jth and ends at the kth colored point, then its minimum length is the sum of the lower bounds from Lemma 2 weighted by the distance between adjacent colored points:

$$P[j, k] = \begin{cases} \sum_{i=1}^{j-1} d_i c_i + \sum_{i=j}^{k-1} d_i s_i + \sum_{i=k}^{n-1} d_i c_i & \text{if } j\text{th point is red} \\ \sum_{i=1}^{j-1} d_i c_i + \sum_{i=j}^{k-1} d_i t_i + \sum_{i=k}^{n-1} d_i c_i & \text{if } j\text{th point is blue} \end{cases}.$$

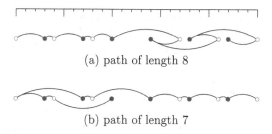

(a) path of length 8

(b) path of length 7

Fig. 3. Alternating paths. (Note the scale – fractional units are used) (Color figure online)

We can find the indices of different colored points $1 \leqslant j < k \leqslant n$ that minimize $P[j,k]$ in $O(n^2)$ time by calculating c_i, s_i, t_i, and d_i for all i in linear time; tabulating the partial sums $\sum_{i=1}^{j-1} d_i c_i$ (for all j) and $\sum_{i=k}^{n-1} d_i c_i$ (for all k) in linear time; and tabulating $\sum_{i=j}^{k-1} d_i s_i$ and $\sum_{i=j}^{k-1} d_i t_i$ (for all pairs $1 \leqslant j < k \leqslant n$) in quadratic time. Once we know the endpoints of the shortest alternating path, we can find the actual path using Lemma 3.

Theorem 2. *Given a set R of n red points and a set B of n blue points such that all points are distinct and collinear, there exists an $O(n^2)$-time algorithm that computes a planar alternating path of minimum length such that no two edges cross and each edge has at most two bends.*

5   Extensions and Generalizations

In this section we discuss how to extend the described approaches to more than two colors and we consider the case where the points are not collinear.

Shortest Alternating Paths and Cycles with More Than Two Colors. Consider 3-colored collinear point-sets. We use the colors red, green and blue, denoted by r, g and b. An alternating cycle is a cycle that connects points in the order rgbrgbr etc. For ease of presentation we consider the cycle oriented in this direction. The following lemma and theorems are the 3-color version of Lemma 1 and of the approaches illustrated in the previous sections for 2-colored point-sets. Proofs are omitted.

Lemma 4. *Let R, G and B be sets of n red, n green and n blue points that are all distinct and lie on the x-axis. Any alternating cycle C on $R \cup G \cup B$, crosses a vertical line, ℓ, between two colored points at least $2 \max\{1, |r - g|, |g - b|, |b - r|\}$ times, when there are r red, g green and b blue points to the left of ℓ.*

Theorem 3. *Let R, G and B be sets of n red, n green and n blue points such that all points are distinct and lie on the x-axis. A shortest planar alternating path (cycle) having $O(n)$ bends per edge can be computed in $O(n^2)$ time.*

It is natural to ask whether one can construct alternating shortest paths and cycles for collinear point-sets having more than three colors with an approach that computes drawings which satisfy generalizations of Lemmas 1 and 4. It is not hard to see that this is not the case even for 4-colored point-sets.

Assume that we have eight points and that we want to embed an alternating cycle with four colors. Figure 4 shows that for any sweep-line there is a solution that crosses this sweep-line exactly twice. So all lower-bounds are 2. However it is not hard to see that there is no embedding that satisfies all lower-bounds simultaneously.

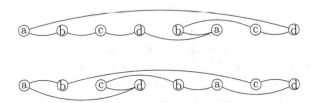

Fig. 4. 4-colored cycles on a set of eight points.

Non-collinear Point Sets. We show that finding the shortest alternating cycle is NP-hard by showing that deciding if there is a shortest alternating cycle of length less than L is NP-hard. Our reduction is from the EXACT COVER problem: Given a family F of subsets of a finite set U, is there a subfamily F' of F, consisting of disjoint sets, such that $\bigcup_{S \in F'} S = U$.

Theorem 4. *Given a k-colored point-set for constant $k \geqslant 1$, it is NP-hard to find the shortest planar alternating cycle.*

Proof. If $k = 1$, shortest planar alternating cycle is Euclidean TSP, which is NP-hard [16]. If $k = 2$, we describe a polynomial time reduction from EXACT COVER that is a slight modification of the reduction by Papadimitriou [16] showing that Euclidean TSP is NP-hard. Let P be the point set obtained from Papadimitriou's reduction from the EXACT COVER instance, rotated slightly so that no points share the same x- or y-coordinate. If the EXACT COVER instance is solvable, the shortest tour of P has length L (see [16]), while if it is not solvable, the shortest tour of P has length at least $L + \sqrt{a^2 + 1} - a$ where $a = 20$. Choose $0 < \epsilon \leqslant (\sqrt{a^2 + 1} - a)/(10(n + 1))$ to be smaller than half the smallest difference between the x- or y-coordinates of points in P. Let $R = (P + (-\epsilon, -\epsilon)) \cup (P + (\epsilon, \epsilon))$ and $B = (P + (-\epsilon, \epsilon)) \cup (P + (\epsilon, -\epsilon))$ (where $P + (x, y) = \{(p_x + x, p_y + y) | p \in P\}$). That is, each point $p \in P$ becomes a cluster of four points (two red and two blue) forming the corners of a square $S(p)$ of side-length 2ϵ centered at p.

If the EXACT COVER instance is solvable, there is a planar alternating tour of $R \cup B$ of length at most $L + 10\epsilon n < L + \sqrt{a^2 + 1} - a$. The alternating tour follows the shortest tour of P from cluster to cluster. Within the cluster for p, it follows three of the four sides of $S(p)$, leaving one side whose endpoints connect to the two neighbors of the cluster. It is not hard to verify that one may choose such a side for each cluster so that the resulting alternating tour is planar. Its length is at most $L + (6 + \sqrt{2})\epsilon n < L + 10\epsilon n < L + \sqrt{a^2 + 1} - a$. If the instance is not solvable, any alternating tour of $R \cup B$ is at least as long as the shortest tour of $R = P$, which has length at least $L + \sqrt{a^2 + 1} - a$. Thus $R \cup B$ has an alternating tour of length at most $L + \sqrt{a^2 + 1} - a$ if and only if the EXACT COVER instance is solvable.

If $k > 2$, the reduction is the same except that inside each square are $2(k - 2)$ points (two of each color other than red and blue). These points lie on the diagonal that connects the red corners of the square, with one point of color i at distance $i\epsilon/(k - 2)$ from each corner, for $i = 1, 2, \ldots, k - 2$. (Red is color 0 and blue is color $k - 1$.) The resulting alternating tour, which uses paths of diagonal points in place of the two red-to-blue sides in each square, has length at most $L + (5\sqrt{2} + 2)\epsilon n < L + \sqrt{a^2 + 1} - a$. $\qquad\qquad\square$

6   Open Problems

The research in this paper suggests several open problems. We conclude the paper by listing some of those that in our opinion are among the most interesting. (1) Can the time complexity of Theorem 2 be improved? (2) Can the bend-complexity of Theorem 3 be improved? (3) The problem of computing shortest alternating cycles on collinear k-colored point-sets is open for $k > 3$. (4) Study the problem of drawing not necessarily alternating shortest bicolored cycles/paths on collinear bicolored point-sets. That is, we are given a cycle/path where any blue (red) vertex may have a neighbor of its same color and we want to draw the cycle/path using the points of $R \cup B$ such that the total edge length is minimized.

References

1. Abellanas, M., Garcia-Lopez, J., Hernández-Peñalver, G., Noy, M., Ramos, P.A.: Bipartite embeddings of trees in the plane. Discr. Appl. Math. **93**(2–3), 141–148 (1999)
2. Akiyama, J., Urrutia, J.: Simple alternating path problem. Discr. Math. **84**, 101–103 (1990)
3. Badent, M., Di Giacomo, E., Liotta, G.: Drawing colored graphs on colored points. Theor. Comput. Sci. **408**(2–3), 129–142 (2008)
4. Bastert, O., Fekete, S.P.: Geometrische Verdrahtungsprobleme. Technical Report 96–247, Universität zu Köln (1996)
5. Chan, T.M., Hoffmann, H.-F., Kiazyk, S., Lubiw, A.: Minimum length embedding of planar graphs at fixed vertex locations. In: Wismath, S., Wolff, A. (eds.) GD 2013. LNCS, vol. 8242, pp. 376–387. Springer, Heidelberg (2013)
6. Frati, F., Glisse, M., Lenhart, W.J., Liotta, G., Mchedlidze, T., Nishat, R.I.: Point-set embeddability of 2-colored trees. In: Didimo, W., Patrignani, M. (eds.) GD 2012. LNCS, vol. 7704, pp. 291–302. Springer, Heidelberg (2013)

7. Di Giacomo, E., Didimo, W., Liotta, G., Meijer, H., Trotta, F., Wismath, S.K.: k-colored point-set embeddability of outerplanar graphs. J. Graph Alg. and Appl. **12**(1), 29–49 (2008)
8. Di Giacomo, E., Liotta, G., Trotta, F.: Drawing colored graphs with constrained vertex positions and few bends per edge. Algorithmica **57**(4), 796–818 (2010)
9. Kaneko, A., Kano, M.: Straight-line embeddings of two rooted trees in the plane. Disc. Comp. Geometry **21**(4), 603–613 (1999)
10. Kaneko, A., Kano, M.: Straight line embeddings of rooted star forests in the plane. Discr. Appl. Math. **101**, 167–175 (2000)
11. Kaneko, A., Kano, M.: Discrete geometry on red and blue points in the plane - a survey-. In: Aronov, B., Basu, S., Pach, J., Sharir, M. (eds.) Discrete and Computational Geometry. Algorithms and Combinatorics, vol. 25. Springer, New York (2003)
12. Kaneko, A., Kano, M., Suzuki, K.: Path coverings of two sets of points in the plane. In: Pach, J. (ed.) Towards a Theory of Geometric Graph, vol. 342. American Mathematical Society, Providence (2004)
13. Kaneko, A., Kano, M., Tokunaga, S.: Straight-line embeddings of three rooted trees in the plane. In: Canadian Conference on Computational Geometry, CCCG 1998 (1998)
14. Kaneko, A., Kano, M., Yoshimoto, K.: Alternating hamilton cycles with minimum number of crossing in the plane. Int. J. Comp. Geometry Appl. **10**, 73–78 (2000)
15. Pach, J., Wenger, R.: Embedding planar graphs at fixed vertex locations. Graphs Comb. **17**, 717–728 (2001)
16. Papadimitriou, C.H.: The Euclidean traveling salesman problem is NP-complete. Theor. Comp. Sci. **4**, 237–244 (1977)

On Embeddability of Buses in Point Sets

Till Bruckdorfer[1]([⊠]), Michael Kaufmann[1], Stephen G. Kobourov[2],
and Sergey Pupyrev[2,3]

[1] Wilhelm-Schickard-Institut für Informatik, Universität Tübingen,
Tübingen, Germany
{bruckdor,mk}@informatik.uni-tuebingen.de
[2] Department for Computer Science, University of Arizona, Tucson, USA
kobourov@cs.arizona.edu
[3] Institute of Mathematics and Computer Science, Ural Federal University,
Yekaterinburg, Russia
spupyrev@gmail.com

Abstract. Set membership of points in the plane can be visualized by connecting corresponding points via graphical features, like paths, trees, polygons, ellipses. In this paper we study the *bus embeddability problem* (BEP): given a set of colored points we ask whether there exists a planar realization with one horizontal straight-line segment per color, called bus, such that all points with the same color are connected with vertical line segments to their bus. We present an ILP and an FPT algorithm for the general problem. For restricted versions of this problem, such as when the relative order of buses is predefined, or when a bus must be placed above all its points, we provide efficient algorithms. We show that another restricted version of the problem can be solved using 2-stack pushall sorting. On the negative side we prove the NP-completeness of a special case of BEP.

1 Introduction

Visualization of sets is an important topic in graph drawing and information visualization and the traditional approach relies on representing overlapping sets via Venn diagrams and Euler diagrams [28]. When more than a handful sets are present, however, such diagrams become difficult to interpret and alternative approaches, such as compact rectangular Euler diagrams are needed [27].

Often the geometric position of the elements of the sets are prescribed as points in the plane. The task is to emphasize the sets where the elements belong to. In visualization approaches for set memberships of items on maps, this is done by connecting points from the same set by corresponding lines (LineSets [2]), tree structures (KelpFusion [24]), and enclosing polygons (BubbleSet [11] or MapSets [13]).

We consider a unified version of the tree-structure approach using a model that has been applied before for drawing orthogonal buses known from VLSI

An extended version including all missing proofs can be found in [7].

© Springer International Publishing Switzerland 2015
E. Di Giacomo and A. Lubiw (Eds.): GD 2015, LNCS 9411, pp. 395–408, 2015.
DOI: 10.1007/978-3-319-27261-0_33

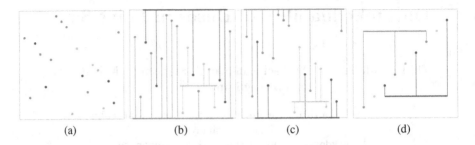

Fig. 1. (a) Fixed positions of points, where points with the same color belong to the same set. (b) A planar bus realization for this setting, while (c) is a non-planar bus realization. (d) A point set without any planar bus realization (Color figure online).

design [22]. Our goal is a membership visualization of points in sets by a tree-structure that consists of a single horizontal segment, called *bus*, to which all the points from the same set are connected by vertical segments, called *connections*; see Fig. 1 for planar and non-planar versions. We assume the sets to be given by single-colored points, such that in the final visualization, called *bus realization*, every point of the same color is connected to exactly one bus associated with this color. The objective is to find a position for each bus, such that crossings of buses with connections are avoided, called *planar* bus realization. We call this the *bus embeddability problem* (BEP). Such a simple visualization scheme makes it very easy to recognize the sets and label them, by placing a label inside each bus (if the bus is drawn thick enough), or directly above/next to the bus.

Related Work. Buses have been used, in a more general form, for visualizing degree-restricted hypergraphs. Ada et al. [1] used horizontal and vertical buses in bus realizations, where the points (representing hypervertices contained in at most four hyperedges) were not predefined in the plane. They asked whether a given hypergraph admits a non-planar bus realizations (allowing connections to cross each other) and showed that the problem is NP-complete. In contrast, if a planar embedding is given, a planar bus realization can be constructed on a $\mathcal{O}(n) \times \mathcal{O}(n)$ grid in $\mathcal{O}(n^{3/2})$ time [6]. These types of problems also have connections to rectangular drawings, rectangular duals and visibility graphs, since the edges of the incidence graph of a hypergraph enforce visibility constraints in the bus realizations [29].

Another related approach is visualization based on graph supports of hypergraphs. Here the goal is to connect the vertices in such a way that each hyperedge induces a connected subgraph [8,20]. Supported hypergraph visualizations inspired edge-bundling and confluent layouts as alternative visualizations for cliques [12,14].

A solution to the BEP problem can be viewed as planar tree support for hypergraphs, and this problem is related to Steiner trees, where the goal is to connect a set of points in the plane while minimizing the sum of edge lengths in the resulting tree; this is a classic NP-complete problem [15]. Hurtado et al. [18]

considered planar supports for hypergraphs with two hyperedges such that the induced subgraph for every hyperedge and the intersection is a Steiner tree. Their objective was to minimize the sum of edge lengths, while allowing degree one or two for the hypervertices. BEP is even more closely related to rectilinear Steiner trees, where the Euclidean distance is replaced by the rectilinear distance; constructing rectilinear Steiner trees is also NP-complete [16]. A single trunk Steiner tree [10] is a path which contains all vertices of degree greater than one. This is a variant that is solvable in linear time. BEP for a single set is the single trunk rectilinear Steiner tree problem, where we ignore the minimization of the sum of the edge lengths. Thus BEP can be seen as a simultaneous single-trunk rectilinear Steiner tree problem. The fact that a bus placement influences the placement of other buses makes the problem hard.

Consider the input to BEP along with a box that encloses all the points. If in BEP the buses extend to the right boundary of this box, or both to the left and right boundary of this box, then this problem corresponds to backbone boundary labeling and can be efficiently solved [4]. In backbone boundary labeling, the problem is to orthogonally connect points by a horizontal backbone segment leading to a label placed at the boundary. In this setting it is always possible to split the problem into two independent subproblems, which is impossible in our case.

BEP is also related to the classical *point set embeddability problem*, where given a set of points along with a planar graph, we need to determine whether there exists a mapping of vertices to points such that the resulting straight-line drawing is planar. The general decision problem is NP-hard [9]. In the variant of orthogeodesic point set embedding, Katz et al. proved that deciding whether a planar graph can be embedded using only orthogonal edge routing is NP-hard [19].

Our Results. In Sect. 2 we solve BEP when the relative order of the buses is prescribed; we also show that BEP is fixed-parameter tractable (FPT) with respect to the number of colors. In Sect. 3 we formulate an integer linear programming (ILP) formulation for BEP and show some experimental results. In Sect. 4 we restrict BEP (when a bus must be above all its points, or a bus must be either at its topmost or bottommost point) and describe efficient algorithms for these settings. Another restricted version of the problem is shown to be equivalent to the problem of sorting a permutation, which is called 2-stack pushall sorting. Finally we prove that BEP is NP-complete, even for just two points per color, if points may not lie on buses.

2   Preliminaries

We begin with some definitions. Suppose we are given a set of points $\mathcal{P} = \{p_1, \ldots, p_n\}$ and colors $\mathcal{C} = \{c_1, \ldots, c_k\}$ together with a function $f : \mathcal{P} \longrightarrow \mathcal{C}, f(p) = c$. For simplicity, we assume that no two points share a coordinate in the input point set, although in some illustrations the input points might violate this assumption. The bus embeddability problem (BEP) asks, whether there is a planar bus realization with one horizontal bus per color. BEP is a decision problem, but in our descriptions whenever the answer is affirmative we also compute a drawing. We refer to such a drawing as a *solution of BEP*. In the negative case, we say that BEP has no solution.

A point p has x-coordinate $x(p)$, y-coordinate $y(p)$, and color $f(p)$. In a bus realization we have connections only between a point p and a bus c of the same color, that is, $c = f(p)$. We denote by $f^{-1}(c)$ the set of points with color c. Bus c naturally extends from the x-coordinate $x_l(c) = \min\{x(p)|p \in f^{-1}(c)\}$ of the leftmost point to the x-coordinate $x_r(c) = \max\{x(p)|p \in f^{-1}(c)\}$ of the rightmost point of $f^{-1}(c)$. We call $[x_l(c), x_r(c)]$ the *span* of c, which is predefined by the input points. The y-coordinate of a bus c is denoted by $y(c)$, which is the only parameter to be determined for a solution for BEP.

Note that BEP is trivial when there are at most two colors: it is always possible to place one bus at the top and the other (if exists) at the bottom of the drawing. Thus in the following we assume $k > 2$. For more than two colors, the relative order of the buses is important; see Fig. 1. Suppose the y-order of the buses is prescribed. The next lemma shows that one can check an existence of a solution for BEP respecting the order.

Lemma 1. *There is a $\mathcal{O}(n \log n)$-time algorithm that, given an order of buses, tests whether there exists a solution for BEP respecting the order.*

Proof. Suppose we are given an order $c_1 < \cdots < c_k$ of the buses from bottom to top. We use discrete values for the y-coordinates increasing from bottom to top, where a unit is $1/n$ of the y-distance of two consecutive points. We first present a simpler $\mathcal{O}(n^2)$-time algorithm, and then describe how to speed it up.

Recall that the span of every bus is defined by an input point set; hence, we only show how to choose y-coordinates of the buses. The first bus, c_1, is placed at y-coordinate $y(c_1) = 0$, and all the points of color c_1 are connected to the bus. Assume that bus c_{i-1} is placed at y-coordinate $y(c_{i-1})$ and is connected to all its points. We place c_i at $y(c_i) = y(c_{i-1}) + 1$ unit and check if the bus crosses a previously drawn (vertical) segment. If it does cross a segment, then we shift c_i one unit upwards by increasing $y(c_i)$ and repeat the procedure. Once the bus is placed without crossings, we connect it to the corresponding points. Consider the vertical segment of a point p of color c_i. It is easy to see that if $y(p) \geq y(c_i)$, then the segment cannot cross a previously placed bus c_j for $j < i$. If $y(p) < y(c_i)$ and the vertical segment crosses a bus, then such a crossing is unavoidable in any solution respecting the given order. Hence, we may stop the algorithm reporting that no solution exists. Otherwise, we proceed with the next color.

The above algorithm can easily be implemented in quadratic time. However, we can do better using the following observation: Every bus is placed at its bottommost "valid" y-coordinate, that is, the one that does not produce crossings with previously placed buses. To find such a y-coordinate efficiently for each color, we store all points of the already processed colors in a data structure D that supports the range operation such as "extracting minimum/maximum on a given range". For every color c_i, we extract a point with the maximum y-coordinate in the range corresponding to the span of c_i. The bus of c_i is placed at the maximum of the extracted y-coordinate and the y-coordinate of bus $y(c_{i-1})$. Then all the points of color c_i are added to D. A balanced tree (e.g., a segment tree) providing logarithmic complexity for insert and extract operations is sufficient for our needs. □

In general the correct order of the buses for a planar bus realization is not known. One can apply Lemma 1 for each of the $k!$ possible bus orders, which yields an $\widetilde{\mathcal{O}}(k!)$-time[1] algorithm for BEP. Next, we improve the running time with an algorithm providing deeper insight into the structure of the problem.

Lemma 2. *There is a $\widetilde{\mathcal{O}}(2^k)$-time algorithm for BEP.*

Proof. We solve a given instance of BEP using dynamic programming. Let us call a *state* a pair (h, B), where $0 \le h \le n + 1$ is an integer and B is a subset of $\mathcal{C} = \{c_1, \ldots, c_k\}$. By a solution for a state (h, B) we mean a (planar) bus realization consisting of buses for every color $c \in B$ such that the topmost bus has y-coordinate h. If such a solution exists, we write $F(h, B) = \text{true}$, and otherwise $F(h, B) = \text{false}$. It is easy to see that a solution for the original BEP problem exists if and only if $F(h, \mathcal{C}) = \text{true}$ for some $0 \le h \le n + 1$.

We reduce the problem to solving it for "smaller" states, that are the states with fewer elements in B. As a base case, we set $F(h, B) = \text{true}$ for all $0 \le h \le n+1$ and $|B| = 1$. To compute a value for a state $F(h, B)$ with $|B| > 1$, we consider a color $c^* \in B$. Let $h^* = \max\{y(p)|f(p) \in B \smallsetminus \{c^*\}$ and $x_l(c^*) \le x(p) \le x_r(c^*)\}$, that is, the largest (topmost) y-coordinate of a point of color $B \smallsetminus \{c^*\}$ laying in the span of c^*. It follows from the proof of Lemma 1 that the bus for c^* should be placed at y-coordinate h^*. Thus, $F(h, B)$ is set to true if (a) $h \ge h^*$ and (b) there exists a solution for a state $(h', B \smallsetminus \{c^*\})$ for some $h' < h$. We stress here that in order to compute $F(h, B)$, one needs to consider every color of B as a potential c^*. There are $n2^k$ different states, and a computation for a single state clearly takes a polynomial number of steps. □

The above result shows that the BEP problem is fixed-parameter tractable with respect to k, that is, it can be efficiently solved for a small number of buses. Note that in Sect. 5 we prove that BEP is NP-complete; hence, it is unlikely that a polynomial-time (in terms of k) algorithm exists.

3 An ILP for BEP

In this section we present an integer linear programming (ILP) formulation for BEP that produces a planar bus realization if one exists. The ILP also minimizes the amount of ink in a solution, that is, the sum of all segment lengths.

Lemma 3. *A solution for BEP can be computed by an ILP.*

---

[1] $\widetilde{O}$ hides polynomial factors.

Proof. In a preprocessing step we compute the span of every bus $c \in C$. As mentioned earlier, it remains to compute the y-coordinate variable $y(c)$ of every bus c. To this end, we introduce a planarity constraint for every point $p \in P$ within the span of bus c having a different color. The pairs $(p, c), c \neq f(p)$ are called *conflicting*. Conflicting pairs (p, c) are stored in a matrix $\mathcal{J}$ and induce the constraint $(y(p) < y(c)$ and $y(f(p)) < y(c))$ or $(y(p) > y(c)$ and $y(f(p)) > y(c))$. The matrix $\mathcal{J}$ can be computed in $\mathcal{O}(kn)$ time, where $n = |P|$ and $k = |C|$. In order to minimize the amount of ink, we sum up the lengths of all connections and ignore the lengths of buses, as those are determined by the input.

$$\min \sum_{c \in C} \sum_{f(p)=c} |y(c) - y(p)|$$

$$s.t. \quad (y(p) < y(c) \vee y(f(p)) > y(c)) \wedge (y(p) > y(c) \vee y(f(p)) < y(c)) \quad \forall (p,c) \in \mathcal{J}$$
$$0 \leq y(c) \leq \max_{p \in P}\{y(p)\} + 1$$

Since absolute value (resp. "or") needs one more variable and 3 constraints for every point (resp. for every conflicting pair)[2], the final ILP has $n+k+2|\mathcal{J}|$ variables and $3n+k+6|\mathcal{J}|$ constraints.   □

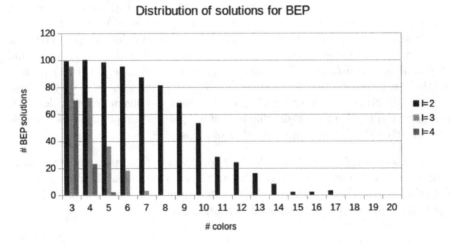

Fig. 2. The percentage of solutions for BEP for a random point set of size $n = kl$ with $l = 2, 3, 4$ points per color out of $k = 3, \ldots, 20$ colors (Color figure online).

In order to get a feeling about the probability that a point set admits a solution of BEP, we ran an experiment with the ILP, implemented with the Gurobi solver [17]. We considered point sets with $k = 3, \ldots, 20$ colors and with $l = 2, 3, 4$ points per color. We randomly placed the points on a 1024×768 area.

---

[2] $\min \sum |a-b| \Leftrightarrow \min \sum e, e \geq a-b, e \geq b-a, e \geq 0$; $(a < b) \vee (c < d) \Leftrightarrow a-b < eM, c-d < (1-e)M, e \in \{0,1\}, M = \infty$.

For each pair (l, k) we counted the number of BEP solutions out of 100 instances; see Fig. 2. The remaining instances were infeasible. For a fixed number of points, l, the number of solutions for BEP decreases with increasing the number of colors, k. It decreases faster the higher l is. On the other hand for a fixed number of colors, k, the number of solutions for BEP also decreases with increasing number of points, l. Hence, studying two points per color promises to be sufficiently interesting. Thus, as the base case for further analysis, we initially consider two points per color, before dealing with the general case, where in real instances solutions rarely exist. It is possible that much more solutions exist if we allow only few crossings, but all non-planar settings are left as open problems.

4   Efficiently Solvable BEP Variants

In this section we consider three variants of BEP, which can be solved in polynomial time. A bus c is called *top* (resp., *bottom*) if all of its points are below (resp., above) the bus, that is, $y(c) \geq y(p)$ (resp., $y(c) \leq y(p)$) for all $p \in f^{-1}(c)$. We distinguish between buses that are above (below) of their points and buses that pass through one of their points. A top-bus is a ⊓-*bus* if $y(c) > y(p)$ for all $p \in f^{-1}(c)$ (Fig. 3(a)), while it is a ⊏-*bus* if $y(c) = y(p)$ for a point p with $y(p) = \max\{y(q)|q \in f^{-1}(c)\}$ (Fig. 3(c)). Similarly we define a ⊔-*bus* and a ⊏-*bus*; see Fig. 3(b) and (d). A bus, whose type is none of the four types from above, is called a *center-bus*. The variant of BEP where only buses of the types in $S \subseteq \{⊓, ⊔, ⊏, ⊏\}$ are allowed to use is denoted by S-BEP.

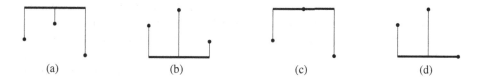

Fig. 3. Illustration of (a) ⊓-bus, (b) ⊔-bus, (c) ⊏-bus, and (d) ⊏-bus.

In Sect. 4.1 we study ⊓-buses and provide an algorithm for ⊓-BEP. The same algorithm obviously solves the ⊔-BEP variant. Next we consider ⊏-buses and ⊏-buses. Note that ⊏-BEP and ⊏-BEP are trivial, since every ⊏-bus (resp., ⊏-bus) is uniquely defined by its span and the topmost (bottommost) point. Hence, we investigate and design an efficient algorithm for the (⊏, ⊏)-BEP variant. Finally in Sect. 4.3, we examine the general BEP for a specific point set, where all points lie on a diagonal. We show that the variant of the problem is equivalent to a longstanding open problem (resolved very recently) of sorting a permutation with a series of two stacks.

4.1 ⊓-BEP

Here, we present an algorithm that decides in polynomial time whether a drawing with ⊓-buses exists for a given input, and constructs such a drawing if one exists.

Theorem 1. *There exists an $\mathcal{O}(n \log n)$-time algorithm for ⊓-BEP.*

Proof. For ease of presentation, we first assume that the input consists of two points per color, that is, $k = n/2$, and provide a simple quadratic-time implementation. Later we generalize the algorithm and improve the running time. Intuitively, the algorithm sweeps a line from bottom to top and processes the points in increasing order of y-coordinates. At every step, we keep all the vertical segments of the "active" colors (the ones without a bus) in the correct left-to-right order. If two vertical segments of the same color are adjacent in the order, then we can draw the corresponding bus and remove the color and its vertical segments. Otherwise, all the active vertical segments have to be "grown" until we reach the next point. It is easy to see that a solution exists if and only if the set of active colors is empty after processing all the points.

More formally, the points are processed one-by-one in increasing order of their y-coordinates. The points are stored in an array sorted by x-coordinate, that is, we have $(p_1, \ldots, p_n)$ with $x(p_1) < \cdots < x(p_n)$. At each iteration, a new point is inserted into the array in the position determined by its x-coordinate. Then the array is modified (or simplified) so that the pairs of points of the same color that are adjacent in the array are removed. That is, if $f(p_i) = f(p_{i+1})$ for some $1 \le i < n$, then we get a new array $(p_1, \ldots, p_{i-1}, p_{i+2}, \ldots, p_n)$. The simplification is performed as long as the array contains monochromatic adjacent points. After this step the algorithm proceeds with the next point. For every color c, we keep the value $y^*(c)$, which is equal to the y-coordinate $y(p), p \in f^{-1}(c')$ of the point of color c', whose insertion into the array induced the removal of points $f^{-1}(c)$ from the array. If the algorithm ends up with a non-empty array, then we report that no solution exists. Otherwise, the y-coordinate of the resulting bus of color c is $y^*(c) + \varepsilon$, where $\varepsilon > 0$ is sufficiently small to avoid overlaps between the buses. An example of the algorithm is illustrated in Fig. 4.

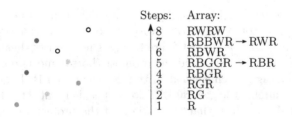

Steps:	Array:
8	RWRW
7	RBBWR → RWR
6	RBWR
5	RBGGR → RBR
4	RBGR
3	RGR
2	RG
1	R

Fig. 4. Running the algorithm from Lemma 1 on a given point set with red (R), green (G), blue (B), and white (W) pairs of points. Since the resulting array is not empty, there is no solution for the instance. Notice that removing any of the colors yields an instance with a solution (Color figure online).

Correctness. The correctness follows from the observation that the algorithm chooses the lowest "available" y-coordinate for every bus, that is, the one that does not induce a crossing between the bus and vertical segments of other colors. Indeed, if at any step of the algorithm we get a color pattern $R, \ldots, B, \ldots, R$ in the array formed by red (R) and blue (B) points and the second blue point p has not been processed yet, then clearly in any solution the red vertical segments reach the y-coordinate of p. Hence, it is safe to "grow" the segments. On the other hand, if processed points form a color pattern RR (that is, two consecutive points of the same color), then there is a solution connecting the corresponding vertical segments at the current y-coordinate. The two points can be removed from consideration, as they cannot create crossings with the subsequent buses. It is also easy to see that the algorithm minimizes ink of the resulting drawing.

Running Time. At every iteration of the algorithm, we need to insert a new point into the sorted array and then run the simplification procedure. Point insertion takes $\mathcal{O}(n)$ time and removal of a pair of points from the array can also be done in $\mathcal{O}(n)$ time. Since every pair is removed only once, the total running time is $\mathcal{O}(n^2)$.

To get down to $\mathcal{O}(n \log n)$ time, we use a balanced binary tree instead of an array to store the points. The tree is sorted by the x-coordinates of the points; hence, insertion/removal of a point takes $\mathcal{O}(\log n)$ time. Note that after inserting/removing a point, the only potential candidate pairs for simplification are the point's neighbors that can be found in $\mathcal{O}(\log n)$ time. Again, every point is inserted/removed only once; thus, the total running time is $\mathcal{O}(n \log n)$.

Finally, we observe that the algorithm can be generalized to handle multiple points per color. To this end, we change the simplification step so that the points are removed only if they form a contiguous subsequence in the array (tree), containing all points of this color. Hence we need to know the number of points for each color, which can be done with a linear-time scan of the input. It is easy to see that the proof of correctness can be appropriately modified and the running time remains the same. □

4.2 (⌐,∟)-BEP

We present an algorithm that decides in polynomial time whether (⌐,∟)-BEP has a solution for a given input, and constructs a drawing if one exists.

Theorem 2. *There exists an $\mathcal{O}(n^2)$-time algorithm for (⌐,∟)-BEP.*

Proof. The span of every bus is predefined by the input, while the y-coordinate has precisely two options. We show that (⌐,∟)-BEP can be modeled by 2-SAT, and thus is efficiently solvable. For ease of presentation, we first assume that the input consists of two points per color and describe a simple quadratic-time algorithm.

The algorithm creates a variable x_c for every color $c \in \mathcal{C}$. The value of x_c is *true* if c is a ⌐-bus, and it is *false* if c is a ∟-bus. Then for every pair of colors c, c', the algorithm creates a clause for the 2-SAT instance when the corresponding buses induce a crossing. Building the clauses with respect to the relative position of points is a straight-forward procedure; 3 examples are illustrated in Fig. 5. We can generalize this idea in a straight-forward manner to the case of more points per color. In the general case the y-coordinate of a bus still has precisely two options. In contrast to the case with two points per color we check several points (not only the leftmost or rightmost point) of color c' for their position with respect to the points of color c, since points lie not necessarily in corners of the enclosing rectangle.

Correctness. The correctness follows from the complete case analysis.

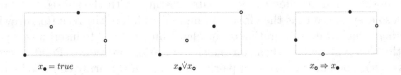

Fig. 5. Three examples for creating clauses for two colors black and white.

Running Time. We remark that for the $n^2/4$ pairs of colors, we create $\mathcal{O}(n^2)$ clauses, each clause in constant time by a case analysis. This results in a 2-SAT instance with k variables $x_c, c \in \mathcal{C}$ and $\mathcal{O}(n^2)$ clauses. We solve this instance in linear time [3] and the solution determines the drawing: c is drawn as a ⌐-bus, if the value of x_c is *true*, otherwise c is drawn as a ⌐-bus.   ▢

4.3   Diagonal BEP

Here we consider a *diagonal* point set in which all points lie on a single diagonal line and there are two points per color. We assume that the point set is *separable*, that is, there is a straight line separating every pair of points having the same color; see Fig. 6. This specific arrangement can be naturally described in terms of permutations. Assuming that the colors are numbered from 1 to k in the order along the diagonal from bottom to top, the input is described by a permutation $\pi = [\pi(1), \ldots, \pi(k)]$ on $\{1, \ldots, k\}$. Such an instance is called *diagonal π-BEP*.

It turns out that this variant of BEP is closely related to the well-studied topic of sorting a permutation with stacks introduced by Knuth in the 1960's [21]. We next show that diagonal π-BEP has a solution if and only if π can be sorted with 2 stacks in series. The problem of deciding whether a permutation is sortable with 2 stacks in series is a longstanding open problem and it has been conjectured to be NP-complete several times [5]. Only very recently a polynomial-time algorithm has been developed [25, 26]. It is an indication that even the restricted variant of BEP is highly non-trivial. Next we prove the equivalence.

First observe that for a diagonal point set with 2 points per color, a top-bus (bottom-bus) can be transformed to a center-bus. For every color c, there are no points of different color within the span of c above the topmost point of c. Hence, we may only consider center-buses in the variant of BEP. For the 2-stack sorting problem, given a permutation π, we want to sort the numbers to the identity permutation $[1, \ldots, k]$ with two stacks S_I, S_{II} using the following operations:

- α_i : read the next element i from input π and push it on the first stack S_I;
- β_i : pop the topmost element i from S_I and push it on S_{II};
- γ_i : pop the topmost element i from S_{II} and print it to the output.

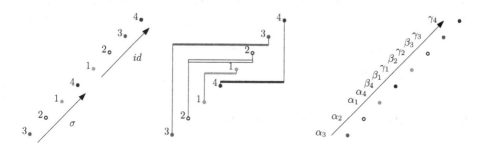

Fig. 6. A diagonal point set with a solution for BEP and the regarding sorting sequence (Color figure online).

To make the equivalence between 2-stack sorting and bus embeddability, we note that the first operation, α_i, corresponds to the left vertical segment of color i, the second one, β_i, is the bus of i, while γ_i corresponds to the right vertical segment of the color; see Fig. 6. A crossing in the drawing correspond to an "invalid" sorting operation in which either a non-topmost element is moved from S_I to S_{II} (a crossing to the "left" of the diagonal), or a non-topmost element is moved from S_{II} to the output (a crossing to the "right" of the diagonal). Hence, sorting sequences of the operations for π are in one-to-one correspondence with planar bus realization for the point set. Since the point set is separable, all the elements of π will be pushed to S_I before any of the elements is popped to the output. This is called 2-stack *pushall* sorting, see [25] for more details.

Theorem 3. *Diagonal π-BEP has a solution if and only if π is 2-stack pushall sortable. This can be checked in $\mathcal{O}(n^2)$ time.*

5 Hardness of BEP

We sketch the idea behind the proof that BEP$^\varepsilon$ for 2 points per color is NP-complete, where BEP$^\varepsilon$ is BEP with minimum distance ε of points to their bus as additional input.

We can easily verify a possible solution using Lemma 1; thus BEP$^\varepsilon$ is in the class NP. To prove the hardness of BEP$^\varepsilon$, we reduce from planar 3-SAT [23], which is 3-SAT, where an instance is represented by a graph whose vertices represent variables and clauses and whose edges represent containment of variables in clauses. We replace the vertices and edges by gadgets. We first restrict ourselves to $(\sqcap, \sqcup)$-BEP and drop the "no points share a coordinate" restriction. Details can be found in [7].

The most important module of the construction is a *chain link*, which is also a gadget for replacing variables. It consists of two points on a common horizontal line that will be connected by a bus. We replace the edges of the graph by chains consisting of nested chain links and replace the clause vertices by a big construction of points, that allows two specific points to be connected via a bus using only one of three choices. We use the input ε to be able to block some choices for this bus. We finally transform the construction into the "no points share a coordinate" setting and allow also center-buses.

Theorem 4. *BEP$^\varepsilon$ for 2 points per color is* NP-*complete.*

6   Conclusion and Future Work

We studied bus embeddability, where a set of colored points is covered by a set of horizontal buses, one per color and without crossings. We described an ILP and an FPT algorithm for the general problem and presented polynomial-time algorithms for several restricted versions. The general problem is shown to be NP-complete even for two points per color when points may not lie on buses.

It is still open to determine the complexity of BEP in the following cases:

- BEP using only center-buses;
- $(\sqcap, \sqcup)$-BEP, that is, BEP without center-buses;
- diagonal BEP with more than 2 points per color;
- general BEP (in our construction, we use an extra ε as a parameter).

A natural generalization would be to allow both horizontal and vertical buses, as in [1,6]. Another variant might be to consider multi-colored points, where a point has to be connected either to all the buses of its corresponding colors, or to at least one of them. For point sets that have no solution for BEP with only one bus per color, we may allow more than one bus or bound the number of crossings. Possible objectives in these scenarios are to minimize the total number of buses over all colors, to minimize the total number of buses, or to minimize the total number of buses if each tree can connect $\leq$ k unicolored points. These objectives are even interesting if a solution to BEP exists.

References

1. Ada, A., Coggan, M., Marco, P.D., Doyon, A., Flookes, L., Heilala, S., Kim, E., Wing, J.L.O., Préville-Ratelle, L.F., Whitesides, S., Yu, N.: On bus graph realizability. In: Canadian Conference on Computational Geometry, pp. 229–232 (2007)
2. Alper, B., Riche, N.H., Ramos, G., Czerwinski, M.: Design study of LineSets, a novel set visualization technique. IEEE Trans. Visual. Comput. Graph. **17**(12), 2259–2267 (2011)

3. Aspvall, B., Plass, M.F., Tarjan, R.E.: A linear-time algorithm for testing the truth of certain quantified Boolean formulas. Inform. Process. Lett. **8**(3), 121–123 (1979)
4. Bekos, M.A., Cornelsen, S., Fink, M., Hong, S.-H., Kaufmann, M., Nöllenburg, M., Rutter, I., Symvonis, A.: Many-to-one boundary labeling with backbones. In: Wismath, S., Wolff, A. (eds.) GD 2013. LNCS, vol. 8242, pp. 244–255. Springer, Heidelberg (2013)
5. Bóna, M.: A survey of stack-sorting disciplines. Electron. J. Comb. **9**(2), 16 (2003)
6. Bruckdorfer, T., Felsner, S., Kaufmann, M.: On the characterization of plane bus graphs. In: Spirakis, P.G., Serna, M. (eds.) CIAC 2013. LNCS, vol. 7878, pp. 73–84. Springer, Heidelberg (2013)
7. Bruckdorfer, T., Kaufmann, M., Kobourov, S., Pupyrev, S.: On embeddability of buses in point sets. CoRR abs/1508.06760 (2015)
8. Buchin, K., van Kreveld, M.J., Meijer, H., Speckmann, B., Verbeek, K.: On planar supports for hypergraphs. J. Graph Algorithms Appl. **15**(4), 533–549 (2011)
9. Cabello, S.: Planar embeddability of the vertices of a graph using a fixed point set is NP-hard. J. Graph Algorithms Appl. **10**(2), 353–366 (2006)
10. Chen, H., Qiao, C., Zhou, F., Cheng, C.K.: Refined single trunk tree: A rectilinear Steiner tree generator for interconnect prediction. In: SLIP, pp. 85–89. ACM (2002)
11. Collins, C., Penn, G., Carpendale, T.: Bubble Sets: Revealing set relations with iso-contours over existing visualizations. IEEE Trans. Visual. Comput. Graph. **15**(6), 1009–1016 (2009)
12. Dickerson, M., Eppstein, D., Goodrich, M.T., Meng, J.Y.: Confluent drawings: Visualizing non-planar diagrams in a planar way. J. Graph Algorithms Appl. **9**(1), 31–52 (2005)
13. Efrat, A., Hu, Y., Kobourov, S.G., Pupyrev, S.: MapSets: visualizing embedded and clustered graphs. In: Duncan, C., Symvonis, A. (eds.) GD 2014. LNCS, vol. 8871, pp. 452–463. Springer, Heidelberg (2014)
14. Gansner, E.R., Koren, Y.: Improved circular layouts. In: Kaufmann, M., Wagner, D. (eds.) GD 2006. LNCS, vol. 4372, pp. 386–398. Springer, Heidelberg (2007)
15. Garey, M.R., Graham, R.L., Johnson, D.S.: The complexity of computing Steiner minimal trees. SIAM J. Appl. Math. **32**(4), 835–859 (1977)
16. Garey, M.R., Johnson, D.S.: The rectilinear Steiner tree problem is NP-complete. SIAM J. Appl. Math. **32**(4), 826–834 (1977)
17. Gurobi Optimization, I.: Gurobi optimizer reference manual (2015). www.gurobi. com
18. Hurtado, F., Korman, M., van Kreveld, M., Löffler, M., Sacristán, V., Silveira, R.I., Speckmann, B.: Colored spanning graphs for set visualization. In: Wismath, S., Wolff, A. (eds.) GD 2013. LNCS, vol. 8242, pp. 280–291. Springer, Heidelberg (2013)
19. Katz, B., Krug, M., Rutter, I., Wolff, A.: Manhattan-geodesic embedding of planar graphs. In: Eppstein, D., Gansner, E.R. (eds.) GD 2009. LNCS, vol. 5849, pp. 207–218. Springer, Heidelberg (2010)
20. Klemz, B., Mchedlidze, T., Nöllenburg, M.: Minimum tree supports for hypergraphs and low-concurrency euler diagrams. In: Ravi, R., Gørtz, I.L. (eds.) SWAT 2014. LNCS, vol. 8503, pp. 265–276. Springer, Heidelberg (2014)
21. Knuth, D.E.: The Art of Computer Programming, Volume 1. Fundamental Algorithms, 3rd edn. Addison Wesley Longman Publishing Co., Inc., Redwood (1997)
22. Lengauer, T.: VLSI theory. In: van Leeuwen, J. (ed.) Handbook of Theoretical Computer Science, Volume A: Algorithms and Complexity (A), pp. 835–868. Elsevier, Amsterdam (1990)

23. Lichtenstein, D.: Planar formulae and their uses. SIAM J. Comput. **11**(2), 329–343 (1982)
24. Meulemans, W., Riche, N.H., Speckmann, B., Alper, B., Dwyer, T.: KelpFusion: A hybrid set visualization technique. IEEE Trans. Visual. Comput. Graph. **19**(11), 1846–1858 (2013)
25. Pierrot, A., Rossin, D.: 2-stack pushall sortable permutations. CoRR abs/1303.4376 (2013)
26. Pierrot, A., Rossin, D.: 2-stack sorting is polynomial. In: Mayr, E.W., Portier, N. (eds.) Symposium on Theoretical Aspects of Computer Science. LIPIcs, vol. 25, pp. 614–626. Schloss Dagstuhl - Leibniz-Zentrum fuer Informatik (2014)
27. Riche, N.H., Dwyer, T.: Untangling Euler diagrams. IEEE Trans. Visual. Comput. Graph. **16**(6), 1090–1099 (2010)
28. Simonetto, P., Auber, D., Archambault, D.: Fully automatic visualisation of overlapping sets. Comput. Graph. Forum **28**(3), 967–974 (2009)
29. Tamassia, R., Tollis, I.G.: A unified approach to visibility representations of planar graphs. Discrete Comput. Geom. **1**, 321–341 (1986)

A Universal Point Set for 2-Outerplanar Graphs

Patrizio Angelini[1], Till Bruckdorfer[1]([✉]), Michael Kaufmann[1],
and Tamara Mchedlidze[2]

[1] Wilhelm-Schickard-Institut für Informatik, Universität Tübingen,
Tübingen, Germany
{angelini,bruckdor,mk}@informatik.uni-tuebingen.de
[2] Institute of Theoretical Informatics, Karlsruhe Institute of Technology,
Karlsruhe, Germany
mched@iti.uka.de

Abstract. A point set $S \subseteq \mathbb{R}^2$ is universal for a class $\mathcal{G}$ if every graph of $\mathcal{G}$ has a planar straight-line embedding on S. It is well-known that the integer grid is a quadratic-size universal point set for planar graphs, while the existence of a sub-quadratic universal point set for them is one of the most fascinating open problems in Graph Drawing. Motivated by the fact that outerplanarity is a key property for the existence of small universal point sets, we study 2-outerplanar graphs and provide for them a universal point set of size $O(n \log n)$.

1 Introduction

Let S be a set of m points on the plane. A *planar straight-line embedding* of an n-vertex planar graph G, with $n \leq m$, on S is a mapping of each vertex of G to a distinct point of S so that, if the edges are drawn straight-line, no two edges cross. Point set S is *universal* for a class $\mathcal{G}$ of graphs if every graph $G \in \mathcal{G}$ has a planar straight-line embedding on S. Asymptotically, the smallest universal point set for general planar graphs is known to have size at least $1.235n$ [11], while the upper bound is $O(n^2)$ [3,8,12]. All the upper bounds are based on drawing the graphs on an integer grid, except for the one by Bannister et al. [3], who use super-patterns to obtain a universal point set of size $n^2/4 - \Theta(n)$ – currently the best result for planar graphs. Closing the gap between the lower and the upper bounds is a challenging open problem [6–8].

A subclass of planar graphs for which the "smallest possible" universal point set is known is the class of *outerplanar* graphs – the graphs that admit a straight-line planar drawing in which all vertices are incident to the outer face. Namely, Gritzmann et al. [10] and Bose [5] proved that any size-n point set in general position is universal for n-vertex outerplanar graphs. Motivated by this result, we consider the class of k-*outerplanar* graphs, with $k \geq 2$, which is a generalization of outerplanar graphs. A planar drawing of a graph is k-outerplanar if

This work has been supported by DFG grant Ka812/17-1. The full version of the paper, including all the missing proofs, can be found in [2].

© Springer International Publishing Switzerland 2015
E. Di Giacomo and A. Lubiw (Eds.): GD 2015, LNCS 9411, pp. 409–422, 2015.
DOI: 10.1007/978-3-319-27261-0_34

removing the vertices of the outer face, called k-th *level*, produces a $(k-1)$-outerplanar drawing, where 1-outerplanar stands for outerplanar. A graph is k-outerplanar if it admits a k-outerplanar drawing. Note that every planar graph is a k-outerplanar graph, for some value of $k \in O(n)$. Hence, in order to tackle a meaningful subproblem of the general one, it makes sense to study the existence of subquadratic universal point sets when the value of k is bounded by a constant or a sublinear function. However, while the case $k = 1$ is trivially solved by selecting any n points in general position, as observed above [5,10], the case $k = 2$ already eluded several attempts of solution and turned out to be far from trivial. In this paper, we finally solve the case $k = 2$ by providing a universal point set for 2-outerplanar graphs of size $O(n \log n)$.

A subclass of k-outerplanar graphs, in which the value of k is unbounded, but every level is restricted to be a chordless simple cycle, was known to have a universal point set of size $O(n(\frac{\log n}{\log \log n})^2)$ [1], which was subsequently reduced to $O(n \log n)$ [3]. It is also known that *planar 3-trees* – graphs not defined in terms of k-outerplanarity – have a universal point set of size $O(n^{5/3})$ [9]. Note that planar 3-trees have treewidth equal to 3, while 2-outerplanar graphs have treewidth at most 5.

Structure of the Paper: After some preliminaries and definitions in Sect. 2, we consider 2-outerplanar graphs in Sect. 3 where the inner level is a forest and all the internal faces are triangles. We prove that this class of graphs admits a universal point set of size $O(n^{3/2})$. We then extend the result in Sect. 4 to 2-outerplanar graphs in which the inner level is still a forest but the faces are allowed to have larger size. Finally, in Sect. 5, we outline how the result of Sect. 4 can be extended to general 2-outerplanar graphs. We also explain how to apply the methods in [3] to reduce the size of the point set to $O(n \log n)$. We conclude with open problems in Sect. 6.

2   Preliminaries and Definitions

A straight-line segment with endpoints p and q is denoted by $s(pq)$. A circular arc with endpoints p and q (clockwise) is denoted by $a(pq)$. We assume familiarity with the concepts of *planar graphs*, *straight-line planar drawings* and their *faces*. A straight-line planar drawing Γ of a graph G determines a clockwise ordering of the edges incident to each vertex u of G, called *rotation at u*. The *rotation scheme* of G in Γ is the set of the rotations at all the vertices of G determined by Γ. Observe that, if G is connected, in all the straight-line planar drawings of G determining the same rotation scheme, the faces of the drawing are delimited by the same edges.

Let $[G, \mathcal{H}]$ be a 2-outerplanar graph, where the outer level is an outerplanar graph G and the inner level is a set $\mathcal{H} = \{G_1, \ldots, G_k\}$ of outerplanar graphs. We assume that $[G, \mathcal{H}]$ is given together with a rotation scheme, and the goal is to construct a planar straight-line embedding of $[G, \mathcal{H}]$ on a point set determining this rotation scheme. Since $[G, \mathcal{H}]$ can be assumed to be connected (as otherwise

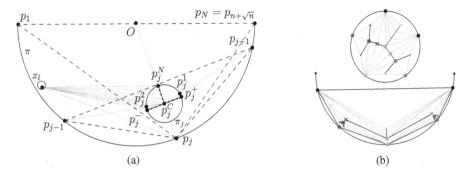

Fig. 1. (a) Illustration of S, focused on S_j of p_j. (b) A cycle-tree graph and its embedding.

we can add a minimal set of dummy edges to make it connected), this is equivalent to assuming that a straight-line planar drawing Γ of $[G, \mathcal{H}]$ is given. We rename the faces of Γ as $F_1, \ldots, F_k$ in such a way that each graph G_h, which can also be assumed connected, lies inside face F_h. Note that, for each face F_h of G, the graph $[F_h, G_h]$ is again a 2-outerplanar graph; however, its outer level F_h is a simple chordless cycle and its inner level G_h consists of only one connected component. In the special case in which G_h is a tree we say that graph $[F_h, G_h]$ is a *cycle-tree* graph. We say that a 2-outerplanar graph is *inner-triangulated* if all the internal faces are 3-cycles. Note that not every cycle-tree graph can be augmented to be inner-triangulated without introducing multiple edges.

3   Inner-Triangulated 2-Outerplanar Graphs with Forest

In this section we prove that there exists a universal point set S of size $O(n^{3/2})$ for the class of n-vertex inner-triangulated 2-outerplanar graphs $[G, \mathcal{H}]$ where $\mathcal{H}$ is a forest.

3.1   Construction of the Universal Point Set

In the following we describe S (Fig. 1(a)). Let π be a half circle with center O and let $N := n + \sqrt{n}$. Uniformly distribute points in $S_{\mathcal{M}} = \{p_1, \ldots, p_N\}$ on π. The points in $S_{\mathcal{D}} = \{p_{i\sqrt{n}+i} : 1 \leq i \leq \sqrt{n}\}$ are called *dense*, while the remaining points in $S_{\mathcal{M}} \setminus S_{\mathcal{D}}$ are *sparse*[1]. For $j = 2, \ldots, N - 1$, place a circle π_j with its center p_j^C on $s(p_j O)$, so that it lies completely inside the triangle $\triangle p_{j-1} p_j p_{j+1}$ and inside the triangle $\triangle p_1 p_j p_N$. Note that the angles $\angle p_j p_j^C p_N$ and $\angle p_j p_j^C p_1$ are smaller than $180°$. Let p_j^N be the intersection point between $s(p_j O)$ and π_j that is closer to O. Also, let p_j^1 (resp. p_j^2) be the intersection point of $s(p_j^C p_{j+1})$ (resp. $s(p_j^C p_{j-1})$) with π_j. Finally, let p_j^3 (resp. p_j^4) be the intersection point of

---

[1] The distribution of the points into dense and sparse portions of the point set is inspired by [1].

π_j with its diameter orthogonal to $s(p_jO)$, such that $a(p_j^3p_j^4)$ does not contain p_j^N. Now, choose a point p_j^+ on the arc $a(p_j^1p_j^3)$, and a point p_j^- on the arc $a(p_j^4p_j^2)$. To complete the construction of S, evenly distribute $\bar{n} - 1$ points on each of the three segments $s_j^N := s(p_j^Cp_j^N)$, $s_j^+ := s(p_j^Cp_j^+)$, and $s_j^- := s(p_j^Cp_j^-)$, where $\bar{n} = n$ if p_j is dense and $\bar{n} = \sqrt{n}$ if it is sparse. We refer to the points on s^N, s^+, s^-, including the points $p_j^N, p_j^C, p_j^+, p_j^-$, as the point set of p_j, and we denote it by S_j. Vertex p_j^C is the center vertex of S_j. The described construction uses $O(n^{3/2})$ points and ensures the following property.

Property 1. For each $j = 1, \ldots, N$, the following visibility properties hold:

(A) The straight-line segments connecting point p_j to: point p_j^-, to the points on s_j^-, to p_j^C, to the points on s_j^+, and to p_j^+ appear in this clockwise order around p_j.

(B) For all $l < j$, consider any point $x_l \in \{p_l\} \cup S_l$ (see Fig. 1); then, the straight-line segments connecting x_l to: p_j^N, to the points on s_j^N, to p_j^C, to the points on s_j^-, to p_j^-, and to p_j appear in this clockwise order around x_l. Also, consider the line passing through x_l and any point in $\{p_j\} \cup S_j$; then, every point in $\{p_q\} \cup S_q$, with $l < q < j$, lies in the half-plane delimited by this line that does not contain the center O of π.

(C) For all $l > j$, consider any point $x_l \in \{p_l\} \cup S_l$; then, the straight-line segments connecting x_l to: p_j^N, to the points on s_j^N, to p_j^C, to the points on s_j^+, to p_j^+, and to p_j appear in this counterclockwise order around x_l. Also, consider the line passing through x_l and any point in $\{p_j\} \cup S_j$; then, every point in $\{p_q\} \cup S_q$, with $j < q < l$, lies in the half-plane delimited by this line that does not contain O.

3.2   Labeling the Graph

Let $[G, \mathcal{H}]$ be an inner-triangulated 2-outerplanar graph where G is an outerplanar graph and $\mathcal{H} = \{T_1, \ldots, T_k\}$ is a forest such that tree T_h lies inside face F_h of G, for each $1 \le h \le k$. The idea behind the labeling is the following: in our embedding strategy, G will be embedded on the half-circle π of the point set S, while the tree $T_h \in \mathcal{H}$ lying inside each face F_h of G will be embedded on the point sets S_j of some of the points p_j on which vertices of F_h are placed. Note that, since π is a half-circle, the drawing of F_h will always be a convex polygon in which two vertices have *small* (acute) internal angles, while all the other vertices have *large* (obtuse) internal angles. In particular, the vertices with the small angle are the first and the last vertices of F_h in the order in which they appear along the outer face of Γ. Since, by construction, a point p_j of F_h has its point set S_j in the interior of F_h if and only if it has a large angle, we aim at assigning each vertex of T_h to a vertex of F_h that is neither the first nor the last. We will describe this assignment by means of a labeling $\ell \colon [G, \mathcal{H}] \to 1, \ldots, |G|$; namely, we will assign a distinct label $\ell(v)$ to each vertex $v \in G$ and then assign to each vertex of T_h the same label as one of the vertices of F_h that is neither the first or the last. Then, the number of vertices with the same label as a vertex of G will determine whether this vertex will be placed on a sparse or a dense point. We formalize this idea in the following.

We rename the vertices of G as $v_1, \ldots, v_{|G|}$ in the order in which they appear along the outer face of Γ, and label them with $\ell(v_i) = i$ for $i = 1, \ldots, |G|$. Next, we label the vertices of each tree $T_h \in \mathcal{H}$. Since trees T_h and $T_{h'}$ are disjoint for $h \neq h'$, we focus on the cycle-tree graph $[F, T]$ composed of a single face $F = F_h$ of G and of the tree $T = T_h \in \mathcal{H}$ inside it. Rename the vertices of F as $w_1, \ldots, w_m$ in such a way that for any two vertices $w_x = v_p$ and $w_{x+1} = v_q$, where $p, q \in \{1, \ldots, |G|\}$, it holds that $p < q$. As a result, w_1 and w_m are the only vertices of F with small internal angles. A vertex of T is a *fork vertex* if it is adjacent to more than two vertices of F (square vertices in Fig. 1(b)), otherwise it is a *non-fork vertex* (cross vertices in Fig. 1(b)). Since $[F, T]$ is inner-triangulated, every vertex of T is adjacent to at least two vertices of F, and hence non-fork vertices are adjacent to exactly two vertices of F. We label the vertices of T starting from its fork vertices. To this end, we construct a tree T' composed only of the fork vertices, as follows. Initialize $T'=T$. Then, as long as there exists a non-fork vertex of degree 3 (namely, with 2 neighbors in F and 1 in T'), remove it and its incident edges from T'. The vertices removed in this step are called *foliage* (small crosses in Fig. 1(b)). All the remaining non-fork vertices have degree 4 (namely 2 in F and 2 in T'); for each of them, remove it and its incident edges from T' and add an edge between the two vertices of T' that were connected to it before its removal. The vertices removed in this step are *branch* vertices (large crosses in Fig. 1(b)). A vertex $w_x \in F$ is called *free* if so far no vertex of T' has label $\ell(w_x)$. To perform the labeling, we traverse T' bottom-up with respect to a root r that is the vertex of T' adjacent to both w_1 and w_m. Since $[F, T]$ is inner-triangulated, this vertex is unique. During the traversal of T', we maintain the invariant that vertices of T' are incident to only free vertices of F. Initially the invariant is satisfied since all the vertices of F are free. Let a be the fork vertex considered in a step of the traversal of T', and let $w_{a_1}, \ldots, w_{a_k}$ be the vertices of F adjacent to a, with $1 \leq a_1 < \cdots < a_k \leq m$ and $k \geq 3$. By the invariant, $w_{a_1}, \ldots, w_{a_k}$ are free. Choose any vertex w_{a_i} such that $2 \leq i \leq k-1$, and set $\ell(a) = \ell(w_{a_i})$. For example, the red fork vertex in Fig. 1(b) adjacent to w_3, w_4, and w_5 in F gets label $\ell(w_4)$. Since vertices $w_{a_2}, \ldots, w_{a_{k-1}}$ cannot be adjacent to any vertex of T' that is visited after a in the bottom-up traversal, the invariant is maintained at the end of each step. When finally $a=r$, then $w_{a_1} = w_1$ and $w_{a_k} = w_m$ are both free.

Now we label the non-fork vertices of T based on the labeling of T'. Let b be a non-fork vertex. If b is a branch vertex, then consider the first fork vertex a encountered on a path from b to a leaf of T; set $\ell(b) = \ell(a)$. Otherwise, b is a foliage vertex. In this case, consider the first fork vertex a' encountered on a path from b to the root r of T. Let $v, w \in F$ be the two vertices of F adjacent to b; assume $\ell(v) < \ell(w)$. If $\ell(a') \leq \ell(v)$, then set $\ell(b) = \ell(v)$; if $\ell(a') \geq \ell(w)$, then set $\ell(b) = \ell(w)$; and if $\ell(v) < \ell(a') < \ell(w)$, then set $\ell(b) = \ell(a')$ (the latter case only happens when a' is the root and b is adjacent to w_1 and w_m). Note that the described algorithm ensures that adjacent non-fork vertices have the same label. We perform the labeling procedure for every $T_h \in \mathcal{H}$ and obtain a labeling for $[G, \mathcal{H}]$. We say that the subgraph of $\mathcal{H}$ induced by all the vertices of $\mathcal{H}$ with label i is the *restricted subgraph* H_i of $\mathcal{H}$ for all $i = 1, \ldots, |G|$ (see Fig. 1(b)).

Lemma 1. *Each restricted subgraph H_i of $\mathcal{H}$, $1 \leq i \leq |G|$, is a tree all of whose vertices have degree at most 2, except for one vertex that may have degree 3.*

Proof Sketch. First, H_i has at most one fork vertex a, which is hence the only one with degree larger than 2. Further, a is incident to at most one path (to no path, if $a = r$) of branch vertices, namely the one connecting it to its parent fork vertex. Finally, a is incident to at most two (if $a \neq r$) or at most three (if $a = r$) paths of foliage vertices, namely the ones whose vertices are incident to the vertex $w \in F$ such that $\ell(w) = i$.   □

3.3   Embedding on the Point Set

We describe an embedding algorithm consisting of three steps (see Fig. 1(b)).

Step a: Let $\omega : G \to \mathbb{N}$ be a weight function with $\omega(v_i) = |\{v \in [G, \mathcal{H}] \mid \ell(v) = i\}|$ for every $v_i \in G$. Note that $\sum_{v_i \in G} \omega(v_i) = n$. We categorize each vertex $v_i \in G$ as *sparse* if $1 \leq \omega(v_i) \leq \sqrt{n}$, and *dense* if $\omega(v_i) > \sqrt{n}$. There are at most $\sqrt{n}$ dense vertices.

Step b: We draw the vertices $v_1, \ldots, v_{|G|}$ of G on the $N := n + \sqrt{n}$ points of π in the same order as they appear along the outer face of Γ, in such a way that dense (resp. sparse) vertices are placed on dense (resp. sparse) points. The resulting embedding $\widetilde{\Gamma}$ of G is planar since Γ is planar. The construction of $\widetilde{\Gamma}$ implies the following.

Property 2. Let $Q = \{p_{j_1}, \ldots, p_{j_m}\} \subseteq \pi$, $j_i < j_{i+1}$, be the polygon representing a face of G. Polygon Q contains in its interior all the point sets $S_{j_2}, \ldots, S_{j_{m-1}}$.

Step c: Finally, we consider forest $\mathcal{H} = \{T_1, \ldots, T_k\}$. We describe the embedding algorithm for a single cycle-tree graph $[F, T]$, where $F = w_1, \ldots, w_m$ is a face of G and $T \in \mathcal{H}$ is the tree lying inside F. We show how to embed the restricted subgraph H_i, for each vertex w_x of F with label $\ell(w_x) = i$, on the point set S_j of the point p_j where w_x is placed. We remark that the labeling procedure ensures that $|H_i| + 1 = \omega(w_x) \leq |S_j|$; also, by Property 2, point set S_j lies inside the polygon representing F, except for the two points where vertices w_1 and w_m have been placed.

  By Lemma 1, H_i has at most one (fork-)vertex a of degree 3, while all other vertices have smaller degree. We place a, if any, on the center point p_j^C of p_j. The at most three paths of non-fork vertices are placed on segments s_j^+, s_j^-, s_j^N starting from p_j^C; namely, the unique path of branch vertices is placed on s_j^N, while the two paths of foliage vertices are placed on s_j^+ or s_j^- based on whether

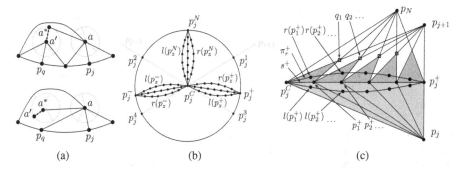

Fig. 2. (a)(top) P contains $a' \neq a$, (a)(bottom) a' is a leaf of T. (b)–(c) Dark-gray triangles are used for construction of petal points $r(p_z^+)$ while light-gray triangles for $l(p_z^+)$.

the vertex of G different from w_x they are incident to is w_{x+1} or w_{x-1}, respectively. If $a = r$, then the path of foliage vertices incident to w_1 and w_m is placed on s_j^N.

We show that this results in a planar drawing of T. First, for every two fork vertices $a \in H_p$ and $a' \in H_q$, with $p < q$, all the leaves of the subtree of T rooted at a have smaller label than all the leaves of the subtree of T rooted at a'. Then, for each $w_x \in F$, with $\ell(w_x) = i$, consider the fork vertex $a \in H_i$, which lies on p_j^C. Let P be any path connecting a to a leaf of T and let a^* be the neighbor of a in P. If P contains a fork vertex other than a (Fig. 2(a)), then let a' be the fork vertex in P that is closest to a (possibly $a'=a^*$) and let p_q^C be the point where a' has been placed. Assume $q < j$, the case $q > j$ is analogous. By definition, the non-fork vertices in the path from a to a' (if any) are branch vertices, and hence lie on s_q^N. Then, Property 1 ensures that the straight-line edge (a, a^*) separates all the point sets S_p with $q < p < j$ from the center of π. Since the vertices on S_p are only connected either to each other or to the vertices on s_j^- and s_q^+, edge (a, a^*) is not involved in any crossing. If P does not contain any fork vertex other than a (Fig. 2(a)), then all the vertices of P other than a are foliage vertices and are placed on a segment s_q^+ or s_q^-, for some q. In particular, if $q < j$, then they are on s_q^-; if $q > j$, then they are on s_q^+; while if $q = j$, then they are either on s_q^+ or on s_q^-. In all the cases, Property 1 ensures that edge (a, a^*) does not cross any edge.

Finally, observe that any path of T containing only non-fork vertices is placed on the same segment of the point set, and hence its edges do not cross. As for the edges connecting vertices in one of these paths to the two leaves of T they are connected to, note that by item (A) of Property 1 the edges between each of these leaves and these vertices appear in the rotation at the leaf in the same order as they appear in the path.

Lemma 2. *There exists a universal point set of size $O(n^{3/2})$ for the class of n-vertex inner-triangulated 2-outerplanar graphs $[G, \mathcal{H}]$ where $\mathcal{H}$ is a forest.*

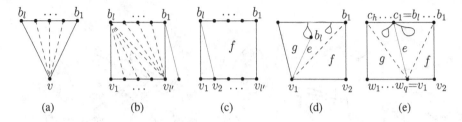

Fig. 3. (a)–(c): Insertion of triangulation edges in (a) a petal face, (b) a non-protected big face, and (c) a big face protected by vertex b_1. (d)–(e) Illustration of the two cases for removing bad faces. Face g is petal in (d) and big in (e). Dummy edges are dashed, the removed edge e is red (Color figure online).

4 2-Outerplanar Graphs with Forest

In this section we consider 2-outerplanar graphs $[G, \mathcal{H}]$ where $\mathcal{H}$ is a forest. Contrary to the previous section, we do not assume $[G, \mathcal{H}]$ to be inner-triangulated. As observed before, augmenting it might be not possible without introducing multiple edges. The main idea to overcome this problem is to first identify the parts of $[G, \mathcal{H}]$ not allowing for the augmentation, remove them, and augment the resulting graph with dummy edges to inner-triangulated (Sect. 4.2); then, apply Lemma 2 to embed the inner-triangulated graph on the point set S; and finally remove the dummy edges and embed the parts of the graph that had been previously removed on the remaining points (Sect. 4.3). To do so, we first need to extend the point set S with some additional points.

4.1 Extending the Universal Point Set

We construct a point set S^* with $O(n^{3/2})$ points from S by adding *petal points* to segments s_j^+, s_j^N, s_j^- of the point sets S_j, for every $j=2,\ldots,N-1$. For simplicity of notation, we skip the subscript j whenever possible. We denote by p_z^σ the z-th point on segment s^σ, with $\sigma \in \{+, -, N\}$ and $z=1,\ldots,\overline{n}$ (where $\overline{n}=\sqrt{n}$ or $\overline{n}=n$, depending on whether p_j is sparse or dense), so that p_1^σ is the point following p^C along s^σ and $p_{\overline{n}}^\sigma = p_j^\sigma$. For each point p_z^σ we add two *petal points* $l(p_z^\sigma)$ and $r(p_z^\sigma)$ to S^*.

We first describe the procedure for s^+, see Fig. 2(c). For each $z=1,\ldots,\overline{n}$, consider the intersection point q_z between segments $s(p_{z-1}^+p_{j+1})$ and $s(p_z^+p_N)$, where $p_{z-1}^+ = p_j^C$ when $z = 1$. By construction, all triangles $\triangle p_{z-1}^+ p_z^+ q_z$ have two corners on s^+, have the other corner in the same half-plane delimited by the line through s^+, and do not intersect each other except at common corners. Hence, there exists a convex arc π_r^+ passing through p_j^C and $p_{\overline{n}}^+ = p_j^+$, and intersecting the interior of every triangle. For each $z = 1,\ldots,\overline{n}$, we place the petal point $r(p_z^+)$ on the arc of π_r^+ lying inside triangle $\triangle p_{z-1}^+ p_z^+ q_z$. For the other petal point $l(p_z^+)$ we use the same procedure by considering triangles $\triangle p_{z-1}^+ p_z^+ p_j$ instead of $\triangle p_{z-1}^+ p_z^+ q_z$. Symmetrically we place the petal points for s^-, using points p_{j-1}

and p_1 to place $l(p_z^-)$ and point p_j to place $r(p_z^-)$, and for s^N, using points p_{j-1} and p_1 to place $l(p_z^N)$ and points p_{j+1} and p_N to place $r(p_z^N)$.

4.2   Modifying and Labeling the Graph

We now aim at modifying $[G, \mathcal{H}]$ to obtain an inner-triangulated graph that can be embedded on the original point set S (**Part A** and **Part B**); in Sect. 4.3 we describe how to exploit this embedding on S to obtain an embedding of the original graph $[G, \mathcal{H}]$ on the extended point set S^* (**Part C**). We describe the procedure just for a cycle-tree graph $[F, T]$ composed of a face F of G and of the tree T inside it.

Part A: We categorize each face f of $[F, T]$ based on the number of vertices of F and of T that are incident to it. Since T is a tree, f has at least a vertex of F and a vertex of T incident to it. If f contains exactly one vertex of F, then it is a *petal face*. If f contains exactly one vertex of T, then it is a *small face*. Otherwise, it is a *big face*. Let $b_1, \ldots, b_l$ be the occurrences of the vertices of T in a clockwise order walk along the boundary of a big face f. If either b_1 or b_l, say b_1, has more than one adjacent vertex in F (namely one in f and at least one not in f), then f is *protected* by b_1. If f is a big face with exactly two vertices incident to F and is not protected, then f is a *bad face*.

The next lemma gives sufficient conditions to triangulate G without introducing multiple edges; we will later use this lemma to identify the "tree components" of T whose removal allows for a triangulation.

Lemma 3. *Let $[F, T]$ be a biconnected simple cycle-tree graph, such that (1) each vertex of F has degree at most four, and (2) there exists no bad face in $[F, T]$. It is possible to augment $[F, T]$ to an inner-triangulated simple cycle-tree graph.*

Proof Sketch. Each petal (small, respectively) face f can be triangulated by adding vertices between the only vertex of F (of T) incident to f and all the other vertices of f. Multiple edges are not created since $[F, T]$ is biconnected and there exists no two petal faces incident to the same vertex v of F, as v has degree at most 4; see Fig. 3(a).

Consider a big face f, with vertex occurrences $v_1, \ldots, v_{l'}, b_1, \ldots, b_l$ (with l, $l' > 1$), where $v_1, \ldots, v_{l'} \in F$ and $b_1, \ldots, b_l \in T$. If f is protected by a vertex, say b_1, then it is triangulated by adding an edge between b_l and every vertex of F, and an edge between $v_{l'}$ and every vertex of T; see Fig. 3(b). The absence of multiple edges is due to the edge connecting b_1 to a vertex of F not incident to f, which implies that $v_{l'}$ is not connected to any vertex of T incident to f other than b_1. Finally, if f is not protected by any vertex, we make it protected by adding an edge (b_l, v_2) and apply the previous case; see Fig. 3(c). Since f is not a bad face, we have $l' > 2$, and hence v_2 is not connected to any vertex of T, which implies that (b_l, v_2) is not a multiple edge. $\square$

We now describe a procedure to transform cycle-tree graph $[F, T]$ into another one $[F, T'']$ that is biconnected and satisfies the conditions of Lemma 3. We do this in two steps: first, we remove some edges connecting a vertex of F and a vertex of T to transform $[F, T]$ into a cycle-tree graph $[F, T'=T]$ that is not biconnected but that satisfies the two conditions; then, we remove the "tree components" of T' that are not connected to vertices of F in order to obtain a cycle-tree graph $[F, T'' \subseteq T']$ that is also biconnected.

To satisfy condition (1) of Lemma 3, we merge all the petal faces incident to the same vertex of F into a single one by repeatedly removing an edge shared by two adjacent petal faces. We refer to these removed edges as *petal edges*, denoted by E_P.

To satisfy condition (2) of Lemma 3, we consider each bad face $f = v_1, v_2, b_1, \ldots, b_l$, where $v_1, v_2 \in F$ and $b_1, \ldots, b_l \in T$. Let g be the face incident to v_1 sharing edge $e = (v_1, b_l)$ with f. We remove e, hence merging f and g into a single face f', that we split again by adding dummy edges, based on the type of face g, in such a way that no new bad face is created. Since f is a bad face, it is not protected by b_l, and hence g is not a small face. If g is a petal face, then f' is still a big face with two vertices of F incident to it, namely v_1 and v_2; see Fig. 3(d). We add edge (v_1, b_1), splitting f' into a petal face $v_1, b_1, \ldots, b_l$ and a triangular face v_1, v_2, b_1. If g is a big face, then f' is a big face; see Fig. 3(e). Let $g = w_1, \ldots, w_q, c_1, \ldots, c_h$, where $w_1, \ldots, w_q \in F$, with $w_q = v_1$, and $c_1, \ldots, c_h \in T$, with $c_1 = b_l$. We add two dummy edges (v_1, c_h) and (v_1, b_1), splitting f' into a small face $w_1, \ldots, w_q, c_h$, a petal face $v_1, b_1, \ldots, b_l = c_1, \ldots, c_h$, and a triangular face v_1, v_2, b_1. The edges removed in this step are *big face edges*, denoted by E_B, and the added edges are *triangulation edges*.

In order to make $[F, T']$ biconnected, note that $[F, T']$ consists of a biconnected component which contains F, called *block-component*, and a set $\mathcal{T}_B$ of subtrees of T', called *tree components*, each sharing a cut-vertex with the block component. We remove the tree components $\mathcal{T}_B$ from $[F, T']$ and obtain an instance $[F, T'' \subseteq T']$, that is actually the block component of $[F, T']$. Since the removal of $\mathcal{T}_B$ does not change the degree of the vertices of F and does not create any bad face, $[F, T'']$ is indeed a biconnected instance satisfying the two conditions of Lemma 3. Thus, by adding further *triangulation edges* we augment it to an inner-triangulated instance $[F, T^\Delta = T'']$.

Lemma 4. *Let $e = (b, v)$ be an edge of $E_P \cup E_B$, where $b \in T$ and $v \in F$. Then, either e is a triangulation edge in $[F, T^\Delta]$ or b belongs to a tree component T_c of $\mathcal{T}_B$ sharing a cut-vertex c with $[F, T'']$. In the latter case, (v, c) is a triangulation edge in $[F, T^\Delta]$.*

Lemma 5. *Let $T_c \in \mathcal{T}_B$ be a tree component such that there exists at least an edge $(b, v) \in E_P \cup E_B$, with $b \in T_c$ and $v \in F$. Then, for each edge in $E_P \cup E_B$ with an endvertex belonging to T_c, the other endvertex is v.*

Performing the above operations for every cycle-tree graph $[F, T]$ yields an inner-triangulated 2-outerplanar graph $[G, \mathcal{H}^\Delta]$ as outcome of **Part A**. We then label $[G, \mathcal{H}^\Delta]$ with the algorithm from Sect. 3.2 and describe next how to extend this labeling to $\mathcal{T}_B$.

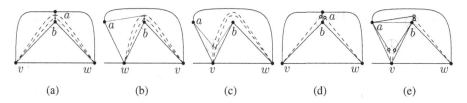

Fig. 4. (a)–(c) Inserting dummy vertices for a tree-component in a face (a, b, v) with $v \in F$ and $a, b \in T^\triangle$, when (a) $\ell(a) = \ell(b)$, (b) $\ell(a) \neq \ell(b)$ and $\ell(w) < \ell(v)$, and (c) $\ell(a) \neq \ell(b)$ and $\ell(w) > \ell(v)$. (d)–(e) Moving dummy vertices to petal points if $\ell(a) = \ell(b)$ and if $\ell(a) \neq \ell(b)$ (Color figure online).

Part B: We consider the tree components $T_c \in \mathcal{T}_B$ for each face F of G; let $[F, T^\triangle]$ be the corresponding inner-triangulated cycle-tree graph. We label the vertices of T_c and simultaneously augment $[F, T^\triangle]$ with dummy vertices and edges, so that $[F, T^\triangle]$ remains inner-triangulated (and hence can be embedded, by Lemma 2) and the vertices of T_c can be later placed on the petal points of the points where dummy vertices are placed. The face of $[F, T'']$ to which T_c belongs might have been split into several faces of $[F, T^\triangle]$ by triangulation edges. We assign T_c to any of such faces f that is incident to the root c of T_c. Then, we label T_c based on the type of f; we distinguish two cases.

Suppose f is a triangular face (c, v, w) with $v, w \in F$ and $c \in T^\triangle$; assume $\ell(v) < \ell(w)$. We create a path P_c containing $|T_c| - 1$ dummy vertices and append P_c at c. Then, we connect every dummy vertex of P_c with both v and w. If $\ell(c) \leq \ell(v)$, then we label the vertices of P_c with $\ell(P_c) = \ell(v)$. If $\ell(c) \geq \ell(w)$, then we label $\ell(P_c) = \ell(w)$.

Suppose f is a triangular face (a, b, v) with $v \in F$ and $a, b \in T^\triangle$, refer to Fig. 4; assume $\ell(a) \leq \ell(b)$. Replace edge (a, b) with a path P_c between a and b with $|T_c| - 1$ internal dummy vertices, and connect each of them to v and to w, where w is the other vertex of F adjacent to both a and b. For each dummy vertex x of P_c, we assign $\ell(x) = \ell(a)$ if $\ell(v) \leq \ell(a)$; we assign $\ell(x) = \ell(b)$ if $\ell(v) \geq \ell(b)$; and we assign $\ell(x) = \ell(v)$ if $\ell(a) < \ell(v) < \ell(b)$. The existence of edge $(a, b) \in T^\triangle$ implies that either a is the parent of b in $T^\triangle$ or vice versa. Suppose the former, the other case is analogous. Then, v and w are the extremal neighbors of b in F, and thus either $\ell(v) \leq \ell(b) \leq \ell(w)$ or $\ell(w) \leq \ell(b) \leq \ell(v)$. Also, if $\ell(a) \neq \ell(b)$, then $\ell(a)$ does not lie strictly between $\ell(v)$ and $\ell(w)$. In fact, this can only happen if $\ell(b)$ strictly lies between $\ell(v)$ and $\ell(w)$, and $\ell(a) = \ell(b)$ (which happens only if a is a non-fork vertex). Since $\ell(a) \leq \ell(b)$, by assumption, this implies that $\ell(a) \leq \ell(v), \ell(w)$. The two observations before can be combined to conclude that, if $\ell(a) = \ell(b)$, then all the tree components lying inside faces (a, b, v) and (a, b, w) have the same label as a and b (Fig. 4(a)). Otherwise, either the tree components inside (a, b, v) have label $\ell(b)$ and those inside (a, b, w) have label $\ell(w)$ (Fig. 4(b)), or the tree components inside (a, b, v) have label $\ell(v)$ and those inside (a, b, w) have label $\ell(b)$ (Fig. 4(c)). All added edges are again *triangulation edges*.

We apply **Part B** to every cycle-tree graph of $[G, \mathcal{H}^\Delta]$, hence creating an inner-triangulated 2-outerplanar graph $[G, \mathcal{H}^A]$ where $\mathcal{H}^A$ is a forest. Since all the dummy vertices of P_c are connected to two vertices $v, w \in F$, they become non-fork vertices. Note that the labeling of the dummy vertices coincides with the one obtained by the algorithm in Sect. 3.2, except for the case when f is a triangular face (a, b, v) with $v \in F$ and $a, b \in T^\Delta$, and $\ell(a) < \ell(v) < \ell(b)$. In this case the algorithm would have labeled either $\ell(P_c) = \ell(a)$ or $\ell(P_c) = \ell(b)$, depending on whether b is the parent of a or vice versa. However, since $\ell(a) < \ell(v) < \ell(b)$ holds in $[F, T^\Delta]$, and since (a, b, v) is a triangular face of $[F, T^\Delta]$, no vertex of $[F, T^\Delta]$ different from v has the same label as v. Hence, graph H_i, for each i, is a tree with at most one vertex of degree 3. We thus apply Lemma 2 to obtain a planar embedding Γ^A of $[G, \mathcal{H}^A]$ on S.

4.3   Transformation of the Embedding

We remove the all the triangulation edges added in the construction, and then restore each tree component T_c, which is represented by path P_c. Since the vertices of P_c are non-fork vertices and have the same label i, by construction, they are placed on the same segment $s \in \{s^+, s^N, s^-\}$ of S_j, where p_j is the point vertex v_i is placed on.

We remove all the internal edges of P_c and move each vertex x of P_c from the point p of s it lies on to one of the corresponding petal points, either $l(p)$ or $r(p)$, as follows. Let v be a vertex of G connected to a vertex of T_c by an edge in $E_P \cup E_B$, if any; recall that, by Lemma 5, all the edges of $E_P \cup E_B$ connecting T_c to G are incident to v. If $\ell(x) < \ell(v)$, then move x to $r(p)$; tree components connected to w in Fig. 4(d) and (e). If $\ell(x) > \ell(v)$, then move x to $l(p)$; tree component connected to v in Fig. 4(e). Otherwise, $\ell(x) = \ell(v)$; in this case $s \neq s^N$, by construction, and hence we have to distinguish the following two cases: If $s = s^+$, then move x to $l(p)$, otherwise move x to $r(p)$ (tree components attached to a and b, respectively, and connected to v in Fig. 4(e)). If no vertex $v \in G$ is connected to T_c, then move x to $r(p)$ if $\ell(c) < \ell(x)$ (tree component attached to a in Fig. 4(e)), and to $l(p)$ otherwise.

We prove that this operations maintain planarity. The internal edges of T_c do not cross since the petal points, together with the point where c lies, form a convex point set, on which it is possible to construct a planar embedding of every tree [4]. As for the edges connecting vertices of T_c to v, by Lemma 4, v has visibility to the root c of T_c, since (v, c) is a triangulation edge; by Property 1, this visibility from v extends to all the segment s where P_c had been placed on; and by the construction of S^*, to all the corresponding petal points. The proof for the edges (a, b) that had been subdivided when merging tree component T_c (green edges in Fig. 4(d) and (e)) is in [2].

Claim 1. *Reinserting every edge (a, b) such that there existed a path P_c between a and b does not introduce any crossing.*

To complete the transformation it remains to insert the edges of $E_P \cup E_B$ which were not inserted in the previous step. Since by Lemma 4 all of these edges were also triangulation edges, their insertion does not produce any crossing.

Lemma 6. *There exists a universal point set of size $O(n^{3/2})$ for the class of n-vertex 2-outerplanar graphs $[G, \mathcal{H}]$ where $\mathcal{H}$ is a forest.*

5 General 2-Outerplanar Graphs

In this section we give a high-level idea of how to extend the result of Lemma 6 to any arbitrary 2-outerplanar graph $[G, \mathcal{H}]$. The complete description can be found in [2].

The idea is to convert every graph $G_h \in \mathcal{H}$ into a tree T_h; embed the resulting graph on S^*; and finally revert the conversion from each T_h to G_h. Each tree T_h is created by substituting each biconnected block B of G_h by a star, centered at a dummy vertex and with a leaf for each vertex of B, where leaves shared by more stars are identified. This results in a 2-outerplanar graph whose inner level is a forest.

The embedding of this graph on S^* is performed similarly as in Lemma 6, with some slight modifications to the labeling algorithm, especially for the vertices of T_h corresponding to cut-vertices of G_h, and to the procedure for merging the tree components. These modifications allow us to ensure that the vertices of each block of G_h lie on a convex portion of S^*, where they can thus be drawn without crossings [5, 10].

We finally reduce the size of S^* to $O(n \log n)$ by using the super-pattern sequence ξ from [3], which is a sequence of integers ξ_j, with $\sum_{j=1,\ldots,n} \xi_j = O(n \log n)$. Sequence ξ majorizes every sequence of integers that sum up to n. We hence assign the size of each point set S_j based on this sequence, instead of using dense or sparse point sets.

Theorem 1. *There exists a universal point set of size $O(n \log n)$ for the class of n-vertex 2-outerplanar graphs.*

6 Conclusions

We provided a universal point set of size $O(n \log n)$ for 2-outerplanar graphs. A natural question is whether our techniques can be extended to other meaningful classes of planar graphs, such as 3-outerplanar graphs. We also find interesting the question about the required area of universal point sets. In fact, while the integer grid is a universal point set for planar graphs with $O(n^2)$ points and $O(n^2)$ area, all known point sets of smaller size, even for subclasses of planar graphs, require a larger area. We thus ask whether universal point sets of subquadratic size require polynomial or exponential area.

References

1. Angelini, P., Di Battista, G., Kaufmann, M., Mchedlidze, T., Roselli, V., Squarcella, C.: Small point sets for simply-nested planar graphs. In: Speckmann, B. (ed.) GD 2011. LNCS, vol. 7034, pp. 75–85. Springer, Heidelberg (2011)

2. Angelini, P., Bruckdorfer, T., Kaufmann, M., Mchedlidze, T.: A universal point set for 2-outerplanar graphs (2015). CoRR abs/1508.05784
3. Bannister, M.J., Cheng, Z., Devanny, W.E., Eppstein, D.: Superpatterns and universal point sets. J. Graph Algorithms Appl. **18**(2), 177–209 (2014)
4. Binucci, C., Di Giacomo, E., Didimo, W., Estrella-Balderrama, A., Frati, F., Kobourov, S., Liotta, G.: Upward straight-line embeddings of directed graphs into point sets. CGTA **43**, 219–232 (2010)
5. Bose, P.: On embedding an outer-planar graph in a point set. CGTA **23**(3), 303–312 (2002)
6. Cabello, S.: Planar embeddability of the vertices of a graph using a fixed point set is NP-hard. J. Graph Algorithms Appl. **10**(2), 353–366 (2006)
7. de Fraysseix, H., Pach, J., Pollack, R.: Small sets supporting fáry embeddings of planar graphs. In: Simon, J. (ed.) STOC '88, pp. 426–433. ACM (1988)
8. Fraysseix, H., Pach, J., Pollack, R.: How to draw a planar graph on a grid. Combinatorica **10**, 41–51 (1990)
9. Fulek, R., Tóth, C.D.: Universal point sets for planar three-trees. J. Discrete Algorithms **30**, 101–112 (2015)
10. Gritzmann, P., Pach, B.M.J., Pollack, R.: Embedding a planar triangulation with vertices at specified positions. Am. Math. Monthly **98**, 165–166 (1991)
11. Kurowski, M.: A 1.235 lower bound on the number of points needed to draw all n-vertex planar graphs. Inf. Process. Lett. **92**(2), 95–98 (2004)
12. Schnyder, W.: Embedding planar graphs on the grid. In: Johnson, D.S. (ed.) SODA '90, pp. 138–148. SIAM (1990)

Linear-Size Universal Point Sets for One-Bend Drawings

Maarten Löffler[1] and Csaba D. Tóth[2]($^{(\boxtimes)}$)

[1] Department of Information and Computing Sciences, Utrecht University,
Utrecht, The Netherlands
m.loffler@uu.nl
[2] Department of Mathematics, California State University Northridge,
Los Angeles, CA, USA
csaba.toth@csun.edu

Abstract. For every integer $n \geq 4$, we construct a planar point set S_n of size $6n - 10$ such that every n-vertex planar graph G admits a plane embedding in which the vertices are mapped to points in S_n, and every edge is either a line segment or a polyline with one bend, where the bend point is also in S_n.

1 Introduction

An *embedding* of a graph $G = (V, E)$ into the Euclidean plane $\mathbb{R}^2$ maps the vertices in V into distinct points in $\mathbb{R}^2$, and the edges in E into interior-disjoint arcs between corresponding vertices. In a *straight-line embedding*, the edges in E are mapped to line segments, and so the embedding of the edges is determined by that of the vertices. It is well known [9] that every planar graph admits a plane straight-line embedding. Finding a compact representation for the straight-line embeddings of planar graphs is a central question in graph drawing.

A point set $P \in \mathbb{R}^2$ is called *n-universal* if every planar graph $G = (V, E)$ with n vertices admits a plane straight-line embedding such that all vertices are mapped to points in P. The quest for finding small n-universal point sets started in the 1990s. De Fraysseix et al. [4] and Schnyder [15] independently showed that there are n-universal sets of size $O(n^2)$. In fact, an $(n-1) \times (n-1)$ section of the integer lattice is n-universal [3,15] for every $n \geq 3$. The current best lower and upper bounds on the minimum size of an n-universal point set are $(1.235 - o(1))n$ by Kurowski [14] and $n^2/4 - \Theta(n)$ by Bannister et al. [1].

In a *polyline* embedding of a graph $G = (V, E)$, the edges are represented by pairwise noncrossing polygonal paths. Everett et al. [8] showed that there is a set S_n of n points in the plane, for every $n \in \mathbb{N}$, such that every n-vertex planar graph has a polyline embedding in which all vertices are mapped into S_n and every edge is a polyline with at most one bend. However, as noted by Dujmović et al. [7], the bend points require a set of size $\Theta(n^3)$ when implementing the embedding method in [8]. By refining this method, Dujmović et al. [7] constructed a point set S_n' of size $O(n^2/\log n)$ for all $n \in \mathbb{N}$ such that every n-vertex planar graph

© Springer International Publishing Switzerland 2015
E. Di Giacomo and A. Lubiw (Eds.): GD 2015, LNCS 9411, pp. 423–429, 2015.
DOI: 10.1007/978-3-319-27261-0_35

has a polyline embedding with at most one bend per edge in which all vertices *and* all bend points along the edges are mapped to S'_n. They also show [7] that if two (resp., three) bends per edge are allowed, a point set of size $O(n \log n)$ (resp., $O(n)$) can accommodate all vertices and bend points in a polyline embedding of every planar graph with n vertices. The main result of this paper is the following.

Theorem 1. *For every integer $n \geq 4$, there exist a set S_n of $6n-10$ points in the plane such that every planar graph with n vertices admits a polyline embedding with one bend per edge such that all vertices and bend points are mapped into S_n.*

Both Everett et al. [8] and Dujmović et al. [7] rely on a *topological book embedding on 2 pages*. Di Giacomo et al. [12,13] proved that every planar graph admits an embedding in the plane such that the vertices are distinct points on the x-axis, and every edge either (i) lies in a closed halfplane above or below the x-axis, or (ii) crosses the x-axis precisely once such that the crossing is on the straight-line segment between the two endpoints so that the half-edge between the left endpoint and the crossing lies below the x-axis and the other half-edge lies above the x-axis; see Fig. 1(left). The edges that lie in a closed halfplane above or below the x-axis are called *arcs*, and the edges that cross the x-axis *biarcs*. Cardinal et al. [2] recently showed that at most $n - 4$ biarcs suffice.

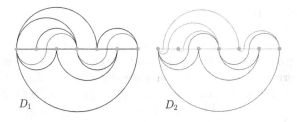

Fig. 1. A biarc diagram D_1. Each arc above the x-axis is deformed into a biarc in D_2.

Proof Technique. If the x-axis is replaced by a convex arc γ, then every edge and half-edge on the convex side of γ can be embedded as a straight-line segment, and every edge and half-edge on the concave side can be embedded as a polyline with one bend. As a first approach, an edge that crossed the x-axis precisely once requires two bend points: one on the concave side of γ and one on the curve γ. Everett et al. [8] carefully arranged n points on a convex curve γ such that the bend points on γ could be eliminated for the embedding of any n-vertex planar graph. Dujmović et al. [7] refined the construction to ensure that the bend points can be chosen from a set of $\Theta(n^2/\log n)$ points on the concave side of γ. We build on the same ideas, but use only $\Theta(n)$ carefully arranged bend points on the concave side of γ.

Fractional Variants. We introduce a new concept here. We say that a point set P is (n, ϱ)-*bend universal*, for $n \in \mathbb{N}$ and $0 \leq \varrho \leq 1$, if every planar graph $G = (V, E)$ with n vertices admits a polyline embedding with one bend per edge such that the vertices are mapped into P and at least $\varrho|E|$ edges are mapped to straight-line segments. There are two variants: the bend points are either required to be in P as well, or they can be chosen freely. In the first case, the problem is equivalent to subdividing at most $(1-\varrho)|E|$ edges to obtain a straight-line embedding. Our proof technique of Theorem 1 can also ensure that at least $\varrho = \frac{1}{3}$ fraction of the edges are mapped to straight-line segments, and we obtain the following.

Theorem 2. *For every* $n \geq 4$, *there exists an* $(n, \frac{1}{3})$-*bend universal set of* $10n-8$ *points in the plane.*

Related Previous Work. The quadratic upper bound for universal point sets is the best possible if the point set is restricted to sections of the integer lattice: Frati and Patrignani [11] showed (based on earlier work by Dolev et al. [6]) that if a rectangular section of the integer lattice is n-universal, then it must contain at least $n^2/9 + \Omega(n)$ points.

Grid drawings have been studied intensively due to their versatile applications. It is known that sections of the integer lattice with $o(n^2)$ points are n-universal for certain classes of graphs. For example, Di Battista and Frati [5] proved that an $O(n^{1.48})$ size integer grid is n-universal for *outerplanar* graphs. Frati [10] showed that 2-trees on n vertices require a grid of size at least $\Omega(n2^{\sqrt{\log n}})$.

2 Construction of a Point Set

For every $k \in \mathbb{N}$, we define a point set T_k together with a partition $T_k = A_k \cup B_k$, where $|A_k| = |B_k| = k$. We label the points by $A_k = \{a_1, \ldots, a_k\}$ and $B_k = \{b_1, \ldots, b_k\}$. Their coordinates are defined by:

$$x(a_i) = -x(b_i) = (1 + \sqrt{2})^{k-i}$$
$$y(a_i) = y(b_i) = i$$

Observe that the points a_i, a_{i+1}, and b_{i+2} (and symmetrically b_i, b_{i+1}, and a_{i+2}) are equidistant and collinear. Figure 2 illustrates the construction.

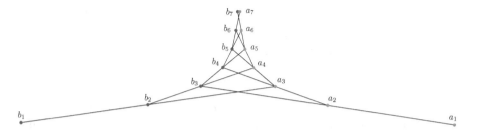

Fig. 2. Illustration of T_7. The figure is scaled horizontally to better fit on this page.

From this construction, we perturb T_k to a new point set $\tilde{T}_k$ to satisfy the following essential property:

Property 1. For every three indices $h < i < j$, the segment $a_h b_j$ passes to the left of a_i, and the segment $b_h a_j$ passes to the right of b_i.

Clearly, this property may be satisfied by an arbitrarily small perturbation of T_k.

Lemma 1. *The polygonal chain* $\alpha = (a_1, \ldots, a_k)$ *is on the convex hull of* A_k *in counter-clockwise order.*

Proof. Property 1 ensures that

$$0 > \text{slope}(a_1 a_2) > \text{slope}(a_2 a_3) > \ldots > \text{slope}(a_{k-1} a_k),$$

consequently $\alpha = (a_1, \ldots, a_k)$ is on the convex hull of A_k in clockwise order. $\square$

Lemma 2. *Consider six indices* h, i, j, h', i', *and* j', *where* $1 \le h < i \le j \le k$, $1 \le h' < i' \le j' \le k$, $i \notin \{h', i', j'\}$, *and* $i' \notin \{h, i, j\}$. *If the segments* $a_h a_i$ *and* $a_{h'} a_{i'}$ *do not cross and the segments* $a_i a_j$ *and* $a_{i'} a_{j'}$ *do not cross, then the polygonal paths* (a_h, b_i, a_j) *and* $(a_{h'}, b_{i'}, a_{j'})$ *do not cross.*

Proof. Since the polygonal chain $\alpha = (a_1, \ldots, a_k)$ is in convex position by Lemma 1, the noncrossing conditions exclude the "interleaving" orders $h < h' < i < i'$, $h' < h < i' < i$, $i < i' < j < j'$, and $i' < i < j' < j$. By Property 1, the segment $a_h b_i$ passes left of $a_{h+1}, \ldots, a_{i-1}$ for all $0 \le h < i$. Since the x-coordinates of $a_i, \ldots, a_k$ are larger than that of b_i, the segment $b_i a_j$ passes left of $a_i, \ldots, a_{j-1}$ for all $i < j \le k$.

By construction, the polygonal paths (a_h, b_i, a_j) and (a'_h, b'_i, a'_j) are y-monotone. Consequently, if the open intervals (h, j) and (h', j') are disjoint, then the two paths cannot cross. Assume now that (h, j) and (h', j') intersect. We distinguish two cases:

Case 1: intervals (h, j) and (h', j') are nested. Without loss of generality, assume $(h', j') \subset (h, j)$. The noncrossing conditions imply that we have either $(h', j') \subset (h, i)$ or $(h', j') \subset (i, j)$. If $(h', j') \subset (h, i)$, then the entire path $(a_{h'}, b_{i'}, a_{j'})$ lies left of segment $a_h a_{i-1}$, hence left of $a_h b_i$. If $(h', j') \subset (i, j)$, then the entire path $(a_{h'}, b_{i'}, a_{j'})$ lies right of segment $b_i a_j$ by Property 1.

Case 2: intervals (h, j) and (h', j') cross. Without loss of generality, assume $h \le h' \le j \le j'$. The constraints on the six indices imply

$$1 \le h \le h' < i' < i < j \le j' \le k.$$

Since both (a_h, b_i, a_j) and $(a_{h'}, b_{i'}, a_{j'})$ are y-monotone, the only possible intersection is between $a_h b_i$ and $b_{i'} a_{j'}$. However, $i' < i$ implies $x(b_{i'}) < x(b_i)$, and Property 1 implies that $b_{i'} a_{j'}$ passes left of a_h. Consequently, (a_h, b_i, a_j) and $(a_{h'}, b_{i'}, a_{j'})$ do not cross, as claimed. $\square$

3   Embedding Algorithm

Let $G = (V, E)$ be a triangulation on n vertices, where $n \geq 4$. Construct a topological book embedding D_1 on 2 pages of Cardinal et al. [2] with at most $n - 4$ biarcs in $O(n)$ time. Assume, without loss of generality, that at most half of the arcs lie above the x-axis (rotate by 180° otherwise).

Modify the embedding D_1 and deform each edge lying above the x-axis into a biarc; refer to Fig. 1 (right). Specifically, for each vertex $v \in V$, let $E_v \subseteq E$ denote the set of edges vw such that v is the left endpoint of vw and vw is an arc lying above the x-axis in the embedding D_1. Apply a homeomorphism in a small neighborhood of v such that all edges in E_v dip below the x-axis, and become biarcs. Let D_2 be the resulting topological book embedding of G. Note that it has at most $(n - 4) + [(3n - 6) - (n - 4)]/2 = 2n - 5$ biarcs, and at least $(3n - 6) - (2n - 5) = n - 1$ arcs.

Consider the set of all vertices and edge-crossings along the x-axis, and denote them by $p_1, \ldots, p_m$, where $m \leq n + (2n - 5) = 3n - 5$. We define a new embedding D_3 into the point set $\tilde{T}_{3n-5}$ such that the vertices are mapped into $\tilde{A}_{3n-5}$ and the bend points into $\tilde{B}_{3n-5}$. For every vertex $v \in V$, if D_2 maps v to point p_i, then let D_3 map it to point a_i. For every edge $uv \in E$, if uv is a biarc that crosses the x-axis at p_i in D_2, then embed uv as a polyline with one bend at b_i; otherwise uv is an arc below the x-axis in D_2, and we embed uv as a straight-line segment. Figure 3 shows the resulting embedding for a small example graph.

Proof of Theorem 1. We set $S_n = \tilde{T}_{3n-5}$, which contains $6n - 10$ points. By Lemma 2, the above embedding algorithm is correct, that is, no two edges cross each other, as required.   □

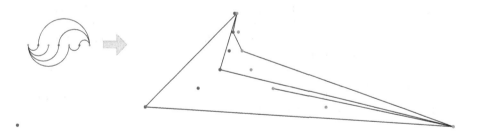

Fig. 3. An example of an embedding produced by our algorithm.

Proof of Theorem 2. Let $G = (V, E)$ be a planar graph on $n \geq 4$ vertices, and let G' be a triangulation of G. Recall that every triangulation admits a Schnyder decomposition into three trees of equal size [15]. Let R be a Schnyder tree of G' that contains at least $|E|/3$ edges from E. Using the embedding algorithm of Di Giacomo et al. [12,13], we obtain a topological book embedding of G' on 2 pages such that all edges of R are mapped to arcs below the x-axis (cf. [2]). This embedding consists of at least $n - 2$ arcs and at most $4n - 4$ biarcs. Similarly to the proof of Theorem 1, we can embed G' on $\tilde{A}_{5n-4}$ such that all edges in R are mapped to straight-line segments, and the remaining edges each have one bend in $\tilde{B}_{5n-4}$. □

We conclude with an observation on the size requirement of our construction.

Lemma 3. S_n *fits on a* $O((1 + \sqrt{2})^{3n})$ *by* $O(n)$ *integer grid.*

Proof. We defined $\tilde{T}_k$ to be an arbitrarily small perturbation of T_k. Alternatively, we can ensure all coordinates are integer by setting $x(b_k) = 1$ and iteratively setting $x(b_i) = \lceil (1 + \sqrt{2})x(b_{i+1}) \rceil$. We observe that the resulting coordinates are upper-bounded by the recurrence $\xi(b_k) = 1$; $\xi(b_i) = 1 + (1 + \sqrt{2})\xi(b_{i+1})$, which solves to

$$\xi(b_i) = \sum_{j=i}^{k}(1 + \sqrt{2})^j = \frac{1}{2}\sqrt{2}(1 + \sqrt{2})^{k-i} - \frac{1}{2}\sqrt{2} \in O((1 + \sqrt{2})^{k-i}).$$

Choosing $k = 3n - 10$ and $i = 1$, the bound on the x-coordinates follows. The y-coordinates are integers ranging from 1 to $3n - 10$. □

Acknowledgements. Research by Tóth was supported in part by the NSF awards CCF-1422311 and CCF-1423615. Research by Löffler was supported in part by the NWO grant 639.021.123.

References

1. Bannister, M.J., Cheng, Z., Devanny, W.E., Eppstein, D.: Superpatterns and universal point sets. J. Graph Algorithms Appl. **18**(2), 177–209 (2014)
2. Cardinal, J., Hoffmann, M., Kusters, V., Tóth, C.D., Wettstein, M.: Arc diagrams, flip distances, and Hamiltonian triangulations. In: Mayr, E.W., Ollinger, N. (eds.) Proceedings of 32nd STACS. LiPIcs, vol. 30, pp. 197–210. Leibniz-Zentrum für Informatik, Dagstuhl (2015)
3. Chrobak, M., Kant, G.: Convex grid drawings of 3-connected planar graphs. Internat. J. Comput. Geom. Appl. **7**, 211–223 (1997)
4. de Frayseix, H., Pach, J., Pollack, R.: How to draw a planar graph on a grid. Combinatorica **10**(1), 41–51 (1990)
5. Di Battista, G., Frati, F.: Small area drawings of outerplanar graphs. Algorithmica **54**(1), 25–53 (2009)
6. Dolev, D., Leighton, F.T., Trickey, H.: Planar embedding of planar graphs. In: Preparata, F. (ed.) Advances in Computing Research, vol. 2, pp. 147–161. JAI Press Inc., London (1984)
7. Dujmović, V., Evans, W., Lazard, S., Lenhart, W., Liotta, G., Rappaport, D., Wismath, S.: On point-sets that support planar graphs. Comput. Geom. Theory Appl. **46**(1), 29–50 (2013)
8. Everett, H., Lazard, S., Liotta, G., Wismath, S.: Universal sets of n points for one-bend drawings of planar graphs with n vertices. Discrete Comput. Geom. **43**(2), 272–288 (2010)
9. Fáry, I.: On straight lines representation of plane graphs. Acta Scientiarum Mathematicarum (Szeged) **11**, 229–233 (1948)

10. Frati, F.: Lower bounds on the area requirements of series-parallel graphs. Discrete Math. Theoret. Comput. Sci. **12**(5), 139–174 (2010)
11. Frati, F., Patrignani, M.: A note on minimum-area straight-line drawings of planar graphs. In: Hong, S.-H., Nishizeki, T., Quan, W. (eds.) GD 2007. LNCS, vol. 4875, pp. 339–344. Springer, Heidelberg (2008)
12. Di Giacomo, E., Didimo, W., Liotta, G., Wismath, S.K.: Curve-constrained drawings of planar graphs. Comput. Geom. Theory Appl. **30**(1), 1–23 (2005)
13. Di Giacomo, E., Didimo, W., Liotta, G.: Spine and radial drawings. In: Tamassia, R. (ed.) Handbook of Graph Drawing and Visualization, Chap. 8, pp. 247–284. CRC Press, Boca Raton (2013)
14. Kurowski, M.: A 1.235 lower bound on the number of points needed to draw all n-vertex planar graphs. Inf. Process. Lett. **92**, 95–98 (2004)
15. Schnyder, W.: Embedding planar graphs in the grid. In: Proceedings of the 1st Symposium on Discrete Algorithms, pp. 138–147. ACM Press, New York, NY (1990)

Contact Representations

Recognizing Weighted Disk Contact Graphs

Boris Klemz[1], Martin Nöllenburg[2]([⊠]), and Roman Prutkin[3]

[1] Institute of Computer Science, Freie Universität Berlin, Berlin, Germany
klemz@inf.fu-berlin.de
[2] Algorithms and Complexity Group, TU Wien, Vienna, Austria
noellenburg@ac.tuwien.ac.at
[3] Institute of Theoretical Informatics, Karlsruhe Institute of Technology,
Karlsruhe, Germany
roman.prutkin@kit.edu

Abstract. Disk contact representations realize graphs by mapping vertices bijectively to interior-disjoint disks in the plane such that two disks touch each other if and only if the corresponding vertices are adjacent in the graph. Deciding whether a vertex-weighted planar graph can be realized such that the disks' radii coincide with the vertex weights is known to be NP-hard. In this work, we reduce the gap between hardness and tractability by analyzing the problem for special graph classes. We show that it remains NP-hard for outerplanar graphs with unit weights and for stars with arbitrary weights, strengthening the previous hardness results. On the positive side, we present constructive linear-time recognition algorithms for caterpillars with unit weights and for embedded stars with arbitrary weights.

1 Introduction

A set of disks in the plane is a *disk intersection representation* of a graph $G = (V, E)$ if there is a bijection between V and the set of disks such that two disks intersect if and only if they are adjacent in G. *Disk intersection graphs* are graphs that have a disk intersection representation; a subclass are *disk contact graphs* (also known as coin graphs), that is, graphs that have a disk intersection representation with interior-disjoint disks. This is also called a *disk contact representation* (DCR) or, if connected, a *circle packing*. It is easy to see that every disk contact graph is planar and the famous Koebe-Andreev-Thurston circle packing theorem [13] dating back to 1936 (see Stephenson [17] for its history) states that the converse is also true, that is, every planar graph is a disk contact graph.

Application areas for disk intersection/contact graphs include modeling physical problems like wireless communication networks [9], covering problems like geometric facility location [16,18], visual representation problems like area cartograms [7] and many more (various examples are given by Clark et al. [4]). Efficient numerical construction of DCRs has been studied in the past [5,15]. Often, however, one is interested in recognizing disk graphs or generating representations that do not only realize the input graph, but also satisfy additional requirements. For example, Alam et al. [1] recently obtained several positive

© Springer International Publishing Switzerland 2015
E. Di Giacomo and A. Lubiw (Eds.): GD 2015, LNCS 9411, pp. 433–446, 2015.
DOI: 10.1007/978-3-319-27261-0_36

and negative results on the existence of balanced DCRs, in which the ratio of the largest disk radius to the smallest is polynomial in the number of disks. Furthermore, it might be desirable to generate a disk representation that realizes a vertex-weighted graph such that the disks' radii or areas are proportional to the corresponding vertex weights, for example, for value-by-area circle cartograms [10]. Clearly, there exist vertex-weighted planar graphs that cannot be realized as disk contact representations, and the corresponding recognition problem for planar graphs is NP-hard, even if all vertices are weighted uniformly [3]. The complexity of recognizing weighted disk contact graphs for many interesting subclasses of planar graphs remained open. Note that graphs realizable as DCRs with unit disks correspond to 1-ply graphs. This was stated by Di Giacomo et al. [6] who recently introduced and studied the ply number concept for graphs. They showed that internally triangulated biconnected planar graphs admitting a DCR with unit disks can be recognized in $O(n \log n)$ time.

In this paper we extend the results of Breu and Kirkpatrick [3] and show that it remains NP-hard to decide whether a DCR with unit disks exists even if the input graph is outerplanar. Our result holds both for the case that arbitrary embeddings are allowed and the case that a fixed combinatorial embedding is specified. The result for the latter case is also implied by a very recent result by Bowen et al. [2] stating that for fixed embeddings the problem is NP-hard even for trees. However, the recognition of trees with a unit disk contact representation remains an interesting open problem if arbitrary embeddings are allowed. For caterpillar-trees we solve this problem in linear time. For vertex weights that are not necessarily uniform we show that the recognition problem is strongly NP-hard even for stars if no embedding is specified. However, for embedded stars we solve the problem in linear time. Some of our algorithms use the *Real RAM* model, which assumes that a set of basic arithmetic operations (including trigonometric functions and square roots) can be performed in constant time.

2   Unit Disk Contact Graphs

In this section we are concerned with the problem of deciding whether a given graph is a *unit disk contact graph* (UDC graph), that is, whether it has a DCR with unit disks. For a UDC graph we also say that it is *UDC-realizable* or simply *realizable*. It is known since 1998 that recognizing UDC graphs is generally NP-hard for planar graphs [3], but it remained open for which subclasses of planar graphs it can be solved efficiently and for which subclasses NP-hardness still holds. We show that we can recognize caterpillars that are UDC graphs in linear time and construct a representation if it exists (Sect. 2.1), whereas the problem remains NP-hard for outerplanar graphs (Sect. 2.2).

2.1   Recognizing Caterpillars with a Unit Disk Contact Representation

Let $G = (V, E)$ be a caterpillar graph, that is, a tree for which a path remains after removing all leaves. Let $P = (v_1, \ldots, v_k)$ be this so-called *inner path* of G.

On the one hand, it is well known that six unit disks can be tightly packed around one central unit disk, but then any two consecutive outer disks necessarily touch and form a triangle with the central disk. This is not permitted in a caterpillar and thus we obtain that in any realizable caterpillar the maximum degree $\Delta \leq 5$. On the other hand, it is easy to see that all caterpillars with $\Delta \leq 4$ are UDC graphs as shown by the construction in Fig. 1a.

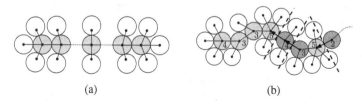

(a)   (b)

Fig. 1. (a) For $\Delta \leq 4$ any caterpillar can be realized. (b) Incremental construction of a DCR. Narrow disks are dark gray and indicated by an outgoing arrow, wide disks are light gray.

However, not all caterpillars with $\Delta = 5$ can be realized. For example, two degree-5 vertices on P separated by zero or more degree-4 vertices cannot be realized, as they would again require tightly packed disks inducing cycles in the contact graph. In fact, we get the following characterization.

Lemma 1. *A caterpillar G with $\Delta = 5$ is a UDC graph if and only if there is at least one vertex of degree at most 3 between any two degree-5 vertices on the inner path P.*

Proof. Consider an arbitrary UDC representation of G and let D_i be the disk representing vertex v_i of the inner path P. Let ℓ_i be the tangent line between two adjacent disks D_{i-1} and D_i on P. We say that P is *narrow* at v_i if some leaf disk attached to D_{i-1} intersects ℓ_i; otherwise P is *wide* at v_i. Let v_i and v_j $(i < j)$ be two degree-5 vertices on P with no other degree 5 vertices between them. The path P must be narrow at the next vertex v_{i+1}, since one of the four mutually disjoint neighbor disks of D_{i-1} except D_i necessarily intersects ℓ_i. If there is no vertex v_k $(i < k < j)$ with $\deg(v_k) \leq 3$ between v_i and v_j we claim that P is still narrow at v_j. If $j = i + 1$ this is obviously true. Otherwise all vertices between v_i and v_j have degree 4. But since the line ℓ_{i+1} was intersected by a neighbor of v_i, this property is inherited for the line ℓ_{i+2} and a neighbor of v_{i+1} if $\deg(v_{i+1}) = 4$. An inductive argument applies. Since P is still narrow at the degree-5 vertex v_j, it is impossible to place four mutually disjoint disks touching D_j for the neighbors of v_j except v_{j-1}.

We now construct a UDC representation for a caterpillar in which any two degree-5 vertices of P are separated by a vertex of degree ≤ 3. We place a disk D_1 for v_1 at the origin and attach its leaf disks *leftmost*, that is, symmetrically pushed to the left with a sufficiently small distance between them. In each subsequent step, we place the next disk D_i for v_i on the bisector of the *free space*, which we define as the maximum cone with origin in D_{i-1}'s center containing no previously inserted neighbors of D_{i-1} or D_{i-2}. Again, we attach the leaves of D_i in a leftmost and balanced way, see Fig. 1b. For odd-degree vertices this leads to a change in direction of P, but by alternating upward and downward bends for subsequent odd-degree vertices we can maintain a horizontal monotonicity, which ensures that leaves of D_i can only collide with leaves of D_{i-1} or D_{i-2}. In this construction P is wide until the first degree-5 vertex is placed, after which it gets and stays narrow as long as degree-4 vertices are encountered. But as soon as a vertex of degree ≤ 3 is placed, P gets (and remains) wide again until the next degree-5 vertex is placed. Placing a degree-5 vertex at which P is wide can always be done.   □

Lemma 1 and the immediate observations for caterpillars with $\Delta \neq 5$ yield the following theorem. We note that the decision is only based on the vertex degrees in G, whereas the construction uses a Real RAM model.

Theorem 1. *For a caterpillar G it can be decided in linear time whether G is a UDC graph if arbitrary embeddings are allowed. A UDC representation (if one exists) can be constructed in linear time.*

2.2   Hardness for Outerplanar Graphs

A planar 3SAT formula φ is a Boolean 3SAT formula with a set $\mathcal{U}$ of variables and a set $\mathcal{C}$ of clauses such that its *variable-clause-graph* $G_\varphi = (\mathcal{U} \cup \mathcal{C}, E)$ is planar. The set E contains for each clause $c \in \mathcal{C}$ the edge (c, x) if a literal of variable x occurs in c. Deciding the satisfiability of a planar 3SAT formula is NP-complete [14] and there exists a planar drawing $\mathcal{G}_\varphi$ of G_φ on a grid of polynomial size such that the variable vertices are placed on a horizontal line and the clauses are connected in a comb-shaped rectangular fashion from above or below that line [12], see Fig. 2a. A planar 3SAT formula φ is *monotone* if

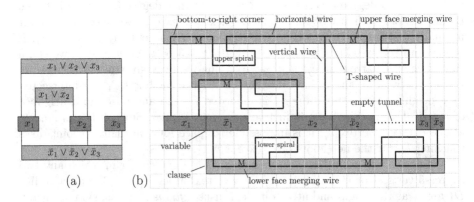

Fig. 2. Sketch of the grid layout $\mathcal{G}_\varphi$ (a) and high-level structure of the construction of G'_φ (b) for the PM3SAT formula $\varphi = (x_1 \vee x_2) \wedge (x_1 \vee x_2 \vee x_3) \wedge (\bar{x}_1 \vee \bar{x}_2 \vee \bar{x}_3)$.

each clause contains either only positive or only negative literals and if G_φ has a planar drawing as described before with all clauses of positive literals on one side and all clauses of negative variables on the other side. The 3SAT problem remains NP-complete for planar monotone formulae [14] and is called Planar Monotone 3-Satisfiability (PM3SAT).

We perform a polynomial reduction from PM3SAT to show NP-hardness of recognizing outerplanar UDC graphs. A graph is *outerplanar* if it has a planar drawing in which all vertices lie on the unbounded outer face. We say that a planar graph G is (combinatorially) *embedded* if we are given for each vertex the circular order of all incident edges as well as the outer face such that a planar drawing respecting this embedding exists. For the reduction we create, based on the planar drawing $\mathcal{G}_\varphi$, an outerplanar graph G'_φ that has a UDC representation if and only if the formula φ is satisfiable.

Arguing about UDC representations of certain subgraphs of G'_φ becomes a lot easier, if there is a single unique geometric representation (up to rotation, translation and mirroring). We call graphs with such a representation *rigid*. Using an inductive argument (see full version [11]), we state the following sufficient condition for rigid UDC structures. All subgraphs of G'_φ that we refer to as rigid satisfy this condition.

Lemma 2. *Let $G = (V, E)$ be a biconnected graph realizable as a UDC representation that induces an internally triangulated outerplane embedding of G. Then, G is rigid.*

The main building block of the reduction is a *wire gadget* in G'_φ that comes in different variations but always consists of a rigid tunnel structure containing a rigid bar that can be flipped into different tunnels around its centrally located articulation vertex. Each wire gadget occupies a square tile of fixed dimensions so that different tiles can be flexibly put together in a grid-like fashion. The bars stick out of the tiles in order to transfer information to the neighboring tiles. Variable gadgets consist of special tiles containing tunnels without bars or with very long bars. Adjacent variable gadgets are connected by narrow tunnels without bars. *Face merging* wires work essentially like normal horizontal wires but their low-level construction differs in order to assert that G'_φ is outerplanar and connected. Figure 2b shows a schematic view of how the gadget tiles are arranged to mimic the layout $\mathcal{G}_\varphi$ of Fig. 2a. The wires connect the positive (negative)

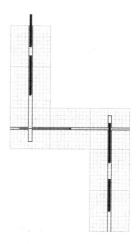

Fig. 3. Variable gadget in state *false* with a positive (left) and a negative literal (right).

clauses to the left (right) halves of the respective variable gadgets. Furthermore, we place a face merging wire (marked by 'M') in the top/bottom left corner of each inner face followed by an *upper (lower) spiral*, which is a fixed 3×4 pattern of wire gadgets. These structures ensure that G'_φ is outerplanar and they limit relative displacements.

The main idea behind the reduction is as follows. Each *variable gadget* contains one thin, long horizontal bar that is either flipped to the left (*false*) or to the right (*true*), see Fig. 3. If the bar is in its left (right) position, this blocks the lower (upper) bar position of the first wire gadget of each positive (negative) literal. Consequently, each wire gadget that is part of the connection between a variable gadget and a clause gadget must flip its entire chain of bars towards the clause if the literal is false. The design of the *clause gadget* depends on its number of literals. Figure 4a illustrates the most important case of a clause with three literals containing a T-shaped wire gadget. The bar of the T-shaped wire needs to be placed in one of the three incident tunnels. This is possible if and only if at least one of the literals evaluates to *true*. A similar statement holds true for clauses with two or one literals; their construction is much simpler: just a horizontal wire gadget or a dead end suffice as clause tile.

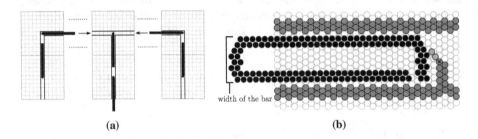

(a)   **(b)**

Fig. 4. (a) Clause gadget with two false inputs (left and right) and one true input. (b) Detailed view of a horizontal wire gadget with a rigid bar (black disks) inside a tunnel (dark gray disks).

All gadgets are realized by combining several rigid UDC subgraphs. As an example, Fig. 4b shows a close-up of the left side of a horizontal wire gadget. Both the black and the dark gray disks form rigid components whose UDC graphs satisfy the precondition of Lemma 2. The black disks implement the bar, the dark gray disks constitute the tunnel. Note how the bar can be flipped or mirrored to the left or the right around the articulation disk (marked 'x') due to the two light gray disks (called *chain* disks) that do not belong to a rigid structure. The width of each bar is chosen such that it differs from the supposed inner width of a tunnel by at most twice the disk diameter, thus admitting some slack. However, we can choose the width of the tunnels/bars (and the gadget tile dimensions) as a large polynomial in the input such that this "wiggle room" does not affect the combinatorial properties of our construction. The description of the face merging wire below discusses this aspect in more detail. Further, we choose the lengths of the bars such that the bars of two adjacent wire gadgets collide if their bars are oriented towards each other. Unlike the bars of wire gadgets, the bars of variable gadgets are not designed to transmit information from tile to tile. Instead they are simply designed to prevent the adjacent vertical wires on either the left or the right side of the variable gadget to be oriented towards it. For this reason, we can choose the width of the variable bars to be very small (e.g., just 2 disks), in order to obtain an overall tighter construction.

Now that we have established how the gadgets work and how they are constructed, consider the properties of the corresponding graph G'_φ that encodes the entire structure. If we would use only the regular wire gadgets as in Fig. 4b for the entire construction, G'_φ would neither be outerplanar nor connected. As illustrated in Fig. 5, for each of the inner faces of $\mathcal{G}_\varphi$ we would obtain a single rigid structure, which we call *face boundary*, with several bars attached to it. These face boundaries, however, would not be connected to each other. Furthermore, the subgraphs that realize the face boundaries would not be outerpla-

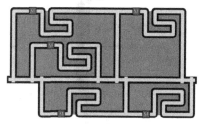

Fig. 5. Schematic of G'_φ if face merging wires (marked 'M') replace some regular wires. Inner faces of $\mathcal{G}_\varphi$ in dark gray, the face 'inside the tunnels' in light gray, the outer face in white and the face boundaries in black

nar. This is why we replace some horizontal wire gadgets in the upper (positive) and lower (negative) part of our construction by *upper* and *lower* face merging wires respectively, which have two purposes. Horizontal wires contain a tunnel that is formed by two face boundaries, called the *upper* and *lower* face boundary of the corresponding gadget tile. These face boundaries are not connected, see Fig. 4b. In a face merging wire, however, the respective face boundaries are connected. Furthermore, a gap is introduced (by removing two disks) to the lower (upper) face boundary in an upper (lower) face merging wire so that the lower (upper) face boundary now becomes outerplanar. Since the face merging wire is supposed to transfer information just like a horizontal wire we cannot connect the two face boundaries rigidly. Instead we create three bars connected to each other with chain disks, see Fig. 6a. The width of the top and the bottom bars are chosen such that they fit tightly inside the narrow cavity in the middle of the tile if placed perpendicularly to the left or right of the respective articulation disk. The third bar ensures that all three bars together are placed either to the left (Fig. 6b) or to the right (Fig. 6a), which allows the desired information transfer.

Together with the incident spiral, a face merging wire ensures that the disks of the lower face boundary deviate from their intended locations relative to the upper face boundary only by up to a small constant distance since (1) the design and the asymmetrical placement of the spirals and the face merging wires preserve the orientations of the respective upper and lower face boundaries, i.e., the left/right/top/bottom sides of these structures are facing as intended in any realization and (2) the width of the tunnels is at most twice the disk diameter larger than the width of the bars and there is at least one bar located in any of the cardinal directions of each spiral. This effect can cascade since the face boundaries might be connected to further face boundaries. However, according to Euler's formula the number of faces in $\mathcal{G}_\varphi$ is linear in the number of clauses

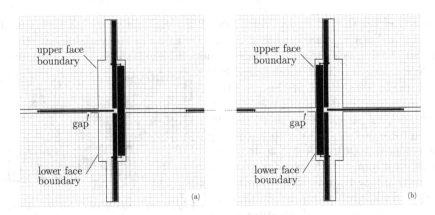

Fig. 6. Upper face merging wire gadget oriented to the right (a) / left (b). It connects the lower and upper face boundaries. The gap causes the faces inside the tunnel and the lower face to collapse.

and variables and, therefore, the total distance by which a disk can deviate from its intended ideal position is also linear in this number. By accordingly adjusting the tile dimensions and bar widths, we can therefore ensure that the wiggle room in our construction does not affect the intended combinatorial properties while keeping the size of G'_φ polynomial. The introduction of face merging wires causes G'_φ to be connected and it causes all inner faces and the face 'inside the tunnels' to collapse. Finally, by introducing a single gap in the outermost rigid structure, G'_φ becomes outerplanar, which concludes our reduction.

This concludes our construction for the case with arbitrary embeddings. Note, however, that the gadgets are designed such that flipping the bars does not require altering the combinatorial embedding of the graph. This holds true even for the face merging wire. Therefore, we can furthermore provide a combinatorial embedding such that G'_φ can be realized with respect to said embedding if and only if φ is satisfiable. Thus, we obtain the following theorem with the remaining arguments of its proof found in [11].

Theorem 2. *For outerplanar graphs the UDC recognition problem is NP-hard. This remains true for outerplanar graphs with a specified combinatorial embedding.*

3   Weighted Disk Contact Graphs

In this section, we assume that a positive weight $w(v)$ is assigned to each vertex v of the graph $G = (V, E)$. The task is to decide whether G has a DCR, in which each disk D_v representing a vertex $v \in V$ has radius proportional to $w(v)$. A DCR with this property is called a *weighted disk contact representation* (WDC representation) and a graph that has a WDC representation is called a *weighted disk contact graph* (WDC graph). Obviously, recognizing WDC graphs is at least as hard as the UDC graph recognition problem from Sect. 2 by setting

$w(v) = 1$ for every vertex $v \in V$. Accordingly, we first show that recognizing WDC graphs is NP-hard even for stars (Sect. 3.1), however, embedded stars with a WDC representation can still be recognized (and one can be constructed if it exists) in linear time (Sect. 3.2).

3.1   Hardness for Stars

We perform a polynomial reduction from the well-known 3-Partition problem. Given a bound $B \in \mathbb{N}$ and a multiset of positive integers $\mathcal{A} = \{a_1, \ldots, a_{3n}\}$ such that $\frac{B}{4} < a_i < \frac{B}{2}$ for all $i = 1, \ldots, 3n$, deciding whether $\mathcal{A}$ can be partitioned into n triples of sum B each is known to be strongly NP-complete [8]. Let $(\mathcal{A}, B)$ be a 3-Partition instance. We construct a star $S = (V, E)$ and a radius assignment $\mathbf{r} : V \to \mathbb{R}^+$ such that S has a WDC representation respecting $\mathbf{r}$ if and only if $(\mathcal{A}, B)$ is a yes-instance.

We create a central disk D_c of radius r_c corresponding to the central vertex v_c of S as well as a fixed number of outer disks with uniform radius r_o chosen appropriately such that these disks have to be placed close together around D_c without touching, creating funnel-shaped *gaps* of roughly equal size; see Fig. 7. Then, a WDC representation of S exists only if all remaining disks can be distributed among the gaps, and the choice of the gap will induce a partition of the integers $a_i \in \mathcal{A}$.

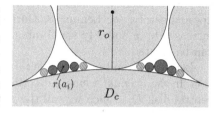

Fig. 7. Reducing from 3-Partition to prove Theorem 3. Input disks (dark) are distributed between gaps. Hatched disks are separators.

We shall represent each a_i by a single disk called an *input* disk and encode a_i in its radius. Each of the gaps is supposed to be large enough for the input disks that represent a *feasible triple*, i.e., with sum B, to fit inside it, however, the gaps must be too small to contain an *infeasible triple*'s disk representation, i.e., a triple with sum $> B$.

While the principle idea of the reduction is simple, the main challenge is finding a radius assignment satisfying the above property and taking into account numerous additional, nontrivial geometric considerations that are required to make the construction work. For example, we require that the lower boundary of each gap is sufficiently flat. We achieve this by creating additional dummy gaps and ensure that they can not be used to realize a previously infeasible instance. Next, we make sure that additional *separator* disks must be placed in each gap's corners to prevent left and right gap boundaries from interfering with the input disks. Finally, all our constructions are required to tolerate a certain amount of "wiggle room", since, firstly, the outer disks do not touch and, secondly, some radii cannot be computed precisely in polynomial time.

Since S is supposed to be a star, the only adjacencies in our construction are the ones with D_c. However, several of the disks adjacent to D_c are required to be placed very close together without actually touching. We shall, whenever we need to calculate distances, handle these barely not touching disks as if

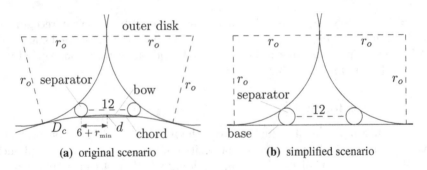

Fig. 8. A gap, bounded in (a) by two outer disks and a bow; in (b) the gap's base replaces its bow. The distance between the separators is 12 in both scenarios.

they were actually touching. We will describe how to compute these distances approximately; see Lemma 8. During this step the radius of the central disk increases by a suitably small amount such that no unanticipated embeddings can be created.

Let $B > 12$ and $n > 6$, and let $m \geq n$ be the number of gaps in our construction. In the *original* scenario described above, a gap's boundary belonging to the central disk D_c, which we call the gap's *bow*, is curved as illustrated in Fig. 8a. We will, however, first consider a *simplified* scenario in which a gap is created by placing two disks of radius r_o right next to each other on a straight line as depicted in Fig. 8a. We refer to this gap's straight boundary as the *base* of the gap. We call a point's vertical distance from the base its *height*. We also utilize the terms *left* and *right* in an obvious manner. Assume for now that we can place two *separator* disks in the gap's left and right corner, touching the base and such that the distance between the rightmost point p_l of the left separator and the leftmost point p_r of the right separator is exactly 12 units. We can assume $B \equiv 0 \mod 4$; see Lemma 3. Thus, we know that $a \in \{B/4 + 1, \ldots, B/2 - 1\}$ for any $a \in \mathcal{A}$. Due to space restrictions, the proofs of the following lemmas are only available in the full version [11].

Lemma 3. *For each $m \geq n$, there exists a 3-Partition instance $(\mathcal{A}', B')$ equivalent to $(\mathcal{A}, B)$ with $|\mathcal{A}'| = 3m$ and $B' = 180B$.*

Our first goal is to find a function $r : \{B/4, B/4 + 1, \ldots, B/2\} \rightarrow \mathbb{R}^+$ that assigns a disk radius to each input integer as well as to the values $B/4$ and $B/2$ such that a disk triple t together with two separator disks can be placed on the base of a gap without intersecting each other or the outer disks if and only if t is feasible. In the following, we show that $r(x) = 2 - (4 - 12x/B)/B$ will satisfy our needs. We choose the radius of the separators to be $r_{min} = r(B/4 + 1) = 2 - (1 - 12/B)/B$, the smallest possible input disk radius. The largest possible input disk has radius $r_{max} = r(B/2 - 1) = 2 + (2 - 12/B)/B$. Note that r is linear and increasing.

Next, we show for both scenarios that separators placed in each gap's corners prevent the left and right gap boundaries from interfering with the input disks.

Lemma 4. *For any $a \in \mathcal{A}$ it is not possible that a disk with radius $r(a)$ intersects one of the outer disks that bound the gap when placed between the two separators.*

For our further construction, we need to prove the following property.

Property 1. Each feasible triple fits inside a gap containing two separators and no infeasible triple does.

It can be easily verified that for x_1, x_2, x_3, $\sum_{i=1}^{3} x_i \leq B$, it is $2 \sum_{i=1}^{3} r(x_i) \leq 12$, implying the first part of Property 1. We define $s_i = 2r_{\min} + 2\sqrt{(r_{\max} + r_{\min})^2 - (r_{\max} - r_{\min})^2}$. In the proof of Lemma 5, we will see that s_i is the horizontal space required for the triple $(r_{\min}, r_{\max}, r_{\min})$, which is the narrowest infeasible triple. Next, let $d(\varepsilon, x) = \sqrt{(r(x) - \varepsilon/2)^2 + (r(x) - r_{\min})^2}$ for $\varepsilon > 0$ and $x \in \{B/4 + 1, \ldots, B/2 - 1\}$. We will see that $d(\varepsilon, x)$ is an upper bound for the distance between the center of a disk $D(x)$ with radius $r(x)$ and the rightmost (leftmost) point of the left (right) separator disk, if the overlap of their horizontal projections is at least $\varepsilon/2$.

Lemma 5. *There exist $\varepsilon > 0$ and $\varepsilon_1, \varepsilon_2, \phi \geq 0$ with $\varepsilon = \varepsilon_1 + \varepsilon_2$ which satisfy the two conditions: (I) $12 + \varepsilon \leq s_i$ and (II) $d(\varepsilon_1, x) \leq r(x) - \phi \forall x \in \{B/4 + 1, \ldots, B/2 - 1\}$. These conditions imply the second part of Property 1 for the simplified scenario.*

So far we assumed that the separators are always placed in the corners of the gap. But in fact, separators could be placed in a different location, moreover, there could even be gaps with multiple separators and gaps with zero or one separator. Since the radius of the separators is $r_{\min}$, which is the radius of the smallest possible input disk, it seems natural to place them in the gaps' corners to efficiently utilize the horizontal space. However, all feasible disk triples (except $(B/3, B/3, B/3)$) require less than 12 units of horizontal space. It might therefore be possible to place a feasible disk triple inside a gap together with two disks that are not necessarily separators but input disks with a radius greater than $r_{\min}$. To account for this problem, we prove the following property.

Property 2. A feasible disk triple can be placed in the gap together with two other disks only if those two disks are separators.

We define $s_f = 2r(B/4) + 2\sqrt{(r(B/2) + r(B/4))^2 - (r(B/2) - r(B/4))^2}$. In the proof of Lemma 6, we will see that s_f is a lower bound for the horizontal space consumption of any feasible triple.

Lemma 6. *There exist $\xi > 0$ and $\xi_1, \xi_2, \psi \geq 0$ with $\xi = \xi_1 + \xi_2$ satisfying the following two conditions: (III) $12 - 24/B^2 + \xi \leq s_f$ and (IV) $d(\xi_1, x) \leq r(x) - \psi \forall x \in \{B/4 + 1, \ldots, B/2 - 1\}$. These conditions imply Property 2 for the simplified scenario.*

We verify in the proofs of Lemmas 5 and 6 in the full version [11] that choosing $\varepsilon_1, \xi_1 = 16/B^2$ and $\varepsilon_2, \phi, \xi_2, \psi = 1/B^2$ satisfies our four conditions.

Intuitively, Conditions (I)–(IV) have the following meaning. By (I), the horizontal space consumption of any infeasible triple is greater than 12 by some fixed buffer. By (III), the horizontal space consumption of any feasible triple is very close to 12. Conditions (II) and (IV) imply that if the overlap of the horizontal projections of a separator and an input disk is large enough, the two disks intersect, implying that triples with sufficiently large space consumption can indeed not be placed between two separators.

In the original scenario, consider a straight line directly below the two separators. We call this straight line the gap's *chord*, see Fig. 8a. The gap's chord has a function similar to the base in the simplified scenario. We still want separators to be placed in the gap's corners. The distance between the rightmost point p_l of the left separator and the leftmost point p_r of the right separator is now allowed to be slightly more than 12. The horizontal space consumption of a disk triple placed on the bow is lower compared to the disk triple being placed on the chord. Moreover, the overlap of the horizontal projections of a separator and an input disk can now be bigger without causing an intersection. However, we show that if the maximum distance d between a gap's bow and its chord is small enough, the original scenario is sufficiently close to the simplified one, and the four conditions still hold, implying the desired properties.

Lemma 7. *In the original scenario, let $d \leq 1/4B^2$, and let the amount of free horizontal space in each gap after inserting the two separators in each corner be between 12 and $12 + 1/4B^2$. Then, Properties 1 and 2 still hold.*

In order to conclude the hardness proof, it therefore remains to describe how to choose the radii for the central and outer disks and how to create the gaps such that $d \leq 1/4B^2$.

Recall that we have a central disk D_c with radius r_c and m outer disks with radius r_o which are tightly packed around D_c such that m equal-sized gaps are created. With basic trigonometry we see that $r_c + r_o = r_o/\sin(\pi/m)$ and, therefore, $r_c = r_o/\sin(\pi/m) - r_o$. Clearly, there always exists a value r_o such that the two separator disks can be placed in each gap's corners and such that the distance between each pair of separators is exactly 12 units. Let $\bar{r}_o$ be this value. Moreover, the maximum distance d between a gap's bow and its chord is of particular importance, see Fig. 8a. Using the Pythagorean Theorem, it can be calculated to be $d = r_c - (\sqrt{(r_c + r_{\min})^2 - (6 + r_{\min})^2} - r_{\min})$. The crucial observation is that we do not necessarily need to choose $m = n$. Instead we may choose any $m \geq n$ and thereby decrease d, as long as we make sure that m is still a polynomial in the size of the input or numeric values and that the $m - n$ additional gaps cannot be used to solve an instance which should be infeasible.

Lemma 8. *There exist constants c_1, c_3, c_4, such that for $m = B^{c_1}$, $\varepsilon_3 = 1/B^{c_3}$ and $\varepsilon_4 = 1/B^{c_4}$, there exist values $\tilde{r}_o$ for r_o and $\tilde{r}_c$ for r_c, for which it holds $\bar{r}_o < \tilde{r}_o \leq \bar{r}_o + \varepsilon_3$ and $\bar{r}_c < \tilde{r}_c \leq \bar{r}_c + \varepsilon_4$ for $\bar{r}_c = \tilde{r}_o/\sin(\pi/m) - \tilde{r}_o$. The constants can be chosen such that $d \leq 1/4B^2$ and such that the amount of free horizontal space in each gap is between 12 and $12 + 1/4B^2$. Finally, $\tilde{r}_o$ and $\tilde{r}_c$ can be computed in polynomial time.*

Lemma 3 already showed how to construct an equivalent 3-Partition instance with $3m \geq 3n$ input integers. We can now prove the main result of this section, see [11] for details. Lemmas 3 and 8 show that the construction can be performed in polynomial time. Properties 1 and 2 let us show that a valid distribution of the input and separator disks among the gaps induces a solution of the 3-Partition instance and vice versa.

Theorem 3. *The WDC graph recognition problem is (strongly) NP-hard even for stars if an arbitrary embedding is allowed.*

3.2 Recognizing Embedded Stars with a Weighted Disk Contact Representation

If, however, the order of the leaves around the central vertex of the star is fixed, the existence of a WDC representation can be decided by iteratively placing the outer disks $D_1, \ldots, D_{n-1}$ tightly around the central disk D_c. A naive approach tests for collisions with all previously added disks and yields a total runtime of $O(n^2)$. However, this can be improved to $O(n)$ by maintaining a list containing only disks that might be relevant in the future. For more details see the full version [11].

Theorem 4. *On a Real RAM, for an embedded, vertex-weighted star S it can be decided in linear time whether S is a WDC graph. A WDC representation respecting the embedding (if one exists) can be constructed in linear time.*

References

1. Alam, M.J., Eppstein, D., Goodrich, M.T., Kobourov, S.G., Pupyrev, S.: Balanced circle packings for planar graphs. In: Duncan, C., Symvonis, A. (eds.) GD 2014. LNCS, vol. 8871, pp. 125–136. Springer, Heidelberg (2014)
2. Bowen, C., Durocher, S., Löffler, M., Rounds, A., Schulz, A., Tóth, C.D.: Realization of simply connected polygonal linkages and recognition of unit disk contact trees. In: Di Giacomo, E., Lubiw, A. (eds.) Graph Drawing (GD'15). LNCS, vol. 9411, pp. 447–459. Springer, Heidelberg (2015)
3. Breu, H., Kirkpatrick, D.G.: Unit disk graph recognitionis NP-hard. Comput. Geom. 9(1–2), 3–24 (1998)
4. Clark, B.N., Colbourn, C.J., Johnson, D.S.: Unit disk graphs. Discrete Math. 86(1–3), 165–177 (1990)
5. Collins, C.R., Stephenson, K.: A circle packing algorithm. Comput. Geom. 25(3), 233–256 (2003)
6. Di Giacomo, E., Didimo, W., Hong, S.H., Kaufmann, M., Kobourov, S., Liotta, G., Misue, K., Symvonis, A., Yen, H.C.: Low ply graph drawing. In: IISA 2015. IEEE (to appear, 2015)
7. Dorling, D.: Area cartograms: Their use and creation. In: Concepts and techniques in modern geography. University of East Anglia: Environmental Publications (1996)
8. Garey, M.R., Johnson, D.S.: Computers and Intractability: A Guide to the Theory of NP-Completeness. W. H. Freeman & Co., New York (1990)

9. Hale, W.: Frequency assignment: theory and applications. Proc. IEEE **68**(12), 1497–1514 (1980)
10. Inoue, R.: A new construction method for circle cartograms. Cartography Geogr. Inf. Sci. **38**(2), 146–152 (2011)
11. Klemz, B., Nöllenburg, M., Prutkin, R.: Recognizing weighted disk contact graphs, September 2015. CoRR arXiv:1509.0072
12. Knuth, D.E., Raghunathan, A.: The problem of compatible representatives. SIAM J. Discrete Math. **5**(3), 422–427 (1992)
13. Koebe, P.: Kontaktprobleme der konformen Abbildung. In: Ber. Sächs. Akad. Wiss. Leipzig, Math.-Phys. Klasse, vol. 88, pp. 141–164 (1936)
14. Lichtenstein, D.: Planar formulae and their uses. SIAM J. Comput. **11**(2), 329–343 (1982)
15. Mohar, B.: A polynomial time circle packing algorithm. Discrete Math. **117**(1), 257–263 (1993)
16. Robert, J.M., Toussaint, G.: Computational geometry and facility location. In: Operations Research and Management Science, pp. 11–15 (1990)
17. Stephenson, K.: Circle packing: a mathematical tale. Not. AMS **50**(11), 1376–1388 (2003)
18. Welzl, E.: Smallest enclosing disks (balls and ellipsoids). In: Maurer, H. (ed.) New Results and New Trends in Computer Science. LNCS, vol. 555, pp. 359–370. Springer, Heidelberg (1991)

Realization of Simply Connected Polygonal Linkages and Recognition of Unit Disk Contact Trees

Clinton Bowen[1], Stephane Durocher[2], Maarten Löffler[3], Anika Rounds[4],
André Schulz[5], and Csaba D. Tóth[1,4(✉)]

[1] Department of Mathematics, California State University Northridge,
Los Angeles, CA, USA
clinton.bowen@my.csun.edu, csaba.toth@csun.edu
[2] Department of Computer Science, University of Manitoba, Winnipeg, MB, Canada
durocher@cs.umanitoba.ca
[3] Department of Information and Computing Sciences, Utrecht University,
Utrecht, The Netherlands
m.loffler@uu.nl
[4] Department of Computer Science, Tufts University, Medford, MA, USA
anika.rounds@tufts.edu, cdtoth@cs.tufts.edu
[5] Theoretical Computer Science, University of Hagen, Hagen, Germany
andre.schulz@fernuni-hagen.de

Abstract. We wish to decide whether a simply connected flexible polygonal structure can be realized in Euclidean space. Two models are considered: polygonal linkages (body-and-joint framework) and contact graphs of unit disks in the plane. (1) We show that it is strongly NP-hard to decide whether a given polygonal linkage is realizable in the plane when the bodies are convex polygons and their contact graph is a tree; the problem is weakly NP-hard already for a chain of rectangles, but efficiently decidable for a chain of triangles hinged at distinct vertices. (2) We also show that it is strongly NP-hard to decide whether a given tree is the contact graph of interior-disjoint unit disks in the plane.

1 Introduction

In this paper, we study the realizability of complex structures that are specified by their local geometry. The complex structures are represented as graphs with constraints on the separation between their vertices, and we ask if these graphs can be embedded in the plane subject to the constraints. We consider two models in the plane; refer to Fig. 1.

1. A **polygonal linkage** is a set $\mathcal{P}$ of convex polygons, and a set H of hinges, where each hinge $h \in H$ corresponds to two or more points on the boundaries of distinct polygons in $\mathcal{P}$. A **realization** of a polygonal linkage is an interior-disjoint placement of congruent copies of the polygons in $\mathcal{P}$ such that the points corresponding to each hinge are identified. A **realization with**

© Springer International Publishing Switzerland 2015
E. Di Giacomo and A. Lubiw (Eds.): GD 2015, LNCS 9411, pp. 447–459, 2015.
DOI: 10.1007/978-3-319-27261-0_37

orientation uses only translated or rotated copies of the polygons in $\mathcal{P}$ (no reflections) and for each hinge, the cyclic order of incident polygons is given. The topology of a polygonal linkage can be represented by the **hinge graph**, a bipartite graph where the vertices correspond to polygons in $\mathcal{P}$ and the hinges in H, and edges represent the polygon-hinge incidences.

2. An (abstract) graph is a **coin graph** if it is the intersection graph of a set of interior-disjoint unit disks in the plane (where the vertices correspond to disks and two vertices are adjacent if and only if the corresponding disks are in contact). A **coin graph with embedding** is a coin graph *together with* a cyclic order of the neighbors for each vertex (i.e., each disk).

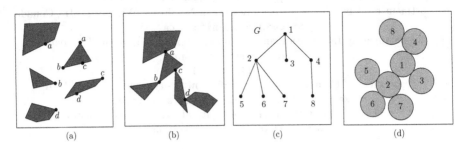

Fig. 1. (a) A set of convex polygons and hinges. (b) A realization of the polygonal linkage (with fixed orientation). (c) A graph G with 8 vertices. (d) An arrangement of interior-disjoint unit disks whose contact graph is G.

The POLYGONAL LINKAGE REALIZABILITY (PLR) problem asks whether a given polygonal linkage admits a realization; and PLR WITH FIXED ORIENTATION asks whether it admits a realization with a given orientation. The COIN GRAPH RECOGNITION (CGR) problem asks whether a given (abstract) graph G is the contact graph of interior-disjoint unit disks in the plane; and CGR WITH FIXED EMBEDDING asks whether a given plane graph G is the contact graph of interior-disjoint unit disks in the plane with the same counterclockwise order of neighbors at each vertex.

These problems, in general, are known to be NP-hard (see details below). However, the hardness reductions crucially rely on graphs with a large number of cycles. We revisit these problems for simply connected topologies, where the hinge graph and the coin graph are trees.

Summary of Results. Our main result is that the realizability problem remains NP-hard for simply connected polygonal linkages, the only exceptions are chains of triangles or rectangles hinged at distinct vertices. In an attempt to identify the most general problem that is not NP-hard, we considered several variants. Some variants are always realizable, some have easy hardness reductions, and some reductions required substantial new machinery. Our most demanding result is the NP-hardness of the recognition of *coin trees* with fixed embedding. We summarize the results here.

1. We start with *chains of polygons*, that is, polygonal linkages in which the hinge graph is a path (Sect. 2). It is easy to see that every chain of triangles or rectangles hinged together at distinct vertices is realizable and a realization can be computed efficiently. However, the problem becomes weakly NP-hard for chains of convex quadrilaterals hinged at distinct vertices or for chains of triangles where one hinge may be at anywhere on the boundary. Our reduction uses PARTITION.

2. We show that PLR (with arbitrary orientation) is strongly NP-hard when the hinge graph is a tree, using an easy reduction from 3SAT with the classic logic engine method (the proof is available in the full paper). The reduction crucially depends on possible reflections of the polygons.

3. We show that PLR with fixed orientation is also strongly NP-hard when the hinge graph is a tree (Sect. 3), using a significantly more involved reduction from PLANAR3SAT. We carefully design gadgets for variables, clauses and a planar graph to simulate PLANAR3SAT.

4. We reduce the recognition of coin trees with fixed embedding to the previous problem (PLR with fixed orientation), by simulating suitable polygons with an arrangement of unit disks (Sect. 4). It would be easy to model a polygon by a *rigid* coin graph (e.g., a section of the triangular grid), but all rigid graphs induce cycles. The main technical difficulty is that when the coin graph is a tree, any realization with unit disks is highly flexible, and simulating a rigid object becomes a challenge. We construct coin trees with "stable" realizations, which may be of independent interest.

Related Previous Work. Previous research has established NP-hardness in several easy cases, but realizability for simply connected structures remained open. Polygonal linkages (or body-and-joint frameworks) are a generalization of classical linkages (bar-and-joint frameworks) in rigidity theory. A linkage is a graph $G = (V, E)$ with given edge lengths. A realization of a linkage is a (crossing-free) straight-line embedding of G in the plane. Based on ideas developed by Bhatt and Cosmadakis [4], who proved that the realizability of linkages is NP-complete on the integer grid, the *logic engine* method [14,15,17,20] has become a standard tool for proving NP-hardness in graph drawing. The logic engine is a graph composed of rigid 2-connected components, where two possible realizations of a 2-connected component encode a binary variable.

However, the logic engine method is **not** applicable to problems with fixed embedding or orientation, where the circular order of the neighbors of each vertex is part of the input. Cabello et al. [7,16] used a significantly more elaborate reduction to show that the realizability of 3-connected linkages (where the orientation is unique by Whitney's theorem [25]) is NP-hard. This problem is efficiently decidable, though, for near-triangulations [7,13].

Note that every *tree* linkage can be realized in $\mathbb{R}^2$ with almost collinear edges. According to the celebrated *Carpenter's Rule Theorem* [10,24], every realization of a path (or a cycle) linkage can be continuously moved (without self-intersection) to any other realization. In other words, the realization space of such a linkage is always connected. However, there are trees of maximum degree 3 with as few as 8 edges whose realization space is disconnected [2]; and deciding whether the realization space of a tree linkage is connected is PSPACE-complete [1]. (Earlier, Reif [22] showed that it is PSPACE-complete to decide whether a polygonal linkage can be moved from one realization to another among polygonal obstacles in $\mathbb{R}^3$.) Cheong et al. [8] consider the "inverse" problems of introducing the minimum number of point obstacles to reduce the configuration space of a polygonal linkage to a unique realization.

Connelly et al. [11] showed that the Carpenter's Rule Theorem generalizes to certain polygonal linkages obtained by replacing the edges of a path linkage with special polygons (called *slender adornments*). Our Theorem 3 indicates that if we are allowed to replace the edges of a linkage with arbitrary convex polygons, then deciding whether the realization space is empty or not is already NP-hard.

Recognition problems for intersection graphs of various geometric object have a rich history [20]. Breu and Kirkpatrick [6] proved that it is NP-hard to decide whether a graph G is the contact graph of unit disks in the plane, i.e., recognizing *coin graphs* is NP-hard; see also [14]. Recognizing outerplanar coin graphs is already NP-hard, but decidable in linear time for caterpillars [21]. It is also NP-hard to recognize the contact graphs of pseudo-disks [20] and disks of bounded radii [5] in the plane, and unit disks in higher dimensions [19,20]. All these hardness reductions produce graphs with a large number of cycles, and do not apply to trees. Note that the contact graphs of disks *of arbitrary radii* are exactly the planar graphs (by Koebe's circle packing theorem), and planarity testing is polynomial. Consequently, every tree is the contact graph of disks of *some* radii in the plane. However, deciding whether a given star is realizable as a contact graph of disks of given radii but arbitrary embedding is already NP-hard [21].

Eades and Wormald [16] showed that it is NP-hard to decide whether a given tree is a *subgraph* of a coin graph. Schaefer [23] proved that deciding whether a graph with given edge lengths can be realized by a straight-line drawing (possibly with crossing edges) has the same complexity as the existential theory of the reals. Both reductions crucially rely on a large number of cycles. Our work is the first to simulate rigid polygons with truly flexible combinatorial structures that have simply connected topology.

2   Chains of Polygons

In this section, we consider polygonal linkages whose hinge graph is a path. We call such a linkage a *chain of polygons*, given by a sequence of convex polygons $(P_1, \ldots, P_n)$, and $n - 1$ hinges, where the ith hinge corresponds to a pair of points on the boundaries of P_i and P_{i+1}, for $i = 1, \ldots, n - 1$. Generalizing an observation by Demaine et al. [12][Lemma 2], we formulate a simple sufficient condition for the realizability of a chain of polygons.

Proposition 1. *Consider a chain of convex polygons $(P_1, \ldots, P_n)$ with $n - 1$ hinges. If P_i admits parallel tangent lines through both of its hinges for $i = 2, \ldots, n - 1$, then the chain of polygons is realizable with fixed orientation. Furthermore, a realization can be computed in $O(n)$ time.*

It follows that every chain of triangles (resp., rectangles) hinged at distinct vertices is realizable with fixed orientation. Surprisingly, the realizability of a chain of arbitrary polygons is already NP-hard, even if the polygons are convex quadrilaterals hinged at vertices, or triangles hinged at arbitrary boundary points. We reduce the problem from PARTITION, which is weakly NP-hard (i.e., NP-hard when the input is a sequence of n integers between 1 and 2^n). We give two NP-hardness proofs for the problem: an easier reduction for the case where reflections of polygons are allowed, and, as an extension, a more technical proof for the case when the orientation of the polygons is fixed. The main idea behind both proofs is that any realization of the chain enforces a bounded rectangular region (frame) in which the remaining polygons have to be fitted. The width of the remaining polygons encode the integers given by the PARTITION instance. Simply speaking, the joint of the first and the last polygon inside the frame have to be vertically aligned to get the last big polygon in. This is possible if and only if we have a yes-instance for PARTITION. See Fig. 2 for an example of the reduction, further details are given in the full paper.

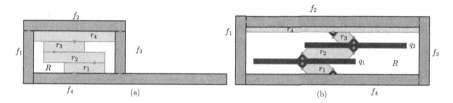

Fig. 2. (a) A chain of 8 rectangles encode PARTITION for 3 integers (a_1, a_2, a_3). Rectangles $f_1, \ldots, f_4$ form a frame around a rectangle R in any realization. (b) A chain of 16 polygons encode PARTITION for 3 integers (a_1, a_2, a_3).

For the second proof we first reduce to instances in which the chained polygons are either triangles, rectangles, or hexagons that are formed by rectangles of height 2 from which an isosceles triangle of side lengths 1 is cut off on every corner. We then replace these polygons by a subchain of triangles whose unique realization redefines these shapes (Fig. 3 depicts this idea).

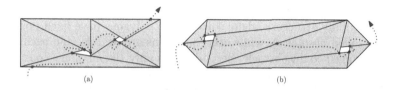

Fig. 3. (a) A rectangle with two hinges on opposite sides is split into a chain of 8 triangles, which has a unique realization (even with reflections). (b) A hexagon with two hinges at opposite vertices is split into a chain of 8 triangles, which has a unique realization with fixed orientation.

Theorem 1. *It is weakly NP-hard to decide whether a chain of rectangles is realizable.*

Theorem 2. *It is weakly NP-hard to decide whether a chain of convex polygons is realizable with fixed orientation. This is already true if the chain of polygons is formed by triangles whose hinges are not restricted to vertices.*

3   Realizability of Polygonal Linkages with Fixed Orientation

Theorem 3. *It is strongly NP-hard to decide whether a polygonal linkage whose hinge graph is a **tree** can be realized with fixed orientation.*

Our proof for Theorem 3 is a reduction from PLANAR-3-SAT (P3SAT): decide whether a given Boolean formula in 3-CNF with a planar associated graph is satisfiable. The *graph associated* to a Boolean formula in 3-CNF is a bipartite graph where the two vertex classes correspond to the variables and to the clauses, respectively; there is an edge between a variable x and a clause C iff x or $\neg x$ appears in C. See Fig. 4(left).

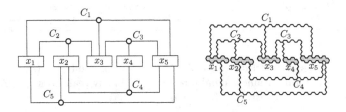

Fig. 4. Left: the associated graph $A(\Phi)$ for a Boolean formula Φ. Right: the schematic layout of the variable, clause, and transmitter gadgets in our construction.

The Big Picture. Given an instance Φ of P3SAT with n variables and m clauses, we construct a simply connected polygonal linkage $(\mathcal{P}, H)$, of polynomial size in n and m, such that Φ is satisfiable iff $(\mathcal{P}, H)$ admits a realization with fixed orientation. We construct a polygonal linkage in two main steps: First, we construct an auxiliary structure where some of the polygons have fixed position in the plane (called *obstacles*), while other polygons are flexible, and each flexible polygon is hinged to an obstacle. Second, we modify the auxiliary construction into a polygonal linkage by allowing the obstacles to move freely, and by adding new polygons and hinges as well as an exterior *frame* that holds the obstacle polygons in place. All polygons in our constructions are regular hexagons or long and skinny rhombi because these are the polygons that we can "simulate" with coin graphs in Sect. 4.

We start with embedding the graph $A(\Phi)$ associated to Φ into a hexagonal tiling, and then replace the vertices by variable and clause gadgets, and the edges by transmitter gadgets (to be described below). A variable gadget corresponds to a cycle in the hexagonal tiling, a clause gadget to single vertex incident to three hexagons, and a transmitted gadget to a path along a sequence of edges and vertices of the tiling. Refer to Fig. 4(right).

The main idea for the auxiliary construction is the following. We thicken the edges of the hexagonal tiling into *corridors* of uniform width, and the vertices of the tiling into regular triangles, which form *junctions* between three corridors. The boundaries of the corridors form regular hexagons, which will be the obstacle polygons in our auxiliary construction. In each corridor, we insert flexible hexagons, with one corner hinged to the boundary of the corridor. Each flexible hexagon has two possible realizations (say, *left* and *right*) that can encode a binary variable: all flexible polygons turn in the same direction along a cycle (clockwise or counterclockwise) with suitable spacing between the hexagons (Fig. 5(a) and (b) and with a small flexible polygon at each junction (Fig. 5(c)). Similarly, the value of a binary variable is transmitted via a chain of corridors and junctions. A clause of Φ is simulated by a single junction (Fig. 7), where a small flexible polygon ensures that hexagons from at most two adjacent corridors enter the junction (i.e., at most two literals are false).

Auxiliary Construction: Flexible Hexagons in a Rigid Frame. Let Φ be a Boolean formula in 3CNF with variables $x_1, \ldots, x_n$ and clauses $C_1, \ldots, C_m$, and let $A(\Phi)$ be the associated planar graph. We modify $A(\Phi)$ to obtain a plane graph $\tilde{A}(\Phi)$ of maximum degree 3 as follows: Replace each *variable* vertex v by a cycle whose length equals the degree of v, and distribute the edges incident to v among the vertices of the cycle.

Embed $\tilde{A}(\Phi)$ into the section of a hexagonal tiling (Fig. 4), contained in a regular hexagon of side length N, where N is a polynomial of n and m [3]. Let $t = 2N^3 + 1$ (t will be the number of flexible hexagons in a corridor). Scale the grid such that the cells become regular hexagons of side length $(5t - 1)/2 + \sqrt{3}$, and then scale each cell independently from its center to a hexagon of side length $(5t - 1)/2$. These large hexagons are considered fixed obstacles in our auxiliary construction. Between two adjacent obstacle hexagons, there is a $\frac{5t-1}{2} \times \sqrt{3}$ rectangle, which we call a *corridor*. Three adjacent corridors meet at a regular triangle, which we call a *junction*. We next describe variable, clause, and transmitter gadgets.

The basic building block of both variable and transmitter gadgets consists of t regular hexagons of side length 1 (*unit hexagons*, for short) attached to a wall of a corridor such that the hinges divide the wall into $t + 1$ intervals of length $(1, 2.5, \ldots, 2.5, 1)$ as shown in Fig. 5(a) and (b) for $t = 3$. Since the height of the corridor is $\sqrt{3}$, each hexagon has exactly two possible realizations: it can lie either *left* or *right* of the hinge in a horizontal corridor. For simplicity, we use the same notation (R and L) in nonhorizontal corridors, too. Hence, the *state* of each flexible hexagon in a realization is either L or R. The following observation describes the key mechanism of a corridor.

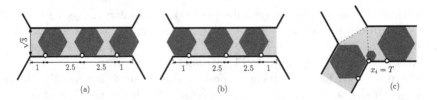

Fig. 5. (a) A corridor when all unit hexagons are in state R. (b) A corridor where all unit hexagons are in state L. (c) A junction where a small hexagon between two corridors ensures that at most one unit hexagon enters the junction from those corridors.

Observation 1

(1) If the leftmost hexagon is in state R, then all t hexagons are in state R, and the rightmost hexagon enters the junction on the right of the corridor.

(2) Similarly, if the rightmost hexagon is in state L, then all t hexagons are in state L, and the leftmost hexagon enters the junction on the left of the corridor.

Each junction is a regular triangle, adjacent to three corridors. In some of the junctions, we attach a small hexagon of side length $\frac{1}{3}$ to one or two corners of the junction (see Fig. 5(c) and Fig. 6). Importantly, if such a small hexagon is attached to a vertex between two corridors, then a unit hexagon can enter the junction from at most one of those corridors.

The **variable gadget** for variable x_i is constructed as follows. Recall that variable x_i corresponds to a cycle in the associated graph $\tilde{A}(\Phi)$, which has been embedded as a cycle in the hexagonal tiling, with corridors and junctions. In each junction along this cycle, attach a small hexagon in the common boundary of the two corridors in the cycle. Observation 1 and the small hexagons ensure that the state of any unit hexagon along the cycle determines the state of all other unit hexagons in the cycle. This property defines the binary variable x_i: If $x_i = T$, then all unit hexagons in the top horizontal corridors are in state R; and if $x_i = F$, they are all in state L.

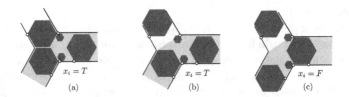

Fig. 6. The common junction of a variable gadget and a transmitter gadget. (a) When $x_i = T$, a hexagon of the transmitter may enter the junction of the variable gadget. (b) When $x_i = T$, the transmitter gadget has several possible realizations. (c) When $x_i = F$, no hexagon from the transmitter enters a junction of the variable gadget.

A **transmitter gadget** is constructed for each edge (x_i, C_j) of the graph $A(\Phi)$. It connects a junction of the variable gadget x_i with the junction representing the clause gadget C_j. The gadget consists of a path of corridors and junctions: at each interior junction, attach a small hexagon in the common boundary of the two corridors in the path (similarly to the variable gadget). At the common junction with the variable gadget x_i, we attach one additional small hexagon to one of the vertices (refer to Fig. 6). If the literal x_i (resp., $\overline{x}_i$) appears in C_j, then we attach a small hexagon to the corner of this junction such that if $x_i = F$ (resp., $\overline{x}_i = F$), then the unit hexagon of the transmitter gadget cannot enter this junction. This ensures that false literals are always correctly transmitted to the clause junctions (and true literals can always transmit correctly).

The **clause gadget** lies at a junction adjacent to three transmitter gadgets (see Fig. 7). At such a junction, we attach a unit line segment to an arbitrary vertex of the junction, and a small hexagon of side length $\frac{1}{3}$ to the other end of the segment. If unit hexagons enter the junction from all three corridors (i.e., all three literals are false), then there is no space left for the small hexagon. But if at most two unit hexagons enter the junction (i.e., one of the literals is true), then the unit segment and the small hexagon are realizable.

The following lemma summarizes our result about the auxiliary construction.

Lemma 1. *For every instance Φ of P3SAT, the above polygonal linkage with flexible and obstacle polygons has the following properties: (1) it has polynomial size; (2) its hinge graph is a forest; (3) it admits a realization such that the obstacle polygons remain fixed if and only if Φ is satisfiable.*

The remaining details of our construction can be found in the full paper.

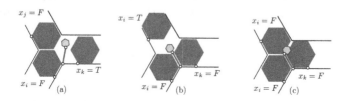

Fig. 7. (a-b) A clause gadget $(x_i \lor x_j \lor x_k)$ is realizable when at least one of the literals is TRUE. (c) The clause gadget cannot be realized when all three literals are FALSE.

4   Recognition of Coin Trees with Fixed Embedding

In this section, we reduce recognition of coin trees with fixed embedding from the realizability of polygonal linkages with cycle-free hinge graphs, which was shown to be strongly NP-hard in Sect. 3.

Theorem 4. *It is NP-hard to decide whether a given plane tree is a coin graph with fixed embedding.*

It is enough to show that the polygons and the hinges used in Sect. 3 can be simulated by disks whose contact graphs are trees. For a constant $\lambda > 0$, we say that a coin graph G with embedding is a λ-*stable approximation* of a polygon P if in every realization of G as the contact graph of interior-disjoint unit disks, the Hausdorff distance between the union of disks and a congruent copy of P is at most λ. In the remainder of this section, we design plane trees that approximate (i) a long and skinny rhombus, and (ii) a regular hexagon. We use these trees and suitable hinges to prove Theorem 4.

Let $|ab|$ denote the Euclidean distance between points a and b in the plane, and note that the distance between the centers of two kissing unit disks is precisely 2. Let $\angle abc \in [0, 2\pi)$ denote the counterclockwise angle that rotates ray $\overrightarrow{ba}$ to $\overrightarrow{bc}$. The following lemma about four unit disks is the key idea for our stability arguments.

Lemma 2. *Let (a, b, c, d) be a polygonal path in the plane such that $|ab| = |bc| = |cd| = 2$ and the unit disks centered at a, b, c, and d are interior-disjoint. Then the sum of angles at the interior vertices on the left (resp., right) of the chain is greater than π.*

Proof. Without loss of generality, consider the two angles on the left side at the two interior vertices, $\angle abc$ and $\angle bcd$. We have $|ab| = |bc| = |cd| = 2$, since the coin graph of the unit disks is P_4. If (a, b, c, d) is a rhombus, then $|ad| = 2$ and $\angle abc + \angle bcd = \pi$. Hence $|ad| > 2$ implies $\angle abc + \angle bcd > \pi$. □

We construct a caterpillar graph on $n = 8 + 4k$ vertices, for any $k \geq 0$, and show that it is a 2-stable approximation of a long and skinny rectangle. Recall that a *caterpillar* is a tree in which all vertices are either on or adjacent to a central path. For $k \geq 0$, let T_k be a plane caterpillar with central path $C = (a_{-k}, \ldots, a_{-1}, a_0, a_1, \ldots, a_k)$ such that the sequence of vertex degrees along the path is

$$1, \underbrace{3, \ldots, 3}_{k-2}, 4, 5, 4, \underbrace{3, \ldots, 3}_{k-2}, 1, \tag{1}$$

and all leaves are attached to the left side of C. Figure 8(left) shows that T_k can be embedded as a subgraph of a triangular grid. This embedding can be perturbed into a coin graph (such that the distance between any two leaves is strictly more than 2).

Lemma 3. *For every integer $k \geq 0$, the plane tree T_k in Fig. 8(left) is a 2-stable approximation of a rhombus of width $2k + 3$ and height $2 + 4\sqrt{3}$.*

The proof of the lemma boils down to a careful estimation of the realizable distances of the tree T_k. With the help of Lemma 2 we can show that for every $j \geq 1$ the centers of a_j and a_{-j} lie at distance at most 2 from their "canonical" position as indicated in Fig. 8(left). The proof goes via induction on j and can be found in the full paper.

We can now extend the tree T_k to a larger tree T_k' with $\Theta(k^2)$ vertices as shown in Fig. 9 that is a 2-stable approximation of a regular hexagon.

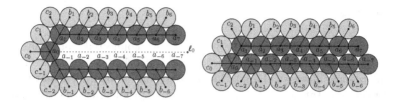

Fig. 8. This caterpillar T_6 consists of two oppositely oriented chains, each of which can only bend towards the other.

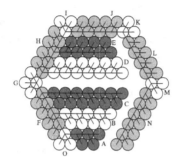

Fig. 9. The embedded tree T'_6 approximates a regular hexagon.

Lemma 4. *For every integer $k \geq 3$, the plane tree T'_k is a 2-stable approximation of a regular hexagon of side length $2k$.*

Proof (Sketch). Let $k \geq 0$ be an arbitrary integer. To construct T'_k, consider five 3-regular caterpillars, each of length k, joined in sequence at vertices of degree four (except one joint vertex of degree five) such that the leaf vertices lie on the outside. See Fig. 9. The joint vertices force five bends, each with a turn of more than $\pi/3$, resulting in a hexagonal shape. The interior of the hexagon is filled with pairs of 3-regular caterpillars aligned symmetrically across the x-axis, similar to the realizations of T_k. Since T_k is a subgraph of T'_k, the vertices in the subgraph are 2-stable by Lemma 3 (c.f., the branches C and D in Fig. 9).

By arguments analogous to those in the proof of Lemma 3, additional horizontal branches are similarly constrained and, therefore, also 2-stable (i.e., the branches A, B, E, and J in Fig. 9). Similarly, the five branches forming the hexagons boundary (i.e., the branches F, H, J, L, and N in Fig. 9) can only move towards the interior of the hexagon. The empty space between any two disks is strictly less than one disk diameter, and the result follows. $\qquad\square$

We now have everything ready to give the reduction for Theorem 4. Roughly speaking, we replace the polygons used in the reduction that is presented in Sect. 3 by stable approximations. The details of the proof of Theorem 4 are given in the full paper.

5   Conclusions

We have shown that deciding whether a simply connected polygonal linkage is realizable in the plane (with or without fixed orientation) is strongly NP-hard. The realizability of a chain of hinged polygons is weakly NP-hard (with or without fixed orientation); and it remains an open problem whether it is strongly NP-hard.

Our hardness proof for the recognition of coin graphs with fixed embedding used subgraphs that "approximate" a regular hexagon. It remains an open problem whether a similar approximation is possible for coin graphs with *arbitrary embedding*. We believe it is, but it would require an approximation of a "dense" packing of unit disks whose contact graph is a tree: this leads to a challenging problem in discrete geometry.

Acknowledgements. Our results in Sect. 4 were developed at the *First International Workshop on Drawing Algorithms for Networks of Changing Entities (DANCE 2014)*, held in Langbroek, the Netherlands, and supported by the NWO project 639.023.208. Research by Rounds and Tóth was supported in part by the NSF awards CCF-1422311 and CCF-1423615. Research by Durocher was supported in part by NSERC.

References

1. Alt, H., Knauer, C., Rote, G., Whitesides, S.: On the complexity of the linkage reconfiguration problem. In: Pach, J. (ed.) Towards a Theory of Geometric Graphs, vol. 342, Contemporary Mathematics, pp. 1–14. AMS, Providence (2004)
2. Ballinger, B., Charlton, D., Demaine, E.D., Demaine, M.L., Iacono, J., Liu, C.-H., Poon, S.-H.: Minimal locked trees. In: Dehne, F., Gavrilova, M., Sack, J.-R., Tóth, C.D. (eds.) WADS 2009. LNCS, vol. 5664, pp. 61–73. Springer, Heidelberg (2009)
3. Biedl, T., Kant, G.: A better heuristic for orthogonal graph drawings. Comput. Geom. **9**(3), 159–180 (1998)
4. Bhatt, S.N., Cosmadakis, S.S.: The complexity of minimizing wire lengths in VLSI layouts. Inform. Process. Lett. **25**(4), 263–267 (1987)
5. Breu, H., Kirkpatrick, D.G.: On the complexity of recognizing intersection and touching graphs of discs. In: Brandenburd, F.J. (ed.) GD 1995. LNCS, vol. 1027, pp. 88–98. Spinger, Heidelberg (1996)
6. Breu, H., Kirkpatrick, D.G.: Unit disk graph recognition is NP-hard. Comput. Geom. **9**, 3–24 (1998)
7. Cabello, S., Demaine, E.D., Rote, G.: Planar embeddings of graphs with specified edge lengths. J. Graph Alg. Appl. **11**(1), 259–276 (2007)
8. Cheong, J.-S., van der Stappen, A.F., Goldberg, K., Overmars, M.H., Rimon, E.: Immobilizing hinged polygons. Int. J. Comput. Geom. Appl. **17**(1), 45–70 (2007)
9. Connelly, R., Demaine, E.D.: Geometry and topology of polygonal linkages. In: Goodman, J.E., O'Rourke, J. (eds.) Handbook of Discrete and Computational Geometry, ch. 9, pp. 197–218. CRC, Boca Raton (2004)
10. Connelly, R., Demaine, E.D., Rote, G.: Straightening polygonal arcs and convexifying polygonal cycles. Discrete Comput. Geom. **30**(2), 205–239 (2003)

11. Connelly, R., Demaine, E.D., Demaine, M.L., Fekete, S.P., Langerman, S., Mitchell, J.S.B., Ribó, A., Rote, G.: Locked and unlocked chains of planar shapes. Discrete Comput. Geom. **44**(2), 439–462 (2010)
12. Demaine, E.D., Eppstein, D., Erickson, J., Hart, G.W., O'Rourke, J.: Vertex-unfoldings of simplicial manifolds. In: 18th Sympos. on Comput. Geom., pp. 237–243. ACM Press, New York (2002)
13. Di Battista, G., Vismara, L.: Angles of planar triangular graphs. SIAM J. Discrete Math. **9**(3), 349–359 (1996)
14. Di Battista, G., Eades, P., Tamassia, R., Tollis, I.G.: Graph Drawing: Algorithms for the Visualization of Graphs. Prentice Hall, Upper Saddle River (1999)
15. Eades, P., Whitesides, S.: The realization problem for Euclidean minimum spanning trees is NP-hard. Algorithmica **16**(1), 60–82 (1996)
16. Eades, P., Wormald, N.C.: Fixed edge-length graph drawing is NP-hard. Discrete Appl. Math. **28**, 111–134 (1990)
17. Fekete, S.P., Houle, M.E., Whitesides, S.: The wobbly logic engine: Proving hardness of non-rigid geometric graph representation problems. In: Di Battista, G. (ed.) GD 1997. LNCS, vol. 1353, pp. 272–283. Springer, Heidelberg (1997)
18. Gregori, A.: Unit-length embedding of binary trees on a square grid. Inform. Process. Lett. **31**, 167–173 (1989)
19. Hliněný, P.: Touching graphs of unit balls. In: Di Battista, G. (ed.) GD 1997. LNCS, vol. 1353, pp. 350–358. Springer, Heidelberg (1997)
20. Hliněný, P., Kratochvíl, J.: Representing graphs by disks and balls (a survey of recognition-complexity results). Discrete Math. **229**(1–3), 101–124 (2001)
21. Klemz, B., Nöllenburg, M., Prutkin, R.: Recognizing weighted disk contact graphs. In: Di Giacomo, E., Lubiw, A. (eds.) GD 2015. LNCS, vol. 9411, pp. 433–446. LNCS, Spinger, Heidelberg (2015)
22. Reif, J.H.: Complexity of the mover's problem and generalizations. In: 20th FoCS, pp. 421–427. IEEE, New York (1979)
23. Schaefer, M.: Realizability of graphs and linkages. In: Pach, J. (ed.) Thirty Essays on Geometric Graph Theory, pp. 461–482. Springer, Heidelberg (2013)
24. Streinu, I.: Pseudo-triangulations, rigidity and motion planning. Discrete Comput. Geom. **34**(4), 587–635 (2005)
25. Whitney, H.: Congruent graphs and the connectivity of graphs. Amer. J. Math. **54**, 150–168 (1932)

Towards Characterizing Graphs with a Sliceable Rectangular Dual

Vincent Kusters[1]([⊠]) and Bettina Speckmann[2]

[1] Department of Computer Science, ETH Zürich, Zürich, Switzerland
`vincent.kusters@inf.ethz.ch`
[2] Department of Computer Science, TU Eindhoven, Eindhoven, The Netherlands
`b.speckmann@tue.nl`

Abstract. Let $\mathcal{G}$ be a plane triangulated graph. A rectangular dual of $\mathcal{G}$ is a partition of a rectangle R into a set $\mathcal{R}$ of interior-disjoint rectangles, one for each vertex, such that two regions are adjacent if and only if the corresponding vertices are connected by an edge. A rectangular dual is sliceable if it can be recursively subdivided along horizontal or vertical lines. A graph is *rectangular* if it has a rectangular dual and *sliceable* if it has a sliceable rectangular dual. There is a clear characterization of rectangular graphs. However, a full characterization of sliceable graphs is still lacking. The currently best result (Yeap and Sarrafzadeh, 1995) proves that all rectangular graphs without a separating 4-cycle are sliceable. In this paper we introduce a recursively defined class of graphs and prove that these graphs are precisely the nonsliceable graphs with exactly one separating 4-cycle.

1 Introduction

Let $\mathcal{G}$ be a plane triangulated graph. A *rectangular dual* of $\mathcal{G}$ is a rectangular partition $\mathcal{R}$ such that *(i)* no four rectangles meet in the same point, *(ii)* there is a one-to-one correspondence between the rectangles in $\mathcal{R}$ and the vertices of $\mathcal{G}$, and *(iii)* two rectangles in $\mathcal{R}$ share a common boundary segment if and only if the corresponding vertices of $\mathcal{G}$ are connected. A graph can have exponentially many rectangular duals [6], but might not even have a single one. Rectangular duals have a variety of applications, for example, as rectangular cartograms in cartography or as floorplans in architecture and VLSI design.

There are several types of rectangular duals that are of particular interest. Often it is desirable to assign certain areas to each rectangle. A recent paper by Eppstein et al. [8] studies *area-universal* rectangular duals, which have the property that any assignment of areas to rectangles can be realized by a combinatorially equivalent rectangular dual. A rectangular dual is *sliceable* if it can be recursively subdivided along horizontal or vertical lines (such duals are also

V. Kusters is partially supported by the ESF EUROCORES programme EuroGIGA, CRP GraDR and the Swiss National Science Foundation, SNF Project 20GG21-134306.

© Springer International Publishing Switzerland 2015
E. Di Giacomo and A. Lubiw (Eds.): GD 2015, LNCS 9411, pp. 460–471, 2015.
DOI: 10.1007/978-3-319-27261-0_38

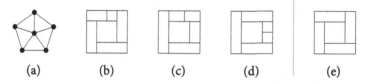

Fig. 1. A graph (a) with rectangular duals (b)-(d) and a rectangular dual of a different graph (e): (b) is not sliceable and not area-universal, (c) is sliceable and not area-universal, (d) is sliceable and area-universal, and (e) is area-universal and not sliceable.

called guillotine floorplans and can be constructed by glass cuts). While it is generally difficult to determine if an area assignment is feasible and to compute the corresponding layout of the rectangles, it is very easy to do so for sliceable duals. Furthermore, sliceable duals more easily facilitate certain layout steps in VLSI layout. Sliceability does not imply area-universality or vice versa (see Fig. 1).

A graph is *rectangular* if it has a rectangular dual and *sliceable* if it has a sliceable rectangular dual. Ungar [20], Bhasker and Sahni [4], and Koźmiński and Kinnen [12] independently gave equivalent characterizations of the rectangular graphs. Eppstein et al. [8] characterized the area-universal rectangular duals. However, despite an active interest in sliceable rectangular duals, a full characterization of sliceable graphs is still lacking. The currently best result by Yeap and Sarrafzadeh [22] from 1995 proves that all rectangular graphs without a separating 4-cycle are sliceable. Dasgupta and Sur-Kolay [7] modified the approach of Yeap and Sarrafzadeh and claimed two sufficient conditions for sliceability. However, Mumford [15] discovered a critical flaw that invalidates their results.[1]

Related Work. Rectangular duals have been studied extensively by the VLSI community. Sliceable layouts more easily facilitate certain steps in the layout process [16]. For instance, the problem of minimizing the perimeter or area of modules in a rectangular layout according to a given measure can be solved in polynomial time for sliceable layouts, but is NP-complete in general [17]. Several papers focus on restricted classes of sliceable and nonsliceable graphs [5,18].

Rectangular duals are also studied in the context of rectangular cartograms, which represent geographic regions by rectangles. The positioning and adjacencies of these rectangles are chosen to suggest their geographic locations and their areas correspond to the numeric values that the cartogram communicates. Van Kreveld and Speckmann [13] gave the first algorithms to compute rectangular cartograms. Eppstein et al. [8] present a numerical algorithm for area-universal rectangular duals which computes a cartogram with approximately the correct areas. For sliceable rectangular duals one can easily compute a combinatorially equivalent rectangular dual with exactly the specified area assignment, if such a rectangular dual exists. Several papers consider *rectilinear duals*: a generalization of rectangular duals which uses simple (axis-aligned) rectilinear polygons instead of rectangles. Every triangulated graph has a rectilinear dual where every

---

[1] Confirmed by Dasgupta and Sur-Kolay, personal communication, 2011-2013.

polygon has eight sides, and eight sides are sometimes necessary [10,14,23]. A series of papers studies the question of how many sides are required to respect all adjacencies and area requirements in general. De Berg, Mumford and Speckmann [3] gave the first bound by showing that forty sides per polygon is always sufficient. After several intermediate results, Alam et al. [2] finally closed the gap by proving that eight sides per polygon is always sufficient.

Sliceable rectangular duals are also called *guillotine partitions* or *guillotine layouts*. In this context a different notion of equivalence is used, which is not based on a dual graph. Specifically, two guillotine partitions are equivalent if they have the same *structure tree* [19]. Yao et al. [21] show that the asymptotic number of guillotine partitions is the nth Schröder number. Ackerman et al. [1] derive the asymptotic number of guillotine partitions in higher dimensions.

Results and Organization. It is comparatively easy to see that the class of sliceable graphs is not closed under minors. Hence we need to explore different approaches to characterize them. In Sect. 3 we introduce a recursively defined class of graphs, so-called *rotating pyramids*, which contain exactly one separating 4-cycle. We conjecture that configurations of rotating pyramids determine if a graph is sliceable. We verify our conjecture for the graphs that contain exactly one separating 4-cycle. The nonsliceable graphs in this class are exactly the graphs that reduce to *rotating windmills*: rotating pyramids with a specific corner assignment. In Sect. 4 we prove that rotating windmills are not sliceable and in Sect. 5 we argue that all other graphs with exactly one separating 4-cycle are sliceable.

2   Preliminaries

An *extended graph* $E(\mathcal{G})$ of a plane graph $\mathcal{G}$ is an extension of $\mathcal{G}$ with four vertices in such a way that the four vertices form the outer face of $E(\mathcal{G})$. These vertices are labeled $t(\mathcal{G})$, $r(\mathcal{G})$, $b(\mathcal{G})$ and $l(\mathcal{G})$ in clockwise order and are called the *poles* of $E(\mathcal{G})$. The vertices of the original graph $\mathcal{G}$ are called the *interior* vertices. Since choosing the extended graph fixes the vertices that correspond to the four corners (and hence the vertices along the four sides) of the rectangular dual, extended graphs are also called *corner assignments* (Fig. 2).

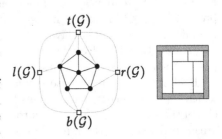

Fig. 2. An extended graph $E(\mathcal{G})$ and the corresponding rectangular dual.

A *separating k-cycle* of an extended graph $E(\mathcal{G})$ is a k-cycle with vertices both inside and outside the cycle. A *triangle* is a 3-cycle. The *outer cycle* of a plane graph is the cycle formed by the edges incident to the unbounded face. An *irreducible triangulation* is a plane graph without separating triangles and where all interior faces are triangles and the outer face is a quadrangle. A graph

Fig. 3. The windmill, the generalized windmill (the hatched shape is an arbitrary graph), and a rectangular dual of the generalized windmill.

$\mathcal{G}$ has a rectangular dual if and only if $\mathcal{G}$ has an extended graph which is an irreducible triangulation [4,12,20].

Sliceable Graphs. A rectangular partition is *sliceable* if it can be recursively subdivided along horizontal or vertical lines. An extended graph $E(\mathcal{G})$ is sliceable if and only if it has a sliceable rectangular dual. A graph $\mathcal{G}$ is sliceable if and only if it has a sliceable extended graph. Since a graph has only polynomially many corner assignments, we consider only extended graphs from now on. The smallest nonsliceable extended graph is the *windmill* depicted in Fig. 3. This extended graph can be generalized to a *generalized windmill* by replacing the center vertex with an arbitrary graph. All generalized windmills are nonsliceable.

A *cut* is a partition of the vertices of a graph in two disjoint subsets. The *cut-set* of the cut is the set of edges whose endpoints are in different subsets of the partition. A cut of $\mathcal{G}$ with cut-set S is *vertical* if the edges dual to S form a path from an interior face incident to $t(\mathcal{G})$ to an interior face incident to $b(\mathcal{G})$. Order the edges in the cut-set $e_1, \ldots, e_m$, according to the order in which they are traversed by the dual path. The *left vertex* of e_i is the endpoint of e_i that is in the same component as $l(\mathcal{G})$ in the graph obtained by deleting $t(\mathcal{G})$, $b(\mathcal{G})$, and S from $E(\mathcal{G})$. The *right vertex* is defined analogously. Let the *left boundary walk* $W_\ell = t(\mathcal{G}), u_1, \ldots, u_\ell, b(\mathcal{G})$ be the sequence of left endpoints of $e_1, \ldots, e_m$ (removing consecutive duplicates), and let the *right boundary walk* $W_r = t(\mathcal{G}), v_1, \ldots, v_r, b(\mathcal{G})$ be the sequence of right endpoints of $e_1, \ldots, e_m$ (removing consecutive duplicates). A walk is a path if it visits every vertex at most once. A path $v_1, \ldots, v_k$ is *chordless* if and only if v_i and v_j are not adjacent for each $1 \le i < j - 1 \le k$. A vertical cut is a *vertical slice* if its boundary walks are chordless paths (Fig. 4). A vertical slice divides $\mathcal{G}$ into $\mathcal{G}_l$ and $\mathcal{G}_r$. Horizontal cuts, top and bottom boundary walks and horizontal slices are defined analogously.

Regular Edge Labelings. The equivalence classes of the rectangular duals of an irreducible triangulation $E(\mathcal{G})$ correspond one-to-one to the *regular edge labelings of $E(\mathcal{G})$*. A regular edge labeling of an extended graph $E(\mathcal{G})$ is a partition of the interior edges of $E(\mathcal{G})$ into two subsets of red (dashed) and blue (solid) directed edges such that: (*i*) around each inner vertex in clockwise order we have four contiguous nonempty sets of incoming blue edges, outgoing red edges, outgoing blue edges, and incoming red

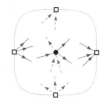

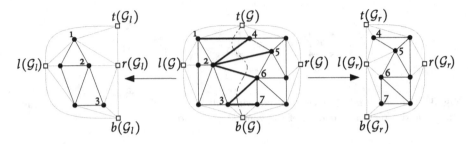

Fig. 4. An extended graph $E(\mathcal{G})$ with a vertical slice indicated by a dash-dotted line and the corresponding $E(\mathcal{G}_\ell)$ and $E(\mathcal{G}_r)$. The edges of the cut-set are bold. The boundary paths are $t(\mathcal{G}), 1, 2, 3, b(\mathcal{G})$ and $t(\mathcal{G}), 4, 5, 6, 7, b(\mathcal{G})$. Both boundary paths are chordless. Figure based on [22].

edges and; *(ii)* $l(\mathcal{G})$ has only outgoing blue edges, $t(\mathcal{G})$ has only incoming red edges, $r(\mathcal{G})$ has only incoming blue edges and $b(\mathcal{G})$ has only outgoing red edges.

A regular edge labeling is *sliceable* if its corresponding rectangular dual is sliceable. One can find a regular edge labeling and construct the corresponding rectangular dual in linear time [11]. A *regular edge coloring* is a regular edge labeling, without the edge directions. A regular edge coloring uniquely determines a regular edge labeling [9, Proposition 2]. A *monochromatic triangle* is a triangle where all edges have the same color. A regular edge labeling (of an irreducible triangulation) induces no monochromatic triangles [9, Lemma 1].

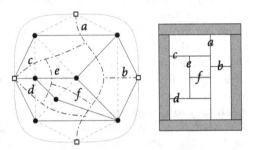

Fig. 5. A regular edge labeling and corresponding rectangular dual. Letters indicate the slices.

Let $\mathcal{R}$ be a rectangular dual of $E(\mathcal{G})$ and let $\mathcal{L}$ be the regular edge labeling that corresponds to $\mathcal{R}$. Any vertical slice in $\mathcal{R}$ has a blue cut-set and red boundary paths in $\mathcal{L}$. Any horizontal slice in $\mathcal{R}$ has a red cut-set and blue boundary paths (see Fig. 5). A slice is a *first slice* of $E(\mathcal{G})$ if it starts and ends at poles of $E(\mathcal{G})$. Slice a is the only first slice in Fig. 5.

k-pyramid Extended Graphs. A *pyramid* is a 4-cycle with exactly one vertex in its interior. A *k-pyramid extended graph* is an irreducible triangulation $E(\mathcal{G})$ such that $\mathcal{G}$ has no cut-vertices, $\mathcal{G}$ has exactly k separating 4-cycles, and all separating 4-cycles in $E(\mathcal{G})$ are pyramids. We argue that it is sufficient for our investigation of sliceability to consider only k-pyramid extended graphs with $k \geq 1$. Firstly, we may assume $\mathcal{G}$ has no cut-vertex (all omitted proofs are in the full version of the paper):

Lemma 1. *Let $E(\mathcal{G})$ be an extended graph such that $\mathcal{G}$ has a cut-vertex v. Then v is adjacent to two opposite poles, say $t(\mathcal{G})$ and $b(\mathcal{G})$. Slice immediately left and immediately right of v. Then $E(\mathcal{G})$ is sliceable if and only if the three extended graphs that result from the two slices are sliceable.*

Secondly, Mumford [15] showed that it is sufficient to consider extended graphs $E(\mathcal{G})$ such that all separating 4-cycles in $\mathcal{G}$ are pyramids. Her proof directly extends to separating 4-cycles in $E(\mathcal{G})$ instead of $\mathcal{G}$, which immediately proves that generalized windmills (Fig. 3) are nonsliceable. Finally, 0-pyramid extended graphs are always sliceable [22].

Yeap and Sarrafzadeh's algorithm. In Sect. 5, we explicitly construct slices in a manner which is based on the algorithm by Yeap and Sarrafzadeh [22]. In Theorem 1 below we give a stronger version of their result and also add a missing case which was overlooked in their original analysis. A cycle C in $E(\mathcal{G})$ splits the plane into two parts: a bounded region and an unbounded region. We say that vertices in the bounded region including C are *enclosed* by C.

Theorem 1. *Let $E(\mathcal{G})$ be a k-pyramid extended graph ($k \geq 0$). Then there exists a vertical cut S such that (i) the left boundary walk P_ℓ of S is a chordless path that contains only vertices with distance 2 to $r(\mathcal{G})$ in $E(\mathcal{G}) \setminus \{t(\mathcal{G}), l(\mathcal{G}), b(\mathcal{G})\}$ and (ii) if the cycle $C_r := \langle r(\mathcal{G}), P_\ell, r(\mathcal{G}) \rangle$ does not enclose a pyramid, then S is a vertical slice. Analogous statements hold for $t(\mathcal{G})$, $l(\mathcal{G})$ and $b(\mathcal{G})$. Consequently, $E(\mathcal{G})$ is sliceable if $k = 0$.*

The following corollary of Lemma 1 gives a final simplification of our problem.

Lemma 2. *Let $E(\mathcal{G})$ be an extended graph with pole p such that p has only one neighbour v in $\mathcal{G}$. Let $E(\mathcal{G}')$ be the extended graph obtained by deleting v from $\mathcal{G}$ and connecting the neighbours of v in $\mathcal{G}$ to p. Then $E(\mathcal{G})$ is sliceable if and only if $E(\mathcal{G}')$ is sliceable.*

Exhaustively applying Lemma 2 to an extended graph $E(\mathcal{G})$ *reduces* $E(\mathcal{G})$ to an extended graph $E(\mathcal{G}')$. We say that $E(\mathcal{G}')$ is *reduced*. The extended graphs $E(\mathcal{G}_\ell)$ and $E(\mathcal{G}_r)$ resulting from a slice in $E(\mathcal{G})$ might not be reduced even if $E(\mathcal{G})$ is. In this sense, Lemma 2 is different from Lemma 1 and Mumford's observation. In the following we focus on the 1-pyramid extended graphs, among which are both sliceable and nonsliceable extended graphs. The smallest nonsliceable one is the windmill in Fig. 3.

3   Rotating Pyramids and Windmills

The graph on the right is the *big pyramid* graph. *Rotating windmills* are recursively defined as follows. The windmill (see Fig. 3) is a rotating windmill. Furthermore, the extended graphs depicted in Fig. 6 are *base rotating windmills*: they are four corner assignments of the big pyramid graph. If $E(\mathcal{G})$ is a rotating windmill other than the windmill, then we can construct another rotating windmill by replacing the

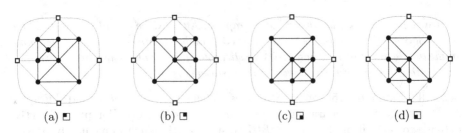

(a) ◨   (b) ◪   (c) ◩   (d) ◧

Fig. 6. The four base rotating windmills.

pyramid in $E(\mathcal{G})$ with a big pyramid using one of three construction steps, labeled ↑, ↖ and ↗, each depicted in Fig. 7.

Intuitively, ↑ extends the rotating windmill in the same direction as the previous extension, ↖ rotates the direction 90° counterclockwise and ↗ rotates the direction 90° clockwise. Note that the construction steps are not allowed to perform a rotation of 180°. We can uniquely identify a rotating windmill by its *construction sequence*. The construction sequence of the windmill is ⊠. The construction sequences of the base rotating windmills are ◨, ◪, ◩ and ◧. If we apply a construction step $s_{k+1} \in \{\uparrow, \nwarrow, \nearrow\}$ to a rotating windmill $bs_1 \cdots s_k$ where $k \geq 0$, $b \in \{\text{◨}, \text{◪}, \text{◩}, \text{◧}\}$, and $s_1, \ldots, s_k \in \{\uparrow, \nwarrow, \nearrow\}$, then the resulting rotating windmill has construction sequence $bs_1 \cdots s_k s_{k+1}$. Figure 8 shows three examples. If $E(\mathcal{G})$ is a rotating windmill, then we call $\mathcal{G}$ a *rotating pyramid*. For a given rotating pyramid $\mathcal{G}$, which is not the pyramid, the *inner graph* $\mathcal{G}'$ is defined as the largest strict subgraph of $\mathcal{G}$ such that $\mathcal{G}'$ is a rotating pyramid.

Drawing Conventions. We draw the edges of the outer cycle of a rotating pyramid $\mathcal{G}$ as a square. The *top side* of $\mathcal{G}$ is the path from the topleft vertex of $\mathcal{G}$ to the topright vertex (including both). The definitions of *right side, bottom side* and *left side* are analogous. Every rotating windmill has two consecutive sides with exactly two vertices, and two consecutive sides with at least two vertices.

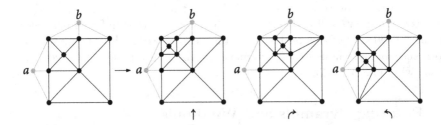

Fig. 7. On the left: the big pyramid in a rotating windmill, along with two of its neighbors in gray. On the right: the results of applying the three construction steps.

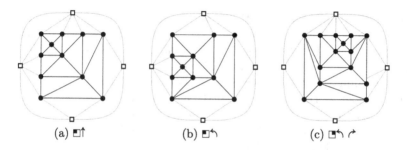

Fig. 8. Three rotating windmills.

Consider the graph $\mathcal{G}$ on the right. The partially drawn edges incident to the vertices on the outer cycle of $\mathcal{G}$ represent connections to vertices not shown in the figure. The inner graph $\mathcal{G}'$ of $\mathcal{G}$ is represented by only its outer cycle; its interior vertices (if any) are not shown. The lines along the top, right, bottom and left sides of $\mathcal{G}'$ contain the $\cdots$ symbol in their center to indicate that there may be zero or more extra vertices on the side. The edges whose color is 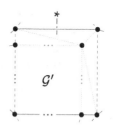 not uniquely determined are gray (dotted). The start of a slice is denoted with $*$, and the end of a slice is denoted with $\times$ (not shown). Every vertex on the top side of $\mathcal{G}'$ is connected to the topleft vertex in the figure, and every vertex on the right side of $\mathcal{G}'$ is connected to the bottomright vertex in the figure. Since $\mathcal{G}'$ is a rotating pyramid, a maximum of two sides of $\mathcal{G}'$ (and they must be consecutive) can have extra vertices.

4   Rotating Windmills are not Sliceable

Before we can prove the main result of this section, we need the following lemma:

Lemma 3. *Let $E(\mathcal{G})$ be an extended graph with a sliceable regular edge labeling $\mathcal{L}$. Let $\mathcal{G}'$ be a subgraph of $\mathcal{G}$ such that the outer cycle of $\mathcal{G}'$ under $\mathcal{L}$ has in clockwise order (i) a nonempty path of red edges followed by a nonempty path of blue edges oriented clockwise, and (ii) a nonempty path of red edges followed by a nonempty path of blue edges oriented counterclockwise. Let $E(\mathcal{G}')$ be the extended graph with labeling $\mathcal{L}'$ induced by coloring the edges of $\mathcal{G}'$ according to $\mathcal{L}$. The labeling $\mathcal{L}'$ is a sliceable labeling for $E(\mathcal{G}')$.*

Proof. The figure shows an example of the labeling of the outer cycle of $\mathcal{G}'$, the induced corner assignment $E(\mathcal{G}')$ and the labeling of $E(\mathcal{G}')$. Observe that the slices in $\mathcal{L}'$ are exactly the slices in $\mathcal{L}$ that cut through edges of $\mathcal{G}'$. Since $\mathcal{L}$ is a sliceable labeling of $E(\mathcal{G}')$, the labeling $\mathcal{L}'$ must also be sliceable.   □

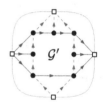

Theorem 2. *Extended graphs that reduce to rotating windmills are not sliceable.*

Proof. Since the reduction operation preserves sliceability, it is sufficient to consider rotating windmills. We will prove the theorem by structural induction on rotating windmills. Our base case is the windmill, which is not sliceable.

Let $E(\mathcal{G})$ be a rotating windmill and assume that all rotating windmills with fewer vertices are nonsliceable. Assume without loss of generality that the construction sequence of $E(\mathcal{G})$ starts with ▫. For the sake of deriving a contradiction, suppose that $E(\mathcal{G})$ is sliceable and consider a sliceable regular edge labeling. We assume wlog that the first slice in $E(\mathcal{G})$ is a vertical slice from $t(\mathcal{G})$ to $b(\mathcal{G})$. We show that any first slice either *(i)* cannot reach $b(\mathcal{G})$ or *(ii)* cuts $E(\mathcal{G})$ in such a way that a smaller graph is forced into a corner assignment that is a rotating windmill. Both cases result in a contradiction.

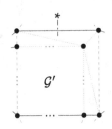

Fig. 9. Graph $\mathcal{G}$.

See Fig. 9. The vertices along the outer cycle of $\mathcal{G}$ are connected to the poles in $E(\mathcal{G})$. Since $t(\mathcal{G})$ has only incoming red edges, the edges along the top side of $\mathcal{G}$ must be blue. A similar reasoning forces the coloring of all edges on the outer cycle of $\mathcal{G}$. Let $\mathcal{G}'$ be the inner graph of $\mathcal{G}$. We distinguish four cases.

Case 1. The first slice does not cut through an edge in the top side of $\mathcal{G}'$, see Fig. 10. As noted previously, the colors of the edges along the outer cycle of $\mathcal{G}$ are forced by the corner assignment. The choice of the slice forces the colors of all dotted edges in Fig. 9. The induced corner assignment of $\mathcal{G}'$ is a rotating windmill $E(\mathcal{G}')$ which is smaller than $E(\mathcal{G})$. By the induction hypothesis, $E(\mathcal{G}')$ is not sliceable. Hence, $E(\mathcal{G})$ is also not sliceable. Contradiction.

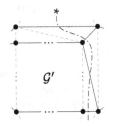

Fig. 10. Case 1.

Case 2. The top side of $\mathcal{G}'$ has at least two edges and the first slice cuts through the rightmost one, as depicted in Fig. 11(a). The induced corner assignment of $\mathcal{G}'$ is not a rotating windmill, so we cannot immediately conclude that $E(\mathcal{G})$ is not sliceable. Let us consider the structure of $\mathcal{G}'$. Note that the top side of $\mathcal{G}'$ has more than two vertices. This means that the construction sequence of $E(\mathcal{G})$ must start with ▫↱.

The slice that enters $\mathcal{G}'$ in Fig. 11(a) continues at the * in Fig. 11(b). Let $\mathcal{G}''$ be the inner graph of $\mathcal{G}'$. Note that the slice must enter $\mathcal{G}''$: if it did not, we would be in Case 1 again. It follows that the slice must enter $\mathcal{G}''$ through some edge on the right side of $\mathcal{G}''$. This forces the colors of all dotted edges in the figure. The slice cannot leave $\mathcal{G}''$ through an edge on the top or bottom side of $\mathcal{G}''$, since the slice cannot continue to $b(\mathcal{G})$ from there. Since the first slice does not reach $b(\mathcal{G})$, it cannot be the first slice. Contradiction.

Case 3. The top side of $\mathcal{G}'$ has at least two edges and the first slice does not cut through the rightmost one, see Fig. 11(c). Hence, the construction sequence of $E(\mathcal{G})$ must start with ▫↱. The first slice continues at * in Fig. 11(d). Let $\mathcal{G}''$ be the inner graph of $\mathcal{G}'$. All edges in $\mathcal{G}'$ incident to the topright vertex in $\mathcal{G}'$ must

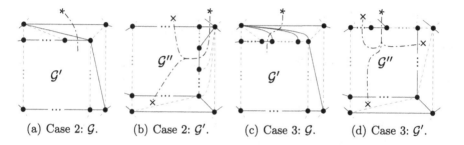

(a) Case 2: $\mathcal{G}$.   (b) Case 2: $\mathcal{G}'$.   (c) Case 3: $\mathcal{G}$.   (d) Case 3: $\mathcal{G}'$.

Fig. 11. (a-b) Graphs $\mathcal{G}$ and $\mathcal{G}'$ in Case 2. (c-d) Graphs $\mathcal{G}$ and $\mathcal{G}'$ in Case 3.

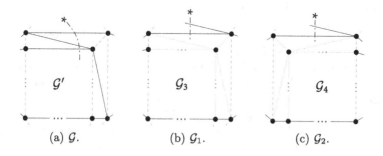

(a) $\mathcal{G}$.   (b) $\mathcal{G}_1$.   (c) $\mathcal{G}_2$.

Fig. 12. Case 4: graph $\mathcal{G}$ and two cases for $\mathcal{G}'$: graphs $\mathcal{G}_1$ and $\mathcal{G}_2$.

be red. This forces the coloring of all remaining edges. So the first slice cannot continue to $b(\mathcal{G})$ after leaving $\mathcal{G}''$: hence it cannot be the first slice. Contradiction.

Case 4. The top side of $\mathcal{G}'$ has exactly one edge e and the first slice cuts through e, see Fig. 12(a). Since $\mathcal{G}'$ has only two vertices on its top side, the construction sequence of $E(\mathcal{G})$ must start with ⊡↑ ($\mathcal{G}' = \mathcal{G}_1$) or ⊡↖ ($\mathcal{G}' = \mathcal{G}_2$). See Fig. 12(b) for $\mathcal{G}' = \mathcal{G}_1$ and Fig. 12(c) for $\mathcal{G}' = \mathcal{G}_2$. The only difference between $\mathcal{G}_1$ and $\mathcal{G}$ (Fig. 9) is that the topright vertex of $\mathcal{G}_1$ has an extra blue edge. Suppose that $E(\mathcal{G})$ is sliceable for $\mathcal{G}' = \mathcal{G}_1$ (the case $\mathcal{G}' = \mathcal{G}_2$ is similar). Let $\mathcal{L}_\mathcal{G}$ be a sliceable regular edge labeling of $E(\mathcal{G})$ and let $\mathcal{L}_\mathcal{G}[\mathcal{G}_1]$ be the restriction of $\mathcal{L}_\mathcal{G}$ to $\mathcal{G}_1$. All edges along the top side and bottom side of $\mathcal{G}_1$ in $\mathcal{L}_\mathcal{G}[\mathcal{G}_1]$ are blue and all the edges along the left side and right side are red. Let $E(\mathcal{G}_1)$ be the corner assignment of $\mathcal{G}_1$ such that $E(\mathcal{G}_1)$ is a rotating windmill. Coloring the edges of $\mathcal{G}_1$ inside $E(\mathcal{G}_1)$ according to $\mathcal{L}_\mathcal{G}[\mathcal{G}_1]$ yields a sliceable regular edge labeling for $E(\mathcal{G}_1)$ by Lemma 3. But since $E(\mathcal{G}_1)$ is a smaller rotating windmill than $E(\mathcal{G})$, it is not sliceable by the induction hypothesis. Contradiction.   □

5   Sliceability of 1-pyramid Extended Graphs

In this section we prove that all reduced 1-pyramid extended graphs other than rotating windmills are sliceable. Given a 1-pyramid extended graph $E(\mathcal{G})$, let C_p be the cycle defined in Theorem 1 for each pole $p \in \{l(\mathcal{G}), b(\mathcal{G}), r(\mathcal{G}), t(\mathcal{G})\}$.

Lemma 4. *Let $E(\mathcal{G})$ be a reduced 1-pyramid extended graph. Suppose that there exists a slice S that splits $E(\mathcal{G})$ into $E(\mathcal{G}_\ell)$ and $E(\mathcal{G}_r)$, such that $E(\mathcal{G}_\ell)$ (or $E(\mathcal{G}_r)$) can be reduced to a rotating windmill. Then we can construct a reduced 1-pyramid extended graph $E(\mathcal{G}')$ such that $E(\mathcal{G}')$ is not a rotating windmill, $\mathcal{G}'$ is a strict subgraph of $\mathcal{G}$ and $E(\mathcal{G})$ is sliceable if $E(\mathcal{G}')$ is sliceable.*

Proof (sketch). One can argue that that $E(\mathcal{G}_\ell)$ (or $E(\mathcal{G}_r)$) is already be a rotating windmill and then locally change S to a slice that does not induce a rotating windmill in the left or right graph. $\square$

Lemma 5. *Let $E(\mathcal{G})$ be a reduced 1-pyramid extended graph. If C_p encloses the pyramid of $\mathcal{G}$ for all poles p, then $E(\mathcal{G})$ is the windmill.*

Proof (sketch). First, the proof argues that since C_ℓ and C_r both enclose the pyramid, there is a cycle C formed by vertices L from P_ℓ and R from P_r that encloses the pyramid. Since $l(\mathcal{G})$ $(r(\mathcal{G}))$ has a path of length two to every vertex on P_ℓ (P_r), one can show that every vertex in $L \setminus R$ must have an edge to a vertex in R. It follows that C is a 4-cycle and since it encloses the pyramid in the 1-pyramid extended graph $E(\mathcal{G})$, the pyramid must be equal to C. Hence, P_ℓ and P_r contain an edge of the outer cycle of the pyramid. By a symmetric argument, P_t and P_b contain an edge of the outer cycle of the pyramid. Next, one can show that every edge of the outer cycle of the pyramid is on a different boundary path. Finally, we can use this property to show that every vertex on the outer cycle of the pyramid is connected to two adjacent poles. It follows that $E(\mathcal{G})$ contains the edges of the windmill. Since $E(\mathcal{G})$ is an irreducible triangulation, no other vertices can be present, which concludes the proof. $\square$

The following algorithm computes a sliceable labeling of a reduced 1-pyramid extended graph that is not a rotating windmill.

1. If $\mathcal{G}$ is a single vertex, we are done.
2. Since $E(\mathcal{G})$ is not a rotating windmill, by Lemma 5, there is a pole p for which C_p does not enclose the pyramid. Use Theorem 1 to compute a slice from p. This slice splits $E(\mathcal{G})$ into $E(\mathcal{G}_\ell)$ and $E(\mathcal{G}_r)$. One of these, say $\mathcal{G}_\ell$, contains the pyramid of $\mathcal{G}$. By Theorem 1, $E(\mathcal{G}_r)$ is sliceable. If $E(\mathcal{G}_\ell)$ can be reduced to a rotating windmill, then proceed to Step 1 with the reduced extended graph $E(\mathcal{G}')$ guaranteed by Lemma 4. Otherwise, reduce $E(\mathcal{G}_\ell)$ using Lemma 2 and go to Step 1 with $E(\mathcal{G}_\ell)$.

The algorithm maintains the invariant that $E(\mathcal{G})$ is a reduced 1-pyramid extended graph that is not a rotating windmill at line 1. Combined with Theorem 2, this concludes the proof of our main result:

Theorem 3. *A 1-pyramid extended graph is sliceable if and only if it cannot be reduced to a rotating windmill.*

References

1. Ackerman, E., Barequet, G., Pinter, R.Y., Romik, D.: The number of guillotine partitions in d dimensions. Inf. Proces. Letters **98**(4), 162–167 (2006)
2. Alam, M.J., Biedl, T., Felsner, S., Kaufmann, M., Kobourov, S.G., Ueckerdt, T.: Computing cartograms with optimal complexity. In: SOCG 2012, pp. 21–30 (2012)
3. de Berg, M., Mumford, E., Speckmann, B.: On rectilinear duals for vertex-weighted plane graphs. Disc. Math. **309**(7), 1794–1812 (2009)
4. Bhasker, J., Sahni, S.: A linear time algorithm to check for the existence of a rectangular dual of a planar triangulated graph. Networks **17**(3), 307–317 (1987)
5. Bhattacharya, B., Sur-Kolay, S.: On the family of inherently nonslicible floorplans in VLSI layout design. In: ISCAS 1991, pp. 2850–2853. IEEE (1991)
6. Buchin, K., Speckmann, B., Verdonschot, S.: Optimizing regular edge labelings. In: Brandes, U., Cornelsen, S. (eds.) GD 2010. LNCS, vol. 6502, pp. 117–128. Springer, Heidelberg (2011)
7. Dasgupta, P., Sur-Kolay, S.: Slicible rectangular graphs and their optimal floor-plans. ACM Trans. Design Automation of Electronic Systems **6**(4), 447–470 (2001)
8. Eppstein, D., Mumford, E., Speckmann, B., Verbeek, K.: Area-universal rectangular layouts. In: SOCG 2009, pp. 267–276 (2009)
9. Fusy, É.: Transversal structures on triangulations: A combinatorial study and straight-line drawings. Disc. Math. **309**(7), 1870–1894 (2009)
10. He, X.: On floor-plan of plane graphs. SIAM J. Comp. **28**(6), 2150–2167 (1999)
11. Kant, G., He, X.: Two algorithms for finding rectangular duals of planar graphs. In: van Leeuwen, J. (ed.) WG 1993. LNCS, vol. 790, pp. 396–410. Springer, Heidelberg (1994)
12. Koźmiński, K., Kinnen, E.: Rectangular duals of planar graphs. Networks **15**(2), 145–157 (1985)
13. van Kreveld, M., Speckmann, B.: On rectangular cartograms. Comp. Geom. **37**(3), 175–187 (2007)
14. Liao, C.C., Lu, H.I., Yen, H.C.: Compact floor-planning via orderly spanning trees. J. Algorithms **48**(2), 441–451 (2003)
15. Mumford, E.: Drawing Graphs for Cartographic Applications. Ph.D. thesis, TU Eindhoven (2008). http://repository.tue.nl/636963
16. Otten, R.: Efficient floorplan optimization. In: ICCAD'83. vol. 83, pp. 499–502 (1983)
17. Stockmeyer, L.: Optimal orientations of cells in slicing floorplan designs. Inf. Control **57**(2), 91–101 (1983)
18. Sur-Kolay, S., Bhattacharya, B.: Inherent nonslicibility of rectangular duals in VLSI floorplanning. In: Kumar, S., Nori, K.V. (eds.) FSTTCS 1988. LNCS, vol. 338, pp. 88–107. Springer, Heidelberg (1988)
19. Szepieniec, A.A., Otten, R.H.: The genealogical approach to the layout problem. In: Proceedings of the 17th Conference on Design Automation, pp. 535–542. IEEE (1980)
20. Ungar, P.: On diagrams representing maps. J. L. Math. Soc. **1**(3), 336–342 (1953)
21. Yao, B., Chen, H., Cheng, C.K., Graham, R.: Floorplan representations: Complexity and connections. ACM Trans. Design Auto. of Elec. Sys. **8**(1), 55–80 (2003)
22. Yeap, G., Sarrafzadeh, M.: Sliceable floorplanning by graph dualization. SIAM J. Disc. Math. **8**(2), 258–280 (1995)
23. Yeap, K.H., Sarrafzadeh, M.: Floor-planning by graph dualization: 2-concave rectilinear modules. SIAM J. Comp. **22**(3), 500–526 (1993)

Pixel and Voxel Representations of Graphs

Md. Jawaherul Alam[1], Thomas Bläsius[2], Ignaz Rutter[2], Torsten Ueckerdt[2], and Alexander Wolff[3]([✉])

[1] University of Arizona, Tucson, USA
[2] Karlsruhe Institute of Technology, Karlsruhe, Germany
[3] Universität Würzburg, Würzburg, Germany
alexander.wolff@uni-wuerzburg.de
http://www1.informatik.uni-wuerzburg.de/wolff

Abstract. We study contact representations for graphs, which we call *pixel representations* in 2D and *voxel representations* in 3D. Our representations are based on the unit square grid whose cells we call pixels in 2D and voxels in 3D. Two pixels are adjacent if they share an edge, two voxels if they share a face. We call a connected set of pixels or voxels a *blob*. Given a graph, we represent its vertices by disjoint blobs such that two blobs contain adjacent pixels or voxels if and only if the corresponding vertices are adjacent. We are interested in the size of a representation, which is the number of pixels or voxels it consists of.

We first show that finding minimum-size representations is NP-complete. Then, we bound representation sizes needed for certain graph classes. In 2D, we show that, for k-outerplanar graphs with n vertices, $\Theta(kn)$ pixels are always sufficient and sometimes necessary. In particular, outerplanar graphs can be represented with a linear number of pixels, whereas general planar graphs sometimes need a quadratic number. In 3D, $\Theta(n^2)$ voxels are always sufficient and sometimes necessary for any n-vertex graph. We improve this bound to $\Theta(n \cdot \tau)$ for graphs of treewidth τ and to $O((g+1)^2 n \log^2 n)$ for graphs of genus g. In particular, planar graphs admit representations with $O(n \log^2 n)$ voxels.

1 Introduction

In Tutte's landmark paper "How to draw a graph", he introduces barycentric coordinates as a tool to draw triconnected planar graphs. Given the positions of the vertices on the outer face (which must be in convex position), the positions of the remaining vertices are determined as the solutions of a set of equations. While the solutions can be approximated numerically, and symmetries tend to be reflected nicely in the resulting drawings, the ratio between the lengths of the longest edge and the shortest edge is exponential in many cases. This deficiency

This work was started at the 2014 Bertinoro Workshop on Graph Drawing. We thank the organizers for creating an inspiring atmosphere and Sue Whitesides for suggesting the problem. A.W. acknowledges support by the ESF EuroGIGA project GraDR (DFG grant Wo 758/5-1).

E. Di Giacomo and A. Lubiw (Eds.): GD 2015, LNCS 9411, pp. 472–486, 2015.
DOI: 10.1007/978-3-319-27261-0_39

triggered research directed towards drawing graphs on grids of small size in both 2D and 3D for different graph drawing paradigms; Brandenburg et al. [12] listed this as an important open problem. In *straight-line grid drawings*, the vertices are at integer grid points and the edges are drawn as straight-line segments. Both Schnyder [36] and de Fraysseix et al. [30], gave algorithms for drawing any n-vertex planar graph on a grid of size $O(n) \times O(n)$. There has also been research towards drawing subclasses of planar graphs on small-area grids. For example, any n-vertex outerplanar graph can be drawn in area $O(n^{1.48})$ [19]. Similar research has also been done for other graph drawing problems, such as *polyline drawings*, where edges can have bends [9], *orthogonal drawings*, where edges are polylines consisting of only axis-aligned segments [9,17], and for drawing graphs in 3D [21,34,35].

A *bar visibility representation* [37] draws a graph in a different way: the vertices are horizontal segments and the edges are realized by vertical line-of-sights between corresponding segments. Improving earlier results, Fan et al. [24] showed that any planar graph admits a visibility representation of size $(\lfloor 4n/3 \rfloor - 2) \times (n-1)$. Generalized visibility representations for non-planar graphs have been considered in 2D [13,23], and in 3D [11]. In all these and many subsequent papers, the size of a drawing is measured as the area or volume of the bounding box.

Yet another approach to drawing graphs are the so-called *contact representations*, where vertices are interior-disjoint geometric objects such as lines, curves, circles, polygons, polyhedra, etc. and edges correspond to pairs of objects touching in some specified way. An early work by Koebe [32] represents planar graphs with touching disks in 2D. Any planar graph can also be represented by contacts of triangles [29], by side-to-side contacts of hexagons [22] and of axis-aligned T-shape polygons [2,29]. 2D-contact representations of graphs with line segments [28], L-shapes [18], homothetic triangles [4], squares and rectangles [15,25] have also been studied. Of particular interest are the so-called *VCPG-representations* introduced by Aerts and Felsner [1]. In such a representation, vertices are represented by interior-disjoint paths in the plane square grid and an edge is a contact between an endpoint of one path and an interior point of another. Aerts and Felsner showed that for certain subclasses of planar graphs, the maximum number of bends per path can be bounded by a small constant.

Contact representations in 3D allow us to visualize non-planar graphs, but little is known about contact representations in 3D: Any planar graph can be represented by contacts of cubes [26], and by face-to-face contact of boxes [14,38]. Contact representations of complete graphs and complete bipartite graphs in 3D have been studied using spheres [6,31], cylinders [5], and tetrahedra [39]. In 3D as well as in 2D, the complexity of a contact representation is usually measured in terms of the *polygonal complexity* (i.e., the number of corners) of the objects used in the representation.

In this paper, in contrast, we are interested in "building" graphs, and so we aim at minimizing the cost of the building material—think of unit-size Lego-like blocks that can be connected to each other face-to-face. We represent each vertex by a connected set of building blocks, which we call a *blob*. If two vertices are adjacent, the blob of one vertex contains a block that is connected (face-to-face) to a block in the blob of the other. The blobs of two non-adjacent vertices are not connected. We call the building blocks *pixels* in 2D and *voxels* in 3D. Accordingly, the 2D and 3D variants of such representations are called *pixel* and *voxel* representations, respectively. We define the *size* of a pixel or voxel representation to be the total number of boxes it consists of. (We use *box* to denote either pixel or voxel when the dimension is not important.)

The same representation was introduced very recently and independently by Cano et al. [16] under the name *mosaic drawings* for interior-triangulated planar graphs and triangular, square, or hexagonal pixels. They want each blob to use a given number of pixels (to represent statistical data, such as population) and to imitate the shapes of given geometric objects (such as countries).

Although pixel representations can be seen as generalizations of VCPG-representations where grid subgraphs instead of grid paths are used, minimizing or bounding the size of such representations has so far been studied neither in 2D nor in 3D.

Our Contribution. We first investigate the complexity of our problem: finding minimum-size representations turns out to be NP-complete (Sect. 2). Then, we give lower and upper bounds for the sizes of 2D- and 3D-representations for certain graph classes:

In 2D, we show that, for k-outerplanar graphs with n vertices, $\Theta(kn)$ pixels are always sufficient and sometimes necessary (see Sect. 3). In particular, outerplanar graphs can be represented with a linear number of pixels, whereas general planar graphs sometimes need a quadratic number.

In 3D, $\Theta(n^2)$ voxels are always sufficient and sometimes necessary for any n-vertex graph (see Sect. 4). We improve this bound to $\Theta(n \cdot \tau)$ for graphs of treewidth τ and to $O((g+1)^2 n \log^2 n)$ for graphs of genus g. In particular, n-vertex planar graphs admit voxel representations with $O(n \log^2 n)$ voxels.

2   Complexity

First, we show that it is NP-hard to compute minimum-size pixel representations. We reduce from the problem of deciding whether a planar graph of maximum degree 4 has a grid drawing with edges of length 1. This problem is known to be NP-hard [7]. The hardness proof still works if the angles between adjacent edges are specified. Note that specifying the angles also prescribes the circular order of edges around vertices (up to reversal). We can only sketch the hardness proof here, details are in the full paper [3].

Theorem 1. *It is NP-complete to minimize the size of a pixel representation of a planar graph.*

Proof Sketch. Clearly the decision problem is in NP. Let G be a planar graph of maximum degree 4 with prescribed angles between edges. Construct a graph H by replacing each vertex by a five-vertex wheel so that the angles between the edges are respected, and subdividing each edge except the ones incident to the wheel centers. Then G has a grid drawing with edge length 1 if and only if H has a representation where each vertex is a pixel. Indeed, from a grid drawing of G one can obtain a drawing of H where two vertices

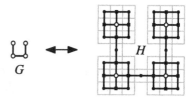

Fig. 1. A graph G drawn with length-1 edges and prescribed angles between adjacent edges, and the resulting graph H drawn with length-1 edges and pixel representation (in gray).

have distance 1 if *and only if* they are adjacent; see Fig. 1. Represent each vertex v of H by a pixel with v at its center. Conversely, if H has a representation where each vertex is a pixel, then for each vertex v of G, the subdivided wheel is a 3×3 square. Placing each vertex v at the center of the square and scaling by $1/4$ yields the grid drawing of G.   □

Next, we reduce computing minimum-size pixel representations to computing minimum-size voxel representations. In our reduction [3], we build a rigid structure around the given graph that forces the given graph to be drawn in a single plane.

Theorem 2. *It is NP-complete to minimize the size of a voxel representation of a graph.*

3   Lower and Upper Bounds in 2D

Here we only consider planar graphs since only planar graphs admit pixel representations. Let G be a planar graph with fixed plane embedding $\mathscr{E}$. The embedding $\mathscr{E}$ is 1-*outerplane* (or simply outerplane) if all vertices are on the outer face. It is k-*outerplane* if removing all vertices on the outer face yields a $(k-1)$-outerplane embedding. A graph G is k-*outerplanar* if it admits a k-outerplane embedding but no k'-outerplane embedding for $k' < k$. Note that $k \in O(n)$, where n is the number of vertices of G.

In Sect. 3.1, we show that pixel representations of an n-vertex k-outerplanar graph sometimes requires $\Omega(kn)$ pixels. As the number of pixels is a lower bound for the area consumption, this strengthens a result by Dolev et al. [20] that says that orthogonal drawings of planar graphs of maximum degree 4 and *width* w sometimes require $\Omega(wn)$ area. As we will see later, width and k-outerplanarity are very similar concepts.

In Sect. 3.2, we show that $O(kn)$ area and thus using $O(kn)$ pixels is also sufficient. We use a result by Dolev et al. [20] who proved that any n-vertex planar graph of maximum degree 4 and width w admits a planar orthogonal drawing of area $O(wn)$. The main difficulty is to extend their result to general planar graphs.

3.1   Lower Bound

Let G be a k-outerplanar graph with a pixel representation Γ. Note that a pixel representation Γ induces an embedding of G. Let Γ induce a k-outerplane embedding of G, which we call a k-*outerplane pixel representation* for short. We claim that the width and the height of Γ are at least $2k - 1$. For $k = 1$ this is trivial as every (non-empty) graph requires width and height at least 1. For $k \geq 2$, let $V_{\text{ext}} = \{v_1, \dots, v_\ell\}$ be the set of vertices incident to the outer face of Γ. Removing V_{ext} from G yields a $(k-1)$-outerplane graph G' with corresponding pixel representation Γ'. By induction, Γ' requires width and height $2(k-1) - 1$. As the representation of V_{ext} in Γ encloses the whole representation Γ' in its interior, the width and the height of Γ are at least two units larger than the width and the height of Γ', respectively.

Clearly, the number of pixels required by the vertices in V_{ext} is at least the perimeter of Γ (twice the width plus twice the height minus 4 for the corners, which are shared) and thus at least $8k - 8$. After removing the vertices in V_{ext}, the new vertices on the outer face require $8(k - 1) - 8$ pixels, and so on. Thus, the representation Γ requires overall at least $\sum_{i=1}^{k}(8i - 8) = 4k^2 - 4k$ pixels, which gives the following lemma.

Lemma 1. *Any k-outerplane pixel representation has size at least $4k^2 - 4k$.*

There are k-outerplanar graphs with n vertices such that $k \in \Theta(n)$. For example, the nested triangle graph with $2k$ triangles (see Fig. 2) has $n = 6k$ vertices and is k-outerplanar for $k \geq 2$. Let G be a graph with c connected components each of which is k-outerplanar and has $\Theta(k)$ vertices. Then each connected component requires $4k^2 - 4k$ pixels (due to Lemma 1) and thus we need at least $(4k^2 - 4k)c$ pixels in total. As G has $n = \Theta(kc)$ vertices, we get $(4k^2 - 4k)c \in \Theta(kn)$, which proves the following.

Theorem 3. *Some k-outerplanar graphs require $\Omega(kn)$-size pixel representations.*

3.2   Upper Bound

In the following two lemmas, we first show how to construct a pixel representation from a given orthogonal drawing and that taking minors does not heavily increase the number of pixels we need. Both lemmas aim at extending a result of Dolev et al. [20] on orthogonal drawings of planar graphs with maximum degree 4 to pixel representations of general planar graphs. As we re-use both lemmas in the 3D case (Sect. 4), we state them in the general d-dimensional setting.

Lemma 2. *Let G be a graph with n vertices, m edges, and an orthogonal drawing of total edge length ℓ in d-dimensional space. Then G admits a d-dimensional representation of size $2\ell + n - m$.*

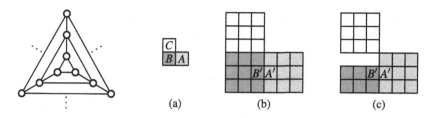

Fig. 2. A nested triangle gra- **Fig. 3.** Constructing a representation of a minor with
ph of outerplanarity $\Omega(n)$.   asymptotically the same number of blocks.

Proof. We first scale the given drawing Γ of G by a factor of 2 and subdivide the edges of G such that every edge has length 1. Denote the resulting graph by G' and its drawing by Γ'. An edge e of length ℓ_e in Γ is represented by a path with $2\ell_e - 1$ internal vertices (the subdivision vertices). Thus, the total number of subdivision vertices is $2\ell - m$. Due to the scaling, non-adjacent vertices in G' have distance greater than 1 in Γ' (adjacent vertices have distance 1). Thus, representing every vertex v by the box having v as center yields a representation of G' with $2\ell + n - m$ boxes (one box per vertex of G'). If we assign the boxes representing subdivision vertices to one of the endpoints of the corresponding edge, we get a representation of G with $2\ell + n - m$ boxes.   □

Lemma 3. *Let G be a graph that has a d-dimensional representation of size b. Every minor of G admits a d-dimensional representation of size at most $3^d b$.*

Proof. Let H be a minor obtained from G by first deleting some edges, then deleting isolated vertices, and finally contracting edges. We start with the representation Γ of G using b boxes and scale it by a factor of 3. This yields a representation 3Γ using $3^d b$ boxes. Then we modify 3Γ, without adding boxes, to represent the minor H. For convenience, we consider the 2D case; the case $d > 2$ works analogously.

Let uv be an edge in G that is deleted. In 3Γ we delete every pixel in the representation of u that touches a pixel of the representation of v. We claim that this neither destroys the contact of u with any other vertex nor does it disconnect the shape representing u. Consider a single pixel B in Γ. In 3Γ it is represented by a square of 3×3 pixels *belonging* to B. If B is in contact to another pixel A in Γ, then there is a pair of pixels A' and B' in 3Γ such that A' and B' are in contact, while all other pixels that touch A' and B' belong to A and B, respectively; see Fig. 3a and b. Assume that we remove in 3Γ all pixels belonging to B that are in contact to pixels belonging to another pixel C touching B in Γ; see Fig. 3c. Obviously, this does not effect the contact between A' and B'. Moreover, the remaining pixels belonging to B form a connected blob. The above claim follows immediately.

Removing isolated vertices can be done by simply removing their representation. Moreover, contracting an edge uv into a vertex w can be done by merging the blobs representing u and v into a single blob representing w. This blob is obviously connected and touches the blob of another vertex if and only if either u or v touch this vertex. □

Now let G be a k-outerplanar graph. Applying the algorithm of Dolev et al. [20] yields an orthogonal drawing of total length $O(wn)$, where w is the *width* of G. The width w of G is the maximum number of vertices contained in a shortest path from an arbitrary vertex of G to a vertex on the outer face. Given the orthogonal drawing, Lemma 2 gives us a pixel representation of G. There are, however, two issues. First, k and w are not the same (e.g., subdividing edges increases w but not k). Second, G does not have maximum degree 4, thus we cannot simply apply the algorithm of Dolev et al. [20].

Concerning the first issue, we note that the algorithm of Dolev et al. exploits that G has width w only to find a special type of separator [20, Theorem 1]. For this, it is sufficient that G is a subgraph of a graph of width w (not necessarily with maximum degree 4; in fact Dolev et al. triangulate the graph before finding the separator).

Lemma 4. *Every k-outerplanar graph has a planar supergraph of width $w = k$.*

Proof. Let G be a graph with a k-outerplane embedding. Iteratively deleting the vertices on the outer face gives us a sequence of deletion phases. For each vertex v, let k_v be the phase in which v is deleted. Note that the maximum over all values of k_v is exactly k. For any vertex v, either $k_v = 1$ or there is a vertex u with $k_u = k_v - 1$ such that u and v are incident to a common face. Thus, there is a sequence $v_1, \ldots, v_{k_v}$ of k_v vertices such that (i) $v_1 = v$, (ii) v_{k_v} lies on the outer face, and (iii) v_i, v_{i+1} are incident to a common face. If the graph G was triangulated, this would yield a path containing k_v vertices from v to a vertex on the outer face. Thus, triangulated k-outerplanar graphs have width $w = k$.

It remains to show that G can be triangulated without increasing k_v for any vertex v. Consider a face f and let u be the vertex incident to f for which k_u is minimal. Let $v \neq u$ be any other vertex incident to f. Adding the edge uv clearly does not increase the value k_x for any vertex x. We add edges in this way until the graph is triangulated. Alternatively, we can use a result of Biedl [8] to triangulate G. Note that we do not need to triangulate the outer face of G. Hence, we do not increase the outerplanarity. □

To solve the second issue (the k-outerplanar graph G not having maximum degree 4), we construct a graph G' such that G is a minor of G', G' is k-outerplanar, and G' has maximum degree 4. Then, (due to Lemma 4) we can apply the algorithm of Dolev et al. [20] to G'. Next, we apply Lemma 2 to the resulting drawing to get a representation of G' with $O(kn)$ pixels. As G is a minor of G', Lemma 3 yields a representation of G that, too, requires $O(kn)$ pixels.

Theorem 4. *Every k-outerplanar n-vertex graph has a size $O(kn)$ pixel representation.*

Proof. Let G be a k-outerplanar graph. After the above considerations, it remains to construct a k-outerplanar graph G' with maximum degree 4 such that G is a minor of G'. Let u be a vertex with $\deg(u) > 4$. We replace u with a path of length $\deg(u)$ and connect each neighbor of u to a unique vertex of this path. This can be done maintaining a plane embedding. We now show that the resulting graph remains k-outerplanar.

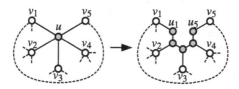

Fig. 4. Replacement of high-degree vertices while preserving k-outerplanarity.

Consider a vertex u on the outer face with neighbors $v_1, \ldots, v_\ell$. Assume the neighbors appear in that order around u such that v_1 is the counter-clockwise successor of u on the outer face; see Fig. 4. We replace u with the path $u_1, \ldots, u_\ell$ and connect u_i to v_i for $1 \leq i \leq \ell$. Call the resulting graph G_u. Note that all u_i in G_u are incident to the outer face. Thus, if G was k-outerplanar, G_u is also k-outerplanar. Moreover, the degrees of the new vertices do not exceed 4 (actually not even 3), and G is a minor of G_u—one can simply contract the inserted path to obtain G.

We can basically apply the same replacement if u is not incident to the outer face. Assume that we delete u in phase k_u if we iteratively delete vertices incident to the outer face. When replacing u with the vertices $u_1, \ldots, u_\ell$, we have to make sure that all these vertices get deleted in phase k_u. Let f be a face incident to u that is merged with the outer face after $k_u - 1$ deletion phases (such a face must exist, otherwise u is not deleted in phase k_u). We apply the same replacement as for the case where u was incident to the outer face, but this time we ensure that the new vertices u_i are incident to the face f. Thus, after $k_u - 1$ deletion phases they are all incident to the outer face and thus they are deleted in phase k_u. Hence, the resulting graph G_u is k-outerplanar. Again the new vertices have degree at most 3 and G is obviously a minor of G_u. Iteratively applying this kind of replacement for every vertex u with $\deg(u) > 4$ yields the claimed graph G'.

The corresponding drawing can then be obtained as follows. Since G' has a supergraph of width $w = k$ by Lemma 4, and G' has maximum degree 4, we use the algorithm of Dolev et al. [20] to obtain a drawing of G' with area (and hence total edge length) $O(nk)$. By Lemma 2, we thus obtain a representation of G' with $O(nk)$ pixels. Since G is a minor of G', Lemma 3 yields a representation of G with $O(nk)$ pixels. □

4   Representations in 3D

In this section, we consider voxel representations. We start with some basic considerations showing that every n-vertex graph admits a representation with $O(n^2)$ voxels. Note that $\Omega(n^2)$ is obviously necessary for K_n as every edge corresponds to a face-to-face contact and every voxel has at most 6 such contacts. We improve on this simple general result in two ways. First, we show that n-vertex graphs with treewidth at most τ admit voxel representations of size $O(n \cdot \tau)$ (see Sect. 4.1). Second, for n-vertex graphs with genus at most g, we obtain representations with $O(g^2 n \log^2 n)$ voxels (see Sect. 4.2).

Theorem 5. *Any n-vertex graph admits a voxel representation of size $O(n^2)$.*

Proof. Let G be a graph with vertices $v_1, \ldots, v_n$. Vertex v_i ($i = 1, \ldots, n$) is represented by three cuboids (see Fig. 5a), namely a vertical cuboid consisting of the voxels centered at the points $(2i, 2, 0), (2i, 3, 0), \ldots, (2i, 2n, 0)$, a horizontal cuboid consisting of the voxels centered at $(2, 2i, 2), (3, 2i, 2), \ldots, (2n, 2i, 2)$, and the voxel centered at $(2i, 2i, 1)$. This yields a representation where every vertex is a connected blob and no two blobs are in contact. Moreover, for every pair of vertices v_i and v_j, there is a voxel of v_i at $(2i, 2j, 0)$ and a voxel of v_j at $(2i, 2j, 2)$ and no voxel between them at $(2i, 2j, 1)$. Thus, one can easily represent an arbitrary edge (v_i, v_j) by extending the representation of v_i to also contain $(2i, 2j, 1)$; see Fig. 5b. Clearly, this representation consists of $O(n^2)$ voxels. $\square$

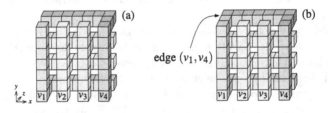

Fig. 5. (a) The basic contact representation without any contacts between vertices. (b) If v_1 and v_4 are adjacent, it suffices to add a single voxel to the representation of v_1 (or to that of v_4).

4.1   Graphs of Bounded Treewidth

Let $G = (V, E)$ be a graph. A *tree decomposition* of G is a tree T where each node μ in T is associated with a *bag* $X_\mu \subseteq V$ such that: (i) for each $v \in V$, the nodes of T whose bags contain v form a connected subtree, and (ii) for each edge $uv \in E$, T contains a node μ such that $u, v \in X_\mu$.

Note that we use (lower case) Greek letters for the nodes of T to distinguish them from the vertices of G. The *width* of the tree decomposition is the maximum bag size minus 1. The *treewidth* of G is the minimum width over all tree decompositions of G. A tree decomposition is *nice* if T is a rooted binary tree, where for every node μ:

- μ is a leaf and $|X_\mu| = 1$ (*leaf node*), or
- μ has a single child η with $X_\mu \subseteq X_\eta$ and $|X_\mu| = |X_\eta| - 1$ (*forget node*), or
- μ has a single child η with $X_\eta \subseteq X_\mu$ and $|X_\mu| = |X_\eta| + 1$ (*introduce node*), or
- μ has two children η and κ with $X_\mu = X_\eta = X_\kappa$ (*join node*).

Any tree decomposition can be transformed (without increasing its width) into a nice tree decomposition such that the resulting tree T has $O(n)$ nodes, where n is the number of vertices of G [10]. This transformation can be done in linear time. Thus, we can assume any tree decomposition to be a nice tree decomposition with a tree of size $O(n)$.

Lemma 5. *Let T be a nice tree decomposition of a graph G. The edges of G can be mapped to the nodes of T such that every edge uv of G is mapped to a node μ with $u, v \in X_\mu$ and the edges mapped to each node μ form a star.*

Proof. We say that a node μ *represents* the edge uv if uv is mapped to μ. Consider a node μ during a bottom-up traversal of T. We want to maintain the invariant that, after processing μ, all edges between vertices in X_μ are represented by μ or by a descendant of μ. This ensures that every edge is represented by at least one node. Every edge can then be mapped to one of the nodes representing it.

If μ is a leaf, it cannot represent an edge as $|X_\mu| = 1$. If μ is a forget node, it has a child η with $X_\mu \subseteq X_\eta$. Thus, by induction, all edges between vertices in X_μ are already represented by descendants of μ. If μ is an introduce node, it has a child η and $X_\mu = X_\eta \cup \{u\}$ for a vertex u of G. By induction, all edges between nodes in X_η are already represented by descendants of μ. Thus, μ only needs to represent the edges between the new node u and other nodes in X_μ. Note that these edges form a star with center u. Finally, if μ is a join node, no edge needs to be represented by μ (by the same argument as for forget nodes). This concludes the proof. $\qquad\square$

We obtain a small voxel representation of G from a nice tree decomposition T of G of treewidth τ roughly as follows. We start with a "2D" voxel representation of the tree T, that is, all voxel centers lie in the x–y plane. We take $\tau + 1$ copies of this representation and place them in different layers in 3D space. We then assign to each vertex v of G a piece of this layered representation such that its piece contains all nodes of T that include v in their bags. For an edge uv, let μ be the node to which uv is mapped by Lemma 5. By construction, the representation of μ occurs multiple times representing u and v in different layers. To represent uv, we only have to connect the representations of u and v. As it suffices to represent a star for each node μ in this way, the number of voxels additionally used for these connections is small.

Theorem 6. *Any n-vertex graph of treewidth τ has a voxel representation of size $O(n\tau)$.*

Proof. Let G be an n-vertex graph of treewidth τ. During our construction, we will get some contacts between the blobs of vertices that are actually not adjacent in G. As G is a minor of the graph that we represent this way, we can use Lemma 3 to get a representation of G. Let T be a nice tree decomposition of G. As a tree, T is outerplanar and, hence, admits a pixel representation Γ with $O(n)$ pixels (by Theorem 4). Let $\Gamma_1, \ldots, \Gamma_k$ be voxel representations corresponding to Γ with z-coordinates $1, \ldots, k = \tau + 1$.

For a vertex v of G, we denote by $\Gamma_i(v)$ the sub-representation of Γ_i induced by the nodes of T whose bags contain v. Now let $c \colon V \to \{1, \ldots, k\}$ be a k-coloring of G with color set $\{1, \ldots, k\}$ such that no two vertices sharing a bag have the same color. Such a coloring can be computed by traversing T bottom up, assigning in every introduce node μ a color to the new vertex that is not already used by any other vertex in X_μ. As a basis for our construction, we represent each vertex v of G by the sub-representation $\Gamma_{c(v)}(v)$.

So far, we did not represent any edge of G. Our construction, however, has the following properties: (i) it uses $O(nk)$ voxels. (ii) every vertex is a connected set of voxels. (iii) for every node μ of T, there is a position (x_μ, y_μ) in the plane such that, for every vertex $v \in X_\mu$, the voxel at $(x_\mu, y_\mu, c(v))$ belongs to the representation of v. Scaling the representation by a factor of 2 ensures that this is not the only voxel for v and that v is not disconnected if this voxel is removed (or reassigned to another vertex).

By Lemma 5 it suffices to represent for every node μ edges between vertices in X_μ that form a star. Let u be the center of this star. We simply assign the voxels centered at $(x_\mu, y_\mu, 1), \ldots, (x_\mu, y_\mu, k)$ to the blob of u. This creates a contact between u and every other vertex $v \in X_\mu$ (by the above property that the voxel $(x_\mu, y_\mu, c(v))$ belonged to v before). Finally, we apply Lemma 3 to get rid of unwanted contacts. The resulting representation uses $O(nk)$ voxels, which concludes the proof. $\square$

Note that cliques of size k require $\Omega(k^2)$ voxels. Taking the disjoint union of n/k such cliques yields graphs with n vertices requiring $\Omega(nk)$ voxels. Note that these graphs have treewidth $\tau = k - 1$. Thus, the bound of Theorem 6 is asymptotically tight.

Theorem 7. *Some n-vertex graphs of treewidth τ require $\Omega(n\tau)$ voxels.*

4.2   Graphs of Bounded Genus

Since planar graphs (genus 0) have treewidth $O(\sqrt{n})$ [27], we can obtain a voxel representation of size $O(n^{1.5})$ for any planar graph, from Theorem 6. Next, we improve this bound to $O(n \log^2 n)$ by proving a more general result for graphs of bounded genus. Recall that we used known results on orthogonal drawings with small area to obtain small pixel representations in Sect. 3.2. Here we follow a similar approach (re-using Lemmas 2 and 3), now allowing the orthogonal drawing we start with to be non-planar.

We obtain small voxel representations by first showing that it is sufficient to consider graphs of maximum degree 4: we replace higher-degree vertices by connected subgraphs as in the proof of Theorem 4. Then we use a result of Leiserson [33] who showed that any graph of genus g and maximum degree 4 admits a 2D orthogonal drawing of area $O((g+1)^2 n \log^2 n)$, possibly with edge crossings. The area of an orthogonal drawing is clearly an upper bound for its total edge length. Finally we turn the pixels into voxels and use the third dimension to get rid of the crossings without using too many additional voxels.

Theorem 8. *Every n-vertex graph of genus g admits a voxel representation of size $O((g+1)^2 n \log^2 n)$.*

Proof. Let G be an n-vertex graph, and let u be a vertex of degree $\ell > 4$. Assume G to be embedded on a surface of genus g, and let $v_1, \ldots, v_\ell$ be the neighbors of u appearing in that order around u (with respect to the embedding). We replace u with the cycle $u_1, \ldots, u_\ell$ and connect u_i to v_i for $1 \leq i \leq \ell$; see Fig. 6a. Clearly, the new vertices have degree 3 and the genus of the graph has not increased. Applying this modification to every vertex of degree at least 5 yields a graph G_4 of maximum degree 4 and genus g. Moreover, G is a minor of G_4 as one can undo the cycle replacements by contracting all edges in the cycles. Thus, we can transform a voxel representation of G_4 into a voxel representation of G by applying Lemma 3.

We claim that the number n_4 of vertices in G_4 is linear in n. Indeed, if m denotes the number of edges in G, then we have $n_4 \leq n + 2m$. Moreover, we can assume without loss of generality that $g \in O(n)$ (otherwise Theorem 5 already gives a better bound). This implies that $m \in O(n)$ and hence, $n_4 \in O(n)$, as we claimed.

We thus assume that G has maximum degree 4. Then G has a (possibly non-planar) orthogonal drawing Γ of total edge length $O(g^2 n \log^2 n)$ [33]. We modify G and Γ as follows. For every bend on an edge e in Γ, we subdivide the edge e once yielding a partition of the edges of the subdivided graph into horizontal and vertical edges. We obtain a graph G' from this subdivision of G by replacing every vertex v by two adjacent vertices v_1 and v_2, and connecting v_1 and w_1 (respectively v_2 and w_2) by an edge if v and w are connected by a horizontal (respectively vertical edge); see Fig. 6b.

We draw G' in 3D space by using the drawing Γ and setting for every vertex v the z-coordinate of v_1 and v_2 to 0 and 1, respectively. The x- and y-coordinates of vertices and edges are the same as in Γ; see Fig. 6b. Note that G is a minor of G': we obtain G from G' by contracting (i) the edge $v_0 v_1$ for every vertex v and (ii) any subdivision vertex. Asymptotically, the total edge length of Γ' is the same as that of Γ, that is, $O((g+1)^2 n \log^2 n)$. By Lemma 2, we turn Γ' into a voxel representation of G' and, by Lemma 3, into a voxel representation of G with size $O((g+1)^2 n \log^2 n)$. $\qquad \Box$

5   Conclusion

In this paper, we have studied pixel representations and voxel representations of graphs, where vertices are represented by disjoint blobs (that is, connected sets

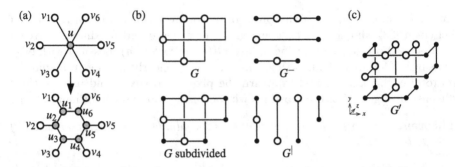

Fig. 6. Constructing voxel representations for bounded-genus graphs: (a) replacing high-degree vertices while preserving the genus, (b) subdividing and decomposing a graph according to a non-planar orthogonal drawing with small area, and (c) constructing a 3D drawing with small total edge length from the decomposition in (b).

of grid cells) and edges correspond to pairs of blobs with face-to-face contact. We have shown that it is NP-complete to minimize the number of pixels or voxels in such representations. Does this problem admit an approximation algorithm?

We have shown that $O((g+1)^2 n \log^2 n)$ voxels suffice for any n-vertex graph of genus g. It remains open to improve this upper bound or to give a non-trivial lower bound. We believe that any planar graph admits a voxel representation of linear size.

References

1. Aerts, N., Felsner, S.: Vertex contact graphs of paths on a grid. In: Kratsch, D., Todinca, I. (eds.) WG 2014. LNCS, vol. 8747, pp. 56–68. Springer, Heidelberg (2014)
2. Alam, M.J., Biedl, T., Felsner, S., Kaufmann, M., Kobourov, S., Ueckerdt, T.: Computing cartograms with optimal complexity. Discrete Comput. Geom. **50**(3), 784–810 (2013)
3. Alam, M.J., Bläsius, T., Rutter, I., Ueckerdt, T., Wolff, A.: Pixel and voxel representations of graphs. Arxiv report (2015). arxiv.org/abs/1507.01450
4. Badent, M., Binucci, C., Di Giacomo, E., Didimo, W., Felsner, S., Giordano, F., Kratochvíl, J., Palladino, P., Patrignani, M., Trotta, F.: Homothetic triangle contact representations of planar graphs. In: Canadian Conference on Computational Geometry (CCCG 2007), pp. 233–236 (2007)
5. Bezdek, A.: On the number of mutually touching cylinders. Comb. Comput. Geom. **52**, 121–127 (2005)
6. Bezdek, K., Reid, S.: Contact graphs of unit sphere packings revisited. J. Geom. **104**(1), 57–83 (2013)
7. Bhatt, S.N., Cosmadakis, S.S.: The complexity of minimizing wire lengths in VLSI layouts. Inform. Process. Lett. **25**(4), 263–267 (1987)
8. Biedl, T.: On triangulating k-outerplanar graphs. Discrete Appl. Math. **181**, 275–279 (2015). arxiv.org/abs/1310.1845

9. Biedl, T.C.: Small drawings of outerplanar graphs, series-parallel graphs, and other planar graphs. Discrete Comput. Geom. **45**(1), 141–160 (2011)
10. Bodlaender, H.L.: Treewidth: algorithmic techniques and results. In: Prívara, I., Ružička, P. (eds.) MFCS 1997. LNCS, vol. 1295, pp. 19–36. Springer, Heidelberg (1997)
11. Bose, P., Everett, H., Fekete, S.P., Houle, M.E., Lubiw, A., Meijer, H., Romanik, K., Rote, G., Shermer, T.C., Whitesides, S., Zelle, C.: A visibility representation for graphs in three dimensions. J. Graph Algorithms Appl. **2**(3), 1–16 (1998)
12. Brandenburg, F.J., Eppstein, D., Goodrich, M.T., Kobourov, S.G., Liotta, G., Mutzel, P.: Selected open problems in graph drawing. In: Liotta, G. (ed.) GD 2003. LNCS, vol. 2912, pp. 515–539. Springer, Heidelberg (2004)
13. Brandenburg, F.J.: 1-visibility representations of 1-planar graphs. J. Graph Algorithms Appl. **18**(3), 421–438 (2014)
14. Bremner, D., et al.: On representing graphs by touching cuboids. In: Didimo, W., Patrignani, M. (eds.) GD 2012. LNCS, vol. 7704, pp. 187–198. Springer, Heidelberg (2013)
15. Buchsbaum, A.L., Gansner, E.R., Procopiuc, C.M., Venkatasubramanian, S.: Rectangular layouts and contact graphs. ACM Trans. Algorithms **4**(1), 8–28 (2008)
16. Cano, R., Buchin, K., Castermans, T., Pieterse, A., Sonke, W., Speckmann, B.: Mosaic drawings and cartograms. Comput. Graph. Forum **34**(3), 361–370 (2015)
17. Chan, T.M., Goodrich, M.T., Kosaraju, S.R., Tamassia, R.: Optimizing area and aspect ratio in straight-line orthogonal tree drawings. Comput. Geom. Theory Appl. **23**(2), 153–162 (2002)
18. Chaplick, S., Kobourov, S.G., Ueckerdt, T.: Equilateral L-contact graphs. In: Brandstädt, A., Jansen, K., Reischuk, R. (eds.) WG 2013. LNCS, vol. 8165, pp. 139–151. Springer, Heidelberg (2013)
19. Di Battista, G., Frati, F.: Small area drawings of outerplanar graphs. Algorithmica **54**(1), 25–53 (2009)
20. Dolev, D., Leighton, T., Trickey, H.: Planar embedding of planar graphs. Adv. Comput. Res. **2**, 147–161 (1984)
21. Dujmović, V., Morin, P., Wood, D.: Layered separators for queue layouts, 3d graph drawing and nonrepetitive coloring. In: Foundations of Computer Science (FOCS 2013), pp. 280–289. IEEE (2013)
22. Duncan, C.A., Gansner, E.R., Hu, Y.F., Kaufmann, M., Kobourov, S.G.: Optimal polygonal representation of planar graphs. Algorithmica **63**(3), 672–691 (2012)
23. Evans, W., Kaufmann, M., Lenhart, W., Mchedlidze, T., Wismath, S.: Bar 1-visibility graphs and their relation to other nearly planar graphs. J. Graph Algorithms Appl. **18**(5), 721–739 (2014)
24. Fan, J.-H., Lin, C.-C., Lu, H.-I., Yen, H.-C.: Width-optimal visibility representations of plane graphs. In: Tokuyama, T. (ed.) ISAAC 2007. LNCS, vol. 4835, pp. 160–171. Springer, Heidelberg (2007)
25. Felsner, S.: Rectangle and square representations of planar graphs. Thirty Essays on Geometric Graph Theory, pp. 213–248 (2013)
26. Felsner, S., Francis, M.C.: Contact representations of planar graphs with cubes. In: Symposium on Computational Geometry (SoCG 2011), pp. 315–320. ACM (2011)
27. Fomin, F.V., Thilikos, D.M.: New upper bounds on the decomposability of planar graphs. J. Graph Theory **51**(1), 53–81 (2006)
28. de Frayseix, H., de Mendez, P.O.: Representations by contact and intersection of segments. Algorithmica **47**(4), 453–463 (2007)
29. de Frayseix, H., de Mendez, P.O., Rosenstiehl, P.: On triangle contact graphs. Comb. Prob. Comput. **3**, 233–246 (1994)

30. de Fraysseix, H., Pach, J., Pollack, R.: How to draw a planar graph on a grid. Combinatorica **10**(1), 41–51 (1990)
31. Hliněný, P., Kratochvíl, J.: Representing graphs by disks and balls (a survey of recognition-complexity results). Discrete Math. **229**(1–3), 101–124 (2001)
32. Koebe, P.: Kontaktprobleme der konformen Abbildung. Berichte über die Verhandlungen der Sächsischen Akademie der Wissenschaften zu Leipzig. Math. Phy. Kla. **88**, 141–164 (1936)
33. Leiserson, C.E.: Area-efficient graph layouts (for VLSI). In: Foundations of Computer Science (FOCS 1980), pp. 270–281. IEEE (1980)
34. Pach, J., Thiele, T., Tóth, G.: Three-dimensional grid drawings of graphs. In: Di Battista, G. (ed.) GD 1997. LNCS, vol. 1353, pp. 47–51. Springer, Heidelberg (1997)
35. Patrignani, M.: Complexity results for three-dimensional orthogonal graph drawing. J. Discrete Algorithms **6**(1), 140–161 (2008)
36. Schnyder, W.: Embedding planar graphs on the grid. In: Symposium on Discrete Algorithms (SODA 1990), pp. 138–148. ACM-SIAM (1990)
37. Tamassia, R., Tollis, I.G.: A unified approach a visibility representation of planar graphs. Discrete Comput. Geom. **1**, 321–341 (1986)
38. Thomassen, C.: Interval representations of planar graphs. J. Comb. Theory B **40**(1), 9–20 (1986)
39. Zong, C.: The kissing numbers of tetrahedra. Discrete Comput. Geom. **15**(3), 239–252 (1996)

User Studies

A Tale of Two Communities: Assessing Homophily in Node-Link Diagrams

Wouter Meulemans[1]($\boxtimes$) and André Schulz[2]($\boxtimes$)

[1] giCentre, City University London, London, UK
wouter.meulemans@city.ac.uk
[2] LG Theoretische Informatik, FernUniversität in Hagen, Hagen, Germany
andre.schulz@fernuni-hagen.de

Abstract. Homophily is a concept in social network analysis that states that in a network a link is more probable, if the two individuals have a common characteristic. We study the question if an observer can assess homophily by looking at the node-link diagram of the network. We design an experiment that investigates three different layout algorithms and asks the users to estimate the degree of homophily in the displayed network. One of the layout algorithms is a classical force-directed method, the other two are designed to improve node distinction based on the common characteristic. We study how each of the three layout algorithms helps to get a fair estimate, and whether there is a tendency to over or underestimate the degree of homophily. The stimuli in our experiments use different network sizes and different proportions of the cluster sizes.

1 Introduction

Networks do not exist without a surrounding context. An object in a network is typically equipped with a set of characteristics (e.g., age, race, or gender in a social network). These characteristics have an influence on the network structure; often, nodes of a network are partitioned into clusters, based on (some) characteristics. Detecting, measuring and understanding network structures and dependencies is an important task in network analysis. In social networks one of these effects is *homophily*. Simply speaking, homophily is a principle that asserts that individuals are more likely to have relationships with similar individuals [10]. There are two main mechanisms behind homophily: (i) individuals with the same characteristics might have a stronger tendency to form relationships (this principle is known as *selection*), (ii) individuals change their behavior to align with their friends (this is known as *socialization*) [3, Sect. 4.2].

Whether homophily is present in a certain network (and to what extent) can be detected by comparing the number of links between nodes of the same cluster (same-cluster) with the number of links between nodes of different clusters (cross-cluster); see Sect. 1.1 for details. This yields a simple formula for the degree of homophily in a network. In this work we study the following questions.

This work was funded by the German Research Foundation (grant SCHU 2458/4-1).

E. Di Giacomo and A. Lubiw (Eds.): GD 2015, LNCS 9411, pp. 489–501, 2015.
DOI: 10.1007/978-3-319-27261-0_40

Question 1. Can an observer detect the degree of homophily in a node-link diagram of a network? Is there a tendency for overestimation or underestimation?

There exist many layout algorithms for node-link diagrams. We expect that the drawing style has a big impact on answering Question 1. Hence, a natural subsequent question is:

Question 2. Which node-link diagram layout is best suitable for detecting homophily? Are there general design principles to improve homophily detection?

We deliberately set the scope to node-link diagrams, since they are probably the most popular style for visualizing networks. Tasks like path tracing can be performed well on node-link diagrams and many users are familiar with their methodology. Other methods for displaying networks (e.g., matrix views, hive plots [9], NodeTrix [6]) or summarizing them (e.g., histograms) may be more effective for enabling homophily assessment, but are out of scope for our study.

For a fair evaluation of the layout methods we include one additional task in our experiments. In particular, we ask for participants to answer shortest-path questions, to detect whether a better homophily assessment diminishes the versatility of the layout and makes other (path-tracing) tasks harder.

Homophily can also occur when there are more than two clusters. More clusters make it harder to detect homophily, simply because there is more information. Moreover, the notion of homophily can be extended to more clusters in slightly different ways. Due to these considerations we decided to restrict our investigations to the most basic case with only two clusters.

Our studies are not necessarily tied to *social* networks. Clusters exist in all kind of networks and it is a natural question to ask whether there exists a bias for cross-cluster or same-cluster links.

1.1 Homophily

Homophily is a natural phenomenon, but it is not always present in (social) networks. In fact, opposite effects might occur, that is, individuals favor to form bonds with individuals that have different characteristics. To understand networks within their surrounding context, we need methods to detect and to measure the effect of homophily with respect to a certain characteristic. We follow the presentation of Easley and Kleinberg [3] to derive such a framework.

Suppose we study a social network in which the individuals are either female or male. We want to decide whether there exists homophily with respect to gender. Assume that a fraction p of the population is male and a fraction $q = 1-p$ is female. In a network without homophily we expect that a random link is male-male with a probability of p^2, female-female with a probability of q^2, and cross-gender with a probability of $2pq$. As a consequence there is evidence for homophily, if the fraction of cross-gender links is considerably less than $2pq$ (and "heterophily" if the fraction is considerably more). We use this "homophily test" to derive a measure for the *degree of homophily* of a network as follows:

Definition 1. *Given a network with two clusters (one with a fraction p of the nodes and the other with a fraction $q = 1 - p$). We say that the degree of homophily is 0 if there are only cross-cluster links, it is 1 if there are no cross-cluster links, and it is 1/2 if there is no homophily, that is we have 2pq cross-cluster links. For all other situations we linearly interpolate between these values.*

1.2   Network Visualization

There exists a vast literature on various methods for network visualization and analysis with clusters, hierarchies or other auxiliary data, e.g. [2,7,12]. Though an extensive review is out of scope, we briefly review a number of methods and describe those we used for our user study.

Force-Directed. The *force-directed* layout (see for example [4]) has been a popular network visualization method since its inception. The core idea of this method is to mimic a physical system in which nodes repel each other and links behave like springs, pulling their ends together. We include this in our study, providing a good baseline visualization method to which to compare. We use the implementation provided by the javascript library D3.js[1].

Polarized. It is straightforward to modify the classical force-directed method to pull the nodes of the two clusters apart. We modified the D3.js layout algorithm by adding a force that moves the nodes left or right, depending on the cluster. We refer to this as the *polarized* layout. This tends to pull the clusters apart, though a clear separation is not guaranteed.

Bipartite. As an extreme form of separating the clusters (*bipartite* layout) we place all vertices of one cluster equidistantly on a vertical line on the left, the other vertices on a vertical line on the right. Cross-cluster links are drawn as straight-line segments forming a 2-layer bipartite drawing of this subnetwork. Same-cluster links are drawn as semi-circles as in an arc diagram. After obtaining an initial ordering of the vertices from a barycentric layout, we use the method of Baur and Brandes [1] to reduce the number of crossings, within one round applying sifts to all nodes of one cluster before the other. We remark that this layout style can also be found as an unchosen design alternative in the interactive visualization system described by Ghani et al. [5]. Their reason for not using this design is that they aim to support many clusters: the same-cluster links drawn as arcs would lead to severe clutter.

Other Methods. Many other methods exist for visualizing (clustered) networks. An example of a method we also considered is by Jusufi et al. [8]. Though potentially useful for assessing homophily, the resulting layouts appear suitable mostly for high-level overview tasks; path tracing is likely to be difficult due to the bundling of the edges. Hence, we did not include this method in our study.

---

[1] http://www.d3js.org/

2   Experimental Design

2.1   Hypotheses

With this user study, we wish to investigate the following three hypotheses:

H1 Homophily assessment is easiest with Bipartite layouts, followed by Polarized layouts and hardest with Force-Directed layouts.
H2 It is harder to assess homophily in networks with differently sized clusters.
H3 Finding a shortest path between two nodes is easiest with Force-Directed layouts, followed by Polarized layouts and hardest with Bipartite layouts.

Underlying our main hypothesis (H1), driving the design of this experiment, is the idea that visually separating the node clusters makes it easier to assess homophily. By pulling the nodes apart, we separate the same-cluster and cross-cluster links, thus potentially making it easier to assess the ratio between them. Cluster separation is stronger for the Bipartite layout than for the Polarized layout; it is not taken into account for Force-Directed layouts at all.

Whether there is homophily, depends not only on the ratio between same- and cross-cluster links, but also on the relative size of the two clusters. We hypothesize that it is easier to assess homophily when the two clusters are of equal size: we may then simply assess whether there are more or less same-cluster links in comparison to cross-cluster links.

Cluster separation may have a negative effect on tasks that are not influenced by the clusters, such as path-tracing tasks. We instantiate this by considering the task of finding the shortest path between two highlighted nodes. The cluster separation may pull neighboring nodes apart, causing longer links in the visualization. Such links become harder to follow. Moreover, switch-backs between the two clusters may be counterintuitive to the idea of a "shortest" path.

2.2   Method

Tasks. We used two Tasks: Bias and Path. The Bias task is targeted at Hypotheses H1 and H2, asking participants to assess the homophily of a network. However, we avoided the use of the term homophily and used an informal description of "Bias" to avoid different behavior between people that knew homophily beforehand and those that did not. In this paper, we shall use Bias to refer to a participant's assessment and homophily for calculated values. For answering Bias trials, participants were given a slider that internally allowed specifying a value between 0 % (only cross-cluster links) and 100 % (only same-cluster links), though no numbers were shown. The aim of the task was for participants to estimate bias, without precisely counting nodes and links; they were instructed accordingly. However, no time limit was given.

The Path task targets Hypothesis H3, asking participants to find the length of a shortest path between two nodes in a given network. To provide their answer, participants were given 5 radio buttons (for 2 to 6 steps). They were instructed

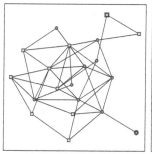 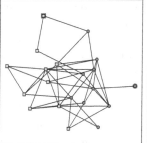 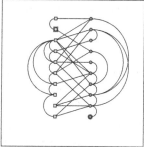

Fig. 1. Three stimuli used, all with Size 1, Balance B and Homophily 50 %. From left to right: force-directed (FD), polarized (P) and bipartite (B). The highlighted nodes were not highlighted for the Bias task.

to balance between answering correctly and answering quickly, ideally within 20 s. But it was mentioned explicitly that the time limit was not enforced.

The two tasks give rise to two sections in the study. To counter learning effects, the order of the sections was determined randomly per participant. Each section was preceded by a page explaining the task and an example question.

Stimuli. We have the following four independent variables for the stimuli:

- **Size.** Four different sizes: (1) 20 nodes, 40 links; (2) 20 nodes, 50 links; (3) 28 nodes, 60 links; (4) 40 nodes, 70 links;[2]
- **Balance.** Two ways to split into two clusters: (B) Balanced, an even split; (U) Unbalanced, one cluster contains 75 % of the nodes.
- **Homophily.** Five levels of degrees of homophily: 25 %; 37.5 %, 50 %; 62.5 %; 75 %;
- **Layout.** Three different layout algorithms (see Sect. 1.2 and Fig. 1): (FD) Force-Directed; (P) Polarized; (B) Bipartite.

To construct a network, we developed a simple random generator that takes as input the number of nodes for each cluster, the total number of links and the desired homophily. The desired homophily gives the fraction of links that should be cross-cluster; the remaining links were divided between the two clusters based on relative cluster sizes. The actual links added to the network were taken randomly (without replacement) from all possible links.

In each network, we also marked two arbitrary nodes (always one in each cluster) for the shortest-path task, controlling for the length of the shortest path. As we did not wish to introduce varying levels of difficultly for the shortest-path task, we need the stimuli to be of comparable difficulty, without always having the same number of steps as answer. We are mainly interested in the effects of Layout on task difficulty; using the same pair for each layout of the same network structure results in perfect balance. Nonetheless, we attempted to balance the

---

[2] Though small for social networks, we purposefully restricted to these sizes to ensure short trials and that path-tracing tasks would not become too difficult.

lengths across different levels of homophily and cluster balance to account for their possible effects on difficulty.

As a result, we have $4 \times 2 \times 5 = 40$ network structures. To each, we applied three Layouts; the resulting drawings were fitted to an SVG canvas, to allow for arbitrary resizing. The two node clusters were drawn using red circles and blue squares, the shade of the color chosen according to ColorBrewer[3]. Links were drawn in black with a small halo to increase separability between crossing links.

Mixed Design. With two tasks and 120 stimuli per task, we would need to give our participants 240 trials for a within-subjects design. This is far beyond what is reasonable for an online study, assuming 20 to 30 s per trial.

A between-subjects design is also not suitable, as we expect performance to be highly dependent on the participant's experience. Moreover, even if different people assess bias differently, it is possible that they are individually consistent: they perceive the homophily levels correctly, but assess the bias strength differently (see Fig. 2). To account for this Layout, Balance and Homophily are unsuitable as between-subjects factors.

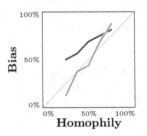

Fig. 2. Bias versus Homophily for two fictive participants. The monotonicity of each line indicates per-participant consistency of bias assessments.

Since network size is not directly of interest for our hypotheses and likely to be an obvious overall factor in increasing difficulty, we decided to use this as a between-subjects measure.

We now have 60 trials per participant. We aimed for a time investment of 20 to 25 min. A pilot study showed that with 60 trials, the actual completion time was around 30 min. Also, one of the pilot participants commented about the monotony of the questions, mentioning that less effort was put into the later trials for each task. We therefore decided to reduce the number of levels in Homophily-Balance interaction. We maintained the five levels of Homophily for the Balanced networks, but reduced it to three levels for Unbalanced networks. Maintaining five levels of Homophily for the Balanced networks provides a good baseline for investigating our main hypothesis on cluster separation and allows us to investigate for individual consistency. This reduced the number of trials to 48 (24 per Task). A second pilot study showed a completion time between 20 and 25 min as desired.

Again, to counter learning effects, the order of the 24 trials for each Task was randomized for each participant. Before each trial the participant was given a pause screen to reduce memory effects and at the same time allow them to pace themselves and reduce the possible impact of interruptions.

Apparatus. We developed our online user study, using a PHP webserver and a MySQL database. As is typical for online studies, we cannot control many aspects of the experimental environment (browser, OS, device, screen size,

---

[3] http://www.colorbrewer2.org/.

interruptions, etc.). We requested participants to use a laptop or desktop, instead of a tablet or phone. They were also asked to avoid or minimize interruptions and indicate at the end of the study any that did occur. To ensure that the browser is appropriate for the user study, we gave them a simple test (setting a slider to a number indicated in a figure) before they could start the actual questions. Some background and preference information was asked after completing the two tasks, though this remained optional for what may be perceived as sensitive information (age, gender, country of residence).

We could not control the screen size, resolution or distance of the participant to his screen. Hence, participants were provided some simple controls to scale the webpage to be comfortably readable and fitting on their screen. Moreover, they were asked to use full-screen mode to reduce distractions.

Recruiting Participants. We recruited volunteers to participate using a mix of mailing lists, social networks and social media. Because we did not know how many people would participate in our study and we have a mixed design, we decided on the following procedure. The four levels of Size were to be filled up to 35 participants, in the following order: 2, 3, 1, 4. Any participants in excess of $4 \cdot 35 = 140$ would be divided equally over Size. If we would fall short of 140 participants, the participants would work on the same or a similar network size.

3   Results

The data set for analysis as well as all stimuli have been made available online[4].

Participants. In total, 105 people volunteered and completed the online questionnaire, which was open for participation for four weeks. We kept close watch at the number of participants and at the end of the second week, we were just in excess of 70 participants and thus decided to disable the last Size group (4).

After two weeks and continuously thereafter, we also inspected all comments left by participants. We excluded from analysis any participants for whom comments or timing indicated a serious interruption, distraction or technical difficulty during a trial, i.e., not during a pause screen or in between sections. Participants were explicitly asked to indicate their effort needed to distinguish nodes from different clusters and finding highlighted nodes; those who indicated having a hard time with this were excluded from analysis. In doing this exclusion while the study was open for participation, our online system assigned new volunteers to fill up the three remaining Size groups evenly. This resulted in three Size groups with 30 participants each.

58 participants are male, 28 are female and 4 did not specify a gender. The men/women ratio was even for Size 1, but the other Sizes have a ratio of approximately $3:1$. Our participants are skewed towards male mathematicians and computer scientists. The average age of our participants is 35.9. Participants with Size 1 where older on average (39.1) and younger with Size 3 (33.9). In terms of country of residence, a majority of the participants live in Europe (61)

---

[4] http://ivv5web01.uni-muenster.de:8013/studyresults.html.

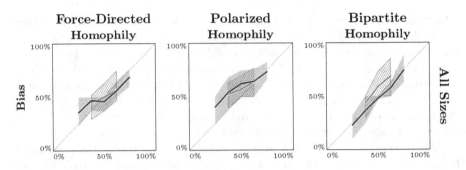

Fig. 3. Bias-Homophily charts. Thick lines indicate the average Bias; shaded blue (Balanced) and hashured red areas (Unbalanced) indicate the 25- and 75-percentile.

with a strong emphasis on Germany (27) and the Netherlands (22). Nine participants live in North America, three in Asia and one in South America; 16 participants did not provide a country of residence.

Hypothesis H1. The results of the bias estimation are summarized in Fig. 3. Each chart shows how the response Bias is correlated to the calculated degree of homophily. It suggests that the Bipartite layout leads to a stronger perception of Bias, with greater agreement (less variability) between the participants. For the Balanced case, the line is close to the diagonal, indicating that the average answer lies close to the calculated homophily. Notably, the Polarized layout and Unbalanced networks have greater variability; Polarized and Unbalanced-Bipartite lead to overestimating homophily (same-cluster links). The results for Balanced-Bipartite are centered on the diagonal, whereas the results for Balanced-Force-Directed are above the diagonal for Homophily below 50 % and below the diagonal for Homophily above 50 %. This suggests that the distinction between different levels of Homophily is clearer for the Bipartite layout.

We cannot simply classify answers as "correct" or "incorrect" as the participants were asked to give an estimation of Bias, without providing them a precise formula of how to determine it. Hence, we score each answer of Bias b for a stimulus with degree of homophily h with a deviation $b - h$. A positive value indicates an overestimation of cross-cluster links, whereas a negative value indicates an overestimation of same-cluster links. The deviation and response times for this task are summarized in Fig. 4. We performed RM-ANOVA on the deviation, to see if there are significant[5] differences of bias estimation for different layouts. The analysis showed a significant effect of Layout on deviation ($p < 0.001$). It also indicated an interaction effect of Layout and Size ($p < 0.001$). A post-hoc Tukey HSD test with Bonferroni adjustment showed a significant difference between Layout FD and P ($p < 0.001$) and between Layout P and B ($p < 0.001$). However, no significant difference between Layout FD and B was found.

---

[5] Significance is reported using a p-value: it indicates the probability that our observation is incorrect, i.e., occurring by chance rather than the different conditions. For an accessible introduction to HCI experimental design and analysis, we refer to [11].

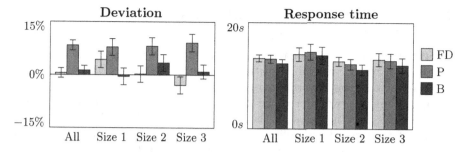

Fig. 4. Average deviation (left) and response time (right) for the Bias task, per Layout and Size. Error bars indicate 95%-confidence intervals.

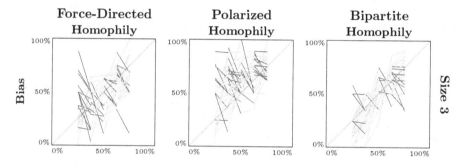

Fig. 5. Bias-Homophily charts for each Layout for Size 3, for the Balanced cases. Each line represents a participant. Monotonicity defects are colored dark purple.

A significant effect of Layout on response time ($p < 0.001$) was found, with the post-hoc test indicating a faster response time for the Bipartite layout.

We may thus partially accept Hypothesis H1: the Bipartite layout indeed outperforms the Polarized layout. However, because the Force-Directed layout outperforms the Polarized layout, it may be that our underlying argument for this hypothesis, i.e., cluster separation, is not the main effect in this difference.

The above focuses on the difference between Bias and Homophily and response time as an indicator of bias assessment. However, this does not readily mean that it is easier for a single participant to consistently assess bias. Let us now briefly turn towards an informal investigation of individual consistency (see also Sect. 2.2). If we chart each participant as a line in a Bias-Homophily plot, we would ideally see only monotonically increasing lines. The reality is of course different: Fig. 5 shows such a plot for Size 3. We observe that there are a lot less defects (decreasing parts of a line) from monotonicity for Layout B than for Layout FD and P. Over all Sizes, there are 90 defects for Layout FD, 113 for P and 60 for B; 24.4% of the participants had no defect for Layout FD, 10% for Layout P and 44.4% for Layout B. This suggests a better homophily perception for the Bipartite layout, but the percentage of people without defects remains rather low. This may in part be explained by the lack of repetitions and training tasks in our study. Unfortunately, this was unavoidable to keep a low time investment of the volunteers.

Hypothesis H2. To investigate the effects of Balance, we again refer to Fig. 3. We observe an increased variability for the Unbalanced cases. For Bipartite layout, we also observe a skew towards overestimating same-cluster links.

For the analysis, we filtered the 25 % and 75 % answers from the data set, as these levels were not used in the Unbalanced case. RM-ANOVA with the resulting data revealed a significant effect of Balance on deviation ($p < 0.05$) and response time ($p < 0.001$). We accept Hypothesis H2.

Hypothesis H3. The performance for the Path task is summarized in Fig. 6. Not surprisingly, RM-ANOVA revealed a significant effect of Size on error rate ($p < 0.001$). To investigate the effects of Layout, we therefore split the data into three subsets, one of each Size. Subsequent analysis of these sets revealed a significant effect of Layout on error rate ($p < 0.001$). The post-hoc test showed that the difference in error rate is significant between the Force-Directed layout and the Bipartite layout for each level of Size ($p < 0.01$). The Force-Directed layout also has a lower error rate for Size 2 ($p < 0.05$) and Size 3 ($p < 0.001$) compared to the Polarized layout. The Bipartite layout has a higher error rate than the Polarized layout for Size 1 ($p < 0.001$) and Size 2 ($p < 0.05$); a hint of a lower error rate was found in Size 3 ($p < 0.1$).

Further investigating this higher error rate, we found that four out of eight stimuli for P in Size 3 had an error rate of 80 % or more (i.e., worse than expected with random answers), suggesting a misleading visualization. After manual inspection of these stimuli, we attribute this to ambiguity of links that pass close by or even through unrelated nodes (see also Sect. 4).

Size does not have a significant effect on response time, but Layout does ($p < 0.001$). The post-hoc test showed a significant difference between the Bipartite layout and the others (both $p < 0.001$) as well as difference between the Force-Directed and Polarized layout ($p < 0.05$).

Combining the results of error rate and response time, we may accept H3: Force-Directed outperforms Polarized, which in turn outperforms Bipartite.

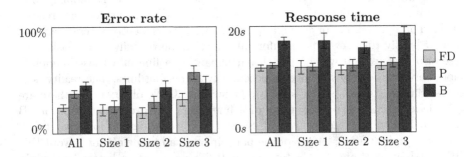

Fig. 6. Average error rate (left) and response time (right) for the Path task, per Layout and Size. Error bars indicate 95 %-confidence intervals.

Preferences. After completing the tasks, we asked the participants to indicate how hard they found it to perform the tasks with each type of network. However, the pilot study indicated that the distinction between the Force-Directed and Polarized layout was not clear while performing the tasks and thus hard to assess afterwards. As we did not want to introduce the three layouts beforehand, we opted to ask participants to rate (1–5) the Force-Directed layout and Bipartite layout only. The overall preference corresponds to the overall performance. For assessing bias, respondents clearly preferred the Bipartite layouts (mean $\mu = 3.5$, standard deviation $\sigma = 0.1$) over the Force-Directed and Polarized ones ($\mu = 2.0$, $\sigma = 0.1$). For finding shortest paths, this was reversed (Bipartite: $\mu = 2.4$, $\sigma = 0.1$; Force-Directed/Polarized: $\mu = 3.8$, $\sigma = 0.1$).

4   Discussion and Conclusion

Our results indicate that we may answer Question 1 positively: observers can indeed assess homophily, but this is affected by the layout and some layouts may in particular lead to an overestimation of homophily. We remark that individual consistency was not very strong, but this was leveled out by taking the average over all participants; see also the discussion below on training and repetition. To answer Question 2, the Bipartite layout performs best, followed by the classic force-directed method. The improved performance of the Bipartite layout, however, must be weighed against a loss in performance for other tasks.

Our results indicate that cluster separation by itself is not a general design principle to improve homophily perception. Future work may investigate such design principles in the context of homophily as well as further explore homophily with multiple clusters and defining a per-cluster homophily degree.

As with any user study, no experiment is flawless. We conclude our paper by discussing some aspects that may undermine our findings.

Bias Estimation. Explaining the Bias task was a difficult thing to do, without making the description overly long. In particular, we chose to go with a simple explanation of bias and not attempt to explain (degree of) homophily in detail. A participant's interpretation of Bias may thus inherently deviate from what is computed with our degree of homophily.

That the Bias task was rather difficult as a result, was also evidenced by some of the participants' comments. In particular, a few participants commented that the Unbalanced condition was hard to assess, further supporting Hypothesis H2.

Visual Representation. By using both color and shape, we tried to make the distinction between the clusters very clear. This supports the Bias task, but may in fact be detrimental for the Path task: participants may have had a tendency to look for connections between same colored nodes before looking at different colored nodes or vice versa. We think that this effect is mitigated by always selecting two nodes of different clusters. Also, network visualizations may simply need to display the different clusters for a variety of possible reasons, while still supporting the task of following paths well.

Whereas the Force-Directed and Polarized layouts use only line segments, the Bipartite layout uses circular arcs for same-cluster links. This difference in graphic encoding likely has a strong effect on distinguishing same-cluster from cross-cluster links, but this is in a large part already done by the clear cluster separation of the layout. However, it may also affect how easily participants can estimate the number of such links and hence affect the bias estimation. Moreover, it is likely to influence how easy it is to follow links for the Path task. As a result, it may be difficult to ascribe the results to the effect of cluster separation alone. This in particular affects for Hypothesis H1, but also means that for the other hypothesis, the effects may be in large part due to the layout as a whole, rather than any single aspect of it.

Training and Repetition. Due to our short intended time investment, there was little time for training the participants. They were given only one example question, before the trials started and each condition was presented only once. As a result, responses to earlier trials may be less accurate (or in case of bias assessment, less consistent). This is countered to some degree by randomizing the order of the trials for each participant. However, as indicated, it undermines evaluating an individual's responses. This is particularly the case for consistency of bias assessment, which is therefore done only informally and treated as an indication. However, plotting the deviation of bias assessment over time did not reveal clear learning effects.

Acknowledgments. The authors would like to thank all anonymous volunteers who participated in the presented user study.

References

1. Baur, M., Brandes, U.: Crossing reduction in circular layouts. In: Hromkovič, J., Nagl, M., Westfechtel, B. (eds.) WG 2004. LNCS, vol. 3353, pp. 332–343. Springer, Heidelberg (2004)
2. Eades, P., Feng, Q., Lin, X., Nagamochi, H.: Straight-line drawing algorithms for hierarchical graphs and clustered graphs. Algorithmica **44**, 1–32 (2006)
3. Easley, D., Kleinberg, J.: Networks, Crowds and Markets: Reasoning About a Highly Connected World. Cambridge University Press, New York (2010)
4. Fruchterman, T.M.J., Reingold, E.M.: Graph drawing by force-directed placement. Softw. Pract. Exp. **21**(11), 1129–1164 (1991)
5. Ghani, S., Kwon, B.C., Lee, S., Yi, J.S., Elmqvist, N.: Visual analytics for multi-modal social network analysis: a design study with social scientists. IEEE TVCG **19**(12), 2032–2041 (2013)
6. Henry, N., Fekete, J.-D., McGuffin, M.J.: Node trix: a hybrid visualization of social networks. IEEE TVCG **13**(6), 1302–1309 (2007)
7. Holten, D.: Hierarchical edge bundles: visualization of adjacency relations in hierarchical data. IEEE TVCG **12**(5), 741–748 (2006)
8. Jusufi, I., Kerren, A., Liu, J., Zimmer, B.: Visual exploration of relationships between document clusters. In: Proceedings of International Conference on Information Visualization Theory and Applications, pp. 195–203 (2014)

9. Krzywinski, M., Birol, I., Jones, S.J.M., Marra, M.A.: Hive plots–rational approach to visualizing networks. Briefings Bioinf. **13**, 627–644 (2011)
10. McPherson, M., Smith-Lovin, L., Cook, J.M.: Birds of feather: homophily in social networks. Ann. Rev. Sociol. **27**, 415–444 (2001)
11. Purchase, H.C.: Experimental Human-Computer Interaction: A Practical Guide with Visual Examples. Cambridge University Press, New York (2012)
12. van den Elzen, S., van Wijk, J.J.: Multivariate network exploration and presentation: from detail to overview via selections and aggregations. IEEE TVCG **20**(12), 2310–2319 (2014)

Shape-Based Quality Metrics for Large Graph Visualization

Peter Eades[1]([✉]), Seok-Hee Hong[1], Karsten Klein[2], and An Nguyen[1]

[1] University of Sydney, Sydney, Australia
{peter.eades,seokhee.hong}@sydney.edu.au, angu5603@uni.sydney.edu.au
[2] Monash University, Melbourne, Australia
karsten.klein@monash.edu

Abstract. We propose a new family of quality metrics for graph drawing; in particular, we concentrate on larger graphs. We illustrate these metrics with examples and apply the metrics to data from previous experiments, leading to the suggestion that the new metrics are effective.

1 Introduction

Several of the earliest papers on Graph Drawing (for example, [21–23]) discussed requirements for a "good" visualization of a graph. For example, Tamassia et al. [22] state:

> *Aesthetics: We use the term aesthetics to denote the criteria that concern certain aspects of readability. A well-admitted aesthetics, valid independently from the graphic standard, is the minimisation of crossings between edges. Also, in order to avoid unnecessary waste of space, it is usual to keep the area occupied by the drawing reasonably small. When the grid standard is adopted, it is meaningful to minimize the number of bends (turns) along the edges, as well as their total length.*

We prefer the term *quality metric* rather than "aesthetics". The underlying and often unstated assumption that these geometric properties of layout measure the "goodness" of a graph drawing was unchallenged until the experiments of Purchase [20]. These experiments showed that task performance is correlated with some of the previously defined quality metrics. A conclusive result was that human task times and error rates were both correlated with the number of *edge crossings*. Subsequent experiments have confirmed and refined these initial results [11,17–19,25]. All these early experiments used relatively small graphs as stimuli; human experiments with larger graphs began recently [13,14]. In particular it has been pointed out that edges and vertices become "blobs" in large graph drawings such as the biological network in Fig. 1; almost all the edge crossings are hidden in the blobs. Any causal relationship between readability and the number of edge crossings seems unlikely. In this paper we propose a quality metric for large drawings such as Fig. 1.

Supported by Australian Research Council grants LP110100519 and DP140100077, and by Tom Sawyer Software and NewtonGreen Technologies.

© Springer International Publishing Switzerland 2015
E. Di Giacomo and A. Lubiw (Eds.): GD 2015, LNCS 9411, pp. 502–514, 2015.
DOI: 10.1007/978-3-319-27261-0_41

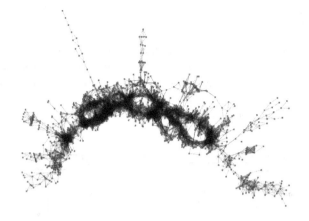

Fig. 1. Crossings can be hidden in a drawing of a large graph.

Although it is seldom explicitly stated as a quality metric for graph drawing, *stress* is often used as such. There are various measures of stress (for example, see [5,6,8,10]); the most commonly used is to define the stress in a drawing D of a connected graph $G = (V, E)$ as $\sum_{u,v \in V} w_{uv} (d_G(u,v) - d_{\Re^2}(D(u), D(v)))^2$, where $d_G(u, v)$ is the graph theoretic distance between u and v, $d_{\Re^2}(D(u), D(v))$ is the Euclidean distance between the locations $D(u)$ and $D(v)$ of u and v, and w_{uv} is a constant. Stress appears to measure the *faithfulness* of a graph drawing [15] rather than its readability. For example, a low value of the stress in a drawing indicates that the Euclidean distances between vertices are (approximately) proportional to the graph-theoretic distances in the graph.

Quality metrics are significant simply because they measure success or failure of a graph drawing method. Most importantly, they are used as optimisation goals in graph drawing algorithms. Methods that aim to draw graphs with a small number of crossings are well established in the literature. Stress minimization algorithms, in one form or another, are by far the most popular methods for drawing undirected graphs.

This paper proposes a new family of quality metrics for graph visualization, especially for large graph drawings. Here, by "large", we mean that the graphs are large enough to make "blobs" such as in Fig. 1 inevitable. This includes dense graphs with at least a few hundred vertices and well as sparse graphs with at least a few thousand vertices.

The proposed metrics are based on the notion of the "shape" of a set of points in $\Re^2$. Our proposal is that a drawing is good if the shape of the set of vertex positions is similar to the original graph.

In Sect. 2, we describe this notion more precisely and illustrate with examples. In Sect. 3 we give some empirical indication that the metrics are valid, based on data sets from previous experiments [1,14]. Section 4 concludes with open problems.

2   Shape-Based Metrics

Figure 2 summarises our proposal. The quality of a drawing D of a graph G is the similarity between G and the "shape" of the set of vertex locations of D. The "shape" is expressed as a graph, called a "shape graph". To make these notions more precise, we need to examine the notion of the shape of a set of points, and the notion of similarity between two graphs.

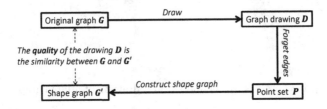

Fig. 2. Shape-based quality metrics.

2.1   Shape as a Graph

Informally, a *shape graph* for a set of points P is a geometric graph $G = (P, E)$ that captures the "shape" of P. The classical example of a shape graph is the α -shape [3]; however, α-shapes capture the shape of the *boundary* of P, and not the *internal structure* of P. Another kind of shape graph is a "proximity graph": an edge is placed between two points $p, q \in P$ if p is "close to" q in some sense. There are many kinds of proximity graphs (see [24]); some examples are below.

- The *k-nearest neighbours graph* has a (directed) edge from point $p \in P$ to point $q \in P$ if the number of points $r \in P$ with $d(p, r) < d(p, q)$ is at most $k - 1$.
- The *Delaunay triangulation*: the dual of the Voronoi diagram on P.
- The *Gabriel graph* (GG) has an edge between distinct points $p, q \in P$ if the closed disc which has the line segment pq as a diameter contains no other elements of P.
- The *relative neighbourhood graph* (RNG) has an edge between distinct points $p, q \in P$ if there is no point $r \in P$ such that $d(p, r) \le d(p, q)$ and $d(q, r) \le d(p, q)$.
- A *Euclidean minimum spanning tree* (EMST) is a minimum spanning tree of P where the weight of the edge between each pair of points is the Euclidean distance.

Each of these shape graphs can be computed in $O(n \log n)$ time using standard methods [16]. In Sect. 3 below, we examine quality metrics based on the Euclidean minimum spanning tree, the Gabriel graph, and the relative neighborhood graph respectively. However, our remarks apply in principle to any shape graph.

2.2   Graph Similarity

Suppose that $G_1 = (V, E_1)$ and $G_2 = (V, E_2)$ are two graphs with the same vertex set. A simple measure for the similarity of G_1 and G_2 is the *Jaccard sum similarity*:

$$JSS(G_1, G_2) = \sum_{u \in V} \frac{|N_1(u) \cap N_2(u)|}{|N_1(u) \cup N_2(u)|}, \tag{1}$$

where $N_i(u)$ is the set of neighbors of u in G_i for $i = 1, 2$. It is straight-forward to compute the Jaccard sum similarity in linear time.

More complex measures for graph similarity include graph edit distance [7], and measures based on the notion that the similarity of two vertices u and u' depends on the similarity of their neighbours (see, for example, [12]). However, these metrics are computationally expensive and do not scale beyond a few thousand vertices; we have found that the Jaccard sum similarity is adequate for our purposes.

2.3   The Metrics

We can now define our proposed metrics. Suppose that D is a drawing of a graph G; we want to measure the quality of D. Let P denote the set of vertex locations of D, and suppose that μ is a shape graph function (that is, μ takes a set of points and produces a shape graph on this set of points). Further, let η be a graph similarity function, that is, η takes two graphs as input and returns a positive real number that indicates the similarity between these two graphs. Then we define the quality metric $Q_{\mu,\eta}$ by $Q_{\mu,\eta}(D) = \eta(G, \mu(P))$.

These metrics are, in spirit, related to the "graph theoretic scagnostics" approach to scatterplots (see [26]).

The "neighborhood inconsistency" [6] and "neighborhood preservation precision" [5,6] metrics used by Gansner et al. are also related, especially when the shape graph μ is a kind of nearest neighbor graph. These two metrics have a different motivation to ours: rather than measure the general notion of shape, they attempt to measure whether neighbours in the layout coincide with neighbours in the graph. Nevertheless, we can regard the "neighborhood inconsistency" as an example of a shape-based metric when the shape graph μ is a k-nearest neighbor graph, and the similarity function η is based on the "stochastic neighbor embedding" of Hinton and Roweis [9].

Throughout this paper we use the Jaccard sum similarity for graph similarity, and so we abbreviate $Q_{\mu,\eta}$ to Q_μ. The time to compute Q_μ depends on the choice of μ; for all such choices μ described in this paper, Q_μ can be computed in time $O(n \log n)$.

2.4   An Example

Although our proposal is aimed at large graphs, this example uses a smaller graph so that it is easier to understand. Consider the graph drawing D_0 in

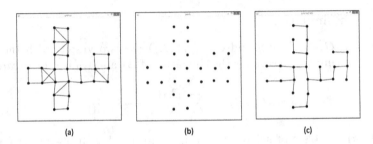

Fig. 3. (a) A graph drawing D_0. (b) The set P_0 of vertex locations of D_0. (c) A Euclidean minimum spanning tree T_0 on P_0.

Fig. 3(a). The set P_0 of vertex locations of D_0 is shown in Fig. 3(b). A Euclidean minimum spanning tree T_0 on P_0 is shown in Fig. 3(c).

Our proposal is that the quality $Q_{EMST}(D_0)$ of the graph drawing D_0 can be measured as the similarity between the (combinatorial) graphs in Fig. 3(a) and (c). Using the Jaccard sum similarity in Eq. (1), we can calculate the value $Q_{EMST}(D_0)$ as 0.61. The comparatively high value of $Q_{EMST}(D_0)$ expresses

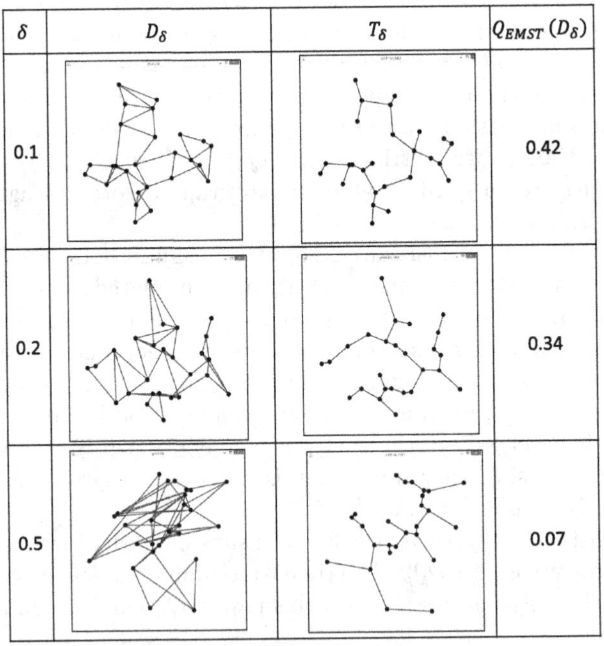

δ	D_δ	T_δ	$Q_{EMST}(D_\delta)$
0.1			0.42
0.2			0.34
0.5			0.07

Fig. 4. The drawing D_δ in the second column is formed from the drawing D_0 in Fig. 3 by moving each vertex in a random direction by a random distance in the range $[0, \delta]$. The graph T_δ in the third column is a Euclidean minimum spanning tree of the vertex locations of D_δ.

the fact that for each vertex u the neighbors of u in the shape graph T_0 overlap considerably with the neighbors of u in G. Intuitively, the graph drawing D_0 is a reasonably faithful representation of the graph G, in that the "shape" of D_0 is similar to G.

Next we examine what happens when we make the drawing progressively bad. Suppose that D_δ is formed from D_0 by moving each vertex in a random direction by a random distance in the range $[0, \delta s]$, where s is the size of the screen. Drawings D_δ for $\delta = 0.1, 0.2$, and 0.5 are shown in Fig. 4. For $\delta = 0.1$, the shape of the drawing is fairly close to the graph, the minimum spanning tree T_δ shares quite a few edges with D_δ, and the value $Q_{EMST}(D_{0.1}) = 0.42$ is reasonably high. As δ increases, the minimum spanning tree T_δ shares fewer edges with D_δ, and the values of $Q_{EMST}(D_\delta)$ fall. For $\delta = 0.5$ the shape of the drawing shows no resemblance to the graph, and $Q_{EMST}(D_\delta)$ is low. Intuitively, as the drawing becomes worse, the shape of the set of points differs more and more from the graph.

Larger examples are shown in Fig. 5, from the data set described in Sect. 3.1. Here the graph drawing (a) has 1160 vertices and 6424 edges, and (b) is a Euclidean minimum spanning tree of (a); the graph drawing (c) has 1749 vertices and 13957 edges, and (d) is a Euclidean minimum spanning tree of (c). In both

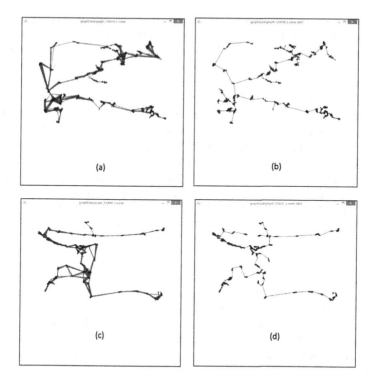

Fig. 5. Two graph drawings from the data set described in Sect. 3.1, and Euclidean minimum spanning trees of the vertex locations.

cases, the Euclidean minimum spanning tree shares many edges with the graph, and expresses the shape of the graph drawing well.

2.5   Remarks

We should point out that the metrics that we have defined above are not normalised across graphs. If D and D' are two drawings of the *same* graph, then $Q_\mu(D) > Q_\mu(D')$ whenever D is a better drawing than D'. However, we make no such claim for two drawings D and D' of two *different* graphs.

Further, note that the Gabriel graph contains the relative neighborhood graph, which in turn contains the Euclidean minimum spanning tree [24]. We expect the Gabriel graph to model the shape of a set of points more precisely than a Euclidean minimum spanning tree.

3   The Experiments

In this section, we describe how the shape-based quality metrics perform on two specific data sets from previous experiments [1,14].

3.1   The "Untangling" Data Set

Marner et al. [14] introduced a new method called *GION* for supporting interaction with graph drawings on large displays. The user study of [14] focussed on the task of *untangling* a graph drawing: subjects were presented with a graph drawing (a Fruchterman-Reingold layout [4]), and were simply asked to untangle the layout. Eight RNA sequence graphs were used, ranging from 1159 to 7885 vertices; there were 16 subjects. The experimental system captured, for each subject and each graph, a snapshot drawing every 5 seconds; the snapshot at time t is denoted by D_t. Two such snapshot graph drawings are shown in Fig. 5(a) and (c). The main result of the experiment was that the GION method is better in several ways than more traditional interaction methods. For more details, see [14].

The experiment gave a large data set of graph drawings (8 graphs × 16 users × 24 snapshot drawings) that we can use to check our shape-based quality metrics. For each snapshot D_t, we computed the number $\chi(D_t)$ of edge crossings, the (scaled) stress $\sigma(D_t)$, and the metrics $Q_{EMST}(D_t)$, $Q_{GG}(D_t)$, and $Q_{RND}(D_t)$, respectively based on Euclidean minimum spanning tree, Gabriel graphs, and relative neighborhood graphs.

Commonly-held graph drawing wisdom is that $\chi(D_t)$ and $\sigma(D_t)$ decrease with the quality of the graph drawing. We expect that quality increases as the graph is untangled, and so we expect that $\chi(D_t)$ and $\sigma(D_t)$ decrease with t. In contrast, the proposed quality metrics $Q_{EMST}(D_t)$, $Q_{GG}(D_t)$, and $Q_{RND}(D_t)$ are expected to increase with t. To make the comparison between these metrics easier, we place them on a comparable scale by inverting and normalising crossings and stress, as follows. We define

$$\bar{Q}_\chi(D_t) = \frac{M_\chi - \chi(D_t)}{M_\chi}, \quad \bar{Q}_\sigma(D_t) = \frac{M_\sigma - \sigma(D_t)}{M_\sigma},$$

where $M_\chi = \max_t \chi(D_t)$ and $M_\sigma = \max_t \sigma(D_t)$. Note that $\bar{Q}_\chi(D_t)$ (respectively $\bar{Q}_\sigma(D_t)$) increases from 0 to 1 as the number of crossings (respectively stress) increases from 0 to the maximum. For the shape-based metrics, we simply linearly normalise Q_{EMST} (respectively Q_{GG} and Q_{RND}) to give $\bar{Q}_{EMST}$ (respectively $\bar{Q}_{GG}$ and $\bar{Q}_{RND}$) so that it increases from 0 to 1 as the quality of the drawing increases.

It is reasonable to assume that the drawing improves in quality as the untangling proceeds. However, the results reported in [14] were counterintuitive in terms of crossings and stress: as the subjects untangled the graph drawings, there was a tendency to *increase* both crossings and stress (that is, both $\bar{Q}_\chi$ and $\bar{Q}_\sigma$ decreased).

In contrast, we found that $\bar{Q}_{EMST}$, $\bar{Q}_{GG}$, and $\bar{Q}_{RND}$ all *increased* as the subjects untangled the drawings. The charts in Fig. 6 show $\bar{Q}_\chi$, $\bar{Q}_\sigma$, $\bar{Q}_{EMST}$, $\bar{Q}_{GG}$, and $\bar{Q}_{RND}$, averaged over all subjects, for the first 3 of the 8 graphs. The horizontal axis is time t; the vertical axis shows the values of the metrics. For graphs #1 and #2, both crossings and stress increase with t (that is, $\bar{Q}_\chi(D_t)$ and $\bar{Q}_\sigma(D_t)$ decrease). In contrast, $\bar{Q}_{EMST}$, $\bar{Q}_{GG}$, and $\bar{Q}_{RND}$ increase. Graphs #4, #5, #6, #7, and #8 showed very similar patterns to graphs #1 and #2. Graph #3 was a little different in that crossings decrease (and thus $\bar{Q}_\chi$ increases), albeit chaotically.

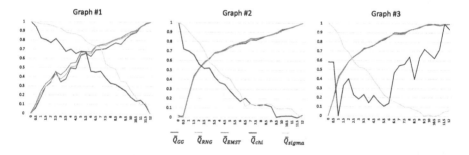

Fig. 6. Metrics against untangling.

Overall, the data from the untangling experiment shows that both crossings and stress metrics became worse as the subjects untangled the graphs, but the shape-based metrics became better. With some provisos (see Sect. 3.3 below), this suggests that the shape-based metrics are better than crossings and stress for measuring untangling.

3.2   The "Preference" Data Set

Chimani et al. [1] report an experiment at the University of Osnabrueck aimed at determining whether human *preferences* in graph drawing correlates with crossings and stress. There were two follow-up experiments, at the Graph Drawing conference in 2014, and at the University of Sydney. The design and results of all three experiments were similar; see [1]. Here we investigate the data from the University of Sydney experiment, aiming to determine whether shape-based metrics are correlated with preference.

This experiment had 40 subjects. Each subject was presented with 20 "instances". Each instance displayed a pair of drawings of the same graph, as in the screenshot in Fig. 7. There is a slider bar at the bottom of the screen, and the subject indicates which of the pair of drawings he/she prefers by sliding to the left or right. The slider bar has a scale on the left from 5 to 1 and on the right from 1 to 5, with zero in the middle. The slider bar is used to give a score to the drawing that the subject prefers, indicating how much the subject prefers it. A score of 5 on the left indicates a strong preference for the drawing on the left, and a score of 5 on the right indicates a strong preference for the drawing on the right.

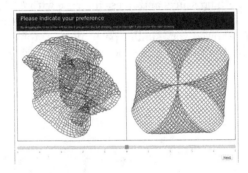

Fig. 7. Example of a typical "instance" (a graph pair shown to participants).

A total of 118 graphs, ranging in size from small (25 vertices and 29 edges) to moderately large (8000 vertices and 15580 edges), were used. Five drawings for each graph were generated, and the instances were chosen randomly. For details, see [1].

The results for a particular quality metric Q are expressed in terms of the "Q-ratio", defined as follows. Consider an instance consisting of two drawings D_{left} and D_{right} of a graph G, such as in Fig. 7. Let $Q(D_{left})$ (respectively $Q(D_{right})$) be the value of the Q metric for D_{left} (respectively for D_{right}). We define the Q-ratio for this instance as

$$\frac{\max(Q(D_{left}), Q(D_{right}))}{\min(Q(D_{left}), Q(D_{right}))}.$$

If the Q-ratio is significantly larger than 1, then we expect that most subjects prefer the drawing with the higher quality (according to the quality metric Q). Further, as the Q-ratio increases, we expect that more and more subjects prefer

the drawing with higher quality. To make this precise, we need to define some
further terms.

For each quality metric Q and each instance I we compute a score $S_Q(I)$ as
follows. Suppose that for this instance, the subject gives a score of x ($0 \leq x \leq 5$).
If the subject chose the drawing with a higher value of the quality metric Q, then
$S_Q(I) = x$; otherwise $S_Q(I) = -x$. The expectation that most subjects prefer the
drawing with the higher quality becomes an expectation that in most instances,
$S_Q(I)$ is positive.

For each metric Q, we chart the median of $S_Q(I)$ over all instances I against
the Q-ratio in Fig. 8. The charts for crossings and stress are shown in Fig. 8(a),
and for EMST, GG, and RNG in Fig. 8(b). For both crossing and stress, there
is adequate data for ratios from 1 to 5; however, the data for ratios larger than
4.5 is small (less than 20 instances) and the results at this end of the spectrum
must be treated with caution.

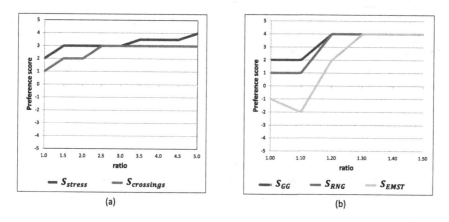

(a)   (b)

Fig. 8. Stress and crossing ratios, shape graph ratios, and preferences.

Crossings. Overall, there is a slight preference for fewer crossings (median over
all instances is $+1$). As the crossing ratio increases, the median preference
for the drawing with fewer crossings increases. When the crossing ratio is
above 2.5 the median preference for the drawing with fewer crossings is $+3$,
and stays steady at $+3$ as the crossing ratio increases beyond 2.5.

Stress. Overall, there is a preference for lower stress (median over all instances
is $+2$). As the stress ratio increases, the median preference for lower stress
rises; it hovers between $+3$ and $+4$ when the stress ratio is above 4.

For EMST, GG, and RNG, there is adequate data for ratios from 1 to 1.5; but
the data for ratios larger than 1.45 is small (less than 20 instances) and the
results at this end of the spectrum must be treated with caution.

EMST. The median preference for the drawing with higher value of $\bar{Q}_{EMST}$ is chaotic when the EMST-ratio is less than 1.2. The preference rises to +4 when the EMST-ratio rises from 1.2 to 1.3, and remains at +4 as the EMST-ratio increases beyond 1.3.

GG. Overall, there is a preference for drawings with a higher value of $\bar{Q}_{GG}$ (median over all instances is +2). The preference for the drawing with higher value of $\bar{Q}_{GG}$ rises smoothly with GG-ratio. When the GG-ratio is above 1.2 the median preference for the drawing with higher value of $\bar{Q}_{GG}$ is +4, and remains at +4 as the GG-ratio increases beyond 1.2.

RNG. Overall, there is a preference for drawings with a higher value of $\bar{Q}_{RNG}$ (median over all instances is +1). The preference for the drawing with higher value of $\bar{Q}_{RNG}$ rises smoothly with RNG-ratio. When the RNG-ratio is above 1.2 the median preference for the drawing with higher value of $\bar{Q}_{RNG}$ is +4, and remains at +4 as the RNG-ratio increases beyond 1.2.

One can conclude that people prefer drawings with fewer crossings, lower stress, and higher values for the shape-based metrics $\bar{Q}_{EMST}$, $\bar{Q}_{GG}$, and $\bar{Q}_{RND}$. Note that the preference for better GG and RNG based metrics appears to be a little stronger than the preference for fewer crossings and lower stress. The overall preference for EMST-based metrics seems unreliable when the EMST-ratio is small.

3.3   Remarks on the Experiments

The data from both the "untangling" experiment and the "preference" experiment support the proposal that the shape-based metrics are good measures of the quality of a graph drawing; there is some indication that the shape-based metrics are better than crossings and stress. However, this support has some limitations:

- Both experiments were designed for other purposes. Neither experiment was designed to test the shape-based metrics. To completely validate the new metrics, further study is needed.
- The "untangling" experiment used a very specific kind of graph: RNA sequence graphs, which are locally dense with a global "tree-like" structure. For more general classes of graphs, further experimentation would be useful.
- The experiments use the notions of "untangledness" and "preference" as proxies for ground truth quality. It would be useful to test the metrics in a *task-oriented* experiment.

4   Conclusion and Open Problems

This paper proposes a new family of metrics, aimed at measuring the quality of large graph drawings. We have some evidence from data in two previous experiments that these metrics are effective.

Our proposal raises several open problems:

- Design experiments to fully validate shape-based metrics. In particular, it would be interesting to know whether time and error of tasks on large graphs (see [13]) is related to shape-based metric values.
- Design algorithms to produce layouts that optimise the metrics. Note that (as with most graph layout problems) optimisation problems of this kind are typically NP-hard (see, for example, [2]), and thus heuristic approaches are in order. In particular, it would be interesting to know whether a stress minimisation algorithm gives a reasonable approximation.

References

1. Chimani, M., et al.: Graph drawings with less stress and fewer crossings are preferable. In: Duncan, C., Symvonis, A. (eds.) GD2014. LNCS, vol. 8871, pp. 523–524. Springer, Heidelberg (2014)
2. Eades, P., Whitesides, S.: The realization problem for euclidean minimum spanning trees in NP-hard. Algorithmica 16(1), 60–82 (1996)
3. Edelsbrunner, H., Kirkpatrick, D.G., Seidel, R.: On the shape of a set of points in the plane. IEEE Trans. Inf. Theory 29(4), 551–558 (1983)
4. Fruchterman, T.M.J., Reingold, E.M.: Graph drawing by force-directed placement. Softw. Pract. Exp. 21(11), 1129–1164 (1991)
5. Gansner, E.R., Hu, Y., Krishnan, S.: COAST: a convex optimization approach to stress-based embedding. In: Wismath, S., Wolff, A. (eds.) GD 2013. LNCS, vol. 8242, pp. 268–279. Springer, Heidelberg (2013)
6. Gansner, E.R., Yifan, H., North, S.C.: A maxent-stress model for graph layout. IEEE Trans. Vis. Comput. Graph. 19(6), 927–940 (2013)
7. Gao, X., Xiao, B., Tao, D., Li, X.: A survey of graph edit distance. Pattern Anal. Appl. 13(1), 113–129 (2010)
8. Hachul, S., Jünger, M.: Drawing large graphs with a potential-field-based multilevel algorithm. In: Pach, J. (ed.) GD 2004. LNCS, vol. 3383, pp. 285–295. Springer, Heidelberg (2005)
9. Hinton, G.E., Roweis, S.T.: Stochastic neighbor embedding. In: NIPS 2002, pp. 833–840 (2002)
10. Hu, Y., Koren, Y.: Extending the spring-electrical model to overcome warping effects. In: Eades, P., Ertl, T., Shen, H.-W. (eds.) PacificVis2009, pp. 129–136. IEEE (2009)
11. Huang, W., Hong, S.-H., Eades, P.: Effects of crossing angles. In: Fujishiro, I., Li, H., Ma, K. L. (eds.) IEEE PacificVis2008, pp. 41–46 (2008)
12. Jeh, G., Widom, J.: Simrank: a measure of structural-context similarity. In: KDD2002, pp. 538–543. ACM (2002)
13. Kobourov, S.G., Pupyrev, S., Saket, B.: Are crossings important for drawing large graphs? In: Duncan, C., Symvonis, A. (eds.) GD 2014. LNCS, vol. 8871, pp. 234–245. Springer, Heidelberg (2014)
14. Marner, M.R., Smith, R.T., Thomas, B.H., Klein, K., Eades, P., Hong, S.-H.: GION: interactively untangling large graphs on wall-sized displays. In: Duncan, C., Symvonis, A. (eds.) GD 2014. LNCS, vol. 8871, pp. 113–124. Springer, Heidelberg (2014)
15. Nguyen, Q.H., Eades, P., Hong, S.-H.: On the faithfulness of graph visualizations. In: Carpendale, S., Chen, W., Hong, S. (eds.) PacificVis2013, pp. 209–216. IEEE (2013)

16. Preparata, F., Shamos, M.: Computational Geometry - An Introduction. Springer, New York (1985)
17. Purchase, H.C.: Which aesthetic has the greatest effect on human understanding? In: Di Battista, G. (ed.) GD 1997. LNCS, vol. 1353, pp. 248–261. Springer, Heidelberg (1997)
18. Purchase, H.C., Allder, J.-A., Carrington, D.A.: Graph layout aesthetics in uml diagrams: user preferences. J. Graph Algorithms Appl. **6**(3), 255–279 (2002)
19. Purchase, H.C., Carrington, D.A., Allder, J.-A.: Empirical evaluation of aesthetics-based graph layout. Empirical Softw. Eng. **7**(3), 233–255 (2002)
20. Purchase, H.C., Cohen, R.F., James, M.I.: Validating graph drawing aesthetics. In: Brandenburg, F.J. (ed.) GD 1995. LNCS, vol. 1027, pp. 435–446. Springer, Heidelberg (1996)
21. Sugiyama, K., Tagawa, S., Toda, M.: Methods for visual understanding of hierarchical system structures. IEEE Trans. Syst. Man. Cybern. **11**(2), 109–125 (1981)
22. Tamassia, R., Batini, C., Di Battista, G.: Automatic graph drawing and readability of diagrams. Technical Report, Universita di Roma La Sapienza, Dipartimento do Informatica e Sistemistica, 01.87 (1987)
23. Tamassia, R., Batini, C., Talamo, M.: An algorithm for automatic layout of entity-relationship diagrams. In: Davis, C.G., Jajodia, S., Ng, P.A., Yeh, R.T. (eds.) ER83, pp. 421–439. North-Holland (1983)
24. Toussaint, G.: Computational Morphology. North Holland, Amsterdam (1988)
25. Ware, C., Purchase, H.C., Colpoys, L., McGill, M.: Cognitive measurements of graph aesthetics. Inf. Vis. **1**(2), 103–110 (2002)
26. Wilkinson, L., Anand, A., Grossman, R.L.: Graph-theoretic scagnostics. In: Stasko, J., Ward, M.O. (eds.) InfoVis2005, p. 21. IEEE (2005)

On the Readability of Boundary Labeling

Lukas Barth[1], Andreas Gemsa[1], Benjamin Niedermann[1]([✉]),
and Martin Nöllenburg[2]

[1] Institute of Theoretical Informatics, Karlsruhe Institute of Technology,
Karlsruhe, Germany
niedermann@kit.edu
[2] Algorithms and Complexity Group, TU Wien, Vienna, Austria

Abstract. Boundary labeling deals with annotating features in images
such that labels are placed outside of the image and are connected by
curves (so-called leaders) to the corresponding features. While boundary
labeling has been extensively investigated from an algorithmic perspective, the research on its readability has been neglected. In this paper
we present the first formal user study on the readability of boundary
labeling. We consider the four most studied leader types with respect to
their performance, i.e., whether and how fast a viewer can assign a feature to its label and vice versa. We give a detailed analysis of the results
regarding the readability of the four models and discuss their aesthetic
qualities based on the users' preference judgments and interviews.

1 Introduction

Creating complex, but comprehensible figures such as maps, scientific illustrations, and information graphics is a challenging task comprising multiple design
and layout steps. One of these steps is labeling the content of the figure appropriately. A good labeling conveys information about the figure without distracting
the viewer. It is unintrusive and does not destroy the figure's aesthetics. At
the same time it enables the viewer to quickly and correctly obtain additional
information that is not inherently contained in the figure. Typically multiple
features are labeled by a set of (textual) descriptions called *labels*. Morrison [15]
estimates the time needed for labeling a map to be over 50 % of the total time
when creating a map by hand. Hence, a lot of research efforts have been made
to design algorithms that automate the process of label placement.

To obtain a clear relation between a feature and its label, the label is often
placed closely to it. However, in some applications this *internal* labeling is not
sufficient, because either features are densely distributed and there are too many
labels to be placed or any extensive occlusion of the figure's details should be
avoided. While in the first case one may exclude less important labels, in the second case even a small number of labels may destroy the readability of the figure.
In either case graphic designers often choose to place the labels outside of the
figure and connect the features with their labels by thin curves, so called *leaders*. This kind of labeling is commonly found in highly detailed scientific figures

E. Di Giacomo and A. Lubiw (Eds.): GD 2015, LNCS 9411, pp. 515–527, 2015.
DOI: 10.1007/978-3-319-27261-0_42

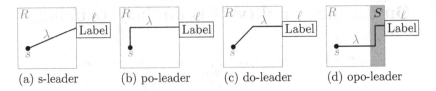

(a) s-leader   (b) po-leader   (c) do-leader   (d) opo-leader

Fig. 1. Illustration of leader types. Type-*opo* leaders use a track routing area S.

as they are used for example in atlases of human anatomy. In the graph drawing community this kind of *external* labeling became well known as *boundary labeling*. Since Bekos et al. [6] have introduced boundary labeling to the graph drawing community, a variety of boundary labeling models have been considered algorithmically. However, they have not been studied concerning their readability from a user's perspective. Here we present the first formal user study on the readability of the four most common boundary labeling models.

Models of Boundary Labeling. The problem of boundary labeling is formalized as follows (refer to Fig. 1). We are given a rectangle R of height h and width w and a finite set P of points in R, which we call *sites*. Each site s is assigned to a text that describes the site. Following traditional map labeling, not the text itself is considered, but its shape is approximated by its axis-aligned bounding box ℓ. We call ℓ the *label* of the site s. The set of all labels is denoted by L.

The boundary labeling problem then asks for the placement of labels such that (1) each label $\ell \in L$ lies outside of R and touches the boundary of R, no two labels overlap, and for each site s and its label ℓ there is a self-intersection-free curve λ in R that starts at s and ends on the boundary of ℓ. We call the curve λ the *leader* of the site s and its label ℓ. The end point of λ that touches ℓ is called the *port* of ℓ. Typically, four main parameters, in which the models differ, are distinguished. The *label position* specifies on which sides of R the labels are placed. The *label size* may be uniform or individually defined for each label. The *port type* specifies whether *fixed ports* or *sliding ports* are used, i.e., whether the position of a port on its label is pre-defined or flexible. Finally, the *leader type* restricts the shape of the leaders. As the leader type is the most distinctive feature of the different boundary labeling models in the literature, we examine how this parameter influences the readability. Regarding the other parameters we restrict our attention to one-sided instances whose labels have unit height, lie on the right side of R and have fixed ports. In the following we list the leader types that are most commonly found in the literature.

Let λ be a leader connecting a site $s \in P$ with a label $\ell \in L$, and let r be the side of R that is touched by ℓ. An *s-leader* consists of a single straight (s) line segment; see Fig. 1(a). A *po-leader* consists of two line segments, the first, starting at s, is parallel (p) to r and the second segment is orthogonal (o) to r; see Fig. 1(b). A *do-leader* consists of two line segments, the first, starting at s, is diagonal (d) at some angle α (typically $\alpha = 45°$) relative to r and the second segment is orthogonal (o) to r; see Fig. 1(c). An *opo-leader* consists of three line segments, the first, starting at s, is orthogonal (o) to r, the second is parallel (p) to r, and the third segment is orthogonal (o) to r; see Fig. 1(d). In case that *opo*-leaders are considered, each leader has its two bends in a strip S next to r whose width is large enough to accommodate all leaders with a minimum pairwise distance of the *p*-segments. The strip S is called the *track-routing area* of R. In the remainder of this paper, we call a labeling based on *s/po/do/opo*-leaders an *s/po/do/opo*-labeling.

Following Tufte's minimum-ink principle [17], the most common objective in boundary labeling is to minimize the total leader length, which means minimizing the total overlay of leaders with the given figure. Further, to increase readability of the labelings, all models usually require that no two leaders cross each other.

Related Work. The algorithmic problem of boundary labeling was introduced at GD 2004 by Bekos et al. [6]. They presented efficient algorithms for models based on *po*-, *opo*- and *s*-leaders. As objective functions they considered minimizing the number of bends and the total leader length. While for *opo*-leaders the labels may lie on one, two, or four sides of R, the labels for *po*-leaders may lie only on one or on two opposite sides of R. In 2005 based on a manual analysis of hand-drawn illustrations (e.g., anatomic atlases), Ali et al. [1] introduced criteria for boundary labeling concerning readability, ambiguity and aesthetics. Based on these they presented force-based heuristics for labeling figures using *s*-leaders and *po*-leaders. In 2006 Bekos et al. considered *opo*-labelings such that labels appear in multiple stacks besides R [4]. Boundary labeling using *do*-leaders has been introduced by Benkert et al. [7] in 2009. They investigated algorithms minimizing a general badness function on *do*- and *po*-leaders and, furthermore, gave more efficient algorithms for the case that the total leader length is minimized. In 2010 Bekos et al. [3] presented further algorithms for *do*-leaders and similarly shaped leaders. Further, Bekos et al. [5] considered *opo*-labelings such that the sites may *float* within predefined polygons in R. Nöllenburg et al. [16] considered *po*-labelings for a setting that supports interactive zooming and panning. In 2011 Gemsa et al. [9] studied the labeling of panorama images using vertical *s*-leaders. Leaders based on Beziér curves and *s*-leaders are further considered in the context of labeling focus regions by Fink et al. [8] (2012). Further, in 2013 Kindermann et al. [11] considered *po*-labelings for the cases that the labels lie on two adjacent sides, or on more than two sides. In 2014 Huang et al. [10] investigated *opo*-labelings with flexible label positions.

Boundary labeling has also been combined in a *mixed model* with internal labels, i.e., labels that are placed next to the sites; e.g., see [14]. *Many-to-one* boundary labeling is a further variant, where each label may connect to multiple sites; e.g., see [13]. Finally, boundary labeling has also been considered in the context of *text annotations*; e.g., see [12]. For a more detailed discussion see [2]. In total we found three papers studying *do*-leaders, nine studying *opo*-leaders, nine studying *po*-leaders, and five papers studying *s*-leaders.

Our Contribution. While boundary labeling has been extensively investigated algorithmically, the research on the readability of the introduced models has been neglected. There exist several user studies on the readability and aesthetics of graph drawings. For example Ware et al. [19] studied how people perceive links in node-links diagrams. However, to the best of our knowledge, there are no studies on the readability of any boundary labeling models. In this paper we present the first user study on readability aspects of boundary labeling. When reading a boundary labeling the viewer typically wants to find for a given site its corresponding label, or vice versa. Hence, a well readable labeling must facilitate this basic two-way task such that it can be performed fast and correctly. We call this the *assignment task*. In this paper we investigate the assignment task with respect to the four most established models, namely models using s-, po-, opo- and do-leaders, respectively. To keep the number of parameters small, we refrained from considering other types of leaders. We conducted a controlled user study with 31 subjects. Further, we interviewed eight participants about their personal assessment of the leader types. We obtained the following main results.

- Type-opo leaders lag behind the other leader types in all considered aspects.
- In the assignment task, do-, po- and s-leaders have similar error rates, but po-leaders have significantly faster response times than do- and s-leaders.
- The participants prefer the leader types in the order do, po, s and opo.

2   Research Questions

As argued before, a well readable boundary labeling must allow the viewer to quickly and correctly assign a label to its site and vice versa. More specifically, the leader λ connecting the label with its site must be easily traceable by a human. We hypothesize that both the response time and the error rate of the assignment task significantly depend on other leaders running close to and parallel to λ in the following sense. *The more parallel segments closely surround λ, the more the response time and the error rate of the assignment task increase.*

However, we did not directly investigate this hypothesis, but we derived from it two more concrete hypotheses that are based on the four leader types. These were then investigated in the user study. To that end, we additionally observe, that in medical figures the density of the sites varies. Both may occur, figures containing a *dense set* of sites, where the sites are placed closely to each other, and figures containing a *sparse set* of sites, where the sites are dispersed. We now motivate the hypothesis as follows.

By definition of the models, the number of parallel leader segments in do-, po- and opo-labelings is quadratic in the number of labels, because each pair of leaders has at least one pair of parallel segments. For opo-labelings each pair of leaders even has up to three pairs of parallel segments. Additionally, the spacing of the first orthogonal segments of opo-leaders is determined by the y-coordinates of the sites rather than by the (more regularly spaced) y-coordinates of the label ports as in po- and do-labelings. In contrast, in an s-labeling the leaders typically have different slopes, so that (almost) no parallel line segments occur. In fact, it is known that the human eye can distinguish angular differences as small as $10'' \approx 0.003°$ [18]. Hence, leaders of do-, po- and opo-labelings, in particular for a dense set of sites, are closely surrounded by parallel segments, while s-leaders for such a set have very different slopes. We therefore propose the next hypothesis.

(H1) For instances containing a dense set of sites,

(a) the assignment task on s-labelings has a significantly smaller response time and error rate than on *do-*, *po-*, and *opo-*labelings.

(b) the assignment task on *do-* and *po-*labelings has a significantly smaller response time and error rate than on *opo-*labelings.

Considering a sparse set of sites, *do-* and *po-*labelings still have many parallel line segments, but this time they are more dispersed. This is normally not true for *opo-*leaders because the actual routing of those leaders occurs in a thin routing area at the boundary of R. Hence, we propose the next hypothesis.

(H2) For instances containing a sparse set of sites, the assignment task on opo-labelings has a significantly greater response time and error rate than on do-, po-, and s-labelings.

In summary, we expect that *opo-*labelings perform worse than the other three, that *do-*and *po-*labelings perform similar, and that *s-*labelings perform best.

3   Design of the Experiment

This section presents the tasks, the stimuli, and the experimental procedure that we used to conduct the user study.

Tasks. In order to test our hypotheses we presented instances of boundary labeling to the participants and asked them to perform the following two tasks.

1. Label-Site-Assignment (T_S): In an instance containing a highlighted label select the related site.
2. Site-Label-Assignment (T_L): In an instance containing a highlighted site select the related label.

Stimuli. We now describe the presented stimuli; for a more detailed description see full version [2]. The stimuli are automatically generated boundary labelings, each using the same basic drawing style. In order to remove confounding effects between background image and leaders we use a plain light blue background. Points, leaders and label texts are drawn in the same style and in black color. Highlighted points are drawn as slightly larger yellow-filled squares with black boundary rather than small black disks. Highlighted labels are shown as white text on a dark gray background. Figure 2 shows four example stimuli.

For all instances we defined R to be a rectangle of 500×750 pixels. In addition to the four leader types as the main factor of interest, we identified three secondary factors that may have an impact on the resulting labelings. This yields four parameters to classify an instance. The first parameter is the *number* $\mathcal{N} = \{15, 30\}$, which allows us to model small instances (15 sites) and large

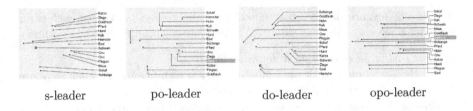

s-leader   po-leader   do-leader   opo-leader

Fig. 2. Examples of stimuli for both tasks and all four leader types.

instances (30 sites). The second parameter is the *distribution* $\mathcal{D} = \{D_U, D_3, D_{10}\}$ that is used for randomly placing the sites in R. We define D_U to be a uniform distribution, which yields dispersed sites. *Dense* and *sparse* sets of sites are modeled by normal distributions with mean $\mu = (250, 375)$ at the center of R, and variance $\sigma = 3000$ and $\sigma = 10000$ in both directions, respectively. The third parameter is the applied *leader type* $\mathcal{T} = \{do, opo, po, s\}$ as defined above. Finally, the fourth parameter $\mathcal{R} = \{0.3, 0.6, 0.9\}$ can be seen as a difficulty level. The parameter $r \in \mathcal{R}$ selects the leader λ whose ink score is the r-quantile among the ink scores of all leaders in the instance, where the *ink score* of a leader specifies how much ink of other leaders is close to it in the drawing.

The parameter space $\mathcal{N} \times \mathcal{D} \times \mathcal{T} \times \mathcal{R}$ gives us the possibility to cover a large variety of different instances. For each of the 72 possible choices of parameters $(n, d, t, r) \in \mathcal{N} \times \mathcal{D} \times \mathcal{T} \times \mathcal{R}$ we have generated two valid boundary labelings I_1 and I_2, one for each task, by minimizing the total leader length via integer linear programming. In each instance each label is randomly chosen from a set of animal names. For *opo*-labelings, the track routing area and the routing of the leaders is chosen such that the *p*-segments of any two leaders have horizontal distance of at least 10 pixels from each other. For examples see full version [2].

It will occur in the instances that leaders lie closely together, e.g., see *opo*-labeling in Fig. 2. However, we do not enforce minimum spacing between leaders because neither any of the studied models nor any of the discussed algorithms enforce minimum spacing explicitly. In fact, a fixed minimum leader spacing may even lead to infeasible instances for certain leader types.

Procedure. The study was run as a within-subject experiment. Four experimental sessions were held in our computer lab at controlled lighting with 12 identical machines and screens using a digital questionnaire in German language. After agreeing to a consent form, each participant first completed a tutorial explaining him or her the tasks T_S and T_L on four instances, each containing one of the four labeling types. Participants were instructed to answer the questions as quickly and as accurately as possible. Afterwards, the actual study started presenting the 144 stimuli to the participant one at a time. Each stimulus was revealed to the participant, after he or she clicked a button in the center of the screen using the mouse. Hence, at the beginning of each task the mouse pointer was always located at the same position. Then he or she performed the task by selecting a label or site using the mouse.

The stimuli were divided into 12 blocks consisting of 12 stimuli each. Each block either contained stimuli only for T_S or only for T_L. For each participant the stimuli were in random order, but in alternating blocks, i.e., after completing a block for T_S a block for T_L was presented, and vice versa. Between two successive blocks a pause screen stated the task for the next block and participants were asked to take a break of at least 15 seconds before continuing.

Especially for professional printings, e.g., for anatomy atlases, not only the figure's readability, but also its aesthetics is of great importance. Further, assigning a label to its site (or vice versa), the viewer should be able to assess whether he or she has done this correctly. We therefore asked all participants about their personal assessment of the aesthetics and readability of the leader types after completing the 144 performance trials. We presented the same four selected instances of the four leader types to each participant. To that end, we selected an instance for each leader type $t \in \mathcal{T}$ based on the 144 instances generated for the tasks T_S and T_L. We score each instance by the sum of its leaders' ink scores. Among all instances with leader type $t \in \mathcal{T}$ and 15 sites, we selected the median instance I with respect to the instance scores of that subset. Hence, for each type of leader we obtain a moderate instance with respect to our difficulty measure. Each participant was asked to rate the different leader types using German school grades on a scale from 1 (excellent) to 6 (insufficient), where grades 5 and 6 are both fail-grades, by answering the following questions.

Q1. How do you rate the appearance of the leader types?
Q2. For a highlighted site, how easy is it for you to find the corresponding label?
Q3. For a highlighted label, how easy is it for you to find the corresponding site?

We further conducted interviews with eight participants after the experiment, in which they justified their grading.

4   Results

In total 31 students of computer science aged between 20 and 30 years completed the experiment, six of them were female and 25 were male. We also asked whether they have fundamental knowledge about labeling figures and maps, which was affirmed by only two participants.

4.1   Performance Analysis

For each of the 144 trials we recorded both the response time and the correctness of the answer, which allows for analyzing two separate quantitative performance measures[1]. Response times were measured from the time a stimulus was revealed until the participant clicks to give the answer. Response times are normalized per participant by his/her median response time to compensate for different reaction times among participants. We split the data into four groups by leader type, and call them $\mathcal{DO}$, $\mathcal{PO}$, $\mathcal{S}$ and $\mathcal{OPO}$, respectively.

---

[1] Raw data at http://i11www.iti.uni-karlsruhe.de/projects/bl-userstudy.

We applied repeated-measures Friedman tests with post-hoc Dunn-Bonferroni pairwise comparisons in SPSS[2] between the four groups to find significant differences in the performance data at a significance level of $p = 0.05$. We chose a non-parametric test since our data are not normally distributed. We now summarize the main findings, while the detailed test results are found in [2].

Response Times. Figure 3a shows the normalized response times broken down into the three considered distributions D_3, D_{10} and D_U, which yield *dense*, *sparse* and *uniform* sets of sites; the corresponding mean and absolute times are found in the full version of this paper [2]. We obtained the following results. Among all leader types, *opo*-leaders have the highest response time. In particular for dense and sparse sets of sites the mean response time is up to a factor 1.8 worse than for the others. For uniform sets we obtain a factor of up to 1.5. Further, for any distribution the measured differences are significant. Comparing the response times of the remaining leader types we obtain the order $po < s < do$ with respect to increasing mean response time. For uniform sets we did not measure any pairwise significant difference between *do*, *po* and *s* leaders. However, for dense and sparse sets we obtained the significant differences as shown in Fig. 3a. We emphasize that for *po*- and *s*-leaders significant differences are measured for sparse, but not for dense sets of sites. In contrast *do*- and *s*-leaders have significant differences for dense sets, but not for sparse sets. Further, *po*- and *do*-leaders have significant differences in both dense and sparse sets. Altogether, this justifies the ranking $po < s < do$ w.r.t. increasing mean response time.

Comparing the instances in terms of T_S and T_L, the mean response time of T_L is slightly lower than that of T_S. Filtering out incorrectly processed tasks does not change the mean response time much and similar results are obtained. The mean response times of *large instances* (any instance with 30 sites and dense, sparse or uniform distribution) are similar to those of dense sets, and the mean response times of *small instances* (any instance with 15 sites and dense, sparse or uniform distribution) are similar to those of uniform sets.

Accuracy. We computed for each leader type and each participant the proportion of instances of that type that the participant solved correctly; see full version for detailed results and figures [2]. For dense and sparse sets of sites we observe that $\mathcal{OPO}$ has success rates around 86 %, while the other groups have success rates greater than 93 %. In particular the differences between success rates of *opo*-leaders and the remaining types are up to 11 % and 13 % for dense and sparse sets, respectively. Any of these differences is significant, while between $\mathcal{PO}$, $\mathcal{DO}$ and $\mathcal{S}$ no significant accuracy differences were measured. For uniform sets of sites, however, no significant differences were measured and any group has a success rate greater than 95 %. Hence, it appears that uniform sets of sites produce well readable labelings with any leader type – unlike dense and sparse instances.

Considering large and small instances separately, the group $\mathcal{OPO}$ has a decreased success rate (81 %), while the other groups remain almost unchanged

---

[2] http://www-01.ibm.com/software/analytics/spss/.

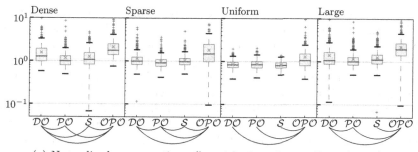

(a) Normalized response times (logarithmic scale, smaller is better).

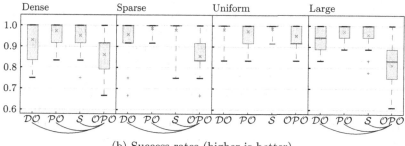

(b) Success rates (higher is better).

Fig. 3. Performance results broken down to dense, sparse and uniform sets as well as to large instances (30 sites). Mean values are indicated by 'x'. Arcs at the bottom show significant differences that were found ($p = 0.05$).

($> 93\%$), which yields for $\mathcal{PO}$ and $\mathcal{OPO}$ a difference of 16 %. For small instances no significant differences were measured. Comparing the instances by tasks T_S and T_L, the success rate of T_S is slightly better than that of T_L except for $\mathcal{OPO}$. For the mean response times the contrary is observed.

4.2 Preference Data

Table 1 shows the average grades given by the participants with respect to the three questions Q1–Q3. Concerning the general aesthetic appeal (question Q1) leaders of type *do* received the best grades (1.8), followed by *po*-leaders (grade 2.3). The participants did not particularly like the appearance of *s*-leaders (grade 3.3) and generally disliked *opo*-leaders (grade 4.6). In the

Table 1. Average grades given by the participants with respect to questions Q1–Q3 (smaller is better).

	do	opo	po	s
Q1	1.8	4.6	2.3	3.3
Q2	2.0	4.6	2.1	2.4
Q3	1.7	4.3	2.3	2.4

full version [2] we list the detailed percentages of participants who graded a particular leader type better, equally, or worse than another type. In addition to the general impression from the average grades it is worth mentioning that between the two most preferred leader types *do* and *po* 48.4 % preferred *do* over *po* and

38.7 % gave the same grades to both leader types. Compared to the s-leaders, a great majority (> 80 %) strictly prefers both do- and po-leaders. In the interviews seven out of eight participants stated that opo-leaders are "confusing, because leaders closely pass by each other". They disliked the long parallel segments of opo-leaders. Further, some participants remarked that opo-leaders "consist of too many bends". For six participants s-leaders were "chaotic and unstructured", unlike do- and po-leaders. Five participants said that they liked the flat bend of do-leaders more than the sharp bend of po-leaders. One participant stated that "po-leaders seem to be more *abstract* than do-leaders". Further, it was said that "the ratio of the segments' lengths is less balanced for po- than do-leaders."

For question Q2 (site-to-label) do- and po-leaders were ranked best (see Table 1), followed by s and more than two grades behind by opo, whereas for question Q3 (label-to-site) do-leaders are further ahead of po- and s-leaders, both of which received similar grades, and are again about two grades ahead of opo-leaders. For questions Q2 and Q3 the most striking observation is that type-s leaders received much better results (almost a full grade point better) than for Q1. This is in strong contrast to the other three leader types, which received grades in the same range as for Q1. This indicates that the participants perceived straight leaders as being well readable during the experiment, but still did not produce very appealing labelings. In the interviews participants stated that "opo-leaders are hard to read because of leaders lying close to each other." They negatively observed that opo-leaders "may not be clearly distinguished", but assessed the "simple shape of s-leaders to be easily legible." Further, they positively noted that "the distances between do-leaders seem to be greater than for other types" and that "po-leaders are easier to follow than other types".

It is remarkable that the participants rated do-leaders best, while they ranked third in our performance test. We conjecture that the participants overestimate the performance of do-leaders, because they like their aesthetics. For s-leaders the reverse is true. In contrast, their assessment on po- and opo-leaders corresponds more closely with the result of our performance test.

In summary, do-leaders obtained the best subjective ratings. The regularly shaped po- and do-leaders both scored better than the irregular and less restricted s-leaders. For any of the three questions opo-leaders were rated a lot worse than the others, which is, according to the interviews, mostly due to the frequent occurrence of many nearby leaders running closely together.

5   Discussion

In Sect. 2 we hypothesized that labelings with many parallel leaders lying close to each other have a significant negative effect on response times and accuracy. Our results from Sect. 4.1 indeed support hypotheses (H1b) and (H2), which said that the assignment task has a significantly smaller response time and error rate for do- and po-labelings than for opo-labelings in dense (H1b) and also sparse sets of sites (H2). Hypothesis (H2) was claimed to also hold for s-labelings versus opo-labelings, which is confirmed by the experiment as well. While greater response times may still be acceptable in some cases, the significantly lower accuracy clearly restricts the usability of *opo*-leaders. Only for small numbers of sites and uniform distributions *opo*-leaders have comparable success rates to the other leader types. This judgment is strengthened further by the preference ratings. On average the participants graded *opo*-leaders between 4 (sufficient) and 5 (poor) in all concerns. The main reason given in the interviews was that *opo*-labelings are confusing due to many leaders closely passing by each other.

However, our results falsified hypothesis (H1a), which claimed that for dense instances type-*s* leaders perform significantly better than the other three leader types. Rather we gained unexpected insights into the readability of boundary labeling. While we had expected that due to their simple shape and easily distinguishable slopes *s*-leaders will perform better than all other types of leaders, we could not measure significant differences between *po*-leaders and *s*-leaders. Interestingly, on average, the participants graded *po*-leaders better than *s*-leaders in all examined concerns, in particular with respect to their aesthetics (Q1). This is emphasized by the statements given by the participants that *po*-labelings appear structured while *s*-labelings were perceived as chaotic. Comparing *do*- and *s*-leaders we measured some evidence for (H1a), namely that the assignment task has significantly smaller response times for *s*- than for *do*-leaders. However, the success rates did not differ significantly.

We summarize our main findings regarding the four leader types as follows:

(1) *do*-leaders perform best in the preference rankings, but concerning the assignment tasks they perform slightly worse than *po*- and *s*-leaders.
(2) *opo*-leaders perform worst, both in the assignment tasks and the preference rankings. They are applicable only for small instances or for uniformly distributed sites.
(3) *po*-leaders perform best in the assignment tasks, and received good grades in the preference rankings.
(4) *s*-leaders perform well in the assignment tasks, but not in the preference rankings. The participants dislike their unstructured appearance.

We can generally recommend *po*-leaders as the best compromise between measured task performance and subjective preference ratings. For aesthetic reasons, it may also be advisable to use *do*-leaders instead as they have only slightly lower readability scores but are considered the most appealing leader type.

An interesting question is why type-*s* leaders (which showed good task performance) are frequently used by professional graphic designers, e.g., in anatomical drawings, although they were not perceived as aesthetically pleasing in our experiment. One explanation may be that our experiment judged all leader types on an empty background, where the leaders receive the entire visual attention of a viewer. In reality, the labeled figure itself is the main visual element and the leaders should be as unobtrusive as possible and not interfere with the figure. It would be necessary to conduct further experiments to assess the influence and interplay of image and leaders on more complex readability tasks.

Another interesting follow-up question is whether the chosen objective function produces actually the most aesthetic and most readable labelings. Despite being the predominant objective function in the literature on boundary labeling, simply minimizing the total leader length most certainly does not capture all relevant quality criteria.

Acknowledgments. We thank Helen Purchase and Janet Siegmund for their advice on the statistical data analysis.

References

1. Ali, K., Hartmann, K., Strothotte, T.: Label layout for interactive 3D illustrations. J. WSCG **13**(1), 1–8 (2005)
2. Barth, L., Gemsa, A., Niedermann, B., Nöllenburg, M.: On the Readability of Boundary Labeling. CoRR, abs/1509.00379 (2015)
3. Bekos, M.A., Kaufmann, M., Nöllenburg, M., Symvonis, A.: Boundary labeling with octilinear leaders. Algorithmica **57**(3), 436–461 (2010)
4. Bekos, M.A., Kaufmann, M., Potika, K., Symvonis, A.: Multi-stack boundary labeling problems. In: Arun-Kumar, S., Garg, N. (eds.) FSTTCS 2006. LNCS, vol. 4337, pp. 81–92. Springer, Heidelberg (2006)
5. Bekos, M.A., Kaufmann, M., Potika, K., Symvonis, A.: Area-feature boundary labeling. Comput. J. **53**(6), 827–841 (2010)
6. Bekos, M.A., Kaufmann, M., Symvonis, A., Wolff, A.: Boundary labeling: models and efficient algorithms for rectangular maps. Comput. Geom. Theory Appl. **36**(3), 215–236 (2007)
7. Benkert, M., Haverkort, H.J., Kroll, M., Nöllenburg, M.: Algorithms for multi-criteria boundary labeling. J. Graph Algorithms Appl. **13**(3), 289–317 (2009)
8. Fink, M., Haunert, J.-H., Schulz, A., Spoerhase, J., Wolff, A.: Algorithms for labeling focus regions. IEEE Trans. Vis. Comput. Graph. **18**(12), 2583–2592 (2012)
9. Gemsa, A., Haunert, J.-H., Nöllenburg, M.: Boundary-labeling algorithms for panorama images. In: ACM GIS 2011, New York, USA, pp. 289–298 (2011)
10. Huang, Z.-D., Poon, S.-H., Lin, C.-C.: Boundary labeling with flexible label positions. In: Pal, S.P., Sadakane, K. (eds.) WALCOM 2014. LNCS, vol. 8344, pp. 44–55. Springer, Heidelberg (2014)
11. Kindermann, P., Niedermann, B., Rutter, I., Schaefer, M., Schulz, A., Wolff, A.: Two-sided boundary labeling with adjacent sides. In: Dehne, F., Solis-Oba, R., Sack, J.-R. (eds.) WADS 2013. LNCS, vol. 8037, pp. 463–474. Springer, Heidelberg (2013)
12. Lin, C., Wu, H., Yen, H.: Boundary labeling in text annotation. In: Banissi, E., et al., (ed.) IV 2009, pp. 110–115. IEEE (2009)
13. Lin, C.-C., Kao, H.-J., Yen, H.-C.: Many-to-one boundary labeling. J. Graph Algorithms Appl. **12**(3), 319–356 (2008)
14. Löffler, M., Nöllenburg, M.: Shooting bricks with orthogonal laser beams: a first step towards internal/external map labeling. CCCG **2010**, 203–206 (2010)
15. Morrison, J.L.: Computer technology and cartographic change. In: Taylor, D. (ed.) The Computer in Contemporary Cartography. Johns Hopkins University Press, Baltimore (1980)
16. Nöllenburg, M., Polishchuk, V., Sysikaski, M.: Dynamic one-sided boundary labeling. In: ACM-GIS 2010, pp. 310–319 (2010)

17. Tufte, E.R.: The Visual Display of Quantitative Information. Graphics Press, Cheshire (2001)
18. Ware, C.: Information Visualization: Perception for Design, 3rd edn. Morgan Kaufmann, San Francisco (2012)
19. Ware, C., Purchase, H., Colpoys, L., McGill, M.: Cognitive measurements of graph aesthetics. Inf. Vis. $1(2)$, 103–110 (2002)

Graph Drawing Contest

Graph Drawing Contest Report

Philipp Kindermann[1]($\boxtimes$), Maarten Löffler[2], Lev Nachmanson[3],
and Ignaz Rutter[4]

[1] Fern Universität in Hagen, Hagen, Germany
philipp.kindermann@fernuni-hagen.de
[2] Utrecht University, Utrecht, The Netherlands
m.loffler@uu.nl
[3] Microsoft, New York, USA
levnach@microsoft.com
[4] Karlsruhe Institute of Technology, Karlsruhe, Germany
rutter@kit.edu

Abstract. This report describes the 22nd Annual Graph Drawing Contest, held in conjunction with the 23rd International Symposium on Graph Drawing (GD'15) in Los Angeles, California, United States of America. The purpose of the contest is to monitor and challenge the current state of graph-drawing technology.

1 Introduction

This year, the Graph Drawing Contest was divided into two parts: the *creative topics* and the *live challenge*.

The creative topics had two graphs: the first one was a graph of inclusion relations between different graph classes, and the second one was a state-transition graph for the game Tic-Tac-Toe. The data sets for the creative topics were published months in advance, and contestants could solve and submit their results before the conference started. The submitted drawings were evaluated according to aesthetic appearance, domain-specific requirements, and how well the data was visually represented.

The live challenge took place during the conference in a format similar to a typical programming contest. Teams were presented with a collection of challenge graphs and had one hour to submit their highest scoring drawings. This year's topic was to minimize the number of crossings in book layouts with a fixed number of pages.

Overall, we received 25 submissions: 13 submissions for the creative topics and 12 submissions for the live challenge.

2 Creative Topics

The two creative topics for this year were a graph of graph classes, and a tic-tac-toe state graph. The goal was to visualize each graph with complete artistic freedom, and with the aim of communicating the data in the graph as well as possible.

© Springer International Publishing Switzerland 2015
E. Di Giacomo and A. Lubiw (Eds.): GD 2015, LNCS 9411, pp. 531–537, 2015.
DOI: 10.1007/978-3-319-27261-0_43

We received 6 submissions for the first topic, and 7 for the second. For each topic, we selected three contenders for the prize, which were printed on large poster boards and presented at the Graph Drawing Symposium. Finally, out of those three, we selected the winning submission. We will now review the top three submissions for each topic (for a complete list of submissions, refer to http://www.graphdrawing.de/contest2015/results.html).

2.1   Graph Classes

The Information System on Graph Classes and their Inclusions (ISGCI)[1] is an initiative to provide a large database of graph classes and their relations, as well as the complexity of several problems that are hard on general graphs. So far, data of 1,511 graph classes and 179,111 inclusions has been collected.

For the first creative topic, participants needed to draw the graph of the graph classes provided by the ISGCI database that are planar. Each node represents a graph class, and each (directed) edge represents an inclusion. For example, the edge from "outerplanar" to "cactus" says that every cactus is outerplanar.

The graph has 65 vertices and 101 edges. The graph is presented in the GraphML File Format[2].

The resulting layout of the graph should contain the label of the vertices, provided as the description of the nodes, and should give a good overview on the hierarchy of the graph classes.

Runner-Up: Evmorfia Argyriou, Michael Baur, Anne Eberle, and Martin Siebenhaller (yWorks). The committee liked the use of edge grouping for nodes with many outgoing edges (such as the "planar" node and the use

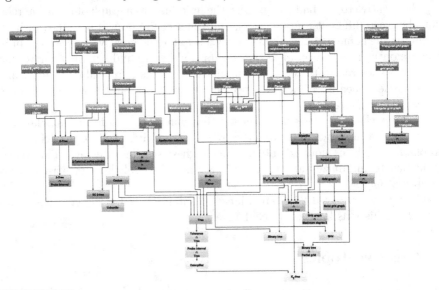

of visual bridges to reduce the visual ambiguity at crossings. Also, the drawing makes good use of the available space and uses color as an additional cue to encode distance from the root.

Runner-Up: Megah Fadhillah (University of Sydney). The committee really liked the idea of having a large "planar" node at the top, which acts as a title for the drawing and immediately explains what is being shown. The drawing uses color to show clusters of nodes and size to encode distance from the root.

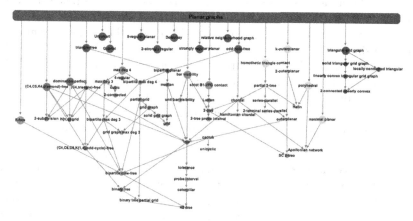

Winner: Tamara Mchedlidze (Karlsruhe Institute of Technology). The committee likes the visual appeal of the drawing. The use of circular arcs for edges makes it easy to follow each individual edge, even when they pass behind

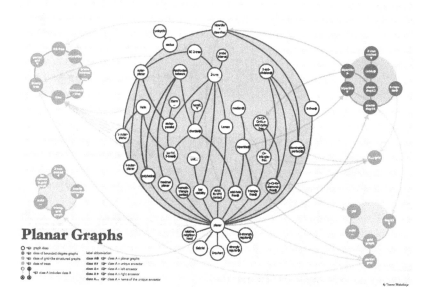

vertices (something which is traditionally considered bad practice in graph drawing). The drawing uses color to single out three meaningful groups of nodes: tree classes, bounded degree classes, and grid-like classes. The author also put a lot of thought into abbreviating node labels to increase the readability of the final drawing.

2.2 Tic-Tac-Toe

Tic-Tac-Toe[3] is a tactical two-player game in which two players take turns entering symbols (X or O) into cells of a three-by-three grid, with the objective of creating a row, column, or diagonal of equal symbols. The game is famous for its relative simplicity.

For the second creative topic, participants were asked to draw the graph of all possible Tic-Tac-Toe game states, in a way that shows as much of the structure and hidden information in the graph as possible. Each node represents a class of symmetric board positions, and there is an edge between two nodes u and v if a board position from v can be reached from a board position from u.

Runner-Up: Remus Zelina, Sebastian Bota, Siebren Houtman, and Radu Balaban (Meurs). The committee really liked visual appeal of the full drawing, which clusters the individual node into groups based on the ply (number of moves played) and the current winner after optimal play. This way, the drawing clearly shows a much smaller meta-graph (25 nodes and 39 edges) with

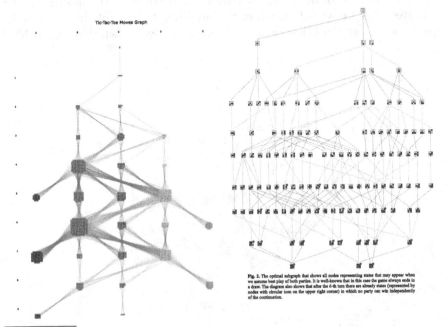

Tic-Tac-Toe Moves Graph

Fig. 2. The optimal subgraph that shows all nodes representing states that may appear when we assume best play of both parties. It is well-known that in this case the game always ends in a draw. The diagram also shows that after the 6-th turn there are already states (represented by nodes with circular icon on the upper right corner) in which no party can win independently of the continuation.

---

[3] http://en.wikipedia.org/wiki/Tic-tac-toe.

colors encoding the game state and "size" of clusters encoding the number of actual board positions that belong to that state. This drawing nicely communicates the global structure of the game.

Runner-Up: Evmorfia Argyriou, Michael Baur, Anne Eberle, and Martin Siebenhaller (yWorks). The second runner-up submitted an interactive visualization of the graph[4]. The committee liked the idea of using an interactive tool to explore the state graph. Using the tool, you can really see the impact of each move, making it very useful for understanding the game. The tool will dynamically show the local neighborhood (all possible paths to get to the situation, and all possible continuations) for any board position. Using small colored disks, the winner after optimal play is visualized for individual nodes.

Winner: Jennifer Hood and Pat Morin (Carleton University). The committee was impressed by the way this drawing manages to illustrate the global structure of the graph while still making it possible to follow individual paths and board positions. The global structure is nicely visualized by presenting the nodes in three columns, indicating which player will win upon optimal play, and nine rows, indicating the ply of the positions. Edge colors distinguish between move types (optimal or non-optimal, which player made the move, and which player wins). The committee especially liked the use of variable node sizes, making them small where necessary without affecting other parts of the

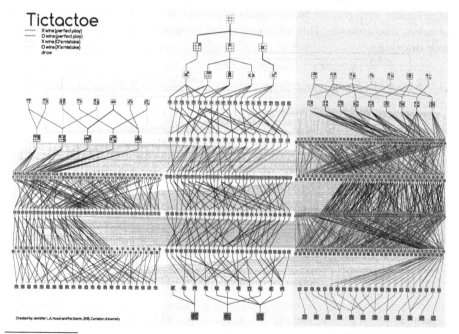

drawing. Similarly, the committee liked the use of solid edges for the (relatively) small set of optimal-play edges and more faded colors for the non-optimal-play edges. Presenting the optimal-play edges (straight) in a different style than the non-optimal-play edges (orthogonal) further enhances their visual distinction.

3   Live Challenge

The live challenge took place during the conference and lasted exactly one hour. During this hour, local participants of the conference could take part in the manual category (in which they could attempt to solve the graphs using a supplied tool), or in the automatic category (in which they could use their own software to solve the graphs). At the same time, remote participants could also take part in the automatic category.

The challenge focused on minimizing the number of crossings in a book embedding with k pages. The input graphs are arbitrary undirected graphs and a maximum number of pages that may be used.

A book with k pages consists of k half-spaces, the pages, that share a single line, the spine of the book. A k-page book embedding of a graph is an embedding of a graph into a book with k pages such that all the vertices lie at distinct positions of the spine and every edge is drawn in one of the pages such that only its endpoints touch the spine.

Note that edges may only cross if they are assigned to the same page. We are looking for drawings that minimize the number of crossings. The results are judged solely with respect to the number of crossings; other aesthetic criteria are not taken into account. This allows an objective way to qualitatively evaluate a given drawing.

3.1   Manual Category

In the manual category, participants were presented with five graphs. These were arranged from small to large and chosen to highlight different types of graphs and graph structures. For illustration, we include the first graph in its initial state and the best manual solution we received (by team snowman). For the complete set of graphs and submissions, refer to the contest website.

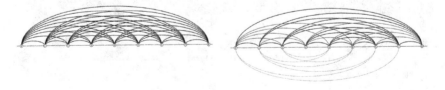

We are happy to present the full list of scores for all teams. The numbers listed are the number of crossings in each graph; the horizontal bars visualize the corresponding scores.

	graph 1	graph 2	graph 3	graph 4	graph 5
To Draw a Mockingbird	18	29	251	44	23
snowman	12	3	147	3	15
Jawaherul	15	3	285	87	95
LVYB	17	20	321	31	
Awesome	14	54	213	74	90
Wouter	15	7	210	6	35
Chipmunks	17	13	269	0	27
Book 'em Dannol	13	0	238	89	101
Super Algos Tübingen	21	6	287	0	15

The winning team is team snowman, consisting of Boris Klemz, Ulf Rüegg, and Fabian Lipp!

3.2 Automatic Category

In the automatic category, participants had to solve the same five graphs as in the manual category, and in addition another five—much larger—graphs. Again, the graphs were constructed to have different structure.

Once more, for illustration, we include the first large graph as it looks in the tool. The graphs themselves can be found on the contest website.

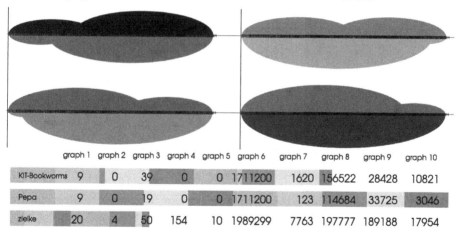

	graph 1	graph 2	graph 3	graph 4	graph 5	graph 6	graph 7	graph 8	graph 9	graph 10
KIT-Bookworms	9	0	39	0	0	1711200	1620	156522	28428	10821
Pepa	9	0	19	0	0	1711200	123	114684	33725	3046
zielke	20	4	50	154	10	1989299	7763	197777	189188	17954

The winning team is team Pepa, consisting of Josef Cibulka!

Acknowledgments. The contest committee would like to thank the generous sponsors of the symposium, Dennis van der Wals for programming most of the tool for the manual category, and all the contestants for their participation. Further details including all submitted drawings and challenge graphs can be found at the contest website: http://www.graphdrawing.de/contest2015/results.html.

Graduate Workshop Report

Graduate Workshop Recent Trends in Graph Drawing: Curves, Graphs, and Intersections

Bernardo M. Ábrego, Silvia Fernández-Merchant, and Csaba D. Tóth[✉]

Department of Mathematics, California State University Northridge,
18111 Nordhoff street, Los Angeles, CA 91330, USA
{Bernardo.Abrego,Silvia.Fernandez,Csaba.Toth}@csun.edu

The Organizing Committee of GD 2015 hosted a gradate workshop, continuing the tradition of previous Symposia, focusing on open problems in graph drawing. The workshop *Recent Trends in Graph Drawing: Curves, Graphs, and Intersections* was held at the California State University Northridge, Los Angeles, CA, September 21–22, 2015, with six invited speakers and 20 registered participants (including 10 students from Cal State Northridge). The following invited talks formed the core of the program:

David Eppstein: *Realizing Graphs as Polyhedra*
Marc van Kreveld: *Quality Ratios in Graph Drawing*
Maarten Löffler: *Drawing What You Do Not Know*
Bettina Speckmann: *Algorithmic Geo-visualization*
Torsten Ueckerdt: *Three Ways to Draw a Graph*
Alexander Wolff: *Simultaneous Drawings with Few Bends*

Each morning, three speakers presented recent results in graph drawing, with an emphasis on open problems. These presentations laid down the groundwork for research in small groups in the afternoons. The participants and the speakers discussed new ideas and observations at the end of each day.

We thank the California State University Northridge, the College of Science and Mathematics, and the Department of Mathematics for their support. In particular, we thank Yen K. Duong and Rebecca E. Say for their invaluable help with hosting the Workshop on campus. We also thank all speakers and all participants for their contributions, and look forward to seeing the results of the collaboration initiated at the Workshop.

E. Di Giacomo and A. Lubiw (Eds.): GD 2015, LNCS 9411, p. 541, 2015.
DOI: 10.1007/978-3-319-27261-0_44

Posters

L-Visibility Drawings of IC-Planar Graphs

Giuseppe Liotta and Fabrizio Montecchiani[(✉)]

Università degli Studi di Perugia, Perugia, Italy
fabrizio.montecchiani@unipg.it

A *visibility drawing* Γ of a planar graph G maps the vertices into non-overlapping horizontal segments (*bars*), and the edges into vertical segments (*visibilities*), each connecting the two bars corresponding to its two end-vertices. Visibilities intersect bars only at their extreme points. In a *strong* visibility drawing, there exists a visibility between two bars if and only if there exists an edge in G between the corresponding vertices. Conversely, in a *weak* visibility drawing, a visibility may not correspond to an edge of the graph. Every planar graph admits a weak visibility drawing [9].

The problem of extending visibility drawings to non-planar graphs has been first addressed by Dean *et al.* [3]. They introduce *bar k-visibility drawings*, which are visibility drawings where each bar can see through at most k distinct bars. In other words, each visibility segment can intersect at most k bars, while each bar can be intersected by arbitrary many visibility segments. The graphs that admit a bar 1-visibility drawing are called 1-*visibile*. Brandenburg and independently Evans *et al.* prove that 1-*planar graphs*, i.e., those graphs that can be drawn with at most one crossing per edge, are 1-visible [1,5]. They focus on a *weak* model, where there is a visibility through at most k bars if there is an edge, while the converse may not be true. In terms of readability, a clear benefit of bar k-visibility drawings is that the crossings form right angles. *Right-angle crossing (RAC) drawings* and their advantages in terms of readability have been extensively studied in the graph drawing literature (see, e.g., [4]). However, in a bar k-visibility drawing crossings involve bars and visibilities, i.e., vertices and edges. These crossings are arguably less intuitive than crossings between edges.

Evans *et al.* introduce a new model of visibility drawings, called *L-visibility drawings* [6]. Their aim is to simultaneously represent two plane st-graphs G_r and G_b (whose union might be non-planar). Each vertex is represented by a horizontal bar and a vertical bar that share an extreme point, i.e. it is an *L-shape* in the set $\{ ∟ , ⌐ , ¬ , Γ \}$. They assume a *strong* model, where two L-shapes are connected by a vertical (horizontal) visibility segment if and only if there exists an edge in G_r (G_b) between the corresponding vertices. Also, no two L-shapes cross one another, and visibilities intersect bars only at their extreme points. The only crossings are between vertical and horizontal visibilites, i.e., between edges of the graph. These crossings form right angles.

In this poster we present results on *weak* L-visibility drawings of non-planar graphs. We focus on *IC-planar graphs*, which are those graphs that admit a drawing where each edge is crossed at most once, and no two crossed edges share an end-vertex (see , e.g., Fig. 1(a)). Their chromatic number is at most five [7],

© Springer International Publishing Switzerland 2015
E. Di Giacomo and A. Lubiw (Eds.): GD 2015, LNCS 9411, pp. 545–547, 2015.
DOI: 10.1007/978-3-319-27261-0_45

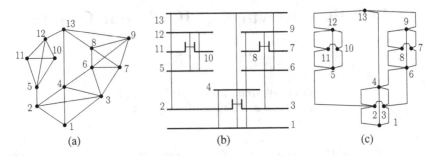

Fig. 1. (a) An IC-plane graph G. (b) An L-visibility drawing of G. (c) A RAC drawing of G.

and they have at most $13n/4 - 6$ edges, which is a tight bound [10]. Recognizing IC-planar graphs is NP-hard [2].

Our main contribution is summarized by the following theorem. See Fig. 1(b) for an example of a L-visibility drawing computed by using Theorem 1.

Theorem 1. *Every n-vertex IC-plane graph G admits an L-visibility drawing in $O(n^2)$ area, which can be computed in $O(n)$ time.*

Theorem 1 contributes to the rapidly growing literature devoted to the problem of drawing "nearly planar" graphs, where only some types of edge crossings are allowed. Brandenburg *et al.* recently described a cubic-time algorithm that computes straight-line RAC drawings of IC-planar graphs. These drawings may require exponential area, which is worst-case optimal [2]. They leave as an open problem to study techniques that compute IC-planar drawings in polynomial area and with good crossing resolution [2]. As a byproduct of Theorem 1, we obtain the following corollary (see Fig. 1(c)).

Corollary 1. *Every n-vertex IC-plane graph G admits a RAC drawing with at most two bends per edge in $O(n^2)$ area, which can be computed in $O(n)$ time.*

For complete proofs of the presented results the reader can refer to [8].

References

1. Brandenburg, F.J.: 1-visibility representations of 1-planar graphs. J. Graph Algorithms Appl. **18**(3), 421–438 (2014)
2. Brandenburg, F.J., Didimo, W., Evans, W.S., Kindermann, P., Liotta, G., Montecchiani, F.: Recognizing and drawing IC-planar graphs. In: Di Giacomo, E., Lubiw, A. (eds.) GD 2015. LNCS, vol. 9411, pp. 295–308. Springer, Heidelberg (2015)
3. Dean, A.M., Evans, W.S., Gethner, E., Laison, J.D., Safari, M.A., Trotter, W.T.: Bar k-visibility graphs. J. Graph Algorithms Appl. **11**(1), 45–59 (2007)
4. Didimo, W., Liotta, G.: The crossing angle resolution in graph drawing. In: Pach, J. (ed.) Thirty Essays on Geometric Graph Theory. Springer, New York (2012)
5. Evans, W.S., Kaufmann, M., Lenhart, W., Mchedlidze, T., Wismath, S.K.: Bar 1-visibility graphs vs. other nearly planar graphs. J. Graph Algorithms Appl. **18**(5), 721–739 (2014)

6. Evans, W.S., Liotta, G., Montecchiani, F.: Simultaneous visibility representations of plane st-graphs using L-shapes. In: Mayr, E.W., (ed.) WG 2015. LNCS. Springer (2015, to appear)
7. Král, D., Stacho, L.: Coloring plane graphs with independent crossings. J. Graph Theory **64**(3), 184–205 (2010)
8. Liotta, G., Montecchiani, F.: L-visibility drawings of IC-planar graphs. arXiv, (2015) http://arxiv.org/abs/1507.08879
9. Tamassia, R., Tollis, I.G.: A unified approach to visibility representations of planar graphs. Discr. Comput. Geom. **1**(1), 321–341 (1986)
10. Zhang, X., Liu, G.: The structure of plane graphs with independent crossings and its applications to coloring problems. Central Europ. J. Math. **11**(2), 308–321 (2013)

On the Relationship Between Map Graphs and Clique Planar Graphs

Patrizio Angelini[1](✉), Giordano Da Lozzo[2](✉), Giuseppe Di Battista[2](✉),
Fabrizio Frati[2](✉), Maurizio Patrignani[2](✉), and Ignaz Rutter[3](✉)

[1] Tübingen University, Tübingen, Germany
angelini@informatik.uni-tuebingen.de
[2] Roma Tre University, Rome, Italy
{dalozzo,gdb,frati,patrigna}@dia.uniroma3.it
[3] Karlsruhe Institute of Technology, Karlsruhe, Germany
rutter@kit.edu

A *map graph* is a contact graph of internally-disjoint regions of the plane, where the contact can be even a point. Namely, each vertex is represented by a simple connected region and two vertices are connected by an edge iff the corresponding regions touch. Map graphs are introduced in [2] to allow the representation of graphs containing large cliques in a readable way.

A *clique planar graph* is a graph $G = (V, E)$ that admits a representation where each vertex $u \in V$ is represented by an axis-parallel unit square $R(u)$ and where, for some partition of V into vertex-disjoint cliques $S = \{c_1, \ldots, c_k\}$, each edge (u, v) is represented by the intersection between $R(u)$ and $R(v)$ if u and v belong to the same clique (*intersection edges*) or by a non-intersected curve connecting the boundaries of $R(u)$ and $R(v)$ otherwise (*link edges*); see Fig. 1(c). Clique planar graphs are introduced in [1], where it is mainly addressed the case where the clique partition S is given.

Figure 2 provides an example of a graph that is both a map graph and a clique planar graph. In [1] it is argued that there are graphs that admit a clique-planar representation while not admitting any representation as a map graph, and vice versa. In this poster we exhibit such counterexamples, establishing that neither of the classes of map graphs and of clique planar graphs is contained in the other.

Lemma 1. *There exists a clique planar graph that is not a map graph.*

Proof. Consider the graph G of Fig. 2(a). Observe that G is not planar since vertices $1, 3, 4, 5$, and 6 (filled red in Fig. 2(a)) form a K_5 subdivision. However, graph G is clique planar (see Fig. 2(b)). If G were also a map graph, some edges could be represented by regions sharing a point. Since only the two triangles $1, 2, 3$ and $4, 5, 6$ could be represented in such a way, this would imply the planarity of a graph G' obtained from G by possibly augmenting one or both of such triangles to wheels, which is not planar as $G' \subseteq G$ (see Fig. 2(a) and (c)). ☐

Work partially supported by MIUR project AMANDA, prot. 2012C4E3KT_001, by DFG grant Ka812/17-1, and by DFG grant WA 654/21-1.

E. Di Giacomo and A. Lubiw (Eds.): GD 2015, LNCS 9411, pp. 548–550, 2015.
DOI: 10.1007/978-3-319-27261-0_46

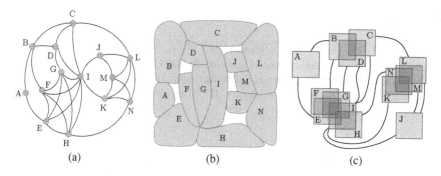

Fig. 1. A graph (a) that is both a map graph (b) and a clique planar graph (c) (Color figure online).

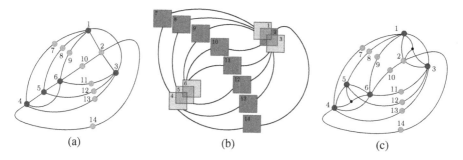

Fig. 2. A graph (a) that is clique planar (b) but not a map graph (c) (Color figure online).

Lemma 2. *There exists a map graph that is not a clique planar graph.*

Proof. Consider a graph $G_h = (V, E)$ composed by three sets V_1, V_2, and V_3 of h vertices each, where the graph induced by $V_1 \cup V_2$ is a clique and the graph induced by $V_2 \cup V_3$ is a clique. Figure 3 shows that G_h is a map graph. Observe that in any partition S of V into vertex-disjoint cliques there are at least $h/2$ vertices in V_2 that do not fall into the same clique with the vertices of V_1 or V_3. The link edges among such vertices induce a $K_{\frac{h}{2},h}$. Hence, for $h = 6$ the clique planarity of G_6 would imply the planarity of $K_{3,3}$. $\square$

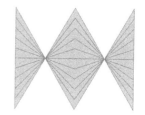

Fig. 3. A map graph that is not clique planar.

By Lemmas 1 and 2 we have the following.

Theorem 1. *Map Graphs $\nsubseteq$ Clique Planar Graphs $\wedge$ Clique Planar Graphs $\nsubseteq$ Map Graphs.*

References

1. Angelini, P., Da Lozzo, G., Di Battista, G., Frati, F., Patrignani, M., Rutter, I.: Intersection-link representations of graphs. In: Di Giacomo, E., Lubiw, A. (eds.) GD 2015. LNCS, vol. 9411, pp. 217–230. Springer, Heidelberg (2015)
2. Chen, Z., Grigni, M., Papadimitriou, C.H.: Map graphs. J. ACM **49**(2), 127–138 (2002)

PED User Study

Till Bruckdorfer[✉], Michael Kaufmann, and Simon Leibßle

Wilhelm-Schickard-Institut für Informatik, Universität Tübingen,
Tübingen, Germany
tillbruck@googlemail.com

Partial edge drawing (PED) is a model for a straight-line drawing of a graph, where edges are subdivided into three parts in order to drop the middle part. The remaining parts are called *stubs* and may not cross. While Burch et al. [1] investigated the usefulness of this model for directed graphs, we focused on undirected graphs and considered PEDs where every stub length is 1/4 of the length of the corresponding edge, so-called 1/4-*SHPEDs* [2,3]. In particular we investigated 1/4-nSHPEDs, i.e., 1/4-SHPEDs where stubs may cross and thus always exist. We developed a controlled on-line user study and claimed that the 1/4-nSHPED model is more readable and understandable than the traditional straight-line model in terms of completion time (CT) and error rate (ER) for tasks testing adjacency and accessibility of vertices.

Design: In our study, we chose graphs randomly ensuring a specific vertex connectivity [4]. We generated *small* graphs with 18 vertices and 30 edges and *midsize* graphs with 25 vertices and 47 edges. Using a force-directed layout algorithm [6] we produced drawings in the traditional straight-line model (*TRA*), and in the 1/4-nSHPED model (*PED*). Tasks were taken from [5] for evaluating the "adjacency" and "accessibility" of a vertex: 1. Which vertex is the farthest Euclidean distance neighbour of v? 2. Do vertices v and w have graph distance 2? To reduce cognitive load each task was considered in a separate block. We randomly chose vertices s.t. (1) the solution of the task is unique, (2) v and w are not on the convex hull, and (3) $\deg(v) \approx 5$ for Task 1, while $\deg(v) \geq 3$ and $\deg(w) \geq 2$ for Task 2. We supported the search for vertices by marking them fat. Participants have been introduced to the model, the intuition behind and introductory examples. The continue-on-demand design allowed users to adjust themselves on the next task. The participants classified themselves into an experience *level* between 1 and 6. Since almost all participants had some experience with graphs, we evaluated only the levels 4 to 6 to get comprehensive results. Our four study variables were the drawing **model** $\in$ {PED, TRA}, the graph **size** $\in$ {18, 25}, the **task** $\in$ {1, 2} and the **level** of experience $\in$ {1, ..., 6}. By repeated-measures design we tested every configuration twice, which resulted in 8 randomly permuted trials per block.

Results: From the 85 participants we averaged ER and CT over the number of participants in each level, see scatterplot in Fig. 1. ER and CT is smaller for most of the results in Task 2 w.r.t. Task 1. We observed that in Task 1 the CT is smallest for midsize TRA and highest for midsize PED. Also ER is smallest for small PEDs and highest for midsize TRAs, similar as midsize PEDs.

© Springer International Publishing Switzerland 2015
E. Di Giacomo and A. Lubiw (Eds.): GD 2015, LNCS 9411, pp. 551–553, 2015.
DOI: 10.1007/978-3-319-27261-0_47

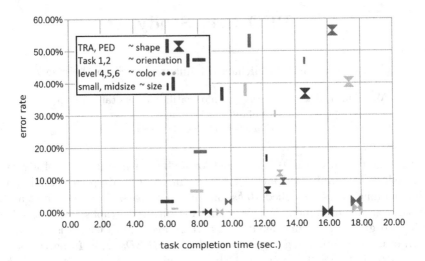

Fig. 1. The scatterplot shows the averaged results w. r. t. the variables (model, task, size, level) in different attributes of graphical visualization (shape, orientation, size, color).

In Task 2 CT was smallest for all TRAs, and midsize PEDs required significantly the highest time, while the ER was small for all but midsize TRAs. Surprisingly in Task 1 TRA supports midsize graphs and PED the small ones. Conclusively the results do not falsify or significantly strengthen our hypothesis.

Discussion: Due to the small study size, we had to trust the self classification of participants and "breaks" between two trials were impossible (otherwise we could have used the same graph in both models). Thus we took graphs with comparable statistical properties, but we did not prevent the possibility to accidentally pick two graphs with significantly different difficulty. Also distractions from environment, handling problems or no unifying screen resolution may have influenced the task completion times. Since most of the participants were used to TRA, it performed sometimes better than expected. In future work we want to compare these TRA layouts with layouts supporting 1/4-SHPEDs. We also like to investigate further graph sizes and stub sizes, as well as to control the error rate with presenting drawings for a fixed number of seconds.

References

1. Burch, M., Vehlow, C., Konevtsova, N., Weiskopf, D.: Evaluating partially drawn links for directed graph edges. In: Speckmann, B. (ed.) GD 2011. LNCS, vol. 7034, pp. 226–237. Springer, Heidelberg (2011)
2. Bruckdorfer, T., Cornelsen, S., Gutwenger, C., Kaufmann, M., Montecchiani, F., Nöllenburg, M., Wolff, A.: Progress on partial edge drawings. In: Didimo, W., Patrignani, M. (eds.) GD 2012. LNCS, vol. 7704, pp. 67–78. Springer, Heidelberg (2013)

3. Bruckdorfer, T., Kaufmann, M.: Mad at edge crossings? break the edges!. In: Kranakis, E., Krizanc, D., Luccio, F. (eds.) FUN 2012. LNCS, vol. 7288, pp. 40–50. Springer, Heidelberg (2012)
4. Barabasi, A.-L., Albert, R.: Emergence of scaling in random networks. Sci. **286**(5439), 509–512 (1999)
5. Lee, B., Plaisant, C., Parr, C.S., Fekete, J.-D., Henry, N.: Task taxonomy for graph visualization. In: BELIV 2006, pp. 1–5. ACM (2006)
6. yWorks GmbH, yFiles graph library. http://www.yworks.com. Accessed February 2015

SVEN: An Alternative Storyline Framework for Dynamic Graph Visualization

Dustin L. Arendt[✉]

Pacific Northwest National Laboratory, Richland, USA
dustin.arendt@pnnl.gov

1 Poster Abstract

The world is a dynamic place, so when we use graphs to help understand real world problems the structure of such graphs inevitably changes over time. Understanding this change is important, but often challenging. Techniques for general purpose dynamic graph visualizations generally fall into one of two broad categories: animation or timeline based techniques [2]. Simple approaches using animation or small multiples experience challenges with change blindness and "preserving the user's mental map" [1]. Storyline visualization techniques [5,7] hold promise, though these techniques were not originally designed as general purpose solutions for dynamic graph visualization.

There are well established criteria for drawing aesthetically pleasing storylines, which are to minimize (1) line crossings, (2) line wiggles, and (3) white space [5]. Past work has approached this problem by using evolutionary or quadratic [3,6] optimization techniques, developing complex ad-hoc solutions [5], or not addressing all of the established aesthetic criteria [7]. Our contribution is a framework that divides the overall storyline drawing problem (including addressing the three aesthetic criteria mentioned above) into relatively simple sub-problems having well-known solutions. We refer to this framework as "Storyline Visualization of Events on a Network" (SVEN).

Input for SVEN can take the form of a contact sequence, which is a list of edges and associated time stamps. Edges in a contact sequence are assumed to represent instantaneous interactions and can repeat at different times. This data is transformed into "interaction sessions" [5] by discretizing time into several windows and finding communities that partitions the nodes into densely connected groups for each time window, similar to [7]. These groups are represented as nodes in a directed acyclic graph whose edges represent the flow of nodes between communities in adjacent time windows. We employ Graphviz's "dot" algorithm, a directed graph layout technique, to determine an ordering for all groups of storylines that has few crossings. Determining which lines will be straightened (without changing the order of groups) is framed as a maximum weighted independent set problem and solved using a simple greedy algorithm [4]. Effective use of whitespace is found by defining the previous ordering and straightening properties as inequality and equality constraints in a linear program. The linear program's objective is to minimize the sum total distance between groups and an optimal solution is found quickly using an off the shelf solver.

© Springer International Publishing Switzerland 2015
E. Di Giacomo and A. Lubiw (Eds.): GD 2015, LNCS 9411, pp. 554–555, 2015.
DOI: 10.1007/978-3-319-27261-0_48

To date, we have demonstrated SVEN by generating visualizations of several benchmark movie datasets: *Star Wars, Inception, The Matrix* (see [6]). A comparison of the runtime performance of our framework against previous work will be important to evaluate the efficiency of the proposed framework. We note that the available literature on storyline visualization leaves much room for further evaluation of the scalability of the algorithms and techniques for storyline visualization, both in terms of usability and runtime performance. For future work, we could employ random dynamic network models to thoroughly evaluate our proposed method.

We believe that scalability challenges inherent to storyline visualizations can be mitigated partially through established interactive visualization patterns such as effective overviews, zooming (temporal), and degree of interest filtering of nodes. Along these lines, we are currently building a standalone web-based application around SVEN that includes capabilities to zoom in on an arbitrary time window, filter out uninteresting nodes, and obtain details about a particular node's interactions over time. The interface also provides alternate views of the data which include node-link and adjacency matrix representations of the interactions within a given time window. We also plan to extend this framework for visualizing many other types of data that can be described simply as "changing group membership."

References

1. Archambault, D., Purchase, H., Pinaud, B.: Animation, small multiples, and the effect of mental map preservation in dynamic graphs. IEEE Trans. Visual. Comput. Graph. **17**(4), 539–552 (2011)
2. Beck, F., Burch, M., Diehl, S., Weiskopf, D.: The State of the Art in Visualizing Dynamic Graphs. EuroVis, STAR (2014)
3. Liu, S., Wu, Y., Wei, E., Liu, M., Liu, Y.: StoryFlow: tracking the evolution of stories. IEEE Trans. Visual. Comput. Graph. **19**(12), 2436–2445 (2013)
4. Sakai, S., Togasaki, M., Yamazaki, K.: A note on greedy algorithms for the maximum weighted independent set problem. Discrete Appl. Math. **126**, 313–322 (2003)
5. Tanahashi, Y., Hsueh, C.H., Ma, K.L.: An efficient framework for generating storyline visualizations from streaming data. IEEE Trans. Visual. Comput. Graph. **6**(1), 1–1 (2015)
6. Tanahashi, Y., Ma, K.L.: Design considerations for optimizing storyline visualizations. IEEE Trans. Visual. Comput. Graph. **18**(12), 2679–2688 (2012)
7. Vehlow, C., Beck, F., Auwärter, P., Weiskopf, D.: Visualizing the evolution of communities in dynamic graphs. Comput. Graph. Forum **34**, 277–288 (2015)

Knuthian Drawings of Series-Parallel Flowcharts

Michael T. Goodrich$^{(\boxtimes)}$, Timothy Johnson, and Manuel Torres

Department of Computer Science, University of California, Irvine, CA, USA
goodrich@acm.org, {tujohnso,mrtorres}@uci.edu

Introduction. In 1963, Knuth published the first paper on a computer algorithm for a graph drawing problem, entitled "Computer-drawn Flowcharts" [8]. In this paper, Knuth describes an algorithm that takes as input an n-vertex directed graph G that represents a flowchart and, using the modern language of graph drawing, produces an *orthogonal drawing* of G. In Knuth's algorithm, every vertex is given the same x-coordinate and every edge has at most $O(1)$ bends, so that drawings produced using his algorithm can be output line-by-line on an (old-style) ASCII line printer and have worst-case area at most $O(n^2)$. Some drawbacks of his approach are that his drawings can be highly non-planar, even if the graph G is planar, and his drawings can have very poor aspect ratios, since every vertex is drawn along a vertical line. Nevertheless, his drawings possess an additional desirable property that has not been specifically addressed since the time of his seminal paper, which we revisit in the present work.

Specifically, inspired by his drawing convention, we say that a directed orthogonal graph drawing is *Knuthian* if there is no vertex having an incident edge locally pointing upwards unless that vertex is a *junction* node, that is, a vertex having in-degree strictly greater than its out-degree. This property (rotated 180 degrees) is related to previously-studied concepts known as "upward" or "quasi-upward" drawing conventions [3–5], where all edges must locally enter a vertex from below and leave going up.

Our Results. In this poster, we announce efficient algorithms for producing Knuthian drawings of degree-three acyclic series-parallel directed graphs, that is, directed orthogonal drawings where vertices are represented as small rectangles or squares and edges are directed paths of horizontal and vertical segments. These are equivalent to the flowcharts of loop-free algorithms. We provide a recursive linear-time algorithm for producing such drawings of degree-three acyclic series-parallel digraphs and we show that such a graph with n vertices has a Knuthian drawing with width $O(n)$ and height $O(\log n)$. We then show how to "wrap" this drawing, while still maintaining it to be Knuthian, to fit within a fixed width, so that the area is $O(n \log n)$ and the aspect ratio is constant. Our drawings strive to achieve few edge bends, both in the aggregate and per edge. Our drawing approach contrasts with previous approaches to drawing series-parallel graphs, including the standard recursive split-join-and-compose method and Knuth's original method [8], as well as more recent methods for drawing series-parallel graphs (e.g., see [1,2,7]). For details, please see the full version of this paper [6].

© Springer International Publishing Switzerland 2015
E. Di Giacomo and A. Lubiw (Eds.): GD 2015, LNCS 9411, pp. 556–557, 2015.
DOI: 10.1007/978-3-319-27261-0_49

Knuthian Drawings of Series-Parallel Flowcharts with $O(n \log n)$ **Area.**
We show that any n-vertex degree-three series-parallel graph has a Knuthian
drawing with $O(n \log n)$ area.

Theorem 1. *A degree-three series-parallel graph with n vertices has a Knuthian
drawing with width $O(n)$ and height $O(\log n)$, such that each edge has at most
two bends and the total number of bends is at most $1.25n$.*

Fixed-Width Drawings. We show how to adapt our $O(n \log n)$-area drawings,
which admittedly have poor aspect ratios, so that they achieve constant aspect
ratios, proving the following theorem.

Theorem 2. *A degree-three series-parallel graph with n nodes has a Knuthian
drawing that can be produced in linear time to have width $O(A + \log n)$ and height
$O((n/A) \log n)$, for any given $A \geq \log n$; hence, the area is $O(n \log n)$. The total
number of bends is at most $3.5n + o(n)$.*

Experimental Results. We tested our Knuth drawing algorithm algorithm on
some sample degree-three series-parallel graphs, based on two distributions used
to create random binary series-parallel decomposition trees.

Acknowledgments. This work was supported in part by the NSF under grants
1011840 and 1228639 and DARPA under agreement no. AFRL FA8750-15-2-0092. The
views expressed are those of the authors and do not reflect the official policy or position
of the Department of Defense or the U.S. Government.

References

1. Bertolazzi, P., Cohen, R.F., Di Battista, G., Tamassia, R., Tollis, I.G.: How to draw
   a series-parallel digraph. Int. J. Comput. Geom. Appl. **04**(04), 385–402 (1994)
2. Biedl, T.: Small drawings of outerplanar graphs, series-parallel graphs, and other
   planar graphs. Discrete Comput. Geom. **45**(1), 141–160 (2011)
3. Chan, T.M., Goodrich, M.T., Kosaraju, S., Tamassia, R.: Optimizing area and
   aspect ratio in straight-line orthogonal tree drawings. Comput. Geom. **23**(2), 153–
   162 (2002)
4. Di Battista, G., Didimo, W., Patrignani, M., Pizzonia, M.: Orthogonal and quasi-
   upward drawings with vertices of prescribed size. In: Kratochvíl, J. (ed.) GD 1999.
   LNCS, vol. 1731, pp. 297–310. Springer, Heidelberg (1999)
5. Garg, A., Goodrich, M.T., Tamassia, R.: Planar upward tree drawings with optimal
   area. Int. J. Comput. Geom. Appl. **06**(03), 333–356 (1996)
6. Goodrich, M.T., Johnson, T., Torres, M.: Knuthian drawings of series-parallel flowcharts. ArXiv ePrint, (2015). http://arxiv.org/abs/1508.03931
7. Hong, S.-H., Eades, P., Lee, S.-H.: Drawing series parallel digraphs symmetrically.
   Comput. Geom. **17**(34), 165–188 (2000)
8. Knuth, D.E.: Computer-drawn flowcharts. Commun. ACM **6**(9), 555–563 (1963)

Gestalt Principles in Graph Drawing

Stephen G. Kobourov[1], Tamara Mchedlidze[2]([⊠]), and Laura Vonessen[1]

[1] Department of Computer Science, University of Arizona, Tucson, USA
[2] Institute of Theoretical Informatics, Karlsruhe Institute of Technology,
Karlsruhe, Germany
mched@iti.uka.de

1 Introduction

Gestalt principles are rules for the organization of perceptual scenes. They were introduced in the context of philosophy and psychology in the 19th century and were used to define principles of human perception in the early 20th century. The *Gestalt* (*form*, in German) principles include, among others: proximity, the grouping of closely positioned objects; similarity, the grouping of objects of similar shape or color; continuation, the grouping of objects that form a continuous pattern; and symmetry, the grouping of objects that form symmetric patterns. Gestalt principles have been extensively applied in user interface design, graphic design, and information visualization.

2 Gestalt Principles in Graph Drawing

Several graph drawing conventions and aesthetics seem to rely on Gestalt principles [1,4,6,8]. In this poster we describe various such relationships; corresponding illustrations can be found in the poster. We believe that such relationships should be further explored and experimentally tested.

Continuation. The principle of continuation suggests that we find it easier to perceive smooth and continuous outlines between points over lines with sudden or irregular changes in direction. That is, we perceive elements as a group when they form a continuous pattern. Moreover, such a pattern will be assumed to continue even if some parts of it are hidden. The principle of continuation appears often in node-link diagrams and is mainly relevant to the drawing and interpretation of edges and paths. The edges of node-link diagrams are often drawn as polylines, using bend points. The principle of continuation suggests that an edge can be more easily followed by the eye when it has few bends which are not sharp. This idea is exploited in slanted orthogonal drawings, which are orthogonal drawings where 90° bends are replaced by diagonal segments.

The continuation principle also seems to be at play in graph drawings with smooth curves, which can be followed by the eye easier than a polyline edge. These curvilinear edges can also be joined together to form edge bundles. Edge bundled drawings and confluent drawings, which represent each (non) edge as a (non) smooth curve, exploit the continuity principle as a human preference

© Springer International Publishing Switzerland 2015
E. Di Giacomo and A. Lubiw (Eds.): GD 2015, LNCS 9411, pp. 558–560, 2015.
DOI: 10.1007/978-3-319-27261-0_50

of smooth curves over curves with abrupt changes of direction. The principle of continuation also allows us to omit parts of a straight-line edge, relying on our perception to "fill in" the missing parts. Such partially drawn edges have been proposed several times as a way to avoid edge crossings [3].

When searching for a path between two nodes, people prefer geodesic paths as these paths are more "continuous" than non-geodesic ones [2]. This notion of geodesic paths underlies greedy, monotone, self-approaching and increasing-chord drawings. The preference of geodesic paths seems to be responsible for the confusion caused by crossings that form small angles. Such crossings trigger extra back-and-forth eye movements [2]. Thus when searching a path from a source to a target node, our eyes try to follow a path towards the target node, and if a crossing edge points towards a potential target it is possible to deviate from the correct path, before reaching the next vertex of the path.

Proximity. The principle of proximity suggests that elements close to each other are perceived as a group. In a node-link diagram this results in nodes which are close to each other being perceived as groups forming clusters. Most of the algorithms to visualize clustered graphs in the form of a node-link diagram rely on this intuition [7]. The principle of proximity is also used by force-directed algorithms, which require that adjacent nodes are close (attractive forces), and that non-adjacent nodes are far apart (repulsive forces). Finally, the proximity principle is at the base of proximity drawings.

Similarity. The similarity principle suggests that objects of similar shape or color are perceived as groups. Nodes in a node-link diagram are often given the same color to indicate they belong to the same cluster. They are given the same size to indicate their equal importance. The same shape implies similar properties. The desire to obtain uniform edge lengths for unweighted graphs seems also to capture the notion of equal importance of the relationships between adjacent nodes. Finally, directed upward drawings indicate similar hierarchical relations between the corresponding pairs of nodes.

Symmetry. The symmetry principle suggests that symmetrical components tend to be grouped together. As a property of node-link diagrams, it is highly preferred by humans [5,6]. However, it is difficult to formalize symmetry of a node-link diagram and to provide a computable measure for it. It has been suggested that straight-line drawings of graphs can become more aesthetically pleasing if the number of edge slopes is relatively small. This kind of aesthetic could be explained by the local symmetries that are created by the similarity of incident edges' slopes around their respective nodes.

References

1. Bennett, C., Ryall, J., Spalteholz, L., Gooch, A.: The aesthetics of graph visualization. In: Computational Aesthetics 2007, pp. 57–64 (2007)
2. Huang, W., Eades, P., Hong, S.-H.: Beyond time and error: a cognitive approach to the evaluation of graph drawings. In: BELIV 2008, pp. 3:1–3:8 (2008)

3. Rusu, A., Fabian, A., Jianu, R., Rusu, A.: Using the Gestalt principle of closure to alleviate the edge crossing problem in graph drawings. In: IV 2011, pp. 488–493, July 2011
4. Sun, D., Wong, K.: On evaluating the layout of UML class diagrams for program comprehension. In: IWPC 2005, pp. 317–326, May 2005
5. Eades, P., Hong, S.-H.: Symmetric graph drawing. In: Tamassia, R. (ed.) Handbook of Graph Drawing and Visualization. Discrete Mathematics and Its Applications. Chapman & Hall/CRC, Boca Raton (2007). ISBN 1584884126
6. van Ham, F., Rogowitz, B.: Perceptual organization in user-generated graph layouts. IEEE Trans. Visual. Comput. Graph. **14**(6), 1333–1339 (2008)
7. Vehlow, C., Beck, F., Weiskopf, D.: The state of the art in visualizing group structures in graphs. In: Eurographics Conference on Visualization (EuroVis) - STARs (2015)
8. Ware, C., Purchase, H., Colpoys, L., McGill, M.: Cognitive measurements of graph aesthetics. Inf. Visual. **1**(2), 103–110 (2002)

Drawing Graphs Using Body Gestures

Yeganeh Bahoo$^{(\boxtimes)}$, Andrea Bunt, Stephane Durocher,
and Sahar Mehrpour

Department of Computer Science, University of Manitoba, Winnipeg, Canada
{bahoo,bunt,durocher,mehrpour}@cs.umanitoba.ca

Abstract. We introduce a new gesture-based user interface for drawing graphs that recognizes specific body gestures using the Microsoft Kinect sensor. Our preliminary user study demonstrates the potential for using gesture-based interfaces in graph drawing.

Traditional input devices for manual data entry of graphs include mice, keyboards, and touch screens. Humans naturally communicate using body gestures. Recent research explores body or *mid-air* gestures as a form of interaction, particularly when using traditional input devices may be unintuitive or undesirable. Inspired by advances in gesture-based input technologies, we investigate the application of mid-air gestures to graph drawing. We created a prototype system called KiDGraD (using **Ki**nect to **D**etect skeletons for **Gra**ph **D**rawing), which uses a Microsoft Kinect to recognize a limited set of body gestures designed to allow the user to manipulate a graph's nodes and edges. We conducted a preliminary user evaluation examining the perceived naturalness of our proposed gesture set and users' attitudes towards our general approach. Feedback from this initial user study suggests that gesture-based graph drawing has a number of potential applications, motivating future research into improved recognition capabilities as well as effective and expressive gesture sets.

Prior research on gesture-based interactions has focused on both gestures on digital surfaces (e.g., multi-touch gestures on digital tables [3]) and on mid-air gestures, where sensors and cameras are used to detect body movements (e.g., [2]). While a number of systems exist for inputting and editing graphs (e.g., [1]), there is limited prior research examining the use of non-traditional user interfaces for drawing or editing graphs from human input. To the authors' knowledge, this work is the first to examine a mid-air gesture-based user interface for graph drawing.

The KiDGraD user interface includes a drawing area, consisting of a grid illustrating the graph and overlayed with a sketch of the user's detected skeleton, a sidebar to access commands, and a header that displays the active command. The system implements five operations: adding nodes, deleting nodes, adding edges, deleting edges, and reset. A user can activate commands in one of two ways: performing the corresponding gesture (see Fig. 1) or using the sidebar.

This research was supported in part by the Natural Sciences and Engineering Research Council of Canada (NSERC).

E. Di Giacomo and A. Lubiw (Eds.): GD 2015, LNCS 9411, pp. 561–562, 2015.
DOI: 10.1007/978-3-319-27261-0_51

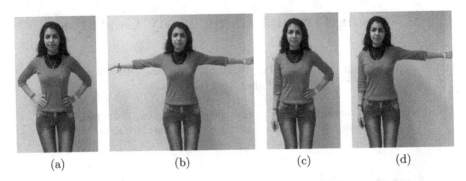

Fig. 1. The Add Node, Add Edge, Delete Node, and Delete Edge gestures.

As a first proof-of-concept exploration of our gesture-based approach to graph drawing, we conducted an informal usability study with ten participants. The goals of the study were to gain initial insight into the intuitiveness and ease of the gestures, as well as to elicit feedback from users on the potential strengths and limitations of this approach. We asked participants to interact with KiDGraD by drawing a number of sample graphs, after which we solicited feedback on both the system concept and the gesture set.

Participants responded quite enthusiastically to the system and the idea of using gestures to draw graphs. With mean responses of 4.0 or greater on a 5-point Likert scale, participants appeared to find the system fun, simple to use, and relatively efficient. Responses for comfort and the system working as participants expected were slightly less positive, and participants suggested a number of potential improvements to both the gesture set and the drawing interface. The post-session interviews revealed that almost all participants felt that the idea of using gestures to draw graphs is interesting, and they were excited to move away from using mice or keyboards.

Once the gesture set and recognition technologies are refined, there are a number of interesting directions to explore in terms of applications. In particular, our participants thought that mid-air graph drawing might be beneficial in educational settings, where the system could be used as an engaging way to teach children about graphs.

References

1. Fröhlich, M., Werner, M.: Demonstration of the interactive graph visualization system da Vinci. In: Tamassia, R., Tollis, I.G. (eds.) GD 1994. LNCS, vol. 894, pp. 266–269. Springer, Heidelberg (1995)
2. Jang, S., Elmqvist, N., Ramani, K.: GestureAnalyzer: visual analytics for pattern analysis of mid-air hand gestures. In: Proceedings of SUI, pp. 30–39 (2014)
3. Pfeuffer, K., Alexander, J., Chong, M.K., Gellersen, H.: Gaze-touch: combining gaze with multi-touch for interaction on the same surface. In: Proceedings of UIST, pp. 509–518 (2014)

Augmenting Planar Straight Line Graphs to 2-Edge-Connectivity

Hugo Alves Akitaya[1], Jonathan Castello[2], Yauheniya Lahoda[2(✉)],
Anika Rounds[1], and Csaba D. Tóth[1,2]

[1] Tufts University, Medford, MA, USA
{hugo.alves_akitaya,anika.rounds}@tufts.edu, cdtoth@acm.org
[2] California State University Northridge, Los Angeles, CA, USA
{jonathan.castello.652,yauheniya.lahoda.428}@my.csun.edu

Abstract. We show that every planar straight line graph (PSLG) with n vertices can be augmented to a 2-edge-connected PSLG with the addition of at most $\lfloor (4n-4)/3 \rfloor$ new edges. This bound is the best possible.

Edge-connectivity augmentation is a classic problem in combinatorial optimization motivated by applications in fault-tolerant network design. Given an undirected graph $G = (V, E)$ and a number $\tau \in \mathbb{N}$, we want to find a set F of new edges of minimum cardinality such that $G' = (V, E \cup F)$ is τ-edge-connected. In this note, we consider edge-connectivity augmentation for planar straight line graphs (PSLG) with n vertices in general position (no three collinear vertices).

Every graph with $t \in \mathbb{N}$ components can be augmented into a connected graph with the addition of $t - 1$ new edges. Every PSLG with n vertices can be augmented to a connected PSLG (encompassing graph) with at most $n - 1$ new edges. Every connected PSLG on n vertices can be augmented to a 2-edge-connected PSLG with at most $\lfloor (2n-2)/3 \rfloor$ new edges [3]. Both bounds are the best possible. The combination of the two bounds implies that every PSLG on n vertices can be augmented to 2-edge-connectivity with the addition of at most $\lfloor 5(n-1)/3 \rfloor$ new edges. However, this bound is not tight. We derive a better bound and show the following.

Theorem 1. *Every PSLG with $n \geq 3$ vertices can be augmented to a 2-edge-connected PSLG with the addition of at most $\lfloor (4n-4)/3 \rfloor$ new edges. This bound is the best possible.*

The upper bound in Theorem 1 is attained for a triangulation on $k \geq 3$ vertices, with an isolated vertex placed in each of the $2k - 5$ bounded faces and 3 vertices in the outer face that pairwise do not see each other (that is, $n = k + (2k - 5) + 3 = 3k - 2$). The proof of the upper bound is constructive and distinguishes between two cases depending on the number of components in the graph. Due to space limitation, we give an outline of the proof here.

Let G be a PSLG on $n \geq 3$ vertices in general position. Let c be the number of components in G. In the first case $c \leq \lfloor (2n+1)/3 \rfloor$, and we augment G to a 2-edge-connected PSLG as follows: first use $c-1$ new edges to obtain a connected

E. Di Giacomo and A. Lubiw (Eds.): GD 2015, LNCS 9411, pp. 563–564, 2015.
DOI: 10.1007/978-3-319-27261-0_52

PSLG, and then use $\lfloor (2n - 2)/3 \rfloor$ edges to make it 2-edge-connected [3]. The total number of new edges is at most

$$(c - 1) + \left\lfloor \frac{2n - 2}{3} \right\rfloor \leq \left\lfloor \frac{2n + 1}{3} \right\rfloor - 1 + \left\lfloor \frac{2n - 2}{3} \right\rfloor \leq \left\lfloor \frac{4n - 4}{3} \right\rfloor. \tag{1}$$

In the second case, when $c \geq \lfloor (2n+1)/3 \rfloor + 1 = \lfloor (2n+4)/3 \rfloor$, we develop an augmentation algorithm that uses a convex subdivision of G. A convex subdivision H is obtained from G by successively shooting rays from the reflex vertices of all nonsingleton components of G, similar to [2]. The isolated vertices of G lie in the interiors of the convex cells of H. For every convex subdivision H constructed in this way, we derive an upper bound for the number of cells h.

Lemma 1. *Let G be a PSLG with n vertices, b bridges, and c components. Then every convex subdivision of G has at most $h \leq 2n - 2c - b + 1$ cells.*

We augment G successively with new edges, and we always denote by G' the *current* graph. Graph G' is a planar straight line multigraph (PSLMG). Let $T \subseteq G'$ denote the set of nonsingleton connected components in G'.

Our augmentation algorithm works as follows:

1. Construct a convex subdivision H of G. Let $C = \{C_i : i = 1 \ldots h\}$ be the set of convex cells. Compute T.
2. For each cell $C_i \in C$: (a) for each nonsingleton component adjacent to C_i select an arbitrary vertex incident to C_i; (b) connect the selected vertices and singleton vertices in the cell C_i into a simple polygon; (c) recompute T.
3. Replace each bridge of G' by a double edge.
4. Transform the multigraph G' into a simple graph.

In step 2 we add $c + h - 1$ edges. Since we do not create any new bridges in step 2, we add b edges in step 3. The total number of new edges e' added is $e' \leq c + h - 1 + b$. By Lemma 1, since $c \geq \lfloor (2n+4)/3 \rfloor$, we obtain:

$$e' \leq c + h - 1 + b \leq 2n - c \leq 2n - \left\lfloor \frac{2n + 4}{3} \right\rfloor \leq \left\lfloor \frac{4n - 4}{3} \right\rfloor. \tag{2}$$

In step 4 we can transform the 2-edge-connected multigraph G' into a 2-edge-connected simple graph without increasing the number of edges by Lemma 2.

Lemma 2. *[1] Let G' be a 2-edge-connected PSLMG and let e be a double edge in G'. Then we can obtain a 2-edge-connected PSLMG from G' by decrementing the multiplicity of e by one and adding at most one new edge of multiplicity 1.*

References

1. Abellanas, M., García, A., Hurtado, F., Tejel, J., Urrutia, J.: Augmenting the connectivity of geometric graphs. Comput. Geom. **40**(3), 220–230 (2008)
2. Bose, P., Houle, M.E., Toussaint, G.T.: Every set of disjoint line segments admits a binary tree. Discrete Comput. Geom. **26**, 387–410 (2001)
3. Tóth, C.D.: Connectivity augmentation in planar straight line graphs. European J. Combin. **33**(3), 408–425 (2012)

Author Index